LOW TEMPERATURE PHYSICS -LT 13

Volume 1: Quantum Fluids

LOW TEMPERATURE PHYSICS-LT 13

LOW TEMPERATURE PHYSICS - LT 13

Edited by

K. D. Timmerhaus

University of Colorado
Boulder, Colorado and
National Science Foundation
Washington, D.C.

W. J. O'Sullivan

University of Colorado
Boulder, Colorado

and

E. F. Hammel

Los Alamos Scientific Laboratory
University of California
Los Alamos, New Mexico

Volume 1: Quantum Fluids

PLENUM PRESS · NEW YORK-LONDON

Library of Congress Cataloging in Publication Data

International Conference on Low Temperature Physics, 13th, University of Colorado, 1972.
    Low Temperature physics—LT 13; [proceedings]

    Includes bibliographical references.
    CONTENTS: v. 1. Quantum fluids.—v. 2. Quantum crystals and magnetism.—v. 3. Superconductivity.—v. 4. Electronic properties, instrumentation, and measurement.
    1. Low temperatures—Congresses. 2. Free electron theory of metals—Congresses. 3. Energy-band theory of solids—Congresses. I. Timmerhaus, Klaus D., ed. II. O'Sullivan, William John, ed. III. Hammel, E. F., 1918-   ed. IV Title.
QC278.I512 1972   536'.56   73-81092
ISBN 0-306-35121-8 (v. 1)

The proceedings of the XIIIth International Conference on Low Temperature Physics, University of Colorado, Boulder, Colorado, August 21-25, 1972, will be published in four volumes, of which this is volume one.

© 1974 Plenum Press, New York
A Division of Plenum Publishing Corporation
227 West 17th Street, New York, N.Y. 10011

United Kingdom edition published by Plenum Press, London
A Division of Plenum Publishing Company, Ltd.
4a Lower John Street, London W1R 3PD, England

Printed in the United States of America

Foreword

The 13th International Conference on Low Temperature Physics, organized by the National Bureau of Standards, Los Alamos Scientific Laboratory, and the University of Colorado, was held in Boulder, Colorado, August 21 to 25, 1972, and was sponsored by the National Science Foundation, the U.S. Army Office of Scientific Research, the U.S. Atomic Energy Commission, the U.S. Navy Office of Naval Research, the International Institute of Refrigeration, and the International Union of Pure and Applied Physics. This international conference was the latest in a series of biennial conferences on low temperature physics, the first of which was held at the Massachusetts Institute of Technology in 1949. (For a complete list of previous LT conferences see p. viii. Many of these past conferences have been coordinated and sponsored by the Commission on Very Low Temperatures of IUPAP. Subsequent LT conferences will be scheduled triennially beginning in 1975.

LT 13 was attended by approximately 1015 participants from twenty five countries. Eighteen plenary lectures and 550 contributed papers were presented at the Conference.

The Conference began with brief introductory and welcoming remarks by Dr. R.H. Kropschot on behalf of the Organizing Committee, Professor J. Bardeen on behalf of the Commission on Very Low Temperatures of the IUPAP, and Professor O.V. Lounasmaa on behalf of the International Institute of Refrigeration. The eighth London Award was then presented by Professor E. Lynton to Professor A.A. Abrikosov (in absentia). The recipient's award address, as delivered by Dr. P. Hohenberg, will surely remain for all who were privileged to hear it as one of the high points of the Conference. We wish to gratefully acknowledge the members of the Fritz London Award Committee: (E.A. Lynton, chairman; J.F. Allen; P. Hohenberg; F. Reif; D. Scalapino, and M. Tinkham) for assuming the responsibility for selection of the Award recipient and for making this timely award.

LT 13, originally scheduled to be held at the University of California at San Diego, was shifted in 1971 to Boulder because facility and accommodation problems developed after the first announcement of the proposed location of LT 13. Nevertheless, the original organizers of LT 13, Professors H. Suhl and B. Matthias, contributed significantly in the initial phases of the LT organization. As originally conceived, E.F. Hammel and the staff of the Los Alamos Scientific Laboratory organized the technical programs from the very beginning. By March of 1971 invitations had been sent to some fifty U.S. low temperature physicists inviting them to serve on the National Organizing Committee of LT 13.

The first meeting of the U.S. National Committee was held in Washington, D.C. on April 25, 1971. By that time arrangements had been made for the shift of the Conference to Boulder, Colorado, and R.H. Kropschot had accepted the General Chairmanship of LT 13. At the April meeting timetables were established, an

International Advisory Committee was appointed, and the general format of the Conference was agreed upon. Strong preferences were expressed for:(a) a general low temperature physics conference; (b) approximately one half of the available time to be allotted to plenary lectures, covering new developments in low temperature physics but presented in such a way as to interest a general audience; (c) somewhat more emphasis on low temperature instrumentation and measurement than in the past, and (d) information (perhaps through a plenary lecture) on recent developments in applied low temperature physics or cryoengineering.

In order to implement these proposals, a small Executive Committee was established (listed on p. viii). It was clear from the outset that LT 13 would be a large as well as a diverse conference, and that proper development of the technical program would require the help of many experts. Consequently, the subject matter of the Conference was divided into six main divisions. Division Chairmen were appointed and were given essentially full authority to arrange both the plenary and the contributed paper sessions in their divisions. In some instances the Division Chairmen appointed small committees to assist with some of this work. A listing of the six divisions and their Chairmen is given below. Certainly the success of LT 13 was in many ways due to the superb work of this group of individuals and those who assisted them.

Division I
   Quantum Fluids—Prof. I.I. Rudnick, *UCLA*
Division II
   Quantum Solids—Dr. N.R. Werthamer, *Bell Telephone Laboratory*
Division III
   Superconductivity—Prof. T.H. Geballe, Stanford and Dr. P. Hohenberg, *Bell Telephone Laboratory*
Division IV
   Magnetism—Dr. S. Foner, *Francis Bitter National Magnet Laboratory, MIT.*
Division V
   Electronic Properties—Dr. P. Marcus, *IBM Laboratory*
Division VI
   Measurements and Instrumentation—Prof. J. Mercereau, *California Institute of Technology*

The International Advisory Committee for LT 13 also contributed significantly to the success of LT 13 by forwarding to the Organizing Committee information on exciting new work in low temperature physics as well as the names of those younger scientists whose presence at LT 13 should be encouraged.

Within this framework many other individuals helped in the organization and execution of this Conference. Special thanks are due to W.E. Keller, R.L. Mills, L.J. Campbell, W.E. Overton, Jr., R.D. Taylor, W.A. Steyert, and E.R. Brilly of the Los Alamos Scientific Laboratory, who helped with the handling of abstracts and manuscripts of the contributed and plenary papers. At Boulder L.K. Armstrong and L.W. Christiansen did an equally fine job with the local arrangements both prior to and during the Conference.

Also at Boulder, the task of preparing and distributing the Call for Papers, the subsequent announcements, and the Conference Program was carried out by the Conference Publications Committee consisting of W. O'Sullivan and K.D. Timmerhaus. Since both these individuals are continuing to serve as the General Editors of the Conference Proceedings, our grateful thanks are hereby extended to them for their contribution to LT 13 before, during and after the Conference. The Editors acknowledge the services of graduate student Michael Ciarvella, University of Colorado, in the process of preparing the manuscript for publication.

The Conference organizers are also deeply indebted to Professors G. Uhlenbeck and S. Putterman for arranging an evening session on "The Origins of the Phenomenological Theories of Superfluid Helium," to Professor J. Allen for the showing of his new motion picture on Liquid Helium II, and to the organizers of the numerous impromptu but extremely valuable sessions that were developed in addition to the regular program. Clearly the organization and operation of a conference as large as LT 13 can succeed only with the help of many dedicated people. We had hoped that LT 13 would be a stimulating and enjoyable conference. The fact that it met both those expectations is due to all those who wanted it to be a fine conference and worked hard to realize it.

Finally, we wish to express our appreciation to the Session Chairmen and to the individuals who reviewed the manuscripts prior to publication in these Proceedings.

E. F. Hammel
R. P. Hudson
R. H. Kropschot

Acknowledgment

The editors take great pleasure in recognizing the outstanding assistance which Mrs. E. R. Dillman of the University of Colorado provided in the preparation of these four volumes comprising the Proceedings of the 13th International Conference on Low Temperature Physics. Words cannot express our gratitude for conscientious and devoted effort in this thankless task.

K. D. Timmerhaus
W. J. O'Sullivan
E. F. Hammel

Biennial Conferences on Low Temperature Physics

1949 LT 1   Massachusetts Institute of Technology, Cambridge, Massachusetts, U.S.A.
1951 LT 2   Oxford University, Oxford, England
1953 LT 3   Rice University, Houston, Texas, U.S.A.
1955 LT 4   University of Paris, Paris, France
1957 LT 5   University of Wisconsin, Madison, Wisconsin, U.S.A.
1958 LT 6   University of Leiden, Leiden, The Netherlands (scheduled to celebrate the 50th Anniversary of the liquefaction of helium by Kamerlingh Onnes)
1960 LT 7   University of Toronto, Toronto, Canada
1962 LT 8   University of London, London, England
1964 LT 9   Ohio State University, Columbus, Ohio, U.S.A.
1966 LT 10   Moscow, U.S.S.R.
1968 LT 11   University of St. Andrews, St. Andrews, Scotland
1970 LT 12   Kyoto, Japan
1972 LT 13   University of Colorado, Boulder, Colorado, U.S.A.

LT 13 Executive Committee

J. Bardeen	R. H. Kropschot	W. J. O'Sullivan	K. D. Timmerhaus
E. F. Hammel	D. G. McDonald	R. N. Rogers	L. K. Armstrong
R. A. Kamper	F. Mohling	H. Snyder	G. J. Goulette

Contents*

* Tables of contents for Volumes 2, 3, and 4 and an index to contributors appear at the back of this volume.

3. Static Films

4. Flowing Films

5. Superfluid Hydrodynamics

6. Helium Bulk Properties

7. Ions and Electrons

8. Sound Propagation and Scattering Phenomena

9. ^{3}He–^{4}He Mixtures

Fritz London Award Recipient
Presented at
13th International Conference on Low Temperature Physics
August 21–25, 1972

A. A. ABRIKOSOV
Landau Institute for Theoretical Physics
Moscow, U.S.S.R.

Fritz London Award Address

A. A. Abrikosov

Landau Institute for Theoretical Physics
Moscow, USSR

First of all I would like to thank the Fritz London Award Committee for the high appraisal of my work expressed by their awarding to me this prize. It was a particular pleasure for me since the last Soviet physicist to receive the London Award was my teacher, Landau, to whom I and many other Soviet physicists are greatly indebted. His early death in a tragic accident in 1968 was a great loss for science.

I would like to recount some memories of a period of approximately a decade which was of great significance for my scientific life, during which time I had the opportunity almost every day to communicate with Landau and to profit by his advice. Maybe this is the reason why this period was so fruitful for me.

In 1950 Ginsburg and Landau wrote their well-known article on superconductivity. Without the microscopic theory the meaning of several quantities entering their treatment remained unclear, above all the meaning of the "superconducting electron wave function" itself. Nevertheless this theory was the first to explain such phenomena as the surface energy at the superconducting–normal phase boundary and the temperature and size dependence of the critical field and current of thin films.

The experimental verification of the predictions of the Ginsburg–Landau theory concerning the critical fields of thin films was undertaken by my friend Zavaritzki, who was at that time a young research student of Shalnikov's. I often discussed the matter with Zavaritzki. Generally his results fitted the theoretical predictions well. He even managed to observe the change in the order of the phase transition with decreasing effective thickness (i.e., the ratio of the thickness to the penetration depth at a given temperature). To do this, he used the hysteresis of the dependence of the resistance $\rho(H)$ on the field. One day Zavaritzki slightly altered his technique of sample preparation. Usually he evaporated a metal drop onto a glass plate and then put such a mirror into the Dewar vessel. Instead of this, he began to carry out the evaporation inside the Dewar vessel, with the glass plate at helium temperature.

Now we know that in this case the atoms reaching the plate are trapped at the sites where they hit the plate and are unable to move and to form a regular structure. Therefore an amorphous substance is produced, which at every effective thickness will be a type II superconductor. But at that time this was not known, of course.

The critical field versus thickness dependence measured by Zavaritski did not follow the formulas given in the article by Ginsburg and Landau. This gave the impression of a paradox. Apart from its beauty, the theory really explained a lot of things and we were surprised to see that suddenly it had failed.

Discussing with Zavaritzki the possible origin of this discrepancy, we came to the idea that the approximation $\kappa \ll 1$ based on the surface tension data (where κ is the Ginsburg–Landau parameter) could be incorrect for objects such as low-temperature films. Particularly one could suppose that $\kappa > 1/\sqrt{2}$. According to Ginsburg and Landau, the surface energy should be negative under these conditions. Intuitively it was felt that in this case the phase transition in a magnetic field would always be of second order, and this was in fact what Zavaritski observed.

When I calculated the dependence of the critical field on the effective thickness with $\kappa > 1/\sqrt{2}$, it appeared that the theory corresponded to the experimental data. This gave me the courage to state in my article of 1952 containing this calculation that apart from ordinary superconductors whose properties were familiar, there exist in nature superconducting substances of another type, which I proposed to call superconductors of the second group (now called type II superconductors). The division between the first and the second group was defined by the relation between the quantity κ and its critical value $1/\sqrt{2}$.

After this I tried to investigate the magnetic behavior of bulk type II super-conductors. The solution of the Ginsburg–Landau equation in the form of an infinitesimal superconducting layer in a normal sea was already contained in their article. Starting from this solution I found that below the limiting critical field, which is the stability limit of every superconducting nucleation, a new and very peculiar phase arose, with a periodic distribution of the Ψ function, magnetic field, and current. I called it the mixed state.

Landau showed a notable interest in this work and wanted me to publish my results for the vicinity of the upper critical field, which I named H_{c2}. But I wanted to understand how the new mixed state looks in the total range of fields.

At this time I became ill and had to stay in bed for almost three months. One day Landau visited me. The conversation, as in most cases, concerned everything but physics, and Landau sipped with great pleasure from a glass of glühwein, which was not at all like him. And then suddenly I destroyed all this paradise by telling him what I had invented for the mixed state, namely, the elementary vortices. As Landau's eyes fell on the London equation with a δ function on the right-hand side, he became furious. But then, remembering that an ill person should not be bothered, he took possession of himself and said, "When you recover we shall discuss it more thoroughly." Then he hastily bade farewell and disappeared.

He did not come to me any more. When I felt better and appeared at the Institute and tried to tell him again about the vortices, he swore rather ingeniously. At that time I was still very young and did not know the temper of my teacher well enough. He had seen in his life many kinds of pseudoscience, and this made him suspicious toward unusual statements. However, by making some effort and disregarding the noise which he made, one could always "drag" him through any reasonable idea. But at that time I sadly put my calculations in my table drawer "until better times."

But in fact the idea was not so bad. Analyzing the solution that I got close to H_{c2}, I saw that in the plane perpendicular to the field there are points where Ψ becomes zero. The phase of the Ψ function changes by 2π along a path around such a point. I thought about why such singularities should appear, and saw that it could not be otherwise. Indeed the Ginsburg–Landau equation contained not the magnetic field but the vector potential. If the magnetic field does not vary in sign over the whole sample, then the vector potential must increase with the coordinate. But the physical state in a uniform field (this is true close to H_{c2}) must be uniform or at most vary periodically in space. So the increase of the vector potential must be compensated by a change of the phase of the Ψ function. Consider Fig. 1. Let the field be along the z axis and let us choose $A_y = Hx$. Consider the (xy) plane. Let the black points be those I noted earlier. If we want to have a unique determination of the phase we must draw cuts in the plane. We draw them through the black points parallel to the y axis. From the figure it is evident that when going around the points the phase increases by $(\Delta\varphi)_1 = \pi y/a$ if we move along the lower path and by $(\Delta\varphi)_2 = -\pi y/a$ if we move along the upper one. That means that at every cut the gradient of the phase $\partial\varphi/\partial y$ undergoes a jump $2\pi/a$. Using ordinary units (at that time I used the dimensionless Ginsburg–Landau units), one sees that the compensation of the increase of A_y demands

$$(2e/c)\,Hb = 2\pi\hbar/a$$

or

$$Hab = \pi\hbar c/e = \Phi_0$$

which is the flux quantum. Since I used dimensionless quantities, I did not mention the flux quantum on the right but I understood that with a decreasing magnetic field the cell dimensions ab must increase, and as a limit one vortex must be considered where the phase of Ψ changes by 2π in going around it. On the z axis one must have $\Psi = 0$ since otherwise the Ψ function is not uniquely defined. Such a picture gave me the possibility of obtaining the lower critical field H_{c1} and the magnetization curve $M(H)$.

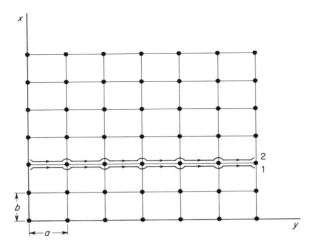

Fig. 1

But as I have said before, at that time (in 1953) after the fuss my professor made I dropped the matter. There was also another reason. Interesting news appeared in a completely different field. I mean in quantum electrodynamics.

After the wonderful work of Schwinger, Feynman, and Dyson, many people were interested in knowing whether it would be possible to sum up all the higher-order corrections and to find formulas for the Green's functions and various physical phenomena without developing in powers of the fine structure constant $e^2\hbar c$. My friend Isaak Khalatnikov as well as myself had an old interest in the problem. At this time an article by S. Edwards was published in which an attempt was made to sum up a ladder sequence of Feynman graphs for the electron–photon vertex part. We studied this article and finally came to the conclusion that Edwards did what he could, but not what was really necessary, since he had no reason to choose this particular sequence. We tried to do something better and wrote, finally, a relation between the electron Green's function and the vertex part which, as soon became clear, was completely wrong. However, when we substituted it into the Dyson equations we obtained various interesting consequences, as, for example, expressions for the electron mass and for the renormalized interaction.

Landau became extremely interested, but being busy with other problems, he had no time to study the new technique of the quantum field theory. So he asked Khalat and me to teach him. I must confess that at that moment we were able to do calculations but had no true understanding of the fundamentals of the theory. Landau swore heavily, but after a month he said that he understood everything. He explained to us that we should find the main sequence of graphs having the highest power of the big logarithm with a given power of the interaction constant. This simple idea put everything in its place. We were indeed successful in calculating the asymptotic expressions for the Green's functions and various physical phenomena at high energies. Moreover, the principle of summing the main Feynman graphs proved afterward to be extremely useful in various statistical problems.

Being occupied with such interesting things, I of course did not turn my mind back to the work on type II superconductors. But here the following happened. Landau had an old interest in the state of He II in a rotating vessel. On the one hand, the helium should not be dragged along by the walls, but on the other hand, this was energetically favorable. In 1955 Landau and E. Lifschitz published an article in which they proposed a layer-type structure with velocity jumps on the layer boundaries. After a year they discovered Feynman's article in which it was shown that in rotating helium elementary vortices appear. Landau immediately said that Feynman was right, and that he and Lifshitz were wrong. Of course it was true. Using our superconducting notation the He II could be considered as an extreme case of a type II superconductor with a correlation length of the order of interatomic distances and with an infinite penetration depth. But at that time this was not so evident.

When Landau began to praise Feynman's work I asked him, "Dau, why are you ready to accept the vortices from Feynman while you flatly rejected the same idea from me?" Landau answered, "You had something different." "Well then, look, please," I said, and produced my calculations from the drawer. This time no objections followed. We discussed the subject very thoroughly and Landau's remarks were very useful.

When everything was put in order I remembered that I had already seen very similar magnetization curves with two critical fields, namely those of superconducting alloys. Digging for the corresponding experimental data, I found the old work (1937) of Shúbnikov, of Khotkévich, Shepelióv, and Riabínin on the magnetization curves of Pb–Tl alloys. The authors prepared their samples very carefully, annealing them for a long time close to the melting temperature. So their samples were probably sufficiently uniform, and this was also confirmed by a rather small hysteresis. But at that time and during the subsequent twenty-five years everybody explained this form of the magnetization curve in terms of the formation of a "Mendelssohn sponge," i.e., of a nonuniform structure with a distribution of critical parameters. It is worth mentioning that even many very good experimentalists finally believed in the mixed state only after they saw the powder figures of a vortex lattice obtained in 1966 by Essmann and Träuble.

So my work was published in 1957 in *Zh. Eksperim. i Teor. Fiz.* In the same year I reported it at a low-temperature conference in Moscow at which some physicists from Oxford and Cambridge took part. Nobody understood a single word. This could be explained, however, by the fact that I had a terrible cold with high temperature and had hardly any idea myself of what I talked about. The translation of the article was then published in the *Journal of Physics and Chemistry of Solids*, but with more than 100 errors in the formulas and text, and this of course did not improve the situation.

In the same year 1957 the famous article by Bardeen, Cooper, and Schrieffer appeared. Everybody became enthusiastic about its ideas, we ourselves among others. Therefore my article did not attract attention. Of course, it would be unjust for me to complain since on the basis of the BCS theory we managed to get a lot of interesting results, and also to develop and to improve greatly the methods of statistical calculations.

Then, in 1961 Kunzler and his colleagues discovered that the alloy Nb_3Sn possesses a critical field of about 100 kOe. Shortly after that, alloys with high critical fields began to be used for constructing superconducting coils. This drew attention to the theory of superconducting alloys.

In my work of 1957 I noted the connection between the quantity κ and the free path length. But as I already said, nobody knew about this article. However, there existed a series of articles by A.B. Pippard in which he qualitatively established the connection between the sign of the surface tension and the ratio of the correlation length to the penetration depth. He also mentioned the decrease of the correlation length with the free path length. On the basis of Pippard's ideas, Bruce Goodman in 1961 rediscovered that alloys with high critical fields have a negative surface tension. Goodman calculated the magnetic properties of such alloys, supposing a simple layer model for the distribution of the normal and superconducting phases. The results were in qualitative agreement with experiment.

I don't know how it happened, but probably somebody told Goodman about my work. What followed was completely incredible. In 1962 Goodman published another article in which he gave a short presentation of my theory and analyzed the experimental data for type II superconductors, comparing them with the predictions of both theories, mine and his own. He came to the conclusion that the vortex model fits experiment much better than the laminar one. So the aim of Goodman's article was to prove that his theory was worse than mine! I have never in my life seen another example of this kind, and took the first opportunity to express my admiration to Goodman.

After this article by Goodman, physicists working on superconductivity finally developed an interest in my work, and I suppose that this was to a considerable extent the cause for the favor done to me by the London Award Committee. Therefore I would like to express once more my gratitude to Bruce Goodman.

So that's the whole story. Of course the fact that the Award has the name of Fritz London is particularly pleasant for me, since he, together with Heinz London, developed the first phenomenological equations of superconductivity. As it eventually turned out, they described the electrodynamics of just the type II superconductors. Also it was Fritz London who introduced the notion of the magnetic flux quantum, which has a direct relation to the subject.

Finally, I would like to say a bit about some of my other activity in the low-temperature field. With one exception it had a much quicker reception. The work, together with that of Gor'kov, on the microscopic theory of superconductivity (high-frequency behavior, Knight shift, and the influence of impurities on the properties of superconductors, particularly magnetic impurities), was accepted rather soon after publication, and has been developed further. It appeared, by the way, that the gapless superconductivity which we predicted for magnetic alloys could also exist under rather different conditions. My work on the Kondo effect and the studies with Khalatnikov of the properties of liquid He^3 were also well treated.

The exception I have mentioned is the theory of semimetals of the Bismuth type, which I constructed together with Falkovski ten years ago. This theory explained such peculiarities as the strange crystal structure, the small number of free carriers, and the large dielectric constant, and gave formulas for the energy spectrum of Bi which fitted experiment very well.

Recently I obtained some new results based on this theory and these will be published soon in the *Journal of Low Temperature Physics*. I hope that they will be interesting to those who work on semimetals and on the metal–insulator transition. I regret that I could not participate in this conference and report these results myself.

In conclusion I would like once more to express my gratitude to the Fritz London Award Committee for the honor it bestowed upon me and also for the opportunity it gave me to prepare this address.

QUANTUM FLUIDS

1
Plenary Topics

Light Scattering As a Probe of Liquid Helium

T.J. Greytak*

Department of Physics and Center for Materials and Engineering Science
Massachusetts Institute of Technology, Cambridge, Massachusetts

Laser light sources and advances in optical spectrometers have made possible the use of inelastic light scattering as a probe of liquid helium. Pure ^{4}He, pure ^{3}He, and isotopic mixtures have been studied in several laboratories. The experiments fall into two categories: Brillouin and Raman scattering. Brillouin scattering measures the spectrum of the equilibrium density fluctuations which contains contributions from all the hydrodynamic modes of the liquid. The velocity and attenuation of high-frequency first and second sound have been obtained by this technique, and it is currently being used to study the dynamics of the critical regions associated with the lambda transition and with the tricritical point. Raman scattering gives information about the elementary excitations in the medium. These experiments have been used to measure roton linewidths, to demonstrate the existence of a bound state of two rotons, and to study optical phonons in solid helium.

The first light scattering experiment to give meaningful information about the nature of liquid helium was reported thirty years ago by Jakovlev.[1]† At that time the Bose–Einstein condensation in a system of noninteracting Bose particles was being considered as a model for the lambda transition in pure ^{4}He. One feature of this model is a divergence of the density fluctuations at the transition temperature. It was known that density fluctuations about thermal equilibrium can give rise to the scattering of light in an otherwise homogeneous medium, an effect generally referred to today as Brillouin scattering. Therefore, if the model were directly applicable, there would be a divergence of the light scattering in the vicinity of the lambda transition, a dramatic phenomenon known as critical opalescence. By using a very powerful mercury arc lamp as a source, Jakovlev was able to see the scattering from ^{4}He below T_λ by eye and to verify estimates[3] that it was of the order of that from dry air at atmospheric pressure. No significant increase in the scattering was observed as the helium was allowed to warm slowly through T_λ. This type of experiment has now been repeated several times with increasing degrees of sophistication,[4–6] but no critical opalescence has ever been observed. These experiments do not rule out all Bose–Einstein models of the transition, however, since it can now be shown that one may have a Bose–Einstein-like transition, without critical opalescence, in a system of *interacting* particles.

---

* Alfred P. Sloan Research Fellow.

† A light scattering experiment in liquid helium was tried in 1932, but no significant difference was noted between the empty and full cells.[2]

Experimentally, the field lay dormant for the next twenty-five years. During this period the Russian theorists were pointing out that a good deal of information could be obtained about superfluid ^{4}He,[3] ^{3}He–^{4}He mixtures,[7] and the Fermi liquid ^{3}He,[8] by studying the spectrum of the scattered light. However, these experiments were not feasible at that time for two reasons. First, the scattering is very weak due to the small value of the atomic polarizability α of the helium atom. It is interesting to note that the smallness of α also contributes to the fact that helium is the only permanent liquid, since the strength of the van der Waals forces between atoms depends upon α. Second, the frequency shifts involved in the spectrum of the scattered light were small, comparable to the linewidths of the best sources then available. Today, however, the use of laser light sources and advances in optical spectrometers have made possible not only Brillouin scattering experiments, which study the thermal fluctuations in helium, but also Raman scattering experiments, which give information about the elementary excitations.

In a modern Brillouin scattering experiment a monochromatic laser beam is passed through the liquid and the spectrum of the scattered light is measured at some fixed scattering angle. The scattering arises from the local inhomogeneities in the dielectric constant caused by the thermal fluctuations in the number density of helium atoms. Therefore, one expects that the angular dependence of the scattering will be governed by the spatial correlation of the fluctuations (diffraction) and the spectrum of the scattered light will contain information about their time evolution (modulation). In particular, the amount of light scattered at a frequency shift $\omega/2\pi$ from the incident frequency is proportional to the spectral function $S(K, \omega)$, which is defined to be the space and time Fourier transform of the density correlation function:

$$S(K, \omega) = \int \int \left[\exp(i\mathbf{K} \cdot \mathbf{R}) \exp(-i\omega\tau) \right] < n(\mathbf{r}, t)\, n(\mathbf{r} + \mathbf{R}, t + \tau) > d\mathbf{R}\, d\tau$$

where $\mathbf{K}$ is the wave vector of the fluctuations being investigated and is fixed, once the scattering angle θ is determined, by the relation $\mathbf{K} = 2\mathbf{k}_0 \sin(\theta/2)$ with $\mathbf{k}_0$ the wave vector of the incident light in the medium. For example, when scattering at $90°$ one investigates those fluctuations whose wavelength is $1/\sqrt{2}$ times the wavelength of the light. Neutron and x-ray scattering in helium also measure $S(K, \omega)$, so in that sense the three scattering probes are similar. The important difference, however, is that neutrons and x rays can study high values of K, where $2\pi/K$ is of the order of the interatomic spacing. In Brillouin scattering, on the other hand, typical values of K are much smaller, about one-thousandth of the wave vector of a roton. For temperatures above about $1°$K the mean free paths of the elementary excitations in helium are small compared to the wavelength of light. Therefore, the dynamics of the density fluctuations studied in Brillouin scattering can be described by hydrodynamic equations. In physical terms, one is not able to study the individual elementary excitations by this technique; rather, one is dealing with their collective behavior.

To investigate the form of $S(K, \omega)$ in the hydrodynamic limit, it is convenient to first decompose the density $n(\mathbf{r}, t)$ into its spatial Fourier components $n(\mathbf{K}, t)$:

$$n(\mathbf{K}, t) = (1/\sqrt{v}) \int \left[\exp(i\mathbf{K} \cdot \mathbf{r}) \right] n(\mathbf{r}, t)\, d\tau$$

This leads to an alternative expression for $S(K, \omega)$:

$$S(K, \omega) = \int e^{-i\omega\tau} \langle n(K, t)\, n^*(K, t + \tau) \rangle \, d\tau$$

which shows that it is the frequency spectrum of the time correlation function for the amplitude of a fluctuation at a single wave vector $\mathbf{K}$. To calculate this correlation function, one uses a hypothesis due to Onsager which is now known to be a result of the fluctuation-dissipation theorem: The time dependence of the equilibrium correlation function is identical to the time evolution of a properly prepared macroscopic disturbance. Imagine that at $t = 0$ a disturbance in the number density, sinusoidal in space, is impressed on the liquid. The liquid is then released and the subsequent motion is recorded. Because one is dealing with small disturbances, the response of the liquid is linear and the spatial form of the disturbance, a sine wave, will not be altered, but its amplitude will change in time and gradually damp out to zero. $S(K, \omega)$ is proportional to the time Fourier transform of the symmetrized form of this behavior. The problem of finding $S(K, \omega)$ is now reduced to a well-posed initial value problem. The appropriate equations which must be solved are the linearized equations of two-fluid hydrodynamics. One is then looking for the solution of a set of coupled linear equations constrained by given initial conditions. The problem can, in principle, be solved exactly and unambiguously; however, it is algebraically tedious. Numerous authors have considered the problem.[3,7-13] Fortunately, one need not go through the mathematics to get an overall picture of the resulting spectra. Similar equations and initial conditions arise when solving for the transient response of an electronic circuit or of a set of coupled mechanical oscillators. The response is a linear combination of the various normal modes of the system. One expects, then, to see in the correlation function for $n(\mathbf{K}, t)$ and in $S(K, \omega)$ the characteristics of each of the normal modes of the two-fluid hydrodynamic equations.

The hydrodynamic normal modes of pure ^{4}He below T_λ are first and second sound. Figure 1(b) is a schematic representation of the spectrum of Brillouin scattering for this case. Part of the response to our imagined initial disturbance of the medium would oscillate at the first-sound frequency and would slowly damp out due to dissipative processes. This causes a pair of lines to appear in $S(K, \omega)$ centered at $\omega_1 = \pm V_1 K$ with half-width at half-height $\Gamma_1 = V_1 \alpha_1$, where V_1 and α_1 are respectively the velocity and amplitude attenuation coefficient of first sound at the frequency ω_1. The rest of the response to the initial disturbance would oscillate at the much lower second-sound frequency, giving rise to a pair of lines in $S(K, \omega)$ at $\omega_2 = \pm V_2 K$, with $\Gamma_2 = V_2 \alpha_2$. Obviously, the observation of such spectra would allow one to measure the velocity and attenuation of first and second sound. The advantage of this technique over conventional accoustic methods is that the measurements are made on very high-frequency sound. The frequency of the first sound studied in a Brillouin experiment is typically ~ 700 MHz and that of second sound could be as high as 100 MHz. As the temperature of the helium is raised toward T_λ, however, the velocity of second sound decreases to zero. Above T_λ this mode is no longer oscillatory, but exhibits a pure exponential decay in time. The corresponding Brillouin spectrum is shown in Fig. 1(a) and is characteristic of simple normal fluids. The second-sound mode below T_λ becomes the entropy diffusion mode above T_λ.

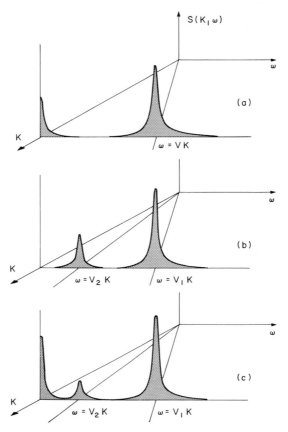

Fig. 1. Schematic representation of $S(K, \omega)$ for three systems: (a) pure ^{4}He above T_λ, (b) pure ^{4}He below T_λ, (c) a superfluid mixture of ^{3}He and ^{4}He.

The first observation of Brillouin scattering from ^{4}He was by St. Peters et al.[14] in 1968 and similar independent work was soon reported by Pike et al.[15] These results were used to study the absorption and dispersion of high-frequency first sound. However, in these experiments second sound did not appear in the spectra. The reason is that although second sound is a normal mode, it does not couple strongly to the density. It is primarily a temperature oscillation and at saturated vapor pressure and below T_λ the thermal expansion coefficient is very small. Looked at another way, one often thinks of second sound in the two-fluid model as an out-of-phase oscillation of the superfluid and normal fluid densities in which the total density does not change appreciably. This effect has been demonstrated very nicely in the experiments by Pike.[16] He was able to see the central line well above T_λ, watch it diminish in amplitude as the temperature was lowered, and observe that it disappeared in the noise near T_λ and below.

Actually, the weakness of the second-sound scattering only pertains near saturated vapor pressure. In 1968 Ferrell et al.[17] pointed out that close to the lambda line at higher pressures the thermal expansion coefficient becomes much larger and the scattering from second sound should become comparable to that from first sound. They also pointed out that close to the lambda line the conventional equations of two-fluid hydrodynamics are not valid, but must be modified to take into account the critical behavior near the transition. Up to now we have thought of Brillouin scattering as a means of measuring the parameters which enter well-known equations, those parameters which are involved in the velocities and attenuations, for example. But now we have the opportunity to get information on the nature of the equations themselves. In particular it is interesting to study the dynamics of the fluctuations when the coherence length in the liquid becomes comparable with the wavelength of the fluctuations being observed.

The scattering from second sound in ^{4}He under pressure has now been observed by Winterling et al.[18] and is being reported at this conference. Figure 2 shows typical traces from this experiment. Notice that as T_λ is approached from below, the second-sound splitting decreases and its intensity increases. The strong attenuation of first sound near T_λ is also noticeable. In these experiments the coherence length becomes comparable to the wavelength of the fluctuations at $T_\lambda - T \sim 0.1$ m°K; therefore, the intensity and shape of the second-sound spectrum are being examined very carefully in this region. Measurements of the temperature dependence of the total intensity of the Brillouin scattering under similar conditions have been made by Palin et al.,[6] and at this conference they are reporting spectral measurements as well.

Now let us turn to the very interesting problem of Brillouin scattering from ^{3}He–^{4}He mixtures. Figure 1(c) shows the spectrum to be expected in the superfluid

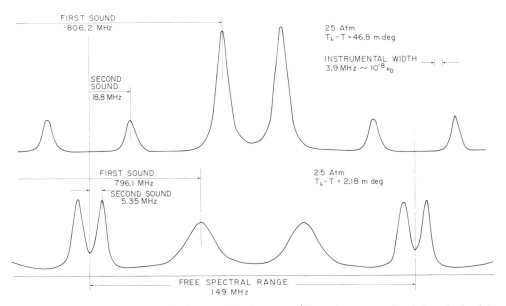

Fig. 2. Experimental traces of Brillouin scattering from pure ^{4}He under pressure just below the lambda temperature. See Ref. 18.

phase of the mixture. The general behavior of the Brillouin doublet due to first sound does not differ markedly from that expected from a simple fluid. Second sound is still a valid normal mode of the system; however, since the ^{3}He atoms have a larger effective volume than the ^{4}He atoms and move only with the normal fluid, the second sound couples more strongly to the density than is the case in pure ^{4}He. This makes the second-sound doublet more prominent in the spectrum. A new line, centered at $\omega = 0$, also appears in the spectrum due to nonpropagating concentration fluctuations. Pike et al.[19] and Palin et al.[20] were the first to observe spectra of this type in the mixtures. Their experiments were done down to 1.25°K. Benjamin et al.[21] at this conference report results obtained at temperatures as low as 0.5°K. Measurements made on such spectra are a valuable source of thermodynamic data on the mixture. There are, however, several regions of the phase diagram which are of particular interest. The transition from the normal to the superfluid phase of the mixture offers another lambda line along which to study the critical behavior of second sound. Also of interest along the lambda line is the behavior of the mass diffusion constant, which determines the width of the central line of the spectrum. At low temperatures in the mixtures the ^{3}He quasiparticles are the dominant elementary excitation and the primary source of the normal fluid density. In this region Brillouin scattering can give significant information about the ^{3}He quasiparticles and their interactions. In particular there is a direct analogy between the Brillouin spectrum of a classical gas[22] and the inner three components (scattering from the gas of quasiparticles) of the mixture spectrum. Finally one has the tricritical point, the junction of the lambda line and the phase separation curve. As this point is approached the concentration fluctuations diverge, giving rise to critical opalescence. No spectral measurements have as yet been made in this region. However, Watts et al.[23] have studied the divergence of the total scattered intensity and their experiments are contributing to the understanding of the critical exponents associated with the tricritical point.

Brillouin scattering is characterized by scattering from fluctuations about thermal equilibrium. There are other experimental methods in which light scattering is used to study first[24,25] and second sound[26-28] in helium; however, these techniques require that the sound be injected into the medium by various means. Unfortunately, there is no space to include them in this review.

Conservation of momentum in a first-order process limits the momentum transfer to the medium to $2\mathbf{k}_0$, or about one-thousandth of the momentum of a roton. In 1968 Halley[29] pointed out that if one considered second-order processes, in which two elementary excitations of nearly equal and opposite momenta are created in the medium, the corresponding spectra would in principle contain contributions from all regions of the dispersion curve. More specifically, the second-order Raman spectrum at an energy loss E is a measure of the density of two excitation states[29,30] $\rho_2(\mathbf{K} = 0, E)$ evaluated at total wave vector $\mathbf{K} = 0$ and total energy E. Second-order Raman spectra have been studied in solids for some time, but their interpretation is often difficult due to the fact that they may contain overlapping contributions from many regions of the Brillouin zone. However, the isotropic nature of a liquid and the simple dispersion curve for the excitations in ^{4}He lead to a much more favorable situation. The primary contribution to the spectrum comes from the rotons (of minimum energy ε_0) and consists of a well-defined line near

$E = 2\varepsilon_0$. There are smaller contributions from the maximum of the dispersion curve and from the plateau region, located at values of E near twice the corresponding dispersion curve energies.

The original motivation for doing "two-roton Raman scattering" was to use the high optical resolution available in such experiments to obtain more accurate values for the roton energies and linewidths than had been found by neutron scattering. In fact, the first Raman scattering experiments in ^{4}He by Greytak and Yan[31] were able to measure roton linewidths in the temperature region from 1.3 to 1.9°K. However, in these experiments the exact form of the spectrum did not quite agree with the theoretical prediction. In particular, the scattering from the maximum in the dispersion curve was weaker than expected. In order to explain the discrepancy, Ruvalds and Zawadowski[32] and Iwamoto[33] showed that the presence of even a weak interaction between the excitations can greatly modify $\rho_2(\mathbf{K} = 0, E)$ from the form that it would have in the absence of interactions. Ruvalds and Zawadowski found that the absence of a peak in the spectra near twice the energy of the maximum of the dispersion curve could be explained by a depletion of the density of pair states in this region caused by an interaction between the excitations. They also showed that the same interaction, if assumed to exist between rotons, would enhance the density of pair states in the vicinity of $2\varepsilon_0$ corresponding to a two-roton resonance or bound state.

In order to investigate the possibility of such a bound state, Greytak *et al.*[34] made higher-resolution studies of the Raman spectrum in the vicinity of the two-roton line. Figure 3 shows a typical experimental trace from those experiments. The maximum in the scattering is seen to occur at an energy shift less that $2\varepsilon_0$. This cannot be attributed to the finite linewidth of the individual rotons or to the finite instrumental profile of the spectrometer. Both of these effects would, in the absence of interaction, cause the maximum of the spectrum to occur at energy shifts greater than $2\varepsilon_0$, as shown by the dashed curve in the figure. The dotted curve in the figure shows the best fit of the data to the theory of Ruvalds and Zawadowski and corresponds to a binding energy for the roton pair of $0.37 \pm 0.10°$K. The primary source of the uncertainty in the value of the binding energy is the uncertainty in the neutron measurements of ε_0. The instrumental width of the spectrometer used in this experi-

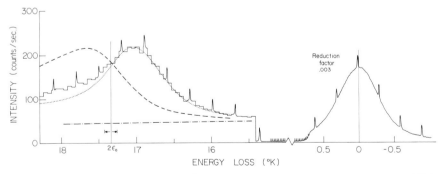

Fig. 3. The Raman spectrum of liquid ^{4}He at 1.2°K. The strong peak at zero energy shift is caused by Brillouin scattering and indicates the instrumental profile. The dotted curve is a theoretical fit to the data based on a two-roton bound state. The dashed curve would correspond to noninteracting rotons. The dot-dashed line is the background level. See Ref. 34.

ment was about twice the binding energy, so the fine details of the intrinsic spectrum are obscured in the experimental trace. For the same reason, reliable measurements of the width of the bound state could not be made. Figure 4 shows the intrinsic spectrum, that which would be seen by a spectrometer of infinite resolution, calculated from the theory of Ruvalds and Zawadowski using the measured binding energy and values of the single-roton linewidth derived from the measurements of Greytak and Yan. The spectrum in the noninteracting case at $T = 0$ is zero for $E < 2\varepsilon_0$ and is proportional to $(E - 2\varepsilon_0)^{-1/2}$ for $E > 2\varepsilon_0$. For finite temperatures the singularity at $E = 2\varepsilon_0$ disappears and the density of unbound roton pair states is represented by a smooth function as shown by the dashed curve in Fig. 4. The result of the interaction is to deplete the unbound states in the vicinity of $E = 2\varepsilon_0$ and to create a new state below the continuum of unbound states. In the theory of Ruvalds and Zawadowski the width of this bound state is twice the single-roton linewidth. The two significant pieces of information obtained from two-roton Raman experiments, the binding energy of the pair and the single-roton linewidths, are now being used together with other experimental data to develop more detailed models for the interaction between rotons.[35-38]

Let us return to the consideration of the Raman spectrum as a whole. At low temperatures the roton peak is the most prominent feature of the spectrum but there are other contributions as well. At slightly larger energy shifts there is a contribution from the creation of two excitations near the maximum of the dispersion curve. This does not form a distinct peak in the spectrum due to the effect of interactions, as mentioned above, but it must still be taken into account when trying to fit the entire measured spectrum. At a shift of twice the energy of the plateau on the dispersion curve there is a broad peak which can be associated with the creation

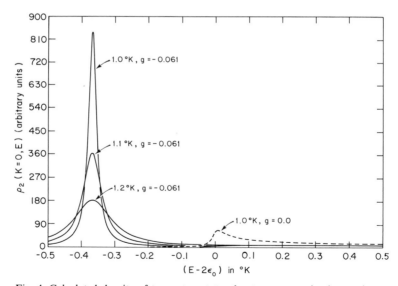

Fig. 4. Calculated density of two-roton states due to an attractive interaction at several temperatures. The dashed curve on the right corresponds to the non-interacting case. Note the well-defined separation of the bound state even at 1.2°K.

of a pair of excitations on the plateau. Beyond this, however, there is a definite long exponential tail to the Raman spectrum which has not, as yet, been explained in terms of any simple processes involving the elementary excitations. This tail seems to be more closely related to the "collision-induced" Raman scattering found in classical monatomic liquids and dense gases.[39] In that case the spectrum of the light is governed by the motion of individual atoms and the dynamics of the collisions with their neighbors. It may be, then, that this portion of the helium Raman spectrum can be more easily understood from an atom–atom correlation approach than from the point of view of elementary excitations.

As the temperature of the helium is raised toward T_λ the two-roton peak in the spectrum broadens due to the increasing single-roton linewidth. Before the lambda temperature is reached the distinction between roton and plateau scattering is lost and the spectrum exhibits only one broad maximum at a finite energy shift. This particular form of the spectrum persists above T_λ[40] even when the pressure is raised[41] almost to the solidification line. Of course the position and exact shape of the broad maximum are pressure and temperature dependent. A similar spectral form has also been found by Slusher and Surko[41] for Raman scattering in liquid ^{3}He. The theoretical interpretation of these broadened spectra, where the contributions from specific elementary excitations are not immediately evident, is a challenging problem. Can the liquid ^{3}He results, for example, give information about short-wavelength zero-sound modes? These questions are still under experimental and theoretical investigation.

Slusher and Surko[41] also studied the Raman scattering from solid ^{3}He and ^{4}He. In the hcp phases of both solids first-order Raman scattering could be observed from the transverse optic phonons. The energy and linewidth which they measured for these phonons have been important to the understanding of the lattice dynamics of quantum crystals. They also observed a broad, shifted peak due to the second-order Raman scattering in the solids which was remarkably similar to the scattering from the liquid state when the crystals were melted. A comparison of the data with theoretical calculations based on two-phonon scattering[42] seemed to indicate that scattering from more than two phonons must be used to explain the intensity and spectral shape of the broad peak in the solids.

One can see from this brief review that light scattering provides valuable new information about the properties of liquid and solid helium, and that the results obtained complement rather than supplant the results of neutron and x-ray scattering.

References

1. I.A. Jakovlev, *J. Phys. USSR* **7**, 307 (1943).
2. J.C. McLennan, H.D. Smith, and J.O. Wilhelm, *Phil. Mag.* **14**, 161 (1932).
3. V.L. Ginsburg, *J. Physics (Moscow)* **7**, 305 (1943).
4. A.W. Lawson and Lothar Meyer, *Phys. Rev.* **93**, 259 (1954).
5. H.Grimm and K. Dransfeld, *Z. Naturforsch.* **22A**, 1629 (1967).
6. C.J. Palin, W.F. Vinen, and J.M. Vaughan, *J. Phys. C* **5**, L139 (1972).
7. L.P. Gorkov and L.P. Pitaevskii, *Soviet Phys.—JETP* **33**, 486 (1958).
8. A.A. Abrikosov and I.M. Khalatnikov, *Soviet Phys.—JETP* **34**, 135 (1958).
9. B.N. Ganguly and A. Griffin, *Can. J. Phys.* **46**, 1895 (1968).
10. A. Griffin, *Can. J. Phys.* **47**, 429 (1969).

11. B.N. Ganguly, *Phys. Lett.* **29A**, 234 (1969).
12. W.F. Vinen, in *Physics of Quantum Fluids* (Tokyo Summer Lectures in Theoretical and Experimental Physics, 1970, R. Kubo and F. Tanako, eds.), Syokabo, Tokyo, Japan.
13. B.N. Ganguly, *Phys. Rev. Lett.* **26**, 1623 (1971); *Phys. Lett.* **39A**, 11 (1972); W.F. Vinen, *J. Phys. C* **4**, L287 (1971).
14. R.L. St. Peters, T.J. Greytak, and G.B. Benedek, *Bull. Am. Phys. Soc.* **13**, 183 (1968); *Opt. Comm.* **1**, 412 (1970).
15. E.R. Pike, J.M. Vaughan, and W.F. Vinen, *J. Phys. C* **3**, L40 (1970).
16. E.R. Pike, *J. Physique* **33**, Cl-25 (1972).
17. R.A. Ferrell, N. Menyhàrd, H. Schmidt, F. Schwabl, and P. Szépfalusy, *Ann. Phys. (N.Y.)* **47**, 565 (1968).
18. G. Winterling, F. Holmes, and T. Greytak, this volume.
19. E.R. Pike, J.M. Vaughan, and W.F. Vinen, *Phys. Lett.* **30A**, 373 (1969).
20. C.J. Palin, W.F. Vinen, E.R. Pike, and J.M. Vaughan, *J. Phys. C* **4**, L225 (1971).
21. R.F. Benjamin, D.A. Rockwell, and T.J. Greytak, this volume.
22. T.J. Greytak and G.B. Benedek, *Phys. Rev. Lett.* **17**, 179 (1966).
23. D.R. Watts, W.I. Goldburg, L.D. Jackel, and W.W. Webb, *J. Physique* **33**, Cl-155 (1972); D.R. Watts and W.W. Webb, this volume.
24. M.A. Woolf, P.M. Platzman, and M.G. Cohen, *Phys. Rev. Lett.* **17**, 294 (1966).
25. W. Heinicke, G. Winterling, and K. Dransfeld, *Phys. Rev. Lett.* **22**, 170 (1969).
26. G. Jacucci and G. Signorelli, *Phys. Lett.* **26A**, 5 (1967).
27. S. Cunsolo, G. Grillo, and G. Jacucci, in *Proc. 12th Intern. Conf. Low Temp. Phys., Kyoto, 1970,* Academic Press of Japan, Tokyo, (1971).
28. D. Petrac and M.A. Woolf, *Phys. Rev. Lett.* **28**, 283 (1972).
29. J.W. Halley, *Bull. Am. Phys. Sov.* **13**, 398 (1968); *Phys. Rev.* **181**, 338 (1969).
30. M.J. Stephen, *Phys. Rev.* **187**, 279 (1969).
31. T.J. Greytak and J. Yan, *Phys. Rev. Lett.* **22**, 987 (1969); in *Proc. 12th Intern. Conf. Low Temp. Phys. Kyoto, 1970,* (Academic Press of Japan, Tokyo, 1971).
32. J. Ruvalds and A. Zawadowski, *Phys. Rev. Lett.* **25**, 333 (1970).
33. F. Iwamoto, *Prog. Theor. Phys. (Japan)* **44**, 1135 (1970).
34. T.J. Greytak, R. Woerner, J. Yan, and R. Benjamin, *Phys. Rev. Lett.* **25**, 1547 (1970).
35. A. Zawadowski, J. Ruvalds, and J. Solana, *Phys. Rev. A* **5**, 399 (1972).
36. J. Solana, V. Celli, J. Ruvalds, I. Tüttö, and A. Zawadowski, to be published.
37. J. Yan and M.J. Stephen, *Phys. Rev. Lett.* **27**, 482 (1971).
38. R.J. Donnelly, to be published.
39. J.P. McTague, P.A. Fleury, and D.B. DuPre, *Phys. Rev.* **188**, 303 (1969).
40. E.R. Pike and J.M. Vaughan, *J. Phys. C* **4**, L362 (1971).
41. R.E. Slusher and C.M. Surko, *Phys. Rev. Lett.* **27**, 1699 (1971).
42. N.R. Werthamer, *Phys. Rev.* **185**, 348 (1969); N.R. Werthamer, R.L. Gray, and T.R. Koehler, *Phys. Rev. B* **4**, 1324 (1971).

Phase Diagram of Helium Monolayers*

J.G. Dash

Department of Physics, University of Washington
Seattle, Washington

Introduction

In the past few years there has been increased interest in the physics of thin films, stimulated largely by the technical achievements and promise in solid state electronics and in superconductivity. Another important stimulus comes from surface science research, which has brought new and highly detailed insights into the properties of the top few atomic layers of solids. The field of adsorption can be considered as a part of "thin films," with a subdivision for physisorption and a further subdivision for adsorbed helium films. But there has been very little communication between the helium film subfield and the other parts of the parent field, even though there appear to be some extremely important areas of overlap and much can be learned from each. The reason for this lack of communication is that those who work on helium films have been primarily interested in the films as special cases of the bulk phases. To some extent they are: Certainly there is a range of film thickness in which a film behaves mainly as a thin slice of the bulk. But when they become thin enough this approximation no longer holds: A film begins to take on such different character that one must treat the layer as a new regime. This change does not occur suddenly; furthermore, the thickness at which it occurs depends on what properties are being considered. But where it does occur, there one must take account of specific properties of the substrate. In the limit of a single monolayer the substrate properties can become supremely important; and one must be at least as concerned with the characteristics of the substrate and the surface–helium interactions as with the helium–helium interactions. We speak as new converts to this view, which was forced on us by work done in the past couple of years.

I will describe the steps leading to our conversion: This may help to make understandable the new results vis à vis older work.

Historical

The heat capacities of high-coverage He monolayers adsorbed on various surfaces have seemed to indicate that they are two-dimensional solids. The surfaces have been of quite different materials: jeweler's rouge[1], porous Vycor glass[2], N_2-plated Vycor[3], Ar-plated Cu[4,5] Ne-plated Cu[6] and bare Cu.[7,8] This has been

* Research supported by The National Science Foundation.

pleasing: One can readily picture a dense monolayer held down by van der Waals forces to the wall, and with an excitation spectrum approaching a two-dimensional solid, with a characteristic temperature of the observed order of magnitude. But trouble develops when one studies lower densities, i.e., monolayers of less than maximum coverage. For we expected that at some moderate fractional coverage the solid would melt, giving some kind of anomaly in the heat capacity, and that at sufficiently low density it would resemble a two-dimensional gas. Now Goodstein and McCormick had seen a broad maximum at intermediate coverage for ^{4}He on Ar-plated Cu, and we speculated that this was a diffuse melting transition.[4] Stewart then undertook a study, in a new cell, of ^{3}He and ^{4}He down to 0.1 monolayer, specifically to detect the two-dimensional gas phase. But he found solidlike behavior persisting down to the lowest coverage.[5] We interpreted this to mean that the film clustered into dense islands with quite large lateral binding energies: The binding could possibly be strongly enhanced by certain substrate interactions. But estimates of the lateral binding[9] and substrate enhancement that were made by Schick and Campbell[10] were an order of magnitude weaker. There was one barely possible alternative model, that the solidlike properties might in some way be due to im-mobile, sitewise adsorption, but detailed calculations by Novaco and Milford[11] removed that possibility: They showed that independent He adatoms on these surfaces should be highly mobile, so that the low-lying single-particle states are virtually the same as for a two-dimensional gas.

We then undertook a more determined search for the missing gas phase. Princehouse[7] extended the range of measurements to still lower coverages and higher temperatures, so that even if the lateral latent heat of sublimation of the solid were as great as $60k_B$ there would be enough sensitivity to detect the 2D vapor. The vapor did not appear, which meant that the latent heat must be even greater. But such strong binding is now beyond what can be attributed to even enhanced He–He interactions: Such magnitudes can only be found in the direct He–surface interaction. And that cast into serious question the whole basis for our models: that the substrates are essentially acting only as attractive regular surfaces. Now, we had not been quite so naive as to believe that our surfaces were perfectly uniform. There is a vast literature on adsorption, all indicating that heterogeneity is the rule, with very few exceptions. In fact that was one of the reasons we chose to preplate the Cu substrates with Ar (suggested by G. D. Halsey) to make the surface more uniform for the adsorption. Now here were results that indicated that a monolayer of Ar plating is not enough. Stewart had found no change with two layers of Ar or Ne.[6] However, although we were being forced to invoke surface heterogeneity, there was no explanation of how that could produce the observed heat capacities, nor how one could obtain similar heat capacities on different surfaces.

Then Roy and Halsey[12] proposed a model which seems both plausible and prom-ising. They proposed that the surface heterogeneities are in the form of long-range variations in the van der Waals attraction, on a scale greater than typical interatomic spacings. They give no explanation for the long ranges but it would not be incon-sistent with what one knows about surfaces. The Roy and Halsey model pictures the binding energy profile on a typical surface as resembling the surface of the moon, with regions of stronger and weaker binding, causing appreciable lateral fields. On such a surface low-coverage films would flow into the most attractive regions where they would force dense clusters, solidified by the strong lateral fields. The density and other properties of these clusters would be characteristic of both the substrate and the helium, and if the surfaces are heterogeneous enough, the films might act as 2D solids having only moderate variations with coverage and substrate.

New experimental results confirmed this view. For some time we had been attempting to develop other substrates suitable for the low-temperature film studies. A number of experimental requirements must be met, some of which tend to be mutually exclusive and it takes some appreciable effort to develop each new material. There was some partial success with a sintered Ag substrate:[13] It gave results qualitatively similar to He on Cu and Ar-plated Cu. Then Bretz successfully developed a graphite substrate, and with his very first results the missing 2D gas phase was found.[14] The rest of this paper will concern the properties of helium on graphite, the phase diagram explored during the past two years by several teams made up of Bretz, Hickernell, Huff, McLean, Vilches, and myself.[14-17] But first we examine some independent evidence concerning the uniformity of graphite.

Graphite Substrates

In the late 1940's and early 1950's it was discovered that graphitized carbon black, used as an adsorbent for various simple gases, gave vapor pressure isotherms and isosteres that were much closer in form to the theoretical shapes for uniform surfaces. In subsequent years many studies were made of adsorption on graphitized carbon black; it has become the prototype of a uniform substrate, and one of the very few exceptions to the rule of heterogeneity. The reasons for its uniformity include crystalline and chemical perfection and a low density of growth steps, strains, and dislocations on the surface. Within the past few years another exceptional substrate has been found, exfoliated graphite. Thomy and Duval[18] have studied the vapor pressures of several gases on exfoliated graphite, and they find isotherms which have even sharper steps than on graphitized carbon black. Furthermore, they find considerable structure in the isotherm of the first layer, indicating 2D gas, liquid, and solid phases. Both graphitized carbon black and exfoliated graphite are virtually pure basal plane surfaces. Thus, there is extensive evidence for the unusual uniformity of these substrates. The evidence is largely based on vapor pressures of gases other than helium. The material that Bretz used in developing a graphite substrate suitable for low-temperature calorimetry is Grafoil,* a commercial product designed for high-temperature gaskets. It appears to consist of exfoliated natural crystals, interleaved as in a roughly stacked deck of cards, rolled into wide sheets of uniform thickness. It has virtually 100% basal plane exposure, considerable surface area, excellent thermal conductivity along the sheet direction, and very high chemical purity.

Since the first indications of success more and more of our laboratory's activities were turned toward work with this substrate. The specific heat work has been carried out during the past two years in three independent calorimeters with exfoliated graphite substrates; the results are summarized in the phase diagram of the first monolayer (Fig. 1).

---

* Trademark of Union Carbide Corp.

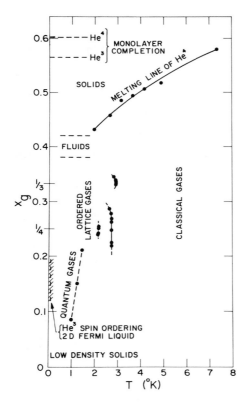

Fig. 1. Phase diagram of the first He monolayer on exfoliated graphite.*

Phases of the First Helium Monolayer on Graphite

Figure 1 shows regions which are 2D analogs of the principal bulk states of matter, as well as several that resemble regimes unique to the heliums. In addition we find some which have no analogs in bulk helium; these can only be understood in terms of the combination of He and detailed surface properties. Most of our evidence is based on specific heats, but some more recent NMR work will be mentioned in support. Recent measurements at California Institute of Technology, which are reported in contributed paper sessions at this conference, appear to be consistent with our observations.[19]

Two-Dimensional Classical Gases. This is the simplest phase to recognize; in this region the specific heats are k_B/atom, independent of T over an appreciable region. Except for the highest and lowest coverages, all films showed this signature at relatively high temperatures, near 4°K. This behavior is consistent with the calculations of Hagen et al.[21] that the independent adatom states of He on the basal plane graphite are highly mobile along the surface.

---

* **Note added in proof:** Further analysis of heat capacity data together with newer measurements have led to different interpretations of several portions of the phase diagram. For details see Bretz et al.[26]

Two-Dimensional Quantum Gases. As the moderately low-coverage films are cooled below about 2°K, the heat capacities of the two isotopes diverge, ^{3}He falling below the classical value and ^{4}He rising to form rounded peaks near 1°K. The ^{3}He behaves as 2D ideal Fermi gases, with Fermi temperatures of a few °K, approximately twice as large as would correspond to noninteracting ^{3}He at the experimental densities. The ^{4}He disagrees qualitatively with a 2D model, for the 2D Bose gas has a heat capacity identical to a 2D Fermi gas the same mass, nuclear spin, and density. But with a minor complication on the 2D gas model, there is reasonable agreement with the results.

If one supposes that the graphite surface is not completely uniform but has a small degree of heterogeneity in the form of a weak lateral field, then it turns out that the heat capacity of a 2D ideal Bose gas develops a pronounced low-temperature cusp even in quite weak fields.[22] The 2D Fermi gas, on the other hand, is much stiffer, and is not appreciably perturbed if the fields are not too large. A semiquantitative fit to the data was obtained for total energy variations of 1 k_B in the He–substrate binding, which is only about 0.6% of the total binding.

Ordered Lattice Gases. As the density is increased, anomalies develop in both isotopes. The shapes are very sensitive to density, growing to very sharp and strong symmetric peaks near 3°K. The densities for the best peaks are approximately 0.6 of a completed monolayer. A better density gauge is the ratio of He atoms to substrate sites. The "critical densities" correspond to one He/three basal plane hexagons on the graphite surface. At this density a classical film can be set into a regular triangular structure. A 2D classical lattice gas can be put into direct correspondence with a 2D Ising magnet. The film specific heat is in fact quite similar to that of the Ising model. In the critical region it has a logarithmic shape. The coefficient is approximately three times that of the known magnetic solutions, consistent with theory, since the number of spins in the magnetic system translates into the number of sites in the lattice gas; thus the factor three corresponds to the critical density one-third. However, the film results are not completely in accord with such a simple model. For example, the behavior near 4°K indicates a 2D mobile gas, while below the peaks there is no evidence of thermal mobility. An NMR study of ^{3}He has just been made by Rollefson.[20] He finds direct evidence, through linewidth measurements, for mobility changes consistent with the heat capacity evidence. But if mobility is changed on passing through the ordering transition, it cannot be understood in terms of single-particle states but is instead a collective property, and may be somewhat analogous to the metal–insulator transition of electronic solids.

We also show additional ordered phases near density $x_g = \frac{1}{4}$. These phase boundaries are traced out by two lines of weak specific heat peaks. The phases may correspond to the two different regular arrays which can exist at this density.

Spin Ordering in ^{3}He. As moderately low-density ^{3}He is cooled below 1°K the heat capacities first fall below the Fermi curves and then rise to form rounded maxima near 0.1°K. Several features indicate that the maxima are due to nuclear spin ordering. There are no low-temperature anomalies in ^{4}He. If one compares specific heat curves of low-coverage ^{3}He monolayers with liquid ^{3}He, the region of the maxima in the films seems quite similar to the "shoulder" region of the liquid at about the same temperature; in this region the liquid experiences a rapid decrease in nuclear susceptibility. The entropy of low-coverage ^{3}He films increases rapidly near 0.1°K, and above this region it exceeds that of ^{4}He by a nearly constant amount, approximately $Nk_B \ln 2$. As the film density is increased toward the critical density $x_g = \frac{1}{3}$ the maxima are gradually weakened and finally disappear. At the critical density there is no appreciable difference between the ^{3}He and ^{4}He heat capacities and entropies. This behavior is consistent with the picture of the ordered state as an immobile array. In this state the ^{3}He spin–spin interaction is probably only dipolar, with a very low characteristic temperature, of order 10^{-6} °K. Hence at the lowest temperatures of the film studies the spins of the critical density film are still completely classical, and a smooth extrapolation to $T = 0$ of the experimental heat capacity gives no contribution to the entropy.

Two-Dimensional Fermi Liquid. On the low-temperature side of the "spin bumps" the heat capacity tends toward a linear dependence on T. This region might be interpreted as a "2D Fermi liquid." In this region the entropy agrees closely with that of ^{3}He adsorbed on liquid ^{3}He–^{4}He solutions, as deduced from measurements of osmotic pressures.[23]

Two-Dimensional Fluids. As the coverage is increased above the critical $x_g = \frac{1}{3}$ the sharp 3°K ordering peaks rapidly diminish in height, become asymmetric, and move to lower temperature. No traces of the ordering peaks are seen above $x \simeq 0.7$; in this region the heat capacity is a smooth sigmoid, varying as low powers of T below ~ 1.5°K and tending toward Nk_B near 4°K. This region is tentatively identified as a dense 2D gas or fluid.

Two-Dimensional Solid and "Melting" Line of ^{4}He. From $x \simeq 0.73$ to $x = 1$ the films form 2D solids. This identification is based on detailed heat capacity results on ^{4}He and NMR measurements on ^{3}He. The heat capacities at low T vary as T^2. The 2D Debye characteristic temperatures are independent of T at $T/\theta \gtrsim 0.07$. The dependence of θ on x is quite different from results on Ar-plated Cu. Here we find a strong dependence on coverage, changing by over a factor three for a density variation of 50%. This sensitivity to density is much more understandable in terms of simple 2D models than previous work on other substrates. We have made a direct comparison between the 2D θ's and those of bulk solid ^{4}He by the use of a common molecular area scale, and find that both films and hcp solid fall on a single curve over the entire common region. There is no theoretical prediction that the two should be identical, but it is plausible that they should be quite similar.

The 2D solid films display pronounced heat capacity anomalies above the T^2 regions. As x is increased the anomalies increase in height, move to higher temperatures, and become quite narrow. If the temperatures of these peaks are compared with melting temperatures[24] of bulk solid ^{4}He on a common molecular area scale, the two sets of data describe parallel curves, strongly suggesting that the film peaks are due to melting. NMR results on ^{3}He films[20] in this coverage range show large changes in linewidth, i.e., mobility, consistent with a melting line comparable to the ^{4}He line.

The shapes of the specific heat peaks show that the melting phenomenon in the films is a continuous transition rather than the first-order melting in regular 3D solids. It appears that the nature of the film transition is related to the absence of long-range order in lower-dimensional systems.[25]

An extensive report on this work is now in preparation.[26]

Acknowledgments

I wish to thank Dr. W. E. Keller and Group P-8 of the Los Alamos Scientific Laboratory for their hospitality during the period this report was written.

References

1. H.P.R. Frederickse, *Physica* **15**, 860 (1949).
2. D.F. Brewer, A.J. Symonds, and A.L. Thompson, *Phys. Rev. Lett.* **15**, 182 (1965).
3. D.F. Brewer, *J. Low Temp. Phys.* **3**, 205 (1970).
4. W.D. McCormick, D.L. Goodstein, and J.G. Dash, *Phys. Rev.* **168**, 249 (1968).
5. G.A. Stewart and J.G. Dash, *Phys. Rev.* **A2**, 918 (1970).
6. G.A. Stewart and J.G. Dash, *J. Low Temp. Phys.* **5**, 1 (1971).
7. D.W. Princehouse, *J. Low Temp. Phys.* **8**, 287 (1972).
8. J.G. Daunt and E. Lerner, *J. Low Temp. Phys.* **8**, 79 (1972).
9. C.E. Campbell and M. Schick, *Phys. Rev. A* **3**, 691 (1971).
10. M. Schick and C.E. Campbell, *Phys. Rev. A* **2**, 1591 (1970).
11. A.D. Novaco and F.J. Milford, *J. Low Temp. Phys.* **3**, 307 (1970).
12. N.N. Roy and G.D. Halsey, *J. Low Temp. Phys.* **4**, 231, (1971).
13. G.B. Huff, unpublished.
14. M. Bretz and J.G. Dash, *Phys. Rev. Lett.* **26**, 963 (1971).
15. M. Bretz and J.G. Dash, *Phys. Rev. Lett.* **27**, 647 (1971).
16. M. Bretz, G.B. Huff, and J.G. Dash, *Phys. Rev. Lett.* **28**, 729 (1972).
17. D.C. Hickernell, E.O. McLean, and O.E. Vilches, *Phys. Rev. Lett.* **28**, 789 (1972).
18. A. Thomy and X. Duval, *J. Chim. Phys. (Paris)* **66**, 1966 (1969); **67**, 286 (1970); **67**, 1101 (1970).
19. D.L. Goodstein, private communication.
20. R.J. Rollefson, *Phys. Rev. Lett.* **29**, 410 (1972).
21. D.E. Hagen, A.D. Novaco, and F.J. Milford, in *Intern. Symp. on Adsorption–Desorption Phenomena, Florence, Italy, 1971* (to be published).
22. C.E. Campbell, J.G. Dash, and M. Schick, *Phys. Rev. Lett.* **26**, 966 (1971).
23. H.M. Guo, D.O. Edwards, R.E. Sarwinski, and J.T. Tough, *Phys. Rev. Lett.* **27**, 1259 (1971).
24. G. Ahlers, *Phys. Rev. A* **2**, 1505 (1970).
25. J.G. Dash and M. Bretz, *J. Low Temp. Phys.* **9**, 291 (1972).
26. M. Bretz, J.G. Dash, D.C. Hickernell, E.O. McLean, and O.E. Vilches, *Phys. Rev.* **A8**, August 1973.

Helium-3 in Superfluid Helium-4*

David O. Edwards

Department of Physics, The Ohio State University
Columbus, Ohio

This paper is not intended to be a review nor is there sufficient time or space to give an adequate introduction to the work which will be described in the many contributed papers at the conference, so I will confine myself to some topics that have interested our group at Ohio State. I must therefore apologize to a lot of people whose important work has not been mentioned.

At the moment the study of the surface of liquid helium seems to be quite active and since the effect of ^{3}He is dominant at low temperatures even in so-called "pure" ^{4}He (which contains a few parts in 10^7 of ^{3}He) the second half of the paper is devoted to the properties of the liquid surface.

As is well known, ^{3}He dissolved in superfluid ^{4}He behaves more or less like a weakly interacting gas of so-called "quasiparticles." Each ^{3}He moves through the superfluid background almost like a free particle, the main effect of the solvent ^{4}He being to bind the ^{3}He to the liquid and to increase the effective mass to about 2.3 times the real mass. For low temperatures (below $\sim 0.5°$K) where the effect of the native ^{4}He excitations, the phonons and rotons, is negligible, the various thermal properties† look like those of an almost ideal Fermi–Dirac gas with a mass $m \cong 2.3m_3$. For a familiar example of this behavior we can look at the specific heat[2] shown in Fig. 1. From this graph we can see that for ordinary ^{3}He concentrations [$X = N_3/(N_3 + N_4) \sim$ a few percent] the Fermi temperature T_F is relatively high. It is about 0.4°K for a solution of 6.5%, which is the saturated concentration at absolute zero, and it varies roughly as the $\frac{2}{3}$ power of the concentration. More precisely T_F is given by the equation

$$kT_F = p_F^2/2m = \hbar^2 (3\pi^2 n_3)^{2/3}/2m \tag{1}$$

where p_F is the Fermi momentum and $n_3 = N_3/V$ is the ^{3}He number density.

If one fits the specific heat data to the curve for an ideal Fermi–Dirac gas, one finds that the two values of the effective mass for the two different concentrations are not exactly the same but differ by an amount of the order of the concentration difference. In other words, the ideal Fermi–Dirac approximation is only valid to order X. To get a better fit, one has to allow for the effect of the dissolved ^{3}He on each other by means of an empirical "effective interaction" as was done by Bardeen, Baym, and Pines (BBP).[3] The interaction can be determined empirically but one must satisfactorily fit the transport properties as well as all the equilibrium properties.

---

* Work supported by a grant from the National Science Foundation.
† See the recent review by Ebner and Edwards.[1]

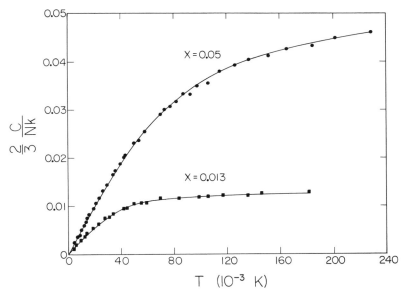

Fig. 1. The specific heat of two solutions of ^{3}He in ^{4}He at the saturated vapor pressure.[2] The curves represent the heat capacity of an ideal Fermi gas with the same number density as the ^{3}He in solutions and with the mass m adjusted to fit the data at each concentration.

For another example we consider the normal fluid density ρ_n. In the low-temperature region (negligible phonon contribution) the normal fluid is composed only of ^{3}He quasiparticles and if these behaved like an ordinary interacting gas, ρ_n should just be proportional to n_3; in fact $\rho_n = n_3 m$. We can turn this around and examine the experimental values of ρ_n/n_3 (called m_i). The values determined from the velocity of second sound[4] are shown in Fig. 2. We see that m_i is almost constant but there is a small systematic dependence on T and on X.

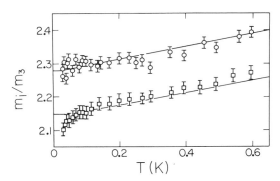

Fig. 2. The inertial mass $m_i = \rho_n/n_3$ for two ^{3}He concentrations at zero pressure: 0.67% (circles) and 4.47% (squares).[4]

The quasiparticle theory was proposed a long time ago by Landau and Pomeranchuk[5] before any of the experimental proofs of that theory were available. Their prediction was based on the idea that the change in the energy on adding a ^{3}He atom to pure ^{4}He in its ground state could be expanded in powers of the momentum given to the system:

$$\varepsilon_0(\mathbf{p}) = -E_3 + p^2/2m + \gamma p^4 + \cdots \tag{2}$$

Neglecting the term in γp^4 gives a free-particle-like dependence of the energy on $\mathbf{p}$. In fact γp^4 is not quite negligible and γ is negative. This explains the slight increase of the inertial mass m_i with the temperature. At higher temperatures we sample more of the high-$\mathbf{p}$ part of the energy spectrum which is slightly flattened compared to the "ideal" parabolic spectrum and which therefore corresponds to a higher average mass. Figure 3 shows the spectrum of Eq. (2) with γ fitted to the inertial mass measurements. Since Eq. (2) is just an expansion of $\varepsilon_0(\mathbf{p})$ in powers of p^2, the term in γp^4 does not become large until p approaches a characteristic momentum of the system, for instance, the momentum of the roton minimum $p_0/\hbar = 2 \text{ Å}^{-1}$.

Equation (2) is obviously valid only for small numbers of added ^{3}He. The theory has been generalized to include the effect of the added ^{3}He on each other in a way which resembles the Landau–Fermi liquid theory. The Fermi liquid theory uses an expansion of the energy of the system about the ground state (e.g., pure liquid ^{3}He at $T = 0$) in powers of the change in quasiparticle distribution function $\delta n_{\mathbf{p}}$. In the "theory of a dilute solution"[1] the ground state is that of pure ^{4}He and we expand the energy as a functional of the momentum distribution of the added ^{3}He atoms $n_{\mathbf{p}}$:

$$\delta(U^{\text{os}}) \equiv \delta(U - N_4\mu_4) = U_0^{\text{os}} + \sum_{\mathbf{p}} \frac{\delta U^{\text{os}}}{\delta n_{\mathbf{p}}} n_{\mathbf{p}} + \sum_{\mathbf{p}\mathbf{p}'} \frac{\delta^2 U^{\text{os}}}{\delta n_{\mathbf{p}}\delta n_{\mathbf{p}'}} n_{\mathbf{p}} n_{\mathbf{p}'} + \cdots \tag{3}$$

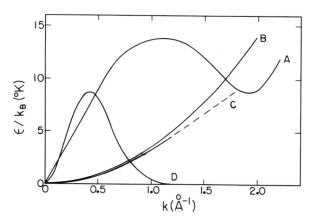

Fig. 3. The excitation spectrum of bulk liquid helium at zero pressure. Curve A is the phonon–roton spectrum. The ^{3}He quasiparticle spectrum is represented by the simple quadratic form (curve B) and with the term in p^4 added (curve C). Curve D (arbitrary units) shows the Maxwell–Boltzmann distribution in momentum for $T = 0.4$.

The functional derivatives* are to be taken at constant volume V and ^{4}He chemical potential μ_4 and at infinite dilution. The combination $U - N_4\mu_4$ (called the osmotic energy) is used because the number of ^{4}He atoms will not remain constant as we change the number of ^{3}He. The condition $\mu_4 = $ const is almost but not quite the same as constant pressure. The original Landau–Pomeranchuk model is equivalent to using only the first two terms on the right of Eq. (3) and writing the quasiparticle energy $\varepsilon_0(\mathbf{p}) = (\delta U/\delta n_{\mathbf{p}})_{V,\mu_4}$ as an expansion in p^2 [Eq. (2)]. If we include the third term, we take into account the effective interaction between quasiparticles as was done by BBP, Emery and others. The quantity

$$U_{\mathbf{p},\mathbf{p}'} = \delta^2 U^{os}/\delta n_{\mathbf{p}} \delta n_{\mathbf{p}'}$$

at a given μ_4 and V can be expected to depend in general on both $\mathbf{p}$ and $\mathbf{p}'$ (and on the relative spin). In the BBP theory, by analogy with the corresponding formula for a real gas of particles interacting via a very weak potential depending only on the interparticle distance, this momentum dependence is simplified so that $U_{\mathbf{pp}'}$ depends only on the relative momentum $\mathbf{q} = \mathbf{p} - \mathbf{p}'$. Unfortunately this simplification does not agree with experiment and $U_{\mathbf{p},\mathbf{p}'}$ is definitely dependent on the center-of-mass momentum as well. This corresponds to an interaction in real space which depends on the particle velocities as well as on their distance apart. Experimentally the velocity dependence of the interaction is shown by the concentration dependence of the inertial mass in Fig. 2. The velocity dependent interaction gives a term in the energy of a distribution of quasiparticles which depends to first order on the square of the particle velocities (and so contributes to the "kinetic energy") and which also depends on the average particle separation or concentration.

Curiously enough, the measurements of the inertial mass and many other properties are in semiquantitative agreement[1,4] with a very simple model which treats the dissolved ^{3}He atoms as tiny spheres moving through the superfluid ^{4}He background. The difference between the effective mass and the real mass is then due to the inertia of the superfluid "back flow" around the little spheres. The velocity-dependent part of the interaction energy is due to the overlap of the dipolar velocity fields around each sphere.[3,6] The sign and magnitude of the γp^4 term are also in rough agreement with this "hydrodynamic" model if the compressibility of the superfluid and of the ^{3}He spheres is taken into account.[4,7]

Of course the hydrodynamic model must not be taken too seriously and in any case it can be expected to be work only for large quasiparticle separations, i.e., low concentrations and temperatures. Probably further progress in understanding the bulk properties of solutions will come when we have made a systematic comparison between theory and experiment without restricting the spin and momentum dependence of the function $U_{\mathbf{p},\mathbf{p}'}$ or of the corresponding scattering amplitude (which appears in the theory of the transport properties). The possible effect of three-body interactions [equivalent to a fourth term in Eq. (3)] must also be taken into account, if necessary by restricting the measurements to sufficiently low concentrations. On the other hand, the comparison should include data at high temperatures where

---

* For simplicity we have not included the spin of the ^{3}He in Eq. (3). See Ref. 1 for a more complete statement of the equation.

$T > T_F$. Although the problem has been studied for many years, such a systematic comparison with experiment is only just beginning.[8]

We now consider the behavior of ^{3}He near the surface of the liquid. It was noticed quite a long while ago that ^{3}He reduces the surface tension of liquid helium and that the reduction becomes much more pronounced as one lowers the temperature below 1°K. Andreev[9] pointed out that this behavior was consistent with the existence of ^{3}He quasiparticle states which are bound to the surface. In such states he proposed that the quasiparticle energy should be

$$\varepsilon(p) = -E_3 - \varepsilon_0 + p^2/2M$$

where the momentum **p** is confined to the plane of the surface and where ε_0 is the additional binding over and above the normal "bulk" binding energy E_3. From experiments on the surface tension[10,11] it appears that $\varepsilon_0/k = (1.95 \pm 0.1)°$K. The mass M is also different from the bulk value m. In Andreev's model the surface quasiparticles form a two-dimensional Fermi–Dirac gas in equilibrium with (i.e., having the same chemical potential μ_3 and temperature T as) the three-dimensional quasiparticle gas in the interior of the liquid. It is the two-dimensional "pressure" exerted by this gas which reduces the surface tension σ below the value for pure ^{4}He, σ_4. At a given bulk concentration X the number of ^{3}He in the surface per unit area N_s grows as the chemical potential increases, so that the depression in the surface tension $\Delta\sigma = \sigma_4 - \sigma$ increases at lower temperatures, as shown in Fig. 4.[11]

The Andreev or independent quasiparticle model can be expected to be valid provided N_s is small compared to an atomic layer. (We rather arbitrarily define

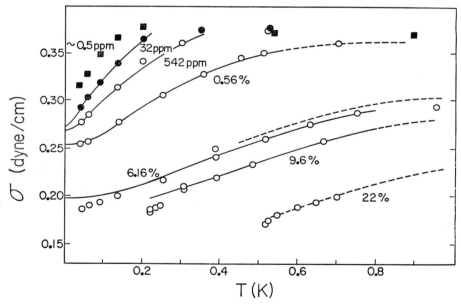

Fig. 4. The surface tension of ^{3}He–^{4}He mixtures as a function of temperature at various ^{3}He concentrations. Dashed lines from Ref. 10, circles and squares from Ref. 11. (The full curves are fits to a particular model and have no significance in the present context).

an atomic layer to have a number per unit area of $(n_3^\circ)^{2/3} = 6.4 \times 10^{14}$ cm^{-2}, where n_3° is the number density in pure liquid ^{3}He.) Before discussing what happens as N_s becomes comparable to an atomic layer or more we ought to describe how N_s can be defined and measured thermodynamically rather than by reference to a particular model.

The definition of N_s depends on a particular choice of the "Gibbs dividing surface."[12,13] The conventional choice of dividing surface which we adopt here makes the number of ^{4}He associated with the surface identically zero. This is shown in Fig. 5. (For simplicity we assume the temperature to be low enough that the vapor phase can be treated as a vacuum.) It should be clear from the diagram that a portion Q_s of any extensive property of the system Q can be associated with the surface using the equation

$$Q_s = Q - V_l q(0)$$

where $q(0)$ is the density of that quantity in the homogeneous interior of the liquid and V_l is the volume under the dividing surface. In this way we can define the amount of ^{3}He on the surface AN_s (A is the area of the surface), the surface entropy AS_s, and the surface free energy* $A\sigma$. Of these in general only the surface tension σ is strictly independent of the choice of the dividing surface, but for dilute ^{3}He–^{4}He solutions the dependence of N_s on the position of the dividing surface is small and the difference between the meaning of N_s in the Andreev model and the thermo-dynamic definition is insignificant.

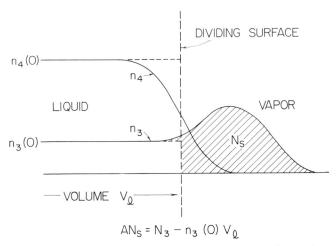

$$AN_S = N_3 - n_3(0)\, V_\varrho$$

Fig. 5. Schematic diagram of the surface of a solution of ^{3}He in ^{4}He showing the ^{3}He and ^{4}He number densities n_3 and n_4 as a function of position, the position of the dividing surface, and the definition of the liquid volume V_l. The shaded area under the curve of n_3 represents the surface density of ^{3}He, N_s.

---

* Actually the part of the potential $U - TS - N_4\mu_4 - N_3\mu_3$ associated with the surface.

It follows from the definition of the surface properties that

$$d\sigma = -N_s \, d\mu_3 - S_s \, dT$$

so that N_s and S_s can be obtained from the surface tension from

$$N_s = -(\partial\sigma/\partial\mu_3)_T \quad \text{and} \quad S_s = -(\partial\sigma/\partial T)_{\mu_3}$$

With these formulas in mind we can examine the results obtained from the surface tension measurements.[11] First we note that there is a finite surface entropy even in pure ^{4}He as shown by the temperature dependence of $\sigma_4(T)$. This is due to the so-called "ripplons"[14] or quantized surface waves which have a dispersion relation $\omega^2 = [\sigma_4(0)/\rho] \, k^3 + gk$. Below about $\sim 0.5°$K their contribution to the entropy becomes negligibly small. Second it is observed that to a fair approximation the contribution to the surface entropy from the ^{3}He, $\partial(\Delta\sigma)/\partial T \equiv [\partial(\sigma_4 - \sigma)/\partial T]_{\mu_3}$, is linear in the temperature as one might expect for a Fermi system. Third, extrapolating the data to $0°$K, we can obtain the graph of $\Delta\sigma^{1/2}$ and N_s as a function of μ_3, as shown in Fig. 6. On this graph the zero of energy or chemical potential is taken at the energy of the lowest bulk quasiparticle state. At $0°$K as one adds ^{3}He to pure ^{4}He no bulk states are occupied at all until one reaches a surface density of just about one atomic layer. In this region N_s is almost linear in μ_3 as one expects for a two-dimensional Fermi gas. (The density of states in energy is independent of energy).

Above one atomic layer the bulk quasiparticle states begin to be filled and at the same time N_s increases more rapidly with μ_3 than in the two-dimensional gas

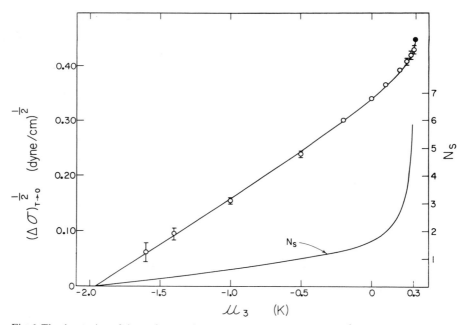

Fig. 6. The depression of the surface tension at $0°$K, $\Delta\sigma$, plotted against the ^{3}He chemical potential. N_s is the corresponding ^{3}He surface density measured in atomic layers.[11]

model. In terms of the μ_3 axis in Fig. 7, the bulk liquid reaches its saturated concentration of 6.5% when $\mu_3/k = 0.31°$K, the chemical potential of pure ^{3}He liquid. We see that as saturation is approached the amount of adsorbed ^{3}He on the surface increases very rapidly and N_s becomes macroscopically large. In other words the surface ^{3}He becomes a macroscopic phase of pure ^{3}He floating above the saturated solution. In principle, therefore, we can study the behavior of this "thin" Fermi system over the complete range of "thickness" from $N_s < 1$ atomic layer to macroscopic thicknesses.

In the limit of very small N_s the Andreev independent, two-dimensional quasi-particle model can be expected to work, while at very large N_s the system can be treated as a macroscopic sample of reduced thickness. The way in which the thickness affects the properties of pure ^{3}He liquid has been studied theoretically by Saam[15] using Fermi liquid theory. He was able to predict successfully the dependence of N_s on μ_3 as saturation is approached.

As an example of these two limits of very small and very large N_s we can look at the low-temperature limit of the entropy ratio S_s/T as a function of N_s (Fig. 7).

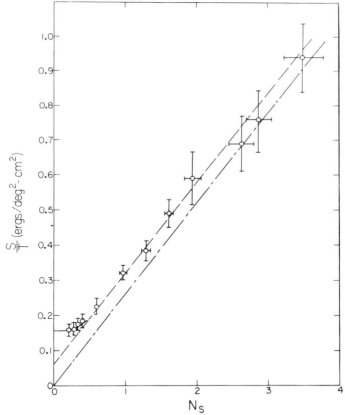

Fig. 7. The ^{3}He entropy per unit temperature as a function of N_s at $0°$K. The broken line through the origin corresponds to the entropy of bulk liquid ^{3}He.[11]

For $N_s \gtrsim 0.4$ atomic layers the entropy per unit area is approximately independent of N_s, as one would expect for a two-dimensional gas approaching infinite dilution. At high N_s the entropy is approximately extensive (∞N_s) and close to the value for pure ³He.

At present very few properties of the surface ³He have been measured but a relatively complete phenomenological theory of the liquid helium surface, including the effect of adsorbed ³He, has recently been published by Andreev and Kompaneets.[16] This work makes clear the very close analogy that exists between the three-dimensional properties of the bulk liquid and the two-dimensional properties of the surface. In this analogy the capillary waves or ripplons are the analog of the phonons and the Andreev surface quasiparticles are the analog of bulk ³He quasiparticles. The analogy has been made even closer by the recent proposal of Reut and Fisher[17] that the ripplon spectrum should exhibit a minimum at high k analogous to the roton minimum in the phonon spectrum.

Andreev and Kompaneets show that the surface obeys a set of two-fluid hydrodynamic equations very like those of bulk liquid helium and the theory predicts the existence of the usual two-fluid phenomena such as second sound. Even before this prediction appeared the propagation of a temperature pulse with a well-defined velocity had been observed in an experiment by Eckardt *et al.*[18] although they had not correctly understood the meaning of the phenomenon. The experiment was carried out at very low temperatures where the effect of the ripplons should be negligible and the "second sound" should just be an adiabatic wave in the two-dimensional ³He gas. The result of a typical experiment is shown in Fig. 8. The speed is given approximately by the familiar sort of formula:

$$u = [2(\Delta\sigma)/N_s M]^{1/2}$$

Measurements of the second-sound velocity should prove to be very useful in determining the effective mass and effective interaction of the two-dimensional quasiparticles. However, the measurement of many other properties of the surface, particularly the surface transport coefficients, is going to be very difficult and should prove a great challenge to experimentalists for the next few years.

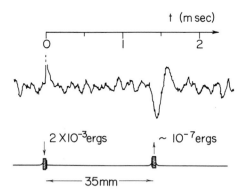

Figure 8. The propagation of a heat pulse along the surface of liquid helium containing a small amount of ³He at 25 m°K. The elements in the liquid surface are a graphite heater and thermometer. The velocity of propagation is in rough agreement with that predicted for "surface second sound" in Ref. 16.

References

1. C. Ebner and D.O. Edwards, *Phys. Rep.* **2c**, 78(1971).
2. A.C. Anderson, W.R. Roach, R.E. Sarwinski, and.J.C. Wheatley, *Phys. Rev. Lett.* **16**, 263 (1966); A.C. Anderson, D.O. Edwards, W.R. Roach, R.E. Sarwinski, and J.C. Wheatley, *Phys. Rev. Lett.* **17**, 367 (1966).
3. J. Bardeen, G. Baym, and D. Pines, *Phys. Rev.* **156**, 207 (1967).
4. N.R. Brubaker, D.O. Edwards, R.E. Sarwinski, P. Seligmann, and R.A. Sherlock, *Phys. Rev. Lett.* **25**, 715 (1970).
5. L.D. Landau and I. Pomeranchuk, *Dokl. Akad. Nauk SSSR* **59**, 669 (1948).
6. W.L. McMillan, *Phys. Rev.* **182**, 299 (1969).
7. C. Ebner, *Phys. Rev.* **A3**, 1201 (1971).
8. Y. Disatnik and H. Brucker, *J. Low Temp. Phys.* **7**, 491 (1972).
9. A.F. Andreev, *Zh. Eksperim. i Teor. Fiz.* **50**, 1415 (1966) [*Soviet Phys.—JETP* **23**, 939 (1966)].
10. K.N. Zinov'eva and S.T. Boldarev, *Zh. Eksperim. i Teor. Fiz.* **56**, 1089 (1969) [*Soviet Phys.—JETP* **29**, 585 (1969)].
11. H.M. Guo, D.O. Edwards, R.E. Sarwinski, and J.T. Tough, *Phys. Rev. Lett.* **27**, 1259 (1971); and to be published.
12. J.W. Gibbs, *Scientific Papers,* Dover, New York (1961), Vol. I, p. 219.
13. L.D. Landau and E.M. Lifshitz, *Statistical Physics,* Pergamon, New York (1958), p. 465.
14. K.R. Atkins, *Can. J. Phys.* **31**, 1165 (1953).
15. W.F. Saam, *Phys. Rev.* **A5**, 335 (1972).
16. A.F. Andreev and Kompaneets, *Zh. Eksperim. i Teor. Fiz.* **61**, 2459 (1971) [*Soviet Phys.—JETP* **34**, 1316 (1972)].
17. L.S. Reut and I.Z. Fisher, *Zh. Eksperim. i Teor. Fiz.* **60**, 1814 (1971) [*Soviet Phys.—JETP* **33**, 981 (1971)].
18. J. Eckardt, D.O. Edwards, F.M. Gasparini, and S.Y. Shen, this volume.

2
Quantum Fluids Theory

Some Recent Developments Concerning the Macroscopic Quantum Nature of Superfluid Helium

S. Putterman

Department of Physics, University of California
Los Angeles, California

Ever since the realization in the late 1930's that the macroscopic dynamics of liquid helium for $T < T_\lambda$ ($\sim 2.2°$K) was quite different from those of ordinary fluids, physicists, led initially by F. London, have been trying to understand its unusual macroscopic behavior from the point of view of the quantum theory. Surely the main reason for interest in He II is that its macroscopic behavior or hydro-dynamics is a direct consequence of the operation of quantum mechanics on a massive scale. In this paper we will discuss those aspects of the macroscopic behavior of He II that are most closely related to the quantum theory.

The first advance in our understanding of the hydrodynamics of superfluid helium was provided by the Landau two-fluid hydrodynamics[1] which roughly pictured He II as a "mixture" of a classical fluid and a superfluid which carried no entropy, experienced no friction, and obeyed the equation of motion

$$(\partial \mathbf{V}_s/\partial t) + (\mathbf{V}_s \cdot \nabla) \mathbf{V}_s = - \nabla \mu - \nabla \Omega \tag{1}$$

where $\mathbf{V}_s$ is the local velocity of the superfluid, Ω is the external potential per unit mass, and μ is the chemical potential. While Landau clearly motivated the two-fluid hydrodynamics from quantum-theoretical considerations, it is not at all obvious from the two-fluid equations by themselves that helium is a quantum fluid since these equations do not contain Planck's constant or, equivalently, based on these equations there is no experiment which can be performed that determines Planck's constant.

This situation changed as a consequence of the ideas put forward by Onsager[2] and Feynman.[3] They proposed that a macroscopic wave function ψ interpreted in a somewhat analogous fashion to the wave function of the ordinary quantum theory be used to describe the superfluid component of He II. Thus, for instance, $\psi^*\psi$ is taken to be the local superfluid density ρ_s or, including a possible phase ϕ,

$$\psi = \sqrt{\rho_s}\, e^{i\phi} \tag{2}$$

The macroscopic velocity of the supercomponent $\mathbf{V}_s$ is the same function of ψ as the probability current in ordinary quantum mechanics, or

$$\mathbf{V}_s = (\hbar/m)\, \nabla \phi \tag{3}$$

* Supported in part by the National Science Foundation.

From (3) one obtains the quantum restriction on $\mathbf{V}_s$:

$$\nabla \times \mathbf{V}_s = 0 \qquad (4)$$

with which, in fact, Landau always supplemented Eq. (1). Because ψ is single-valued, we find that in multiply connected geometries the circulation (line integral of $\mathbf{V}_s$ around a closed contour) must be quantized, or

$$\oint \mathbf{V}_s \cdot d\mathbf{l} = nh/m \qquad (5)$$

where n is an integer which is clearly analogous with the Bohr–Sommerfeld quantization of the old quantum theory or, perhaps, also with the angular momentum quantization of the ordinary quantum theory. Finally, again in analogy with the usual quantum theory, the wave function is taken to vanish at a boundary:

$$\psi \to 0 \quad (\rho_s \to 0) \qquad \text{at a boundary} \qquad (6)$$

The quantum conditions (4) and (5) are *consistent* with the Landau two-fluid theory and in particular Eq. (1). This follows from noting that by taking the curl of Eq. (1) one obtains

$$\partial \nabla \times \mathbf{V}_s / \partial t = \nabla \times [\mathbf{V}_s \times (\nabla \times \mathbf{V}_s)]$$

of which (4) is a solution. Also from Eq. (1) one finds the Kelvin circulation theorem

$$(D_s/Dt) \oint \mathbf{V}_s \, d\mathbf{l} = 0$$

$(D_s/Dt \equiv \partial/\partial t + \mathbf{V}_s \cdot \nabla)$, which states that the circulation of a contour is a constant of its motion so that, indeed, one can quantize it.

The Landau two-fluid equations supplemented by restrictions (4) and (5) constitute the most complete, self-consistent description of superfluid hydrodynamics presently known.

Equation (6), on the other hand, goes beyond the Landau two-fluid picture and strictly speaking is *inconsistent* with it. This happens because in the two-fluid theory eight variables (the five needed to describe a classical liquid plus the three components of $\mathbf{V}_s$) are a complete set. Also, in a vessel at rest in equilibrium the chemical potential, temperature, and normal and superfluid velocities must all be constant; thus, eight variables are constant. Through the equation of state, ρ_s must then also be constant, in disagreement with (6). Nevertheless the quantum restriction (6) as well as conditions (4) and (5) are verified experimentally, as will be discussed below. Our considerations so far are summarized in the top three lines of Table 1.

The restriction (4) of irrotational superfluid flow has been tested by a number of people and here we discuss the work of van Alphen,[4] in which the He II level in standpipes attached to a converging flow was measured. In a classical fluid, application of Bernoulli's law with $\Omega = gz$ along a streamline leads to the well-known fact that the standpipe level after the constriction (i.e., where the liquid flows faster) should be lower. But in He II the restriction (4) implies that one can apply Bernoulli's theorem throughout the fluid (i.e., $\mu + \tfrac{1}{2}V_s^2 + gz = \text{const}$, throughout the fluid,

Table I. Some of the Most Striking Macroscopic Quantum Effects in He II Compared with Their Microscopic Counterparts in the Ordinary Quantum Theory*

Macroscopic quantum effect	Microscopic analog	Consistency of the macroscopic quantum effect with the Landau two-fluid theory
$\mathbf{V} \times \mathbf{V}_s = 0$	—	Consistent
$\oint \mathbf{V}_s \cdot d\mathbf{l} = nh/m$	$\oint p\,dq = nh$	Consistent
$\rho_s \rightarrow 0$ at a boundary	$\psi \rightarrow 0$ at a boundary	Inconsistent
$\Delta V_{s,i}\,\Delta r_i \sim \hbar/m$	$\Delta p\,\Delta q \sim \hbar$	Inconsistent

* In the last column the question of consistency with the fundamental two-fluid hydrodynamics is raised.

and not just on streamlines). In this manner application of Bernoulli's law to the tops of the standpipes leads to the requirement that their levels be equal, which was observed by van Alphen. There is of course a critical velocity of $\mathbf{V}_s$ above which the flow becomes turbulent and the situation more complicated.

The quantization of circulation (5) is an extraordinary condition because through it Planck's constant enters the hydrodynamics. It implies that a careful measurement of the superfluid velocity can be used to determine Planck's constant. In order to appreciate the restrictions imposed by (5), consider the case where there is a vortex line in the superfluid. Then the velocity field in cylindrical coordinates is

$$\mathbf{V}_s = (\kappa/2\pi r)\,\hat{e}_\theta$$

where by (5) κ cannot assume any value as in classical liquids but must be nh/m.

The clearest demonstration of the quantization of vorticity comes from the experiments of Rayfield and Reif.[5] They investigated the energy and velocity of vortex rings in the superfluid. Due to (5) there is an extra relation between these quantities which was observed experimentally. Alternatively, the experiments yielded values for h/m in agreement with previous determinations based upon application of the ordinary quantum theory to microscopic situations.

According to the macroscopic wave function description, the superfluid density ρ_s should vanish at the boundary of the helium [Eq. (6)]. One expects this drop in ψ to occur over very short distances so that the quantum boundary effects will become important in very narrow geometries such as are achieved in thin films or superleaks with very narrow channels.

Experimentally these quantum boundary effects are investigated by measuring quantities that are functions of the average amount of superfluid in the liquid. For instance, in the adsorbed He II films which form on any surface (substrate) in contact with the helium vapor a wave mode (third sound) analogous to long gravity waves propagates with a speed

$$C_3 = [(\bar{\rho}_s/\rho)\,fd]^{1/2} \tag{7}$$

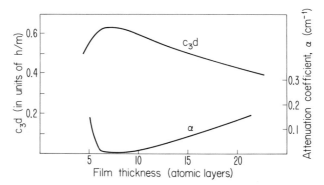

Fig. 1. Graph of C_3d and excess attenuation α at 200 Hz versus film thickness minus one atomic layer (solid layer). The superfluid onset thickness is 4.4 atomic layers and is indicated in the figure; $T = 1.3°K$.

where d is the thickness of the film, $f = 3\alpha/d^4$ is the force of attraction between the surface of the film and substrate, and α is the van der Waals coefficient. From Eq. (7) we see that C_3 depends on the average fraction $\bar{\rho}_s/\rho$ of superfluid in the film (ρ is the total fluid density). If ρ_s vanishes at the boundaries, C_3 will be less than if there were no quantum boundary effect.

From thermodynamics and the experimentally measured pressure and known van der Waals coefficient one determines d, and hence from measuring the speed of third sound one finds $\bar{\rho}_s/\rho$. This has been done for films as thin as four atomic layers. The results[6] indicate that there is a healing length a over which ρ_s drops from its bulk value, given by the known equation of state $\rho_s(\mu, T)$, to zero at a boundary. At low temperatures a is about one atomic layer and at 2°K is about three atomic layers. Experiments suggest the empirical relation $a \approx 1.5T/(\rho_s/\rho) T_\lambda$ valid between 1.0°K and T_λ.

Determinations of the healing length have been made in other experimental arrangements. For instance, from measurements of the angular momentum of a persistent current in He II films Reppy *et al.*[7] have obtained values of a in fairly good agreement with the third-sound data.

In order to explain the experiments which determine a, one has combined two inconsistent theories, namely, the Landau two-fluid equations and Eq. (6). That such a procedure leads to a simple interpretation of the experiments can be regarded as fortunate. One has only to consider the absorption of third sound in order to realize that the problem of unifying the macroscopic quantum ideas [e.g., (6)] with the two-fluid theory is really quite deep.

In Fig. 1 is indicated the attenuation α of third sound[8] in excess of that which follows from a detailed consideration of the two-fluid equations.[9] At the thicknesses shown α is two to three orders of magnitude larger than follows from the two-fluid theory.

It has been proposed[10] that one can understand this absorption in terms of an additional quantum restriction, namely a macroscopic quantum uncertainty principle:

$$\Delta V_{s,i} \Delta r_i \sim \hbar/m \qquad (8)$$

or that the quantum uncertainty in the superfluid velocity field multiplied by the uncertainty in the location of a fluid particle is on the order of $\hbar/m$. The motivation for condition (8) is by analogy with the transition from the old quantum theory to the Schrödinger theory. In that situation the Bohr–Sommerfeld quantization condition $\oint p\,dq = nh$ was replaced by a theory including the wave–particle duality as especially expressed by the uncertainty principle $\Delta p\,\Delta q \sim \hbar$. Since in He II the Onsager–Feynman condition (5), which is the analog of the Bohr–Sommerfeld condition, has been verified, we are led to expect that in a more complete theory of He II our description will include Eq. (8). From Table I this can be seen directly by the substitutions $mV_s \leftrightarrow p, r \leftrightarrow q$. We remark that Eq. (8), like Eq. (6), is inconsistent with the two-fluid theory. In this case the reason lies in the fact that the two-fluid equations are a *deterministic* set of equations, which is clearly inconsistent with the concept of uncertainties.

In any case, by use of (8) and the Landau two-fluid theory we will now try to interpret the third-sound absorption. In thinner and thinner films the fluid particles are more and more localized and hence by (8) the uncertainty in V_s perpendicular to the substrate (taken as the y direction) increases. At a certain thickness, ΔV_{sy} will be as large as C_3 and one certainly expects at this point to have difficulty in propagating third sound. Assuming that all locations of a fluid particle are equally likely,

$$\Delta y = d/\sqrt{12} \tag{9}$$

Then by the above argument we find that when

$$\hbar\sqrt{12}/m \gtreqless C_3 d \tag{10}$$

there should be difficulty in propagating third sound. Experimentally it is found that $C_3 d$ becomes close to or less than $\hbar\sqrt{12}/m$ for the thinnest and thickest films. In each case, as seen in Fig. 1, the absorption of third sound greatly increases. For the thinnest films it is found that there is an onset thickness below which third sound cannot be observed. This turns out to be within 10% of the value of d for which $C_3 d = \hbar\sqrt{12}/m$.

Much attention has been given to the onset phenomenon in which superfluid properties disappear below a sharply defined onset thickness. Many experiments have yielded onset thicknesses in agreement with those of third sound (see Fig. 5 of Ref. 6). Perhaps in terms of the ideas presented here one could interpret it as being the film thickness below which the zero-point oscillations about the average thickness d (as caused by ΔV_{sy}) are large enough to reduce the entire film to normal fluid. This calculation[11] yields an onset thickness in agreement with that discussed above.

Following a suggestion of M. Chester and R. Maynard that it is the nonlinearities inherent in the hydrodynamics combined with the quantum effects which produce the anomalous absorption of third sound, we now attempt to present a calculation of α as a function of thickness.

Due to the uncertainty in V_s, there will be an uncertainty in the thickness or a zero-point motion in the surface of the film which will have a magnitude δy determined by

$$(3\alpha/d^4)\,\delta y \sim (\Delta V_s)^2 \sim 12\hbar^2/m^2 d^2 \tag{11}$$

where we have assumed that the variations in $V_s^2/2$ are balanced by variations in the external field $\Omega = \alpha/d^3$ for a film. That this is the case is perhaps suggested (though not proved) by the fact that $\mu + V_s^2/2 + \Omega$ is constant in equilibrium. Now the third sound which is introduced is not propagated with respect to a flat, quiescent film but with respect to one with large variations in thickness. Hence in calculating the absorption one should take into account the deformation of the wave due to nonlinear effects. Here we get an order of magnitude estimate of α, we assume that the contribution to α due to these effects is the same as the absorption of a saw-tooth shock wave. At quadratic order one finds from the Landau two-fluid equations supplemented by (6) that at a point of the film of thickness y' greater than d the speed of third sound differs from C_3 given by (7) by the amount

$$C_3' = Ay' \tag{12}$$

where

$$A = \frac{C_3}{2(\bar{\rho}_s/\rho)d}\left(\frac{4D}{d} - 1\right) \tag{13}$$

where D is the thickness of the nonsuperfluid part of the film (generally taken to be a solid layer one atom thick adjoining the substrate plus $2\sqrt{2}\,a$). The absorption coefficient suggested by considering a saw-tooth shock wave is

$$\alpha = \omega|A|\,y'/\pi C_3^2 \tag{14}$$

where ω is the frequency. It is this quantity with y', including contributions from the quantum δy [Eq. (11)] and the impressed third sound wave, with which we want to compare the observed absorption of third sound. Actually the quantum δy dominates and we find

$$\alpha = \frac{6\omega}{\pi C_3}\frac{\hbar^2}{m^2\,d^2\,C_3^2}\left|\frac{4D}{d} - 1\right| \tag{15}$$

Experimentally a $\sqrt{\omega}$ dependence seems to be suggested rather than the linear dependence in (15). Nevertheless (15) gives an α very close to the experimental values (e.g., for the same parameters as Fig. 1, $\alpha = 0.15$ cm^{-1} for $d = 5.20$ atomic layers) and in particular there is a value of $d\,(= 4D)$ for which the nonlinearities vanish, $A = 0$, and hence α vanishes also in agreement with experiment.

The calculation just presented is only intended to be suggestive of the quantum mechanisms which in our opinion dominate the absorption of third sound. Certainly the situation must be considered unsatisfactory until there is a single self-consistent theory from which all the properties of third sound can be deduced. Still, based upon the above a prediction can be made that in He II "puddles" below a thickness

$$d \sim (12\hbar^2/m^2g)^{1/3} \sim 6.6 \times 10^{-4}\quad \text{cm}$$

the propagation of long gravity waves will be hindered due to the uncertainty principle. Such an observation at this relatively large thickness would demonstrate that these quantum effects were indeed macroscopic.

In the ordinary quantum theory there are two aspects to the time development of the wave function: first is the causal development governed by the Schrödinger equation, and second is the reduction of the wave packet resulting from a measurement. Helium is unusual in that quantum theory becomes important on the macroscopic level, yet the only features of this unusual property to be directly experimentally or theoretically investigated have to deal with this first aspect of the quantum theory. No one has investigated how to modify the Landau two-fluid theory to include the interaction between macroscopic quantum system and observer, and perhaps the macroscopic quantum uncertainty principle is related to this question.

The situation with regard to our understanding of He II hydrodynamics is comparable to the days of the transition from the old quantum theory to wave mechanics, in that our best theory is a *deterministic* theory from which a quantization condition is used to pick out a discrete set of allowed states. Yet there exist much experimental data indicating the shortcomings of this theory. The major problem is that of finding the macroscopic wave-hydrodynamics which are a generalization of the Landau two-fluid theory plus the quantum ideas such as (5), (6), and (8).

References

1. L.D. Landau, *J. Phys. USSR* **5**, 71 (1941); L.D. Landau, and E.M. Lifshitz, *Fluid Mechanics,* Pergamon, London (1959), Chapter 16.
2. L. Onsager, *Nuovo Cimento (Suppl.)* **6**, 249 (1949).
3. R.P. Feynman, in *Progress in Low Temperature Physics,* C.J. Gorter, ed., North-Holland, Amsterdam (1955), Vol. 1, p. 36.
4. W.M. van Alphen, Flow Phenomena in Superfluid Helium, Thesis, Leiden, 1969.
5. G.W. Rayfield and F. Reif, *Phys. Rev. Lett.* **11**, 305 (1963).
6. I. Rudnick and J.C. Fraser, *J. Low Temp. Phys.* **3**, 225 (1970).
7. R.P. Henkel, E.N. Smith, and J. Reppy, *Phys. Rev. Lett.* **23**, 1276 (1969).
8. T. Wang, Ph.D. Thesis, University of California, Los Angeles, 1971.
9. D. Bergman, *Phys. Rev.* **A3**, 2058 (1971).
10. S.J. Putterman, R. Finkelstein, and I. Rudnick, *Phys. Rev. Lett.* **27**, 1697 (1971).
11. T. Wang and I. Rudnick, *J. Low Temp. Phys.* (to appear).

A Neutron Scattering Investigation of Bose–Einstein Condensation in Superfluid ^{4}He*

H.A. Mook, R. Scherm,† and M.K. Wilkinson

Solid State Division, Oak Ridge National Laboratory
Oak Ridge, Tennessee

London[1] suggested in 1938 that many of the strange properties of superfluid ^{4}He result from the fact that below the λ point some of the ^{4}He atoms undergo Bose–Einstein condensation to a zero-momentum state. Hohenberg and Platzman[2] suggested that this proposal of a zero-momentum state could be checked by high-energy neutron inelastic scattering from ^{4}He. The cross section for such a process is given by

$$d^2\sigma/dE\, d\Omega = b^2 (k'/k_0)\, S(\mathbf{Q}, \omega)$$

where b is the scattering amplitude for ^{4}He; k_0 and k' are the incoming and outgoing wave vectors, respectively; and $S(\mathbf{Q}, \omega)$ is the scattering law. For high-energy neutrons

$$S(\mathbf{Q}, \omega) = \sum_{\mathbf{p}} n(\mathbf{p})\, [h\omega - (h^2/2M)(Q^2 + 2\mathbf{Q}\cdot\mathbf{p})]$$

where $n(\mathbf{p})$ is the ^{4}He momentum distribution, $\mathbf{Q}$ is the neutron momentum transfer, and M is the ^{4}He mass.

The result of such a measurement would then be a distribution centered at $(h^2/2M)\, Q^2$ whose shape reflects the momentum of the ^{4}He atoms. If some fraction of the atoms are in a zero-momentum state, they will manifest themselves in the scattering experiment as a sharper peak on top of the main distribution that reflects the momentum of the normal ^{4}He atoms.

The two-component distribution has been searched for by Cowley and Woods[3,4] and Harling.[5,6] Neither experiment employed sufficient statistical accuracy or resolution to see the condensate directly and thus neither could give a reliable estimate of the condensate fraction.

Our measurements were made on a triple-axis spectrometer at a constant scattering angle of 135° and a momentum transfer of 14.33 Å^{-1}. The results of the measurements are shown in Fig. 1 for ^{4}He at 1.2 and 4.2°K. The data have all been normalized to one run which represents about 20 min counting time per point. Of course, many runs were performed, especially in the area near the peak top where evidence for a condensate is expected.

* Research sponsored by the U. S. Atomic Energy Commission under contract with the Union Carbide Corporation.
† Also affiliated with the Institut Max von Laue—Paul Langevin, Grenoble, France.

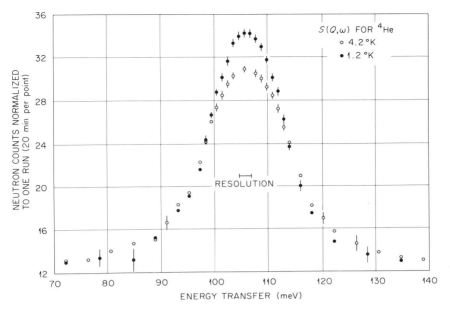

Fig. 1. $S(Q, \omega)$ for ^{4}He at 4.2 and 1.2°K.

The resolution energy width for the constant angle scan used in the experiment was about 2.1 meV. The four-dimensional resolution ellipsoid for the triple-axis spectrometer was calculated by methods similar to those suggested by Cooper and Nathans.[7] The dimensions of the ellipsoid were then checked by measurements with single crystals around the elastic scattering position for several different energies. Good agreement was obtained between the measured and calculated resolution ellipsoids. The value of the resolution width quoted was then calculated: (1) by passing the four-dimensional resolution ellipsoid in a series of steps through an infinitely thin ^{4}He dispersion surface along the trajectory in energy–momentum space corresponding to the constant scattering angle scan; and (2) by obtaining the intersected area for each step. The resolution width quoted is the full-width at half-maximum of the distribution so obtained.

The measurements were alternated between ^{4}He at 1.2 and 4.2°K so that the 4.2°K data could serve as a check on the experiment. The two-component distribution indicative of a condensate is visible at 1.2°K, although the effect is quite subtle. The absolute value of the slopes of the two peaks plotted from the point on the sides of the peaks where the slope is greatest is shown in Fig. 2. It is quite clear that the 1.2°K distribution undergoes a change in slope representative of a condensate. Since the changes in the slope mark the positions on the peak where the condensate begins, it is easy to draw curves representing the normal and condensate distributions and thus estimate a condensate fraction. This type of estimate gives a condensate fraction of about 2%.

To obtain a more reliable estimate, a Gaussian analysis of the data similar to that used by Harling[5,6] and Puff and Tenn[8] was undertaken. This proved to be

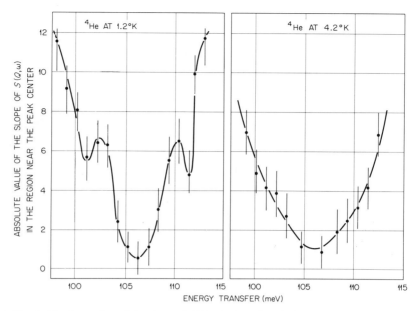

Fig. 2. Absolute value of the slope of $S(Q, \omega)$ taken directly from the measured data plotted from the point on the sides of the distribution where the slope is near its greatest value. The lines are merely smooth curves drawn through the data points and not the result of any analytical fit to the data.

completely unsatisfactory since the measured distributions are decidedly non-Gaussian. $S(\mathbf{Q}, \omega)$ is much too steep on the sides and has too large a wing contribution to be well fit by any Gaussian distribution. However, the distribution could be fit satisfactorily by adding a higher-order term to the Gaussian distribution and thus three more parameters. A Gaussian distribution appears to be satisfactory to represent the condensate contribution. It was thus assumed that the measured distribution could be fit by the following function:

$$I(E) = A_0 + A_1 \exp\left[(E - A_2)^2/A_3^2\right] + A_4 \exp\left[(E - A_5)^4/A_6^4\right]$$
$$+ A_7 \exp\left[(E - A_8)^2/A_9^2\right]$$

where the first two terms fit the main momentum distribution and the last term describes the condensate. The parameters were found by a least squares analysis of the data. A good fit to the data at 4.2°K was found with parameter A_7 equal to zero; however, the 1.2°K data required a nonzero A_7 parameter that was representative of a condensate fraction of 2.4 $\pm$ 1%. This is considerably smaller than the theoretical estimates of the condensate fraction, which range from 6% to 25%.

Acknowledgments

The authors would like to acknowledge a number of helpful discussions with Dr. H. A. Gersch and valuable technical assistance from Mr. J. L. Sellers.

References

1. F. London, *Nature* **141**, 643 (1938).
2. P.C. Hohenberg and P.M. Platzman, *Phys. Rev.* **152**, 198 (1966).
3. R.A. Cowley and A.D.B. Woods, *Phys. Rev. Lett.* **21**, 787 (1968).
4. R.A. Cowley and A.D.B. Woods, *Can. J. Phys.* **49**, 177 (1971).
5. O.K. Harling, *Phys. Rev. Lett.* **24**, 1046 (1970).
6. O.K. Harling, *Phys. Rev. A* **3**, 1073 (1971).
7. M.J. Cooper and R. Nathans, *Acta Cryst.* **23**, 357 (1967).
8. R.D. Puff and J.S. Tenn, *Phys. Rev. A* **1**, 125 (1970).

Temperature and Momentum Dependence of Phonon Energies in Superfluid ^{4}He

Shlomo Havlin and Marshall Luban

Department of Physics, Bar-Ilan University
Ramat-Gan, Israel

One of the challenging theoretical problems relating to superfluid ^{4}He is the detailed form of the phonon portion of the elementary excitation spectrum ε_p. For wave vectors $p/\hbar$ in the range 0.2–1.0 Å^{-1}, the inelastic neutron scattering data[1] yield $\varepsilon_p = c_0 p(1 - \gamma p^2 - \delta p^4 - \cdots)$, where $\gamma = (0 \pm 2) \times 10^{36}$ g^{-2} cm^{-2} sec^2 and $\delta = (2.4 \pm 0.2) \times 10^{75}$ g^{-4} cm^{-4} sec^4. For these wave vectors the term γp^2 can be neglected and thus $1 - \delta p^4$ are the leading terms for $\varepsilon_p/c_0 p$. To date no theoretical work yields $1 - \delta p^4$ as the leading terms for $\varepsilon_p/c_0 p$.

For temperatures between 0.2 and 0.5°K the velocity of sound[2] *decreases* with increasing frequencies in the range 12–84 MHz. The only theoretical results which are in accord with this behavior are those of Maris[3] for the single temperature 0.35°K. His results are obtained by numerical solution of the Landau–Khalatnikov kinetic equations and the phonon Boltzmann equation.

We report here the results of our calculation of the excitation spectrum ε_p of superfluid ^{4}He based on the Landau quantum hydrodynamics model. We obtain an analytic expression for the sound velocity $c(\omega, T)$, $\hbar\omega/k_B T \lesssim 0.1$, which is in excellent agreement with experiment[2] as a function of *both* temperature T and angular frequency ω. For the portion $c_0 p \gtrsim 4k_B T$ of the phonon regime we find, as the dominant contribution, $\varepsilon_p = c_0 p(1 - \delta p^4 + \cdots)$, where $\delta = +0.9 \times 10^{75}$ g^{-4} cm^{-4} sec^4. The quartic dependence, $-\delta p^4$, including its sign, as well as the temperature independence of ε_p in this regime, are in agreement with the neutron data.[1] Our result for δ is of the same order of magnitude, although smaller by a factor of 2.5 than the experimental value.[1]

The phonon energy is given by $\varepsilon_p = c_0 p + \Delta\varepsilon_p$, where $\Delta\varepsilon_p \equiv \text{Re} \sum_p(\varepsilon_p)$ is the real part of the phonon self-energy function. We find $\Delta\varepsilon_p$ as the solution of a self-consistent integral equation obtained by: (i) utilizing the Landau hydrodynamic model; (ii) incorporating scattering, decay, and annihilation three-phonon processes; (iii) choosing the phonon energies to be given by their self-consistent renormalized values; and (iv) noting that the renormalized three-phonon interaction should only allow processes on the energy shell for the dressed phonons. A formal theoretical basis for (iv) has been provided very recently by Eckstein *et al.*[4] generalizing a method due to Fröhlich.[5] This procedure (iv) automatically eliminates the divergence difficulties that had plagued previous theoretical efforts. The excellent agreement between the present theory and the experimental data for ε_p for a wide range of values of p and T in the two regimes $\varepsilon_p/k_B T \lesssim 0.1$ and $\varepsilon_p/k_B T \gtrsim 4$ constitutes confirmation of the basic importance and correctness of (i)–(iv).

We write $\Delta\varepsilon_p = \Delta\varepsilon_p^{(1)} + \Delta\varepsilon_p^{(2)} + \Delta\varepsilon_p^{(3)}$, where the $\Delta\varepsilon_p^{(1-3)}$, arising from scattering, decay, and annihilation processes respectively, satisfy the following self-consistent integral equations:

$$\Delta^{(1)}\varepsilon_p = 2 \int \frac{d^3 p'}{(2\pi\hbar)^3} V_{pp'|\mathbf{p}+\mathbf{p}'|}^2 \frac{n(\varepsilon_{p'}) - n(\varepsilon_{\mathbf{p}+\mathbf{p}'})}{\varepsilon_p + \varepsilon_{p'} - \varepsilon_{\mathbf{p}+\mathbf{p}'}} \tag{1a}$$

$$\Delta^{(2)}\varepsilon_p = \int \frac{d^3 p'}{(2\pi\hbar)^3} V_{pp'|\mathbf{p}-\mathbf{p}'|}^2 \frac{n(\varepsilon_{p'}) - n(-\varepsilon_{\mathbf{p}-\mathbf{p}'})}{\varepsilon_p - \varepsilon_{p'} - \varepsilon_{\mathbf{p}-\mathbf{p}'}} \tag{1b}$$

$$\Delta^{(3)}\varepsilon_p = -\int \frac{d^3 p'}{(2\pi\hbar)^3} V_{pp'|\mathbf{p}+\mathbf{p}'|}^2 \frac{n(\varepsilon_{p'}) - n(-\varepsilon_{\mathbf{p}+\mathbf{p}'})}{\varepsilon_p + \varepsilon_{p'} + \varepsilon_{\mathbf{p}+\mathbf{p}'}} \tag{1c}$$

where $n(\varepsilon) = (e^{\beta\varepsilon} - 1)^{-1}$. The quantity $V_{pp'p''}$ is the renormalized interaction between the three dressed phonons and it is nonzero only on the energy shell for the renormalized phonon energies, having there the value of the bare interaction[6] $V_{pp'p''}^2 = (u + 1)^2 c_0 pp'p''/4\rho$, where $\rho = 0.145$ g cm^{-3} is the density, $c_0 = 2.38 \times 10^4$ cm sec^{-1} is the zero-frequency velocity of sound, and $u = (\rho/c)(\partial c/\partial\rho)_T = 2.84$. The appropriate width of each energy shell is accommodated when the integration variable p' in Eqs. (1a)–(1c) satisfies the following restrictions, respectively:

$$|\varepsilon_p + \varepsilon_{p'} - \varepsilon_{\mathbf{p}+\mathbf{p}'}| \lesssim 2\bar{\Gamma}, \qquad |\varepsilon_p - \varepsilon_{p'} - \varepsilon_{\mathbf{p}-\mathbf{p}'}| \lesssim \Gamma_p, \qquad |\varepsilon_p + \varepsilon_{p'} + \varepsilon_{\mathbf{p}+\mathbf{p}'}| \lesssim 2\bar{\Gamma}$$

$\bar{\Gamma}$ is the thermal phonon width and Γ_p that of a phonon of momentum p. The upper bounds are the largest phonon widths for the relevant processes. Equations (1a)–(1c) are solved assuming that $\Delta\varepsilon_p$, $\Gamma_p \ll c_0 p$, an assumption which is borne out by the final results. The third restriction cannot be satisfied and thus $\Delta\varepsilon_p^{(3)}$ does not contribute to $\Delta\varepsilon_p$.

In the portion $c_0 p/k_B T \gtrsim 4$ of the phonon regime the dominant contribution to ε_p arises from decay processes [Eq. (1b)]. We utilize the following form[7] for Γ_p,* appropriate to this regime, $\Gamma_p = (u + 1)^2 p^5/480\pi\rho\hbar^3$. We obtain $\Delta\varepsilon_p = -c_0 \delta p^5$ and thus $\varepsilon_p = c_0 p(1 - \delta p^4)$, where $\delta = 0.356(u + 1)^2/480\pi\rho\hbar^3 c_0 = 0.9 \times 10^{75}$ g^{-4} cm^{-4} sec^4. The quartic dependence $-\delta p^4$ and its sign for the leading term to $\Delta\varepsilon_p/c_0 p$ are in agreement with Woods and Cowley's measurements; however, our value for δ is smaller by a factor 2.5 than the experimental value.[1] The predicted temperature independence of this result is in fact confirmed by experiment; see, in particular, Fig. 16 of Cowley and Woods[1] showing that up to 1.8°K, ε_p is temperature independent. The relevance of our result for such temperatures follows since for decay processes on the energy shell a phonon can only decay into two other phonons. Thus the rotons which exist for these temperatures play no role.

For the portion $\hbar\omega/k_B T \lesssim 0.1$ ($\hbar\omega \equiv c_0 p$) of the phonon regime[†] the dominant contribution to $\Delta\varepsilon_p$ arises from $\Delta\varepsilon_p^{(1)}$, that is, from the "scattering" processes [Eq.

---

* A calculation valid for all p of Γ_p, the imaginary part of the phonon self-energy function, has been performed by Havlin.[7] This quantity describes the sound attenuation, and the theoretical results are in excellent agreement with existing experimental data, i.e., in the regime $\hbar\omega/k_B T \lesssim 0.1$.

† Our theory for the sound velocity is valid only in the collisionless regime $\omega\tau = \hbar\omega/2\bar{\Gamma} \gg 1$. This requirement is in addition to the requirement $\hbar\omega/k_B T \lesssim 0.1$. Both requirements are met for the values of ω and T used in the experiment of Abraham et al. We expect that $\Delta c/c_0$ as a function of ω for given T will have a maximum for $\omega\tau \simeq 1$.

(1a)]. The calculation of $\Delta \varepsilon_p^{(1)}$ utilizes our result for ε_p in the regime $c_0 p/k_B T \gtrsim 4$, as well as Jäckle and Kehr's[8] expression $\bar{\Gamma} = 0.45 \times 10^{-20} T^5$. For temperatures below 0.6°K, where there are no rotons, we find

$$\Delta c(\omega, T)/c_0 = \Delta \varepsilon_p/c_0 p = A T^4 \ln\left[D(T^2/\omega^2) + 1\right] \tag{2}$$

where $\Delta c(\omega, T) = c(\omega, T) - c_0$, $c(\omega, T)$ is the sound velocity for the indicated parameters, $A = \pi^2 (u + 1)^2 k_B^4 / 60 \rho \hbar^3 c_0^5$, and $D = \bar{\Gamma}^2 / (5\delta \hbar \bar{p}^4 T)^2$, where $\bar{p} = 4.2 k_B T/c_0$ is the thermal momentum, the value of p maximizing the intergrand in (1a). In the following we use the Woods and Cowley value for δ yielding $D = 4.07 \times 10^{18}$ deg^{-2} sec^{-2}. In Fig. 1 we display the experimental results[2] for $\{10^6 (\Delta c/c_0)/ \ln[D(T^2/\omega^2) + 1]\}^{1/4}$ as a function of T for frequencies $\omega/2\pi = 12, 36, 60$ MHz. According to Eq. (2), for all ω the experimental values for the given quantity are predicted to lie on the straight line passing through the origin $(10^6 A)^{1/4} T = 5.1T$. The experimental data are clearly seen to conform with our prediction. By way of

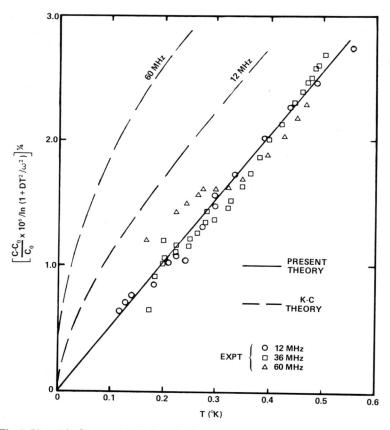

Fig. 1. Plot of $\{10^6 (c - c_0)/c_0 \ln[(DT^2/\omega^2) + 1]\}^{1/4}$ as a function of T. The present theory predicts that for all ω a plot of this quantity is given by the straight line $5.1T$ shown as the solid line. The dashed curves are predictions of the Khalatnikov–Chernikova theory for 12 and 60 MHz.

comparison with previous theoretical work, in Fig. 1 we have also displayed the predictions of the Khalatnikov–Chernikova theory,[9] according to which the data should lie on a family of curves differing according to the value of ω. We hope that experimentalists will soon clarify whether or not the apparent shoulder for the 60-MHz data for $T \simeq 0.3°$K is real.

We remark that even a 20–30% departure from the Woods and Cowley value for δ would drop the experimental points in Fig. 1 noticeably below the theoretical curve 5.1T. Besides confirming the experimental value for δ, this fact emphasizes the intimate connection between the behavior of ε_p in the two widely separated frequency regimes where neutron scattering and sound velocity measurements are performed.

We also note that our result (2) is consistent with a very recent experiment of Roach et al.[10]

Acknowledgments

We wish to thank S. G. Eckstein, D. Friedländer, C. G. Kuper, and B. B. Varga for several very profitable discussions.

References

1. A.D.B. Woods and R.A. Cowley, *Phys. Rev. Lett.* **24**, 646 (1970); R.A. Cowley and A.D.B. Woods, *Can. J. Phys.* **49**, 177 (1971).
2. B.M. Abraham, Y. Eckstein, J.B. Ketterson, M. Kuchnir, and J.P. Vignos, *Phys. Rev.* **181**, 347 (1969).
3. H.J. Maris, *Phys. Rev. Lett.* **28**, 277 (1972).
4. S.G. Eckstein, D. Friedländer, and C.G. Kuper, *Phys. Rev. Lett.*, **30**, 78 (1973).
5. H. Fröhlich, *Proc. Roy. Soc.* **A215**, 291 (1952).
6. I.M. Khalatnikov, *An Introduction to the Theory of Superfluidity,* Benjamin, New York (1965).
7. S. Havlin, M.S. Thesis, Tel-Aviv University, 1970; also to be published.
8. J. Jäckle and K.W. Kehr, *Phys. Lett.* **37A**, 205 (1971).
9. I.M. Khalatnikov and D.M. Chernikova, *Zh. Eksperim. i Teor. Fiz.* **49**, 1957 (1965); **50**, 411 (1966) [*Soviet Phys.—JETP* **22**, 1336; **23**, 274 (1966)].
10. P.R. Roach, B.M. Abraham, J.B. Ketterson, and M. Kuchnir, *Phys. Rev. Lett.* **29**, 32 (1972).

Does the Phonon Spectrum in Superfluid ⁴He Curve Upward?

S.G. Eckstein, D. Friedlander, and C.G. Kuper

Department of Physics, Technion—Israel Institute of Technology
Haifa, Israel

There has recently been some controversy about the shape of the spectrum of phonons in superfluid ⁴He.[1] In this note we use the theory of canonical transformations to demonstrate that the spectrum must be concave at low momenta. It is shown that if the spectrum were convex, there could be no three-phonon vertex in the interaction; *a fortiori*, there can be no three-phonon process in the attenuation of sound. However, the experimental dependence[2] of the attenuation on frequency and temperature is clearly that of a three-phonon process. Hence the spectrum must be concave.

In many-body theory the goal of a canonical transformation is to diagonalize the Hamiltonian

$$H = H_0 + \lambda V \tag{1}$$

$$H_0 = \sum_k \varepsilon_k^0 a_k^\dagger a_k \tag{2}$$

as far as possible, while preserving the notion of "quasiparticles." Full diagonalization under this condition is not possible. Under the unitary transformation

$$b = TaT^\dagger \tag{3}$$

the Hamiltonian transforms to

$$H[a] = T^\dagger H[b] \, T = \bar{H}_0 + U[b] \tag{4}$$

where

$$\bar{H}_0 = \sum_k \varepsilon_k b_k^\dagger b_k \tag{5}$$

Here ε_k are the renormalized quasiparticle energies, and $U[b]$, the renormalized interaction, does not contain any terms in $b_k^\dagger b_k$.

To carry out this program, we expand T as a perturbation product:

$$T = T_1 T_2 T_3 T_4 \cdots, \qquad T_v = \exp(-i\lambda^v S_v) \tag{6}$$

and look for a $U[b]$ in the form

$$U = \sum_v \lambda^v U_v \tag{7}$$

We wish to express the results in terms of the renormalized energies ε_k and not the bare energies ε_k^0 since the former are known experimentally. We therefore write

$$\bar{H}_0 - H_0 \equiv \sum_k (\varepsilon_k - \varepsilon_k^0)\, b_k^\dagger b_k = \sum_\nu \lambda^\nu h_\nu \tag{8}$$

assuming that it has a power-series expansion* in λ. Substituting Eqs. (1), (6), and (8) into Eq. (4) gives the U_ν. The result is the set of equations

$$U_1 = V - h_1 + i[S_1, \bar{H}_0] \tag{9a}$$

$$U_2 = -h_2 + i[S_2, \bar{H}_0] + i[S_1, V - h_1] - \tfrac{1}{2}[S_1, [S_1, \bar{H}_0]] \tag{9b}$$

$$U_\nu = -h_\nu + i[S_\nu, \bar{H}_0] + F_\nu(\bar{H}_0 ; S_1, \ldots, S_{\nu-1} ; h_1, \ldots, h_{\nu-1}) \tag{9c}$$

where F_ν is a known function of its arguments. If a description in terms of non-interacting quasiparticles were possible, all U_ν would vanish for appropriate S_ν and h_ν. In general this is not possible; but we choose S_1 so that as many as possible of the matrix elements of U_1 vanish.† To be explicit, consider the matrix elements of U_1 between the eigenstates $|n\rangle$ and $|m\rangle$ of H_0:

$$\langle n|U_1|n\rangle = \langle n|V - h_1|n\rangle \tag{10a}$$

$$\langle n|U_1|m\rangle = \langle n|V|m\rangle + i(E_n - E_m)\langle n|S_1|m\rangle, \qquad n \neq m \tag{10b}$$

The requirement that U_1 shall have no term in $b_k^\dagger b_k$ determines h_1 uniquely. If $E_n \neq E_m$, choose $\langle n|S_1|m\rangle = i\langle n|V|m\rangle/(E_n - E_m)$. Then $\langle n|U_1|m\rangle = 0$. Obviously, for $E_n = E_m$ no choice of S_1 can make $\langle n|U_1|m\rangle$ vanish. If the spectrum E_n is continuous, it will be possible to make all matrix elements of U_1 vanish except for those connecting states on the energy shell

$$|E_n - E_m| < \Delta \tag{11}$$

where Δ is arbitrarily small.‡ We choose $\langle n|S_1|m\rangle = 0$ on the energy shell.

In exactly the same way we now use Eq. (9b) to determine h_2 and S_2. Like S_1, S_2 will have nonzero matrix elements connecting states off the energy shell only; U_2 will connect states on the energy shell, but will have no term of the form $b_k^\dagger b_k$. In the same way Eq. (9c) determines H_3 and S_3; and so on to all orders.

In phenomenological theories the starting point is a Hamiltonian in which the experimentally determined quasiparticle energies appear—i.e., the starting point is the renormalized Hamiltonian. We have seen that in such a Hamiltonian all interaction terms must be on the energy shell of finite (but arbitrarily small) thickness. Returning now to the problem of phonons in ⁴He, we note that in the three-phonon process $|E_n - E_m| = |\varepsilon_1 + \varepsilon_2 - \varepsilon_3|$, where ε_i are the phonon energies. If the spectrum is convex, then $|E_n - E_m| > \eta$, where η is finite. Choosing $\Delta < \eta$, our renormalization removes all vertices corresponding to the three-phonon process. But the experimental dependence[2] of the acoustic attenuation on temperature and

---

* This is the only restriction on the generality of the method: it can fail only if $\bar{H}_0 - H_0$ is nonanalytic in λ as $\lambda \to 0$.

† This is a generalization of a method due to Fröhlich.[3]

‡ If Δ is arbitrarily small, some matrix elements of S_1 become arbitrarily large. Nevertheless they remain finite, and the exponential series for T_1 is always convergent.

frequency is ωT^4, which is what we expect from a three-phonon vertex. Hence, the spectrum cannot be convex.

The earlier argument of Landau and Khalatnikov[4] asserted that the three-phonon vertex is suppressed for a convex spectrum since the energy-conserving δ-function of the "golden rule" vanishes. It has been shown[5] that this argument is not sufficient to forbid the three-phonon process. The present argument is much stronger—the three-phonon vertex does not contribute because for a convex spectrum there is no such vertex in the renormalized Hamiltonian.

References

1. H. Maris, *Phys. Rev. Lett.* **28**, 277 (1972), and references quoted therein.
2. B.M. Abrahams, Y. Eckstein, J.B. Ketterson, M. Kuchnir, and J. Vigner, *Phys. Rev.* **181**, 347 (1969).
3. H. Fröhlich, *Proc. Roy. Soc.* **A215**, 291 (1952).
4. L.D. Landau and I.M. Khalatnikov, *Zh. Eksperim. i Teor. Fiz.* **19**, 637, 709 (1949); *Collected papers of Landau,* Oxford, Pergamon Press, (1965), Nos. 69, 70, pp. 494, 511.
5. K. Kawasaki, *Prog. Theor. Phys.* **26**, 795 (1961); P.C. Kwok, P.C. Martin, and P.B. Miller, *Sol. St. Comm.* **3**, 181 (1965); C.J. Pethick and D. ter Haar, *Physica* **32**, 1905 (1966).

Model Dispersion Curves for Liquid ^{4}He*

James S. Brooks and Russell J. Donnelly

*Physics Department, University of Oregon, Eugene, Oregon
and The Niels Bohr Institute, Copenhagen, Denmark*

Introduction

We have developed a method which gives a continuous representation of the excitation spectrum (energy vs. momentum) for superfluid helium as a function of temperature and density, using neutron scattering data[1-3] to determine the behavior of the various parameters which describe the shape of the spectrum. We have applied this model spectrum to the calculation of various thermodynamic properties of the liquid.

The Three-Curve Representation

Our representation of the excitation spectrum consists of three connected branches: the linear phonon branch, the parabolic roton minimum, and a parabola which connects the phonon and roton branches. The linear phonon branch is given by

$$E_{\text{phonon}} = u_1 p \tag{1}$$

where E is the excitation energy, u_1 is the speed of first sound, and p is the momentum of the excitation. The roton region is described in terms of the Landau parameters Δ, μ, and p_0, which have been recently measured for numerous temperatures and pressures.[3] The roton branch is given by

$$E_{\text{roton}} = \Delta + (p - p_0)^2/2\mu \tag{2}$$

Our parameterization of Δ, μ, and p_0 is given in the appendix. The connecting region between the phonon and roton branches is represented by a "phonon parabola" in analogy to the roton parabola. Hence we have

$$E_{\text{phonon parabola}} = \Delta_p - (p - p_{0p})^2/2\mu_p \tag{3}$$

where Δ_p is the maximum energy, p_{0p} is the momentum of the maximum, and μ_p is the curvature of the phonon parabola. The values of these parameters are also given in the appendix.

To construct a model curve, we connect the three branches in the following way: Equation (1) is used until $E_{\text{phonon}} = E_{\text{phonon parabola}}$. Equation (2) is then used

* Research supported by the National Science Foundation, Grant number NSF-GP-26361, and by the Air Force Office of Scientific Research, Grant number AF-AFOSR-71-1999.

until $E_{\text{phonon parabola}} = E_{\text{roton}}$. (For temperatures greater than 1.1°K and pressures greater than the vapor pressure there are two possible solutions to $E_{\text{phonon parabola}} = E_{\text{roton}}$ since the phonon parabola overlaps the roton parabola. The first intersection, for smallest p, is always taken.) Equation (3) is then used until $p/\hbar = 2.8 \text{ Å}^{-1}$.

Thermodynamic Calculations

The most extensive neutron measurements have been taken at 1.1°K at the vapor pressure.[2] Here the excitation spectrum, including the multiphonon branch, is most exactly known. We have performed a standard statistical thermodynamic calculation using this data to determine the entropy due to the elementary excitations and have found the calculated value to be about 5% lower than the experimental value.[4] The multiphonon branch makes less than 1% contribution to the entropy. Using the neutron data at 2.1°K at the vapor pressure,[2] we have performed a similar calculation and have found the calculated entropy again to be lower than the experimental value.[4] Our model curves give similar results. In general, when our curves are used at the vapor pressure the calculated values of entropy are lower than measured entropies[5] by $\sim 5\%$ between 1.15 and 2.05°K. For the temperature and pressure ranges 1.15–1.17°K and 0–25 atm the calculated entropy rises as a function of increasing pressure relative to the experimental results.[5] At 1.70°K the error in the calculated entropy goes from -5% at the vapor pressure to a maximum of $+15\%$ at 25 atm. This increase in error as a function of pressure is less severe at temperatures lower than 1.70°K. For instance, the error at 1.15°K and 25 atm is only about $+5\%$.

Entropy calculations by Dietrich et al.[3] show a similar increase in error with increasing pressure for temperatures above 1.4°K. Since our model curves are in good agreement with the neutron data, the statistical thermodynamic assumptions used in our calculations may be the possible source of error.

The difference between calculations for the entropy using the Landau model for the spectrum (involving simplified integrals over the linear phonon and parabolic roton branches separately) and calculations involving numerical integration over our three-branch curve becomes apparent at higher temperatures where the phonon parabola starts to contribute to the entropy. For example, at 2.0°K at the vapor pressure the simple Landau model gives a value of 0.896 J g^{-1} deg^{-1} for the entropy, whereas our three-branch model gives 0.932 J g^{-1} deg^{-1}. (The experimental value[5] is 0.963 J g^{-1} deg^{-1}.)

Conclusion

We have developed useful relations for the parameters which determine the shape of the excitation spectrum as a function of temperature and density, and have produced a three-branch model dispersion curve which gives a continuous representation of this spectrum and can be used for calculations involving elementary excitations. Thermodynamic quantities such as the entropy can be calculated from this model to reasonable accuracy. Systematic deviations between calculated and measured entropies suggest that some improvements in the theoretical expressions for thermodynamic quantities may be needed.

Appendix: Parabolic Parameters

A more complete description of the arguments leading to our expressions for the Landau parameters is given by Donnelly.[6] Briefly, the parabolic parameters are given as follows.

1. Density dependence of the Landau parameters. The position of the roton minimum has been empirically determined to be a function of density only. Dietrich et al.[3] found that

$$p_0/\hbar = 3.64\rho^{1/3} \quad \text{Å}^{-1} \tag{4}$$

The density dependence of the energy gap and effective mass were determined by extrapolation of the neutron data[3] back to zero temperature. The results are[6]

$$\Delta(\rho, 0)/k = (16.99 - 57.25\rho) \, ^\circ\text{K} \tag{5}$$

$$\mu(\rho, 0) = (0.32 - 1.102\rho) \, M_{\text{He}} \tag{6}$$

(The exact values of the coefficients depend on the density table used.)

2. Temperature dependence of the Landau parameters. The decrease in the roton energy gap as a function of temperature has been found to be reasonably well represented by the semiempirical relation

$$\delta(\Delta)/k = (\rho_{nr}/\rho) \, T[1 - (a/T) \, N_r] \, ^\circ\text{K} \tag{7}$$

where the best fit to the data[3] gives $a = 8.75 \times 10^{-23} \, ^\circ\text{K cm}^3$. ($\rho_{nr}$ is the roton density of the normal fluid and N_r is the roton number density.)

We have found[6] that the relation

$$\delta\Delta/\Delta = \delta\mu/\mu \tag{8}$$

agrees well with the observed experimental behavior of Δ and μ. This has enabled us to determine the temperature dependence of the roton effective mass in terms of that of the energy gap. Hence, the total changes in Δ and μ as a function of density and temperature are given by

$$\frac{\Delta(\rho, T)}{k} = \frac{\Delta(\rho, 0)}{k} - \frac{\rho_{nr}}{\rho} T\left(1 - \frac{a}{T} N_r\right) \, ^\circ\text{K} \tag{9}$$

$$\mu(\rho, T) = \mu(\rho, 0)\frac{\Delta(\rho, T)}{\Delta(\rho, 0)} \tag{10}$$

Since

$$\rho_{nr} = (P_0^2/3kT) \, N_r \tag{11}$$

and

$$N_r = [2(\mu kT)^{1/2} \, p_0^2/(2\pi)^{3/2} \, \hbar^3] \exp(-\Delta/kT) \tag{12}$$

we must solve $\Delta(\rho, T)$ and N_r self-consistently. An iterative procedure works satisfactorily if μ in N_r is taken to be $\mu(\rho, 0)$.

3. Phonon parabola parameters. The parameters used in the phonon parabola are

$$p_{0p}/\hbar = 1.122 \quad \text{Å}^{-1}, \qquad \mu_p = 0.58 M_{\text{He}}, \qquad \text{and} \qquad \Delta_p(\rho)/k = (8.45 + 36.90\rho) \, ^\circ\text{K}$$

where the density dependence of Δ_p was taken from the data of Henshaw and Woods.[1]

In the absence of a general equation of state for helium II, values for the speed of sound and the density can be taken from tables,[7] as we did for the calculations in this paper. However, at temperatures below 0.5°K the equation of state developed by Abraham et al.[8] can be used.

Note Added in Proof

A continuous representation of the excitation spectrum which includes phonon dispersion is described in a thesis by J. S. Brooks, University of Oregon, 1973.

Acknowledgments

We would like to express our gratitude to Dr. O.W. Dietrich for several valuable discussions and for access to the neutron scattering data. We would also like to thank Mr. Shannon Davis for his assistance in developing and running the computer programs necessary for our calculations.

References

1. D.G. Henshaw and A.D.B. Woods in Proc. 7th Intern. Conf. Low Temp. Phys., G.M. Graham and A.C. Hollis Hallet, eds. University of Toronto Press, Toronto (1961), p. 539.
2. R.A. Cowley and A.D.B. Woods, *Can. J. Phys.* **49**, 177 (1971).
3. O.W. Dietrich, E.H. Graf, C.H. Huang, and L. Passell, *Phys. Rev.* **A5**, 1377 (1972).
4. J. Wilks, *The Properties of Liquid and Solid Helium,* Oxford Univ. Press, Oxford (1967), Table Al, p. 666.
5. C.J.N. van den Meijdenberg, K.W. Taconis, and R. de Bruyn Outober, *Physica* **27**, 197 (1961).
6. R.J. Donnelly, *Phys. Lett.* **A39**, 221 (1972).
7. J. Wilks, *The Properties of Liquid and Solid Helium,* Oxford Univ. Press, Oxford (1967), Table A2, p. 667; Table A6, p. 670.
8. B.M. Abraham, Y. Eckstein, J.B. Ketterson, M. Kuchnir, and P.R. Roach, *Phys. Rev.* **A1**, 250 (1970).

Analysis of Dynamic Form Factor S(k,ω) for the Two-Branch Excitation Spectrum of Liquid ⁴He

T. Soda, K. Sawada, and T. Nagata

Department of Physics, Tokyo University of Education
Tokyo, Japan

In previous work[1] we explained the experimental intensity curves for the two-branch peaks of an experimental excitation spectrum from the two-branch spectrum and the experimental form factor S_k consistently with sum rules by assuming that the sum rules are exhausted by real one-phonon and two-phonon excitations. We want now to explain the intensity curves of the dynamic form factor $S(k, \omega)$ obtained by the neutron scattering experiments of Cowley and Woods[2] in terms of the energies of one- and two-phonon excitations and their decay rate for small momenta.

We introduce a density propagation function $G(k, z)$ defined by

$$G(k, z) = \langle \rho_k^\dagger (z - H)^{-1} \rho_k \rangle \tag{1}$$

where ρ_k is a Fourier transform of the density operator, H is the total Hamiltonian, and the angular brackets in Eq. (1) means a thermal average over eigenstates for the system. $G(k, z)$ is related to $S(k, \omega)$ by

$$S(k, \omega) = -\pi^{-1} \operatorname{Im} G(k, \omega) \tag{2}$$

We define the mean excitation energy $\langle H \rangle_k$ by

$$\langle H \rangle_k = \langle \rho_k^\dagger H \rho_k \rangle \langle \rho_k^\dagger \rho_k \rangle^{-1} = (k^2/2m) S_k \tag{3}$$

where the last equality comes from the f sum rule and the definition of S_k.

We expand $G(k, z)$ in a similar manner as done by Huang-Klein.[3]

$$
\begin{aligned}
G(k, z) &= \frac{\langle \rho_k^\dagger \rho_k \rangle}{z - \langle H \rangle_k} + \frac{1}{z - \langle H \rangle_k} \left\langle \rho_k^\dagger \frac{H - \langle H \rangle_k}{z - H} \rho_k \right\rangle \\
&= \frac{\langle \rho_k^\dagger \rho_k \rangle}{z - \langle H \rangle_k} + \frac{1}{z - \langle H \rangle_k} \left\langle \rho_k^\dagger \left[\frac{H - \langle H \rangle_k}{z - H} - \frac{H - \langle H \rangle_k}{z - \langle H \rangle_k} \right] \rho_k \right\rangle \\
&= \frac{\langle \rho_k^\dagger \rho_k \rangle}{z - \langle H \rangle_k} + \frac{1}{(z - \langle H \rangle_k)^2} \left\langle \rho_k^\dagger \frac{(H - \langle H \rangle_k)^2}{z - H} \rho_k \right\rangle \\
&= \langle \rho_k^\dagger \rho_k \rangle \left\{ z - \langle H \rangle_k \right. \\
&\qquad \left. - \frac{\langle \rho_k^\dagger [(H - \langle H \rangle_k)^2/(z - H)] \rho_k \rangle \langle \rho_k^\dagger \rho_k \rangle^{-1}}{1 + (z - \langle H \rangle_k)^{-1} \langle \rho_k^\dagger [(H - \langle H \rangle_k)^2/(z - H)] \rho_k \rangle \langle \rho_k^\dagger \rho_k \rangle^{-1}} \right\}^{-1}
\end{aligned} \tag{4}
$$

where we subtracted a zero identity, $\langle \rho_k^\dagger [(H - \langle H \rangle_k)/(z - \langle H \rangle_k)] \rho_k \rangle$, in the second equality.

$G(k, z)$ and $G(k, z)^{-1}$ have the following asymptotic expansions for high z:

$$G(k, z)\Big|_{z \to \infty} = \frac{\langle \rho_k^\dagger \rho_k \rangle}{z} + \frac{\langle \rho_k^\dagger H \rho_k \rangle}{z^2} + \cdots \tag{5}$$

and

$$G(k, z)^{-1}\Big|_{z \to \infty} = \frac{z}{\langle \rho_k^\dagger \rho_k \rangle}\left(1 - \frac{1}{z}\langle H \rangle_k + \cdots \right) = \frac{z - \langle H \rangle_k}{\langle \rho_k^\dagger \rho_k \rangle} + O(z^{-1}) \tag{6}$$

In Eq. (6) we note that $(z - \langle H \rangle_k)/\langle \rho_k^\dagger \rho_k \rangle$ is nonzero except on the real positive axis and $G(k, z)^{-1} - (z - \langle H \rangle_k)(\langle \rho_k^\dagger \rho_k \rangle)^{-1}$ approaches zero as $z \to \infty$. By Cauchy's theorem, we have the following:

$$G(k, z)^{-1} - \frac{z - \langle H \rangle_k}{\langle \rho_k^\dagger \rho_k \rangle} = \frac{1}{2\pi i} \int_C \frac{dz'}{z' - z}\left[G(k, z')^{-1} - \frac{z' - \langle H \rangle_k}{\langle \rho_k^\dagger \rho_k \rangle}\right] \tag{7}$$

where the contour of integration is shown in Fig. 1. The contribution from the circle vanishes if we make the radius of the circle infinite. In the remaining integration around the branch cut the last term in the integrand does not contribute because it is continuous across the branch cut. Therefore, we obtain

$$G(k, z)^{-1} = \frac{z - \langle H \rangle_k}{\langle \rho_k^\dagger \rho_k \rangle} + \pi^{-1} \int_0^\infty \frac{d\omega'}{\omega' - z} \operatorname{Im} G(k, \omega')^{-1} \tag{8}$$

and

$$G(k, z) = \left[z - \langle H \rangle_k + \pi^{-1} \langle \rho_k^\dagger \rho_k \rangle \int_0^\infty \frac{d\omega'}{\omega' - z} \operatorname{Im} G(k, \omega')^{-1}\right]^{-1} \langle \rho_k^\dagger \rho_k \rangle \tag{9}$$

We define the following function:

$$M_{\text{off}}(k, z) = -\pi^{-1} \langle \rho_k^\dagger \rho_k \rangle \int_0^\infty \frac{d\omega'}{\omega' - z} \operatorname{Im} G(k, \omega')^{-1} \tag{10}$$

which from Eq. (4) is also given by

$$M_{\text{off}}(k, z) = \frac{\langle (H - \langle H \rangle_k)^2/(z - H) \rangle_k}{1 + (z - \langle H \rangle_k)^{-1} \langle H - \langle H \rangle_k/(z - H) \rangle_k} \tag{11}$$

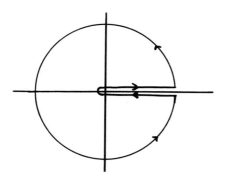

Fig. 1. The contour of integration in the z plane.

where $\langle A \rangle_k$ means $\langle \rho_k^\dagger A \rho_k \rangle / \langle \rho_k^\dagger \rho_k \rangle$. From Eq. (11) we obtain the following asymptotic expansions for $M_{\text{off}}(k, z)$ and $M_{\text{off}}(k, z)^{-1}$ for high z:

$$M_{\text{off}}(k, z)\Big|_{z \to \infty} = \frac{\langle (H - \langle H \rangle_k)^2 \rangle_k}{z} + \frac{\langle (H - \langle H \rangle_k)^2 H \rangle_k}{z^2} + O(z^{-3}) \tag{12}$$

and

$$M_{\text{off}}(k, z)^{-1}\Big|_{z \to \infty} = \frac{z}{\langle (H - \langle H \rangle_k)^2 \rangle_k} \left[1 - \frac{\langle (H - \langle H \rangle_k)^2 H \rangle_k}{z \langle (H - \langle H \rangle_k)^2 \rangle_k} \right] + O(z^{-1}) \tag{13}$$

We also note that the first term in Eq. (13) vanishes only on the real positive axis. In a similar way as we did for $G(k, z)^{-1}$, we express $M_{\text{off}}(k, z)^{-1}$ in a Cauchy integral in the contour given by Fig. 1. We then have

$$\begin{aligned}
M_{\text{off}}(k, z)^{-1} &= \frac{z}{\langle (H - \langle H \rangle_k)^2 \rangle} - \frac{\langle (H - \langle H \rangle_k)^2 H \rangle_k}{[\langle (H - \langle H \rangle_k)^2 \rangle_k]^2} \\
&\quad + \pi^{-1} \int_0^\infty d\omega' \frac{1}{\omega' - z} \operatorname{Im} M_{\text{off}}(k, \omega')^{-1} \\
&= \left[z - \langle H \rangle_k - \frac{\langle (H - \langle H \rangle_k)^3 \rangle}{\langle (H - \langle H \rangle_k)^2 \rangle} \right. \\
&\quad \left. + \pi^{-1} \langle (H - \langle H \rangle_k)^2 \rangle \int_0^\infty d\omega' \frac{1}{\omega' - z} \operatorname{Im} M_{\text{off}}(k, \omega')^{-1} \right] \\
&\quad \times \left[\langle (H - \langle H \rangle_k)^2 \rangle_k \right]^{-1}
\end{aligned} \tag{14}$$

Therefore, we can write $G(k, \omega + i\delta)$ as

$$\begin{aligned}
G(k, \omega + i\delta) &= \langle \rho_k^\dagger \rho_k \rangle \Bigg\{ \omega + i\delta - \langle H \rangle_k \langle (H - \langle H \rangle_k)^2 \rangle_k \\
&\quad \times \left[\omega + i\delta - \langle H \rangle_k - \frac{\langle (H - \langle H \rangle_k)^3 \rangle_k}{\langle (H - \langle H \rangle_k)^2 \rangle_k} \right. \\
&\quad \left. + \pi^{-1} \langle (H - \langle H \rangle_k) \rangle_k^2 \int_0^\infty \frac{d\omega'}{\omega' - \omega - i\delta} \operatorname{Im} M_{\text{off}}(k, \omega')^{-1} \right]^{-1} \Bigg\}^{-1}
\end{aligned} \tag{15}$$

If we introduce the following quantities

$$\langle \Delta \omega^2 \rangle_k = \langle (H - \langle H \rangle_k)^2 \rangle_k \tag{16}$$

$$\Delta_k^{(1)} = \langle (H - \langle H \rangle_k)^3 \rangle_k / \langle (H - \langle H \rangle_k)^2 \rangle_k \tag{17}$$

$$\Delta_k'(\omega) + i\Gamma_k(\omega) = \langle (H - \langle H \rangle_k)^2 \rangle_k \pi^{-1} \int_0^\infty \frac{d\omega'}{\omega' - \omega - i\delta} \operatorname{Im} M_{\text{off}}(k, \omega')^{-1} \tag{18}$$

and

$$\Delta_k(\omega) = -\Delta_k'(\omega) + \Delta_k^{(1)} \tag{19}$$

then $G(k, \omega)$ becomes

$$G(k, \omega) = \frac{\langle \rho_k^\dagger \rho_k \rangle [\omega - \langle H \rangle_k - \Delta_k(\omega) - i\Gamma_k(\omega)]}{[(\omega - \langle H \rangle_k)(\omega - \langle H \rangle_k - \Delta_k(\omega)) - \langle \Delta \omega^2 \rangle_k] - (\omega - \langle H \rangle_k) i\Gamma_k(\omega)}$$

(20)

and $S(k, \omega)$ is given by

$$S(k, \omega) = \frac{\pi^{-1} \langle \rho_k^\dagger \rho_k \rangle \Gamma_k(\omega) \langle \Delta \omega^2 \rangle_k}{[(\omega - \langle H \rangle_k)^2 - (\omega - \langle H \rangle_k) \Delta_k(\omega) - \langle \Delta \omega^2 \rangle_k]^2 + (\omega - \langle H \rangle_k)^2 \Gamma_k(\omega)^2}$$

(21)

Here $\langle H \rangle_k$ is the mean excitation energy, but is exhausted by the one-phonon energy for small momenta as discussed in our previous paper.[1] $\Delta_k^{(1)}$ is given by

$$\Delta_k^{(1)} = \frac{\langle (H - \langle H \rangle_k) H (H - \langle H \rangle_k) \rangle_k}{\langle (H - \langle H \rangle_k)^2 \rangle_k} - \langle H \rangle_k$$

(22)

Since $H - \langle H \rangle_k$ is proportional to the two-phonon density fluctuation $\rho_k^\dagger \rho_{k'}$, $\Delta_k^{(1)} + \langle H \rangle_k$ is the two-phonon energy. On the other hand, $\Delta_k'(\omega)$ has the energy of the rest of the elementary excitations and $\Gamma_k(\omega)$ contains the decay rates of mainly one- and two-phonon excitations for small momenta, and also one of the rest of the elementary excitations for higher momenta.

The expression for $S(k, \omega)$ in Eq. (21) contains the three unknown parameters $\Delta_k(\omega)$, $\Gamma_k(\omega)$, and $\langle \Delta \omega^2 \rangle_k$. Both $\Delta_k(\omega)$ and $\Gamma_k(\omega)$ usually depend on the energy ω. However, we assume they depend weakly on ω for a fixed small momentum. Using the experimental form factor S_k to give the values of $\langle H \rangle_k$ and energies of the extreme points (one minimum and two maximum points) in the intensity curves, we determine the values of the above three parameters.

We take the experimental curves of $S(k, \omega)$ for $k = 0.3 \text{ Å}^{-1}$ from the work of Woods et al.[4] and for $k = 0.6 \text{ Å}^{-1}$ from the work of Cowley and Woods.[2] We put

$$f_k(\omega) = [(\omega - \langle H \rangle_k)^2 - (\omega - \langle H \rangle_k) \Delta_k - \langle \Delta \omega^2 \rangle_k]^2 + (\omega - \langle H \rangle_k)^2 \Gamma_k^2$$

(23)

and set

$$df_k/d\omega = 0$$

for the three extreme points $\omega = \omega_1$, ω_2, and ω_3, where ω_2 corresponds to a minimum and ω_1 and ω_3 correspond to two maxima. Then we determine the value of the parameters given in Table 1. We find that $\langle \Delta \omega^2 \rangle_k$ and Γ_k^2 increases with momentum k, indicating that higher momenta yield more fluctuations and decay rates of elementary excitations.

Table I

k, Å^{-1}	Δ_k,* °K	$\langle \Delta \omega^2 \rangle_k$, °K^2	Γ_k^2, °K^2
0.3	13.34	2.63	3.08
0.6	3.64	18.75	3.67

* $\Delta_k = \Delta_k^{(1)} - \Delta_k'$.

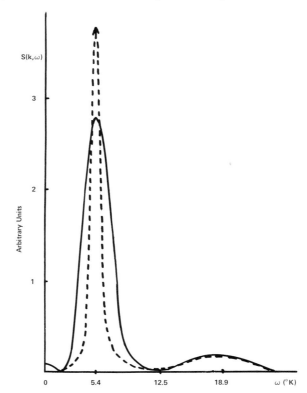

Fig. 2. The curves of $S(k,\omega)$ for $k = 0.3 \text{ Å}^{-1}$. The solid curve is experimental and the broken curve is calculated.

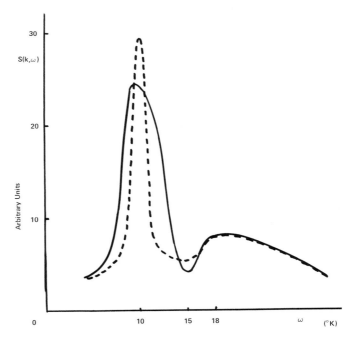

Fig. 3. The curves of $S(k,\omega)$ for $k = 0.6 \text{ Å}^{-1}$. The solid curve is experimental and the broken curve is calculated.

Inserting these values of the parameters in Eq. (21), we plot the shape of the curve of $S(k, \omega)$ in Figs. 2 and 3 for $k = 0.3$ and $0.6\,\text{Å}^{-1}$, respectively. In Fig. 2 we fit the calculated intensities at the smaller maximum with the experimental value. The larger maximum at ω_1 gives a very sharp maximum of a δ-function type due to the fact that ω_1 is very close to the mean excitation energy. We believe that the experiment misses some of the high points near ω_1 and we also have some ambiguities in our calculation for the experimental form factor S_k which is given by different values in the experiments of Achter and Meyer[5] and Hallock,[6] so that we do not expect an exact fit with the experiment for the intensity curve $S(k, \omega)$ near ω_1. In Fig. 3 we also fit the calculated intensities at the smaller maximum ω_3 with the experimental value. For this case we have a slim intensity curve around ω_1. This is due to the background effect and errors in the experimental values of $S(k, \omega)$ as well as the existence of ambiguities in the experimental values of the form factor S_k[5,6] used in the calculation.

A similar calculation for momenta higher than $k = 0.6\,\text{Å}^{-1}$, for example, for $k = 0.8\,\text{Å}^{-1}$, gives rise to imaginary values of Γ_k^2, showing that our assumptions of no energy dependence of $\Delta_k(\omega)$ and $\Gamma_k(\omega)$ does not hold for high momenta.

References

1. T. Soda, K. Sawada, and T. Nagata, *Prog. Theor. Phys.* **44**, 860 (1970).
2. R.A. Cowley and A.D.B. Woods, *Can. J. Phys.* **49**, 177 (1971).
3. K. Huang and A. Klein, *Ann. Phys.* (*N.Y.*) **30**, 203 (1964).
4. A.D.B. Woods, E.C. Svensson, and P. Martel, preprints (1972).
5. E.K. Achter and L. Meyer, *Phys. Rev.* **188**, 291 (1969).
6. R.B. Hallock, *Phys. Rev.* **A5**, 370 (1971).

Excitation Spectrum for Weakly Interacting Bose Gas and the Liquid Structure Function of Helium*

Archana Bhattacharyya and Chia-Wei Woo†

Department of Physics, Northwestern University
Evanston, Illinois

In this paper the excitation spectrum of a Bose gas characterized by a pairwise Fourier-transformable potential $\lambda V(r)$ is calculated to three leading orders in the strength parameter λ, first by the Hugenholtz–Pines prescription,[1] then by the method of correlated basis functions (CBF).

In the Hugenholtz–Pines theory the energies of the low-lying excitations are given by the poles of the Green's function ($\hbar = 1, m = 1$):

$$G(\mathbf{k}, \varepsilon) = \frac{\varepsilon + \frac{1}{2}k^2 - \mu + \Sigma_{11}^-}{[\varepsilon - \frac{1}{2}(\Sigma_{11}^+ - \Sigma_{11}^-)]^2 - [\frac{1}{2}k^2 - \mu + \frac{1}{2}(\Sigma_{11}^+ + \Sigma_{11}^-)]^2 + \Sigma_{02}^2} \tag{1}$$

where Σ_{11}^+, Σ_{11}^-, and Σ_{02} are proper self-energy parts and can be evaluated by perturbation theory. The resulting spectrum is

$$\varepsilon(k) = \frac{k^2}{2} + \lambda n v_k + \lambda^2 \left[\frac{-n^2 v_k^2}{k^2} + \frac{n}{(2\pi)^3} \int d\mathbf{h} \, \frac{v_h^2}{\mathbf{h} \cdot (\mathbf{k} - \mathbf{h})} \frac{\mathbf{k} \cdot \mathbf{h}}{h^2} \right.$$

$$\left. + \frac{n}{(2\pi)^3} \int d\mathbf{h} \, \frac{v_h v_{\mathbf{k}-\mathbf{h}}}{\mathbf{h} \cdot (\mathbf{k} - \mathbf{h})} \right] + \cdots \tag{2}$$

where

$$v_k = \int v(r) \exp(i\mathbf{k} \cdot \mathbf{r}) \, d\mathbf{r} \tag{3}$$

Equation (2) is exact to the order shown.

For applying the method of CBF three different sets of correlated basis functions are constructed, each consisting of density fluctuation operators acting on a different correlating factor $F(1, 2, ..., N)$. The basis functions

$$|\mathbf{k}) = \rho_{\mathbf{k}} F / [N S_F(k)]^{1/2} \tag{4}$$

$$|\mathbf{k} - \mathbf{h}, \mathbf{h}) = \rho_{\mathbf{k}-\mathbf{h}} \rho_{\mathbf{h}} F / [N^2 S_F(|\mathbf{k} - \mathbf{h}|) S_F(h)]^{1/2} \tag{5}$$

$$|\mathbf{k} - \mathbf{h} - \mathbf{l}, \mathbf{h}, \mathbf{l}) = \rho_{\mathbf{k}-\mathbf{h}-\mathbf{l}} \rho_{\mathbf{h}} \rho_{\mathbf{l}} F / [N^3 S_F(|\mathbf{k} - \mathbf{h} - \mathbf{l}|) S_F(h) S_F(l)]^{1/2} \tag{6}$$

* Work supported in part by the National Science Foundation through Grant No. GP-29130 and through the Materials Research Center of Northwestern University.
† Alfred P. Sloan Research Fellow.

etc., are then partially orthogonalized and normalized to yield $|k\rangle$, $|k - h, h\rangle$, $|k - h - l, h, l\rangle$, etc. The function $S_F(k)$ denotes the liquid structure function defined in terms of F,

$$S_F(k) = 1 + \rho \int [g_F(r) - 1] \exp(i\mathbf{k} \cdot \mathbf{r}) \, d\mathbf{r} \tag{7}$$

where

$$g_F(r_{12}) = [N(N - 1)/\rho^2] \int F^2 \, d\mathbf{r}_3 \cdots d\mathbf{r}_N / \int F^2 \, d\mathbf{r}_1 \cdots d\mathbf{r}_N \tag{8}$$

and

$$\rho = N/\Omega \tag{9}$$

The three correlating factors employed are: (i) the optimized Jastrow function $\hat{\Psi}_J \equiv \prod_{i<j} \exp[\frac{1}{2}\hat{u}(r_{ij})]$, (ii) an unoptimized Jastrow function Ψ_J, and (iii) the ground-state eigenfunction of the system Ψ_0.

Case (i): $F = \hat{\Psi}_J$

The expectation value of the Hamiltonian with respect to $|k\rangle$ yields the Feynman-like spectrum

$$\hat{\varepsilon}_F(k) \equiv \langle k|H|k \rangle - \langle 0|H|0 \rangle = k^2/2\hat{S}(k) \tag{10}$$

Treating the matrix elements $\langle k|H|k - h, h \rangle$ and $\langle k - h, h|H - \langle k|H|k \rangle|k - l, l \rangle$, which represent phonon decaying, coalescing, and scattering, as perturbations, we obtain perturbative corrections to $\hat{\varepsilon}_F(k)$. To order λ^2 the only term that contributes is shown in Fig. 1(a) and yields:

$$\hat{\varepsilon}_a(k) = \frac{1}{2} \sum_{h \neq 0, k} \frac{|\langle k|H|k - h, h \rangle|^2}{\hat{\varepsilon}_F(k) - \hat{\varepsilon}_F(|k - h|) - \hat{\varepsilon}_F(h)} \tag{11}$$

(a)   (b)

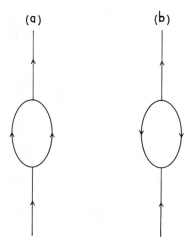

Fig. 1. Second-order perturbation corrections to the Feynman spectrum.

where

$$\langle \mathbf{k}|H|\mathbf{k} - \mathbf{h}, \mathbf{h} \rangle = [\mathbf{k} \cdot (\mathbf{k} - \mathbf{h}) \hat{S}(h) + \mathbf{h} \cdot (\mathbf{k} - \mathbf{h}) \hat{S}(k) + \mathbf{k} \cdot \mathbf{h}\hat{S}(|\mathbf{k} - \mathbf{h}|)$$

$$- k^2 \hat{S}(h) \hat{S}(|\mathbf{k} - \mathbf{h}|) + |\mathbf{k} - \mathbf{h}|^2 \hat{S}(k) \hat{S}(h) + h^2 \hat{S}(k) \hat{S}(|\mathbf{k} - \mathbf{h}|)$$

$$- (k^2 + h^2 - \mathbf{k} \cdot \mathbf{h}) \hat{S}(k) \hat{S}(h) \hat{S}(|\mathbf{k} - \mathbf{h}|)]$$

$$\times [16N\hat{S}(k) \hat{S}(h) \hat{S}(|\mathbf{k} - \mathbf{h}|)]^{-1/2} \tag{12}$$

The excitation spectrum, to order λ^2, given by $\hat{\varepsilon}_F(k) + \hat{\varepsilon}_a(k)$ agrees with Eq. (2) as expected.

Case (ii): $F = \Psi_J$

The exact $S(k)$ can be evaluated with the help of the Hugenholtz–Pines prescription. We now employ a correlating factor Ψ_J which possesses an $S_J(k)$ that agrees with $S(k)$ to $O(\lambda)$ only. The major difference between this treatment and case (i) is that $\langle \mathbf{k}|H|\mathbf{k} \rangle$ now contains an extra term:

$$\varepsilon_F^J(k) \equiv \langle \mathbf{k}|H|\mathbf{k} \rangle - \langle 0|H|0 \rangle = \frac{k^2}{2S_J(k)} + \left\{ \frac{S_J'(k)}{S_J(k)} + \frac{k^2}{4} \left[1 - \frac{1}{S_J(k)} \right] \right\} \tag{13}$$

where $S_J'(k)$ is a complicated integral defined in Ref. 2. It was shown in Ref. 2 that optimizing Ψ_J results in

$$S_J'(k) = -\tfrac{1}{4}k^2 [S_J(k) - 1] \tag{14}$$

thus the simplification of the expression for $\langle \mathbf{k}|H|\mathbf{k} \rangle$.

Perturbative corrections to order λ^2 again arise only from Fig. 1(a). Here $\varepsilon_a^J(k)$ is given by an expression similar to Eq. (11). The spectrum to order λ^2 again agrees with Eq. (2).

Case (iii): $F = \Psi_0$

$$\varepsilon_F(k) \equiv \langle \mathbf{k}|H|\mathbf{k} \rangle - E_0 = k^2/2S(k) \tag{15}$$

Evaluation of the second-order perturbation correction $\varepsilon_a(k)$ is now simplified by the fact that Ψ_0 is an eigenfunction of the Hamiltonian. The matrix element $\langle \mathbf{k}|H|\mathbf{k} - \mathbf{h}, \mathbf{h} \rangle$ in this case reduces to

$$\langle \mathbf{k}|H|\mathbf{k} - \mathbf{h}, \mathbf{h} \rangle = [\mathbf{k} \cdot (\mathbf{k} - \mathbf{h}) S(h) + \mathbf{k} \cdot \mathbf{h}S(|\mathbf{k} - \mathbf{h}|) - k^2 S(h) S(|\mathbf{k} - \mathbf{h}|)]$$

$$\times [4NS(k) S(h) S(|\mathbf{k} - \mathbf{h}|)]^{-1/2} \tag{16}$$

Again $\varepsilon_F(k) + \varepsilon_a(k)$ yields Eq. (2) exactly to $O(\lambda^2)$.

The above calculations have been carried out to demonstrate the flexibility of the method of CBF. Such flexibility can actually be utilized in the study of other model and realistic Bose systems, such as the charged Bose gas and liquid ^{4}He. For example, by carrying out parallel calculations using the correlating factors (i) and (iii), with $S(k)$ known to order λ^n and $\hat{S}(k)$ known to order λ^{n+1}, it is possible to extract the $(n + 1)$th order contribution to $S(k)$ by identifying $[\varepsilon_F(k) + \varepsilon_a(k)]$ with $[\hat{\varepsilon}_F(k) + \hat{\varepsilon}_a(k)]$. In the case of a high-density charged Bose gas the expansion parameter is $r_s^{3/4}$, where r_s denotes half the average interparticle spacing measured in units of the Bohr radius. The leading term in the excitation spectrum is the plasmon frequency ω_{pl}. The next term is proportional to $\omega_{pl} r_s^{3/4}$, and has contributions from the Feynman term as well as from second-order perturbation. The latter now includes both Fig. 1(a) and Fig. 1(b), with

$$\hat{\varepsilon}_b(k) = -\tfrac{1}{2} \sum_{\mathbf{h} \neq 0, \mathbf{k}} \frac{|\langle \mathbf{k}, \mathbf{h}, -\mathbf{k} - \mathbf{h} | H | 0 \rangle|^2}{\hat{\varepsilon}_F(k) + \hat{\varepsilon}_F(h) + \hat{\varepsilon}_F(|k + h|)} \tag{17}$$

and $\varepsilon_b(k)$ given by a similar formula. The matrix element $\langle \mathbf{k}, \mathbf{h}, -\mathbf{k} - \mathbf{h} | H | 0 \rangle$, just as in the case of $\langle \mathbf{k} | H | \mathbf{k} - \mathbf{h}, \mathbf{h} \rangle$, differs in the two cases. Substituting

$$S(k) = \hat{S}_0(k) + r_s^{3/4} \hat{S}_1(k) + \cdots \tag{18}$$

and

$$S(k) = S_0(k) + r_s^{3/4} S_1(k) + \cdots \tag{19}$$

into $[\hat{\varepsilon}_F(k) + \hat{\varepsilon}_a(k) + \hat{\varepsilon}_b(k)]$ and $[\varepsilon_F(k) + \varepsilon_a(k) + \varepsilon_b(k)]$, respectively, we find

$$S_0(k) = \hat{S}_0(k) \tag{20}$$

$$S_1(k) = \hat{S}_1(k) + \Delta S_1(k) \tag{21}$$

where $\Delta S_1(k)$ denotes an expression which can be evaluated readily. Results of this calculation were reported in an earlier note,[3] the details of which are contained in a forthcoming paper.[4] We note that a direct calculation of $\Delta S_1(k)$ constitutes a formidable task for the usual diagrammatic perturbation techniques.

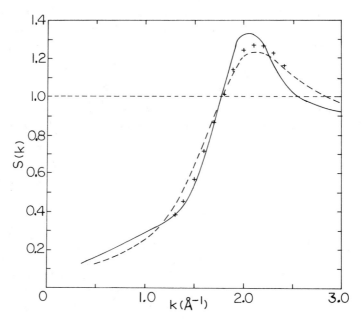

Fig. 2. Improved liquid structure function for ⁴He. (– – –) Optimum Jastrow; (——) experimental (x-ray scattering) + present work.

In Ref. 3 we suggested a similar calculation for $S(k)$ in liquid helium. For helium there is no convenient expansion parameter. A number of diagrams besides those in Fig. 1 may well contribute significantly to the perturbative corrections. However, Fig. 1 must clearly play a major role. Taking the optimized Jastrow $\hat{S}(k)$ as a starting point in the role of $\hat{S}_0(k) + \hat{S}_1(k)$, we proceed to solve the second order Brillouin–Wigner equation in search for an improvement $\Delta S_1(k)$ in $S(k)$. It is expected that the peak will be sharpened, bringing the theoretically calculated liquid structure function into better agreement with experiment. Numerical work leads to the results shown in Fig. 2. Even at this preliminary stage of the work we find the improvement significant and most encouraging.

References

1. N.M. Hugenholtz and D. Pines, *Phys. Rev.* **116**, 489 (1959).
2. E. Feenberg, *Theory of Quantum Fluids,* Academic, New York (1969), Chapter 5.
3. A. Bhattacharyya and C.-W. Woo, *Phys. Rev. Lett.* **28**, 1320 (1972).
4. A. Bhattacharyya and C.-W. Woo, *Phys. Rev.* (to be published).

Elementary Excitations in Liquid Helium*

Chia-Wei Woo†

Department of Physics, Northwestern University
Evanston, Illinois

Recently certain controversies have arisen over the form and direction of the phonon dispersion and the nature of the two-roton excitation in liquid helium. From the standpoint both of theoretical elegance and of satisfactorily interpreting experimental data in a way consistent with sum rules, it is highly desirable to develop a theory which accounts for the two branches of excitations on equal footing, at least in the limit of long wavelengths. We discuss here such an attempt. It is a microscopic theory based on a generalization of Feynman's treatment of the phonon–roton spectrum.

The low-lying excited states of Bose liquids such as liquid helium are best described in the "multiphonon representation." Independent but generally non-orthogonal basis functions are constructed by applying density fluctuation operators $\rho_{\mathbf{k}} = \sum_i \exp(i\mathbf{k} \cdot \mathbf{r}_i)$ on the ground-state wave function Ψ_0 of the system, resulting in $\rho_{\mathbf{k}}\Psi_0$, $\rho_{\mathbf{k}}\rho_{\mathbf{l}}\Psi_0$, $\rho_{\mathbf{k}}\rho_{\mathbf{l}}\rho_{\mathbf{h}}\Psi_0$, etc. We wish then to simultaneously diagonalize the Hamiltonian and the overlap matrix in this space.

In practice we have to settle for a far less ambitious program, and restrict our attention for the moment to the subspace spanned by just the one- and two-phonon states:

$$|\mathbf{k}\rangle = \rho_{\mathbf{k}}\Psi_0/(NS_k)^{1/2} \tag{1}$$

and

$$|\mathbf{k} - \mathbf{l}, \mathbf{l}\rangle = \rho_{\mathbf{k}-\mathbf{l}}\rho_{\mathbf{l}}\Psi_0/(N^2 S_{|\mathbf{k}-\mathbf{l}|}S_l)^{1/2} \tag{2}$$

where S_k denotes the zero-temperature liquid structure function. Included in this treatment are three-phonon (roton) and four-phonon vertices—all of the phonon processes normally considered. (The decay of one phonon into three and its inverse process are omitted. They will be reintroduced when the space is extended to include three-phonon states.) From previous work by Jackson and Feenberg[1] and Lai et al.[2] it appears that the phonon–roton spectrum can be accurately obtained in such a treatment. We thus consider it an adequate starting point.

Taking as a trial wave function

$$\Psi_{\mathbf{k}} = \zeta_{\mathbf{k}}|\mathbf{k}\rangle + \tfrac{1}{2} \sum_{\mathbf{l} \neq 0} \xi_{\mathbf{k}}(\mathbf{l}) |\mathbf{k} - \mathbf{l}, \mathbf{l}\rangle \tag{3}$$

* Work supported in part by the National Science Foundation through Grant No. GP-29130 and through the Materials Research Center of Northwestern University.
† Alfred P. Sloan Research Fellow.

we evaluate and minimize the energy expectation value

$$E_k \equiv (\Psi_\mathbf{k}|H|\Psi_\mathbf{k})/(\Psi_\mathbf{k}|\Psi_\mathbf{k}) \equiv E_0 + \varepsilon_k \tag{4}$$

the necessary matrix elements of H and 1 having been evaluated in earlier work.[2] Two coupled integral equations are obtained:

$$(\varepsilon_k - \varepsilon_k^0)\,\zeta_k = \tfrac{1}{2}\sum_\mathbf{l} \xi_\mathbf{k}(\mathbf{l})\,(\mathbf{k}|W(\varepsilon_k)|\mathbf{k} - \mathbf{l}, \mathbf{l}) \tag{5}$$

and

$$(\varepsilon_k - \varepsilon_{|\mathbf{k} - \mathbf{h}|}^0 - \varepsilon_h^0)\,\xi_\mathbf{k}(\mathbf{h}) = \zeta_k(\mathbf{k}|W(\varepsilon_k)|\mathbf{k} - \mathbf{h}, \mathbf{h})$$
$$+ \tfrac{1}{2}\sum_{\mathbf{l}(\neq\,\mathbf{h})}' \xi_\mathbf{k}(\mathbf{l})\,(\mathbf{k} - \mathbf{h}, \mathbf{h}|W(\varepsilon_k)|\mathbf{k} - \mathbf{l}, \mathbf{l}) \tag{6}$$

where ε_k^0 represents the Feynman spectrum, and

$$W(\varepsilon_k) = H - \varepsilon_k 1 \tag{7}$$

and the prime over the summation indicates omission of $\mathbf{l} = 0$ or $\mathbf{k}$ from the sum. By combining these equations, an eigenvalue equation is formed, the solution of which gives rise to both branches of the excitation spectrum.

For orientation purpose, we formally iterate Eq. (6), solving for $\xi_\mathbf{k}(\mathbf{h})$ in terms of ζ_k, and then substitute the resulting series into Eq. (5). The first iteration yields Brillouin–Wigner's second-order perturbation correction, which was investigated by Jackson and Feenberg.[1] On further iterations, phonon–phonon scattering processes enter into the correction terms, some of which were included in Ref. 2. A direct solution of the eigenvalue equation that arises from Eqs. (5) and (6) corresponds to the summation of diagrams shown in Fig. 1 to all orders. Likewise, the elimination of ζ_k from Eqs. (5) and (6) and a subsequent iterative solution give rise to Fig. 2, which depicts the summation of repeated phonon–phonon (roton–roton) scattering to all orders.

Some of the more interesting results obtained by applying this theory to liquid helium are summarized below. In all cases the Hamiltonian

$$H = -(\hbar^2/2m)\sum_i \nabla_i^2 + \sum_{i<j} v(r_{ij}) \tag{8}$$

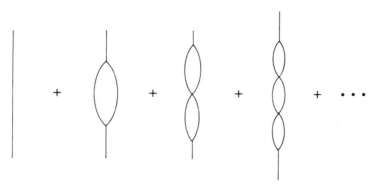

Fig. 1. Phonon renormalization.

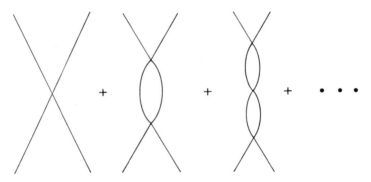

Fig. 2. Roton–roton scattering.

with $v(r)$ given by deBoer and Michels (Lennard-Jones 6–12 form) is used. There are no adjustable parameters in the calculation.

At long wavelengths we express the phonon–roton spectrum in the form

$$\varepsilon_k = \varepsilon_k^0 (1 + c_1\chi + c_2\chi^2 + c_3\chi^3 + \cdots) \tag{9}$$

where

$$\chi \equiv \hbar k/ms \tag{10}$$

s being the velocity of first sound. For $c_1 \neq 0$, Eqs. (5) and (6) reduce to an identity which cannot be satisfied. Also, preliminary numerical work indicates that $c_2 \approx -1.46$, as reported earlier in Refs. 2 and 3 as the result of a low-order perturbation calculation, is reasonably accurate. Armed with sum rule arguments, the fact that linear dispersion is totally negligible,[4] and the more reliable parts of x-ray and neutron scattering data, namely Hallock's best fit[5] of S_k and Cowley and Wood's location (but not the strength) of the multiphonon peak,[6] we obtain with *no* further assumption the following results:

$$\varepsilon_k = \hbar s k (1 + 0.17\chi^2 + 1.28\chi^3 - 4.5\chi^4 + \cdots) \tag{11}$$

$$Z_k = (\hbar k/2ms)(1 - 1.63\chi^2 - 1.28\chi^3 + 0.51\chi^4 - 4.0\chi^5 + \cdots) \tag{12}$$

$$S_k^{\Pi} = 1.1\chi^4 + 2.0\chi^6 + \cdots \tag{13}$$

$$G_k = (1/2ms^2)(1 - 1.80\chi^2 - 2.56\chi^3 + \varepsilon\chi^4 + \cdots), \qquad 0 \leq \varepsilon < 1 \tag{14}$$

$$\langle\omega^3\rangle^{1/3} = 41°K \tag{15}$$

Equations (11)–(15) represent long-wavelength expansion formulas for the phonon spectrum ε_k, the single-phonon and multiphonon structure functions Z_k and S_k^{Π}, the ω^{-1} moment G_k of the dynamic structure function, and the ω^3 moment $\langle\omega^3\rangle$ of the multiphonon contribution. Equations (11)–(14), when plotted and compared to Cowley and Wood's data,[5] all fall well within experimental uncertainties. Equation (11) indicates a slightly concave phonon dispersion. The coefficient of χ^2 at 0.17 is well below Maris's 2.00 or Phillips *et al.*'s 1.03. One must, however, note that there exists an important χ^3 term. The coefficients in Eqs. (12)–(13) differ from Cowley and Woods's own fits, which emphasizes the fact that it is difficult to separate out single-phonon and multiphonon contributions to $S(k, \omega)$. The implications of these

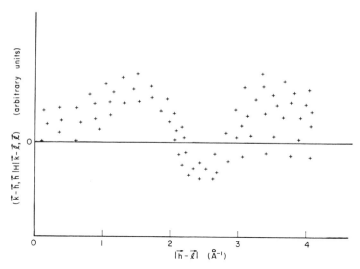

Fig. 3. Noncentral, momentum-dependent roton–roton interaction.

present findings are under investigation. At this moment we have nothing further to add.

The search for the two-roton bound state turns out to be a difficult numerical task. The main reason is that the unrenormalized roton–roton interaction (thus the renormalized vertex as well) is noncentral and strongly momentum dependent. Figure 3 shows samples of $(\mathbf{k} - \mathbf{h}, \mathbf{h} | H | \mathbf{k} - \mathbf{l}, \mathbf{l})$ plotted against the momentum transfer $|\mathbf{h} - \mathbf{l}|$ at $\mathbf{k} \to 0$ for a range of $\mathbf{l}$ and $\mathbf{k} - \mathbf{l}$. Clearly no single curve can pass through the wide-ranging points displayed. Continued work in progress will be reported in detail elsewhere.

Acknowledgments

I am grateful to R. B. Hallock, J. B. Ketterson, and A. D. B. Woods for valuable information and helpful discussions. Parts of the work described here are in collaboration with D. Pines.

References

1. H.W. Jackson and E. Feenberg, *Rev. Mod. Phys.* **34**, 686 (1962).
2. H.-W. Lai, H.-K. Sim, and C.-W. Woo, *Phys. Rev.* **A1**, 1536 (1970).
3. D. Pines and C.-W. Woo, *Phys. Rev. Lett.* **24**, 1044 (1970).
4. P.R. Roach, B.M. Abraham, J.B. Ketterson, and M. Kuchnir, *Phys. Rev. Lett.* **29**, 32 (1972).
5. R.B. Hallock, *Phys. Rev.* **A5**, 320 (1972); and private communications.
6. R.A. Cowley and A.D.B. Woods, *Can. J. Phys.* **49**, 177 (1971).

Long-Wavelength Excitations in a Bose Gas
And Liquid He II at $T = 0$

H. Gould

Department of Physics, Clark University
Worcester, Massachusetts

and

V. K. Wong

Department of Physics, University of Michigan
Ann Arbor, Michigan

There are many assumptions that have been made, both in theoretical[1] and experimental[2] investigations, about the nature of the long-wavelength excitations of liquid He II. As an example, it is common to assume that the excitation spectrum $\omega(k)$ and various structure functions $S(k)$ can be expanded in a power series in the wave vector k. A theoretical justification of the various assumptions is an important problem, but in the absence of a tractable microscopic theory of a strongly interacting Bose system we must look for other ways to test these assumptions. One way is by a qualitative analysis coupled with the use of plausible sum rule arguments. Another way is by a rigorous microscopic analysis of a solvable model of a Bose gas.

Information about the excitations of the system is contained in the well-known dynamic structure function $S(k, \omega)$. The function $S(k, \omega)$ can be measured by a particle scattering experiment, e.g., neutron scattering. At $T = 0$, $S(k, \omega)$ can be expressed in terms of the density–density response function $F(k, \omega)$ by

$$S(k, \omega) = - (1/\pi) \operatorname{Im} F(k, \omega)$$

$S(k, \omega)$ satisfies some well-known sum rules which will be useful in the following discussion.

We have

$$\int_0^\infty d\omega S(k, \omega) = S(k) \qquad \text{and} \qquad \int_0^\infty d\omega\, \omega S(k, \omega) = k^2/2m$$

$S(k)$ is the static structure function; the second relation is referred to as the f sum rule.

Let us first consider a simple qualitative analysis of the structure of $S(k, \omega)$. Consider a simple multiphonon diagram for $F(k, \omega)$, as shown in Fig. 1(a). The imaginary part $S(k, \omega)$ is found by the usual procedure of cutting the diagram for $F(k, \omega)$, with the result that $S(k, \omega)$ can be represented by three terms Fig. 1 (b–d). Figure 1(b) represents the production of a real phonon that exhausts the f-sum rule. What is left over is represented by Fig. 1(c), which is equal to the contribution from the

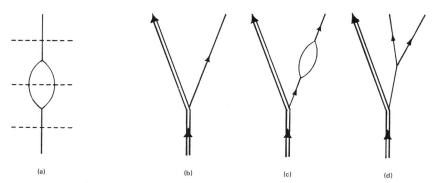

Fig. 1. First order multiphonon processes. The structure of the vertices is ignored. (a) The density response function $F(k, \omega)$. Solid lines represent excitations and dashed lines denote cuts needed to obtain the imaginary part $S(k, \omega)$. (b)–(d) Neutrons (double lines) producing elementary excitations (single line). Excitations represented by directed lines are on the energy shell. Parts (b)–(d) correspond to coherent one-phonon scattering and coherent and incoherent two-phonon processes, respectively.

imaginary part of the double pole due to the coalescence of the single poles of the two isolated propagators in Fig. 1(a). Figure 1(c) corresponds to the production of a real phonon that decays into two virtual phonons that in turn recombine into a real phonon. Thus this process corresponds to the physical picture of backflow. The last diagram can be interpreted as the production of a virtual phonon that decays into two real phonons. The first two processes can be described as "coherent"; that is, both represent the production of a real phonon. The third process is incoherent. More complicated diagrams would lead to coherent and incoherent n-phonon processes in $S(k, \omega)$, that is, a coherent backflow with n virtual phonons and an incoherent production of n real phonons, respectively. Thus we are led to write in the long-wavelength limit

$$S(k, \omega) = Z(k)\,\delta(\omega - \omega(k)) + Y(k)\,\delta(\omega - \omega(k)) + X(k, \omega), \qquad Z(k) = k^2/2m\omega(k)$$

We refer to the three terms in $S(k, \omega)$ as the one-phonon, the backflow, and the background terms, respectively. We know that as $k \to 0$ only the one-phonon term contributes and thus from the first sum rule the relation $S(k) = Z(k) = k^2/2m\omega(k)$ becomes exact. This relation is commonly referred to as the Feynman one-phonon relation. The usefulness of the above separation of $S(k, \omega)$ into three terms is that the corrections to the Feynman relation are isolated into two terms, the backflow Y and the background X. If we appeal to some simple sum rule arguments, we can show that the leading k dependence of the background contribution to $S(k)$ is order k^4, as first done by Miller et al.[3] In the same spirit, we can show that the leading k dependence of the backflow is order k^3. The main assumptions are that the backflow current is nonzero from hydrodynamic considerations and, of course, our separation of the coherent processes into two parts. We thus conclude that the one-phonon Feynman relation for $S(k)$ breaks down at order k^3.

Let us return to Fig. 1(a) and investigate the real part of $F(k, \omega)$ which contributes to $\omega(k)$. By counting powers of k, we see that the one-loop part of the diagram has a long-wavelength logarithmic singularity. If we calculate the diagram more carefully, we find that its contribution to $\omega(k)$ as $k \to 0$ is $k^5 \ln(1/k)$. We can extend the calculation to all orders in perturbation theory by using a Ward identity to determine the vertices. We conclude that $\omega(k)$ is a nonanalytic function of k due to the logarithmic singularity associated with the long-wavelength limit of a propagator pair.

These qualitative arguments are confirmed by a rigorous microscopic analysis of a dilute Bose gas at $T = 0$. Our microscopic calculations of $S(k, \omega)$, $S(k)$, and $\omega(k)$ were made using a dielectric formulation[4] cast into a convenient form by a judicious use of Ward-type identities based on the continuity equation. In contrast to the usual Green's function formulation, our approach enables us to achieve perturbation approximations which obey local number conservation and related sum rules as well as give an excitation spectrum that is without a gap. The approach also allows us to make detailed calculations beyond the Bogoliubov approximation with relative ease and without spurious divergences.

A careful and detailed discussion for our microscopic calculations will be reported in a review article.[5] Some of the main results [6,7] are summarized here. Let g be a dimensionless coupling constant that summarizes the interaction between two bosons. We find that $S(k, \omega)$ can be separated into three parts, as depicted in Fig. 1(b–d):

$$S(k, \omega) = Z^{(0)} \delta(\omega - \omega_k^{(0)} - g\omega_k^{(1)}) + g[Z^{(1)}(k) + Y^{(1)}(k)] \delta(\omega - \omega_k^{(0)})$$
$$+ gX^{(1)}(k, \omega) + O(g^2)$$

In zeroth order we have the Bogoliubov result: a delta function at Bogoliubov spectrum $\omega_k^{(0)}$ with intensity $Z^{(0)} = k^2/2m\omega_k^{(0)}$. To the first order beyond Bogoliubov we find backflow and background contributions to $S(k, \omega)$ and, in addition, a shift in the spectrum ω_k and intensity $Z(k)$. The excitation spectrum has the nonanalytic form

$$\omega_k/k = c_0 + c_2 k^2 + c_{L4} k^4 \ln(1/k).$$

The mth ω-moment of $S(k, \omega)$ can be written as

$$S_m(k) = (\omega_k^{(0)} + g\omega_k^\omega)^m Z^{(0)}(k) + g\omega_k^{(0)m}[Z^{(1)}(k) + Y^\omega(k)] + gX_m(k) \tag{1}$$
$$X_m(k) = \int_0^\infty d\omega\, \omega^m X(k, \omega)$$

The term $Z^{(1)}(k) + Y^{(1)}(k)$ does not have logarithmic terms through order k^5; the leading k dependences of $X_m(k)$ are: X_{-1} is proportional to $k^4 \ln(1/k)$ and X_0, X_1, X_2 are proportional to k^4. Explicit calculation shows that $S_m(k)$ is consistent with the compressibility and f-sum rules. $S_m(k)$ has a term proportional to k^{5+m} $\ln(1/k)$ for $m = -1, 2$. For $m = 0$ the first term in Eq. (1) does not contribute a $k^5 \ln(1/k)$ term and the static structure function $S_0(k)$ has the form

$$S_0(k) = (k/2c_0)[1 + s_2 k^2 + s_3 k^3 + \cdots]$$

The numerical coefficients given in Ref. 6 are correct, but the cancellation of the $k^5 \ln(1/k)$ terms in $S_0(k)$ was overlooked. $S_0(k)$ has both even and odd powers and displays an inflection point. The background $X_0(k)$ gives rise to the $s_3 k^3$ term and the backflow contributes to the $s_2 k^2$ term. Thus, because of backflow, the Feynman relation for $S_0(k)$ breaks down at the $s_2 k^2$ term.

In closing, we note that some of the common assumptions have not fared well under close scrutiny. It may be well to take other assumptions (e.g., assuming the low-T contribution to the specific heat[8] is due entirely to the elementary excitation spectrum) with some caution until proven otherwise. In fact, an investigation of the multiphonon contribution to the specific heat is now in progress.

References

1. D. Pines and C.-W. Woo, *Phys. Rev. Lett.* **24**, 1044 (1970).
2. R.A. Cowley and A.D.B. Woods, *Can. J. Phys.* **49**, 177 (1971).
3. A. Miller, D. Pines, and P. Nozieres, *Phys. Rev.* **127**, 1452 (1962).
4. S.K. Ma, H. Gould, and V.K. Wong, *Phys. Rev.* **A3**, 1453 (1971).
5. V.K. Wong and H. Gould, *Ann. Phys. N.Y.* (to be submitted).
6. H. Gould and V.K. Wong, *Phys. Rev. Lett.* **27**, 301 (1971).
7. V.K. Wong and H. Gould, *Phys. Lett.* **39A**, 331 (1972).
8. N.E. Phillips, C.G. Waterfield, and J.K. Hoffer, *Phys. Rev. Lett.* **25**, 1260 (1970).

Bose–Einstein Condensation in Two-Dimensional Systems

Y. Imry and David J. Bergman

Department of Physics and Astronomy, Tel Aviv University, Tel Aviv, Israel
and Soreq Nuclear Research Center, Yavne, Israel

and

L. Gunther

Department of Physics, Tufts University
Medford, Massachusetts

It is a well-known and accepted fact that two- and one-dimensional systems of bosons do not undergo Bose–Einstein condensation at any finite temperature.[1] We would like to point out that while this is true when the system is kept at a fixed volume, a quite different behavior is obtained if the system is kept instead at a fixed pressure. In the latter case an ideal Bose gas undergoes a continuous phase transition (i.e., not a first-order transition) to a state where all the particles are condensed in the lowest-lying single particle (s.p.) state. A modified form of this transition may remain even in realistic systems of interacting bosons.

In order to see how this transition arises in two dimensions, we write down the usual expressions for the density n and pressure P of a two-dimensional ideal Bose gas:

$$n \equiv N/V = (1/\lambda_{\text{Th}}^2)\, F_1(\alpha) \tag{1}$$

$$P = (kT/\lambda_{\text{Th}}^2)\, F_2(\alpha) \tag{2}$$

where

$$\lambda_{\text{Th}} \equiv (h^2/2\pi mkT)^{1/2} = \text{thermal de Broglie wavelength} \tag{3}$$

$$F_\sigma(\alpha) = \sum_{s=1}^{\infty} s^{-\sigma} e^{-s\alpha} \tag{4}$$

When $\alpha \to 0$, $F_2(\alpha)$ is clearly bounded, and reaches its maximum value of $\pi^2/6$ when $\alpha = 0$. It would thus appear from (2) that P cannot be larger than the critical value $P_c(T)$, defined by

$$P_c(T) \equiv (kT/\lambda_{\text{Th}}^2)\, \pi^2/6 \tag{5}$$

This is wrong for the usual reason: Eq. (2) was obtained by using the integration approximation for the sum on states, which breaks down at least for the lowest state when condensation occurs.

A more careful evaluation of N and P, which treats the lowest s.p. state, with an energy ε_0 and an occupation number N_0, separately, leads to the following modified equations:

$$N = N_0 + (1/\lambda_{Th}^2) V F_1 [\alpha + (\varepsilon_0/kT)] \tag{6}$$

$$P = (N_0 \varepsilon_0/V) + (kT/\lambda_{Th}^2) F_2 [\alpha + (\varepsilon_0/kT)] \tag{7}$$

$$N_0 = [(\varepsilon_0/kT) + \alpha]^{-1} \tag{8}$$

The ground-state contribution to the pressure $N_0 \varepsilon_0/V$ depends on the boundary conditions at the walls and on the exact shape of the box. Periodic boundary conditions would lead to $\varepsilon_0 = 0$, but the more realistic vanishing condition leads to

$$\varepsilon_0 = O(h^2/mV) \tag{9}$$

If we now attempt to solve (6) and (7) for $P > P_c(T)$, we can arrive fairly easily at the following results:

$$N_0 = N = O(V^2) \tag{10}$$

$$\alpha = (1/N_0) - (\varepsilon_0/kT) = O(1/V^2) - O(1/V) \tag{11}$$

$$N - N_0 = (1/\lambda_{Th}^2) V F_1 (1/N_0) = O(V \log V) \ll N \tag{12}$$

$$P = NP_0 + P_c(T) \tag{13}$$

where $P_0 \equiv \varepsilon_0/V$ is the pressure of a single particle in the ground state. These results represent a Bose-condensed phase at a temperature T and pressure $P > P_c(T)$, in which essentially all the particles are in the ground state and the volume (as well as the entropy) is subextensive, i.e., $V = O(N^{1/2})$. The rigorous proof for nonexistence of Bose condensation in two dimensions[1] does not apply in this case because the density n is infinite in the thermodynamic limit* [i.e., $n = O(N^{1/2})$].

As one approaches the condensation temperature $T_c(P)$ [the inverse of the function $P_c(T)$] from above at a fixed pressure, the volume decreases to zero (actually, to a quantity which is $o(N)$) continuously according to the following asymptotic expression:

$$V \to (\pi h^2/12mP)^{1/2} (N/|\log \varepsilon|) \to o(N)$$

where

$$\varepsilon \equiv T/T_c(P) - 1 \tag{14}$$

The entropy exhibits a similar behavior. Other thermodynamic quantities also have peculiar forms of singular behavior. Some examples are the following expressions:

$$c_P/k \to \pi^2/3\varepsilon (\log \varepsilon)^2 \to \infty \tag{15}$$

$$c_V/k \to \pi^2/3 |\log \varepsilon| \to 0 \tag{16}$$

$$kT\chi = N_0 \to 3 |\log \varepsilon|/\pi^2 \varepsilon \to O(N) \tag{17}$$

In the last equation χ is the response function (i.e., generalized susceptibility) of the condensate order parameter $\langle \psi(\mathbf{r}) \rangle$ to its thermodynamically conjugate field [χ is equal to N_0/kT only above $T_c(P)$].

As one approaches $T_c(P)$ from below at a fixed P the volume (as well as the en-

---

* This restriction was first emphasized by Rehr and Mermin.[2]

tropy) has the following behavior:

$$V \sim \{N/[T_c(P) - T]\}^{1/2} \tag{18}$$

while $N_0 = N - O(N^{1/2} \log N)$.

Thus, there are no jumps in the usual extensive thermodynamic variables V and S at $T_c(P)$, both of them going continuously to zero in the thermodynamic limit. Their derivatives, however, are singular, e.g., c_P. In addition to these, there is a discontinuous change in N_0 (i.e., in the order parameter of the condensed phase) at $T_c(P)$: On the scale of N (that is, viewed as an extensive quantity) N_0 jumps from essentially zero to essentially N.

Without going into further details (they will be published elsewhere) we would like to point out that similar considerations may be made for one-dimensional systems, where they also lead to the possibility of Bose condensation at finite T and P. In the three-dimensional boson gas it has long been known that if one cools it at a fixed P below some critical temperature $T_c(P)$, it undergoes a first-order transition to a condensed state which is also collapsed (i.e., has a vanishing volume in the thermodynamic limit).[3,4]

The results up to this point can be derived fairly rigorously due to the great simplicity of the ideal boson gas. What consequences can we deduce from them for realistic systems?

When the bosons interact only by means of a strong, short-range repulsion we expect the effects of the interaction to be felt only when the density is so high that the average interparticle distance is comparable to the hard-core diameter. At these high densities the pressure will be forced upward by the interactions. At a fixed pressure they will prevent the system from actually collapsing to a "zero-volume" condensed state. While in three dimensions the phase transition to a Bose condensed state is still expected to take place, at least whenever $\lambda_{\mathrm{Th}} \gg$ the hard-core radius, in two dimensions we expect it to disappear. However, since the deviations from ideal gas behavior depend on n, which increases only as $P^{1/2}|\log \varepsilon|$, while c_P of the ideal gas increases as $1/\varepsilon(\log \varepsilon)^2$, i.e., much faster, we can expect to see a peak developing in c_P for small ε if the pressure is low enough. A similar type of remnant from the ideal gas transition in two dimensions would be an anomalous rise in N_0, which increases as $|\log \varepsilon|/\varepsilon$ as long as the interactions are unimportant.

When attractive interactions are also brought into play we still expect to see a first-order transition into a condensed state in three dimensions. As before, when the range of interaction is much less than λ_{Th} we expect the transition to be to a (partially) condensed state. This is in agreement with the properties of the gas-to-liquid transition in ${}^4\mathrm{He}$ at low temperatures. In two dimensions there may or may not be a phase transition and it may or may not be first order. But even if there is a transition it will not be to a Bose-condensed state, since the density now remains finite, and the theorem of Ref. 1 is applicable.

More light can be shed on the interacting boson gas by reviewing some results that have been obtained for quantum lattice gases.[9,11] These results lead us to expect that when the attractive interactions are strong enough there will be a phase transition even in two dimensions.

Some of the results on two-dimensional boson systems may be useful in the attempt to understand the experimental properties of thin films of ${}^4\mathrm{He}$: (a) In their experiments on submonolayer ^{4}He films Bretz and Dash[5] found a peak in the specific heat which was later attributed to incipient Bose–Einstein condensation brought about by nonuniformities in the substrate–helium potential energy.[6] The mechanism we have discussed before would also lead to such a peak. (b) In the somewhat thicker ^{4}He films where superfluid flow and third sound can occur (i.e., films that are at least a few atoms thick) the superfluid-to-normal transition seems to be accompanied by a jump in ρ_s,[7,8] even though none of the usual extensive quantities such as V or S have any observable discontinuity. The jump in N_0 at the condensation point of the two-dimensional boson gas which we have found and the lack of any associated jump in other extensive quantities could be connected with these experimental phenomena.

References

1. P.C. Hohenberg, *Phys. Rev.* **158**, 383 (1967).
2. J.J. Rehr and N.D. Mermin, *Phys. Rev.* **B1**, 3160 (1970).
3. B. Kahn, Doctoral Thesis; in *Studies in Statistical Mechanics,* J. de Boer and G.E. Uhlenbeck, eds., North-Holland, Amsterdam (1965), Vol. III.
4. F. London, *Superfluids,* Dover, New York (1954), Vol. II.
5. M. Bretz and J.G. Dash, *Phys. Rev. Lett.* **26**, 963 (1971).
6. C.E. Campbell, J.G. Dash, and M. Schick, *Phys. Rev. Lett.* **26**, 966 (1971).
7. R.S. Kagiwada, J.C. Fraser, I. Rudnick, and D.J. Bergman, *Phys. Rev. Lett.* **22**, 338 (1969).
8. R.P. Henkel, G. Kukich, and J.D. Reppy, in *Proc. 11th Intern. Conf. Low Temp. Phys.* 1968, St. Andrews University Press, Scotland (1969), Vol. I, p. 178.
9. T. Matsubara and H. Matsuda, *Prog. Theor. Phys.* **16**, 416, 569 (1956).
10. N.D. Mermin and H. Wagner, *Phys. Rev. Lett.* **17**, 1133 (1966).
11. M.E. Fisher, *Rep. Prog. Phys.* **30**, 615 (1967).

Phonons and Lambda Temperature in Liquid ^{4}He As Obtained by the Lattice Model

Paul H. E. Meijer

Physics Department, The Catholic University of America
Washington, D. C.

and

W. D. Scherer

Center for Experimental Design and Data Analysis
National Oceanic and Atmospheric Administration, Boulder, Colorado

1. Consider a face-centered lattice with lattice constants d and define a set of operators a_i^* and a_i which create and annihilate Bose particles at the ith site of the lattice. Imposing the usual commutation rules plus a constraint against multiple occupation of the same site leads to a set of hard-core boson operators that can be mapped on a set of spin-$\frac{1}{2}$ operators.[1]

We introduce a kinetic energy term that allows for hopping from one site to another and a potential energy $-v_0$ between the nearest neighbors. This leads to the following Hamiltonian:

$$H = (h^2/2md^2) \sum_{\langle ij \rangle} (a_i^* - a_j^*)(a_i - a_j) - v_0 \sum_{\langle ij \rangle} a_i^* a_i a_j^* a_j - \mu \sum_i a_i^* a_i \quad (1)$$

where m is the mass of the ^{4}He atoms. In this expression we also introduced the chemical potential by adding a term $-\mu N$. If we replace a_i^* by S_i^+, we obtain an anisotropic Heisenberg Hamiltonian given by

$$H = -H \sum_i S_i^z - \tfrac{1}{2} J \sum_{\langle ij \rangle} (S_i^+ S_j^- + S_j^+ S_i^-) - J' \sum_{\langle ij \rangle} S_i^z S_j^z \quad (2)$$

where $J = h/md^2$, $J' = v_0$ and $H = \mu - (z/2)(J - J')$. Here z is the number of nearest neighbors. For liquid helium we find $J > J'$, i.e., the coupling is dominated by the x–y term, the symmetry is oblate.

2. A prolate form can be obtained by performing a rotation in spin space; the resulting Hamiltonian will, however, contain additional terms. This rotation implies the choice of three Euler angles. We choose the first and the last as being fixed and the middle angle as being variable. This angle θ is determined by the request that all terms linear in S^+ and S^- be suppressed. Introducing a decoupling in the new z direction, we find a temperature-dependent rotation angle and a Hamiltonian consisting of two pieces. One part is similar to (2) with $H = 0$ and a different set of

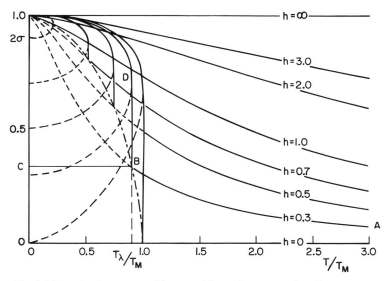

Fig. 1. Magnetization measured in rotated frame 2σ versus reduced temperature T/T_M for constant magnetic field h.

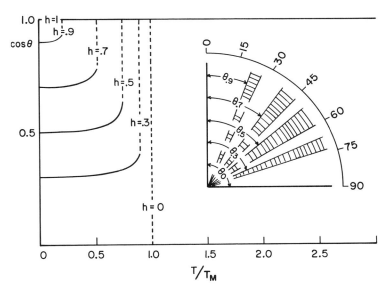

Fig. 2. Rotation angle at constant magnetic field θ_h versus reduced temperature T/T_M.

coupling constants $\tilde{J}$ and $\tilde{J}'$ instead of J and J', and another part given by

$$H_c = (i/8)\sum_{i,j}(J_{ij} - J'_{ij})(\sin^2\theta)(S_i^{+\prime}S_j^{+\prime} - S_i^{-\prime}S_j^{-\prime}) \qquad (3)$$

The first part is a prolate Heisenberg Hamiltonian since $\tilde{J} < \tilde{J}'$. The second part represents a coupling that leads to a set of equations of motion where the S^+ equation depends on the S^- equation and vice versa. These equations are solved by a Fourier transform in space followed by a Bogoliubov transformation[2] between the operators. The result is similar to the transformation in superconductivity but with hyperbolic functions. The spectrum has a linear k dependence as long as the rotation angle in the transformation preceding the two last transformations is finite. Hence, this linear spectrum is modified at higher temperatures since the "constants" in the secular equation are temperature dependent.

To determine this dependence, we evaluate the self-consistence equation for the magnetization[3] in the rotated frame (see Fig. 1), which in turn determined the expectation value of the magnetization in the original frame of quantization, that is, the average number of particles at a given chemical potential. We find that just below the lambda temperature the system changes very rapidly from the zero rotation angle (above T_c) to the finite value attained at $T = 0$ (compare Fig. 2.) We would like to stress that despite the jump in the rotation angle the transition is of second order, since it is associated with a kink in the chemical potential.

We finally obtain two equations relating the density, the temperature, and the chemical potential: one for above the lambda temperature and one for below the lambda temperature. The lambda temperature lies somewhat below 2.18°K depending on the density chosen. The density in a lattice model is an absolute density since it is compared to the completely filled lattice. The choice is therefore determined by the choice of the lattice distance d. There are no adjustable parameters.

References

1. T. Matsubara and H. Matsuda, *Prog. Theor. Phys.* **16**, 569 (1956).
2. N.N. Bogoliubov, *J. Phys. (USSR)* **11**, 23 (1947).
3. D.N. Zubaref, *Soviet Phys.—Uspekhi* **3**, 320 (1960).

Equation of State for Hard-Core Quantum Lattice System

T. Horiguchi and T. Tanaka*

Department of Physics, Ohio University
Athens, Ohio

It is well known that the melting curves for ^{3}He and ^{4}He have a negative slope for a while until a minimum melting temperature is reached, and increase at higher temperatures.[1] For ^{3}He the negative slope is explained by taking into account a large entropy of spin disorder in solid phase and a small entropy in the ordered Fermi fluid. For ^{4}He the entropy from phonon excitations plays an important role in accounting for this phenomenon.

We shall attack these fascinating phenomena from a quite different viewpoint. By the order–disorder theory of the classical lattice model a satisfactory explanation, at least in the first approximation, has been given for the change of phase from fluid to solid in terms of the interatomic potential.[2] The melting curve starts from zero pressure at the absolute zero temperature and increases as the temperature in P–T plane. We shall apply the order–disorder theory to the Matsubara and Matsuda quantum lattice model[3] and show that in the lower-temperature region the melting curve has a negative slope within the pair approximation of the cluster variation method.

In the present paper, as the first step of our approach to this problem, it is assumed that N atoms which are Bose particles are distributed over L lattice sites of a simple cubic lattice with lattice constant d. Following Matsubara and Matsuda,[3] the Hamiltonian of our quantum lattice system is given by

$$H = Kz \sum_i a_i^* a_i - K \sum_{\langle ij \rangle} (a_i^* a_j + a_j^* a_i) + \sum_{\langle ij \rangle} V_{ij} a_i^* a_j^* a_i a_j \qquad (1)$$

where $K = \hbar^2/2md^2$ (d is the lattice constant) and z is the number of nearest-neighbor lattice sites. Here a_i^* and a_i are the creation and annihilation operators for Bose particle at the ith lattice site. We assume the commutation relations for $i \neq j$, and to exclude the multiple occupation of atoms at the same lattice site, we impose the anticommutation relations for $i = j$. The hopping of atoms is assumed to occur from an occupied site to one of its unoccupied nearest-neighbor sites. This effect is represented by the second term of the Hamiltonian. The potential energy is assumed as

$$V_{ij} = \begin{cases} \infty & \text{if the two atoms are at the same lattice site, or at nearest-neighbor lattice sites} \\ -\varepsilon & \text{if the two atoms are at next-nearest-neighbor lattice sites} \\ 0 & \text{otherwise} \end{cases}$$

---

* On leave from Catholic University of America.

Then the third summation in Eq. (1) is taken over nearest- and next-nearest-neighbor pairs.

Following Lennard-Jones and Devonshire,[2] we introduce the so-called normal sites (α sites) and the abnormal sites (β sites) in the basic simple cubic lattice as shown in Fig. 1. The solid state is the state in which α sites are more abundantly populated by atoms than β sites and the fluid state corresponds to the state in which all the lattice sites are equally populated. In the present theory the fcc lattice is considered as the lattice structure in the solid state but in a more realistic treatment of ^{4}He the hexagonal close-packed structure must be considered as the solid state at low temperatures. The number density is defined by

$$\rho = \langle a_i^* a_i \rangle = N/L = 1/v \tag{2}$$

where $v = V/Ld^3$ and $V = Nd^3$. Denoting the number densities of α sites and β sites by ρ_α and ρ_β, we have

$$\rho_\alpha = (1/L_\alpha) \left\langle \sum_{i \in (\alpha \text{ sites})} a_i^* a_i \right\rangle = N_\alpha/L_\alpha \tag{3}$$

$$\rho_\beta = (1/L_\beta) \left\langle \sum_{i \in (\beta \text{ sites})} a_i^* a_i \right\rangle = N_\beta/L_\beta \tag{4}$$

where L_α and L_β are the numbers of the α and β sites, respectively, and N_α and N_β are the numbers of atoms on α and β sites, respectively. The long-range order parameter is defined as follows:

$$R = (\rho_\alpha - \rho)/(1 - \rho) \tag{5}$$

Since the completely ordered state is defined as the state in which all α sites are occupied, ρ_α is equal to one and hence we have $R = 1$. In the disordered state the atoms are equally populated on α sites and β sites, then $\rho_\alpha = \rho$ and we have $R = 0$. The number densities on α sites and β sites are expressed in terms of the long-range

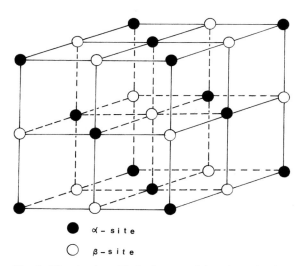

● α - s i t e

○ β - s i t e

Fig. 1. The simple cubic lattice containing the α-site fcc sublattice.

order parameter R as follows:

$$\rho_\alpha = \rho + R(1 - \rho) \tag{6}$$

$$\rho_\beta = \rho - R(1 - \rho) \tag{7}$$

In the cluster variation method[4] the free energy F for a system of N particles in thermal equilibrium is calculated from

$$F = \min \mathscr{F} \tag{8}$$

where $\mathscr{F}$ is the trial free energy function and is defined by

$$\mathscr{F} = \operatorname{tr} \rho_t(H + kT \ln \rho_t) \tag{9}$$

Here H is the Hamiltonian of the system, k is Boltzmann's constant, and T is the absolute temperature. ρ_t is the trial density matrix, normalized to unity,

$$\operatorname{tr} \rho_t = 1 \tag{10}$$

and determined so that $\mathscr{F}$ is minimized. By introducing the reduced trial density matrices by

$$\rho_t^{(n)}(i_1, \ldots, i_n) = \operatorname{tr}_{\{L\} - i_1 - \ldots - i_n}\rho_t, \qquad n = 1, 2, \ldots, L \tag{11}$$

where the trace symbol on the right-hand side indicates that the trace is taken over all variables except for $i_1, \ldots, i_n$, Eq. (9) is rewritten as follows:

$$\mathscr{F} = \sum_i \operatorname{tr} \{\rho_t^{(1)}(i) h^{(1)}(i)\} + \sum_{i>j} \operatorname{tr} \{\rho_t^{(2)}(i, j) h^{(2)}(i, j)\}$$

$$- T\left[\sum_i S^{(1)}(i) + \sum_{i>j} S^{(2)}(i, j) + \cdots + S^{(L)}(i_1, \ldots, i_L)\right] \tag{12}$$

$h^{(1)}(i)$ is the energy operator for one particle at the ith site and $h^{(2)}(i, j)$ is the interaction energy between particles at the ith and jth sites. $S^{(1)}(i), S^{(2)}(i,j), \ldots$, are the entropy terms and are given as follows:

$$S^{(1)}(i) = -k \operatorname{tr} \{\rho_t^{(1)}(i) \ln \rho_t^{(1)}(i)\} \tag{13}$$

$$S^{(2)}(i, j) = -k[\operatorname{tr} \{\rho_t^{(2)}(i, j) \ln \rho_t^{(2)}(i, j)\} - \operatorname{tr} \{\rho_t^{(1)}(i) \ln \rho_t^{(1)}(i)\}$$

$$- \operatorname{tr} \{\rho_t^{(1)}(j) \ln \rho_t^{(1)}(j)\}] \tag{14}$$

Keeping terms up to $S^{(2)}(i, j)$ is equivalent to the Bethe approximation and is called the pair approximation in the cluster variation method. Instead of minimizing the trial free energy function with respect to the trial density matrices, we have an alternative and easier method in which the $\mathscr{F}$ is considered to be a function of correlation functions for particles and the minimization is achieved with respect to these quantities. To do this, the reduced trial density matrices are expressed element by element in terms of the expectation value of certain operators. This is accomplished by requiring that these matrices give the proper expectation values and at the same time satisfy all the reducibility and normalization conditions. Then we have

$$\rho_\alpha^{(1)} = \begin{pmatrix} \rho_\alpha & \\ & 1 - \rho_\alpha \end{pmatrix}, \qquad \rho_\beta^{(1)} = \begin{pmatrix} \rho_\beta & \\ & 1 - \rho_\beta \end{pmatrix} \tag{15}$$

$$\rho_{\alpha\beta}^{(2)}(1) = \begin{pmatrix} \sigma_{\alpha\beta}(1) & & \\ \rho_\alpha - \sigma_{\alpha\beta}(1) & W & \\ W & \rho_\beta - \sigma_{\alpha\beta}(1) & \\ & & 1 - (\rho_\alpha + \rho_\beta) + \sigma_{\alpha\beta}(1) \end{pmatrix} \tag{16}$$

$$\rho_{\alpha\alpha}^{(2)}(2) = \begin{pmatrix} \sigma_{\alpha\alpha}(2) & & \\ \rho_\alpha - \sigma_{\alpha\alpha}(2) & & \\ & \rho_\alpha - \sigma_{\alpha\alpha}(2) & \\ & & 1 - 2\rho_\alpha + \sigma_{\alpha\alpha}(2) \end{pmatrix} \tag{17}$$

$$\rho_{\beta\beta}^{(2)}(2) = \begin{pmatrix} \sigma_{\beta\beta}(2) & & \\ \rho_\beta - \sigma_{\beta\beta}(2) & & \\ & \rho_\beta - \sigma_{\beta\beta}(2) & \\ & & 1 - 2\rho_\beta + \sigma_{\beta\beta}(2) \end{pmatrix} \tag{18}$$

where $\rho_\alpha^{(1)}$ and $\rho_\beta^{(1)}$ replace $\rho_t^{(1)}(i)$, and $\rho_{\mu\nu}^{(2)}(k)$ replace $\rho_t^{(2)}(i, j)$, where i and j are the kth neighbors of each other and i is on a μ site and j is on a ν site. W is the expectation value of the exchange term and we assume that $\langle a_i^* a_j \rangle = \langle a_j^* a_i \rangle$. The $\sigma_{\mu\nu}(k)$ are the pair-correlation functions of the two particles, where one is on a μ site, the other is on a ν site, and they are the kth neighbors of each other. However, due to the nearest-neighbor hard-core potential we have $\sigma_{\alpha\beta}(1) = 0$. Thus, we have the trial free energy function for our quantum lattice system within the pair approximation as follows:

$$\begin{aligned} \mathscr{F}/L = 6K\rho &- 6KW - 3\varepsilon[\sigma_{\alpha\alpha}(2) + \sigma_{\beta\beta}(2)] \\ &+ kT[-(17/2)[\rho_\alpha \ln \rho_\alpha + (1-\rho_\alpha)\ln(1-\rho_\alpha) + \rho_\beta \ln \rho_\beta + (1-\rho_\beta)\ln(1-\rho_\beta)] \\ &+ 3(\{\rho + [R^2(1-\rho)^2 + W^2]^{1/2}\}\ln\{\rho + [R^2(1-\rho)^2 + W^2]^{1/2}\} \\ &+ \{\rho - [R^2(1-\rho)^2 + W^2]^{1/2}\}\ln\{\rho - [R^2(1-\rho)^2 + W^2]^{1/2}\} \\ &+ (1-2\rho)\ln(1-2\rho) + \sigma_{\alpha\alpha}(2)\ln\sigma_{\alpha\alpha}(2) + 2[\rho_\alpha - \sigma_{\alpha\alpha}(2)]\ln[\rho_\alpha - \sigma_{\alpha\alpha}(2)] \\ &+ [1 - 2\rho_\alpha + \sigma_{\alpha\alpha}(2)]\ln[1 - 2\rho_\alpha + \sigma_{\alpha\alpha}(2)] + \sigma_{\beta\beta}(2)\ln\sigma_{\beta\beta}(2) \\ &+ 2[\rho_\beta - \sigma_{\beta\beta}(2)]\ln[\rho_\beta - \sigma_{\beta\beta}(2)] + [1 - 2\rho_\beta + \sigma_{\beta\beta}(2)]\ln[1 - 2\rho_\beta + \sigma_{\beta\beta}(2)])] \end{aligned} \tag{19}$$

By minimizing the trial free energy function with respect to R, W, $\sigma_{\alpha\alpha}(2)$, and $\sigma_{\beta\beta}(2)$ at given volume and temperature, we obtain the free energy of our system and corresponding equilibrium quantities. According as the obtained free energy corresponds to $R \neq 0$ or $R = 0$, our system is either in the ordered state or in the disordered state, respectively, at given volume and temperature.

From the thermodynamic relations the expression for the pressure is given by

$$Pd^3 = -F/L + \rho\left[\frac{\partial(F/L)}{\partial\rho}\right]_{T,L} \tag{20}$$

Inserting the expression (19) with the equilibrium quantities R, W, $\sigma_{\alpha\alpha}(2)$, and $\sigma_{\beta\beta}(2)$ into Eq. (20), one gets the curve for the pressure vs. volume at given temperature. In the coexisting region the Maxwell construction is taken into account. The resulting P–T phase diagram is shown in Fig. 2. The curve for $K = 0$ is the one for the

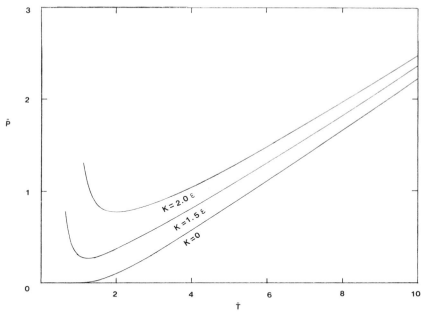

Fig. 2. P–T phase diagram. $\tilde{T} = T/\varepsilon$ and $\tilde{P} = Pd^3/\varepsilon$.

classical lattice model within the same approximation. As seen in Fig. 2, the melting curve for the quantum lattice model has negative slope at lower temperatures. If we use the true mass of ^{4}He (i.e., $m = 6.648 \times 10^{-24}$ g) and choose a lattice constant $d = 2.015$ Å, then for $K = 1.5\varepsilon$ we have $T_m = 1.24°$K, $P_m = 2.99$ atm as the minimum point of the melting curve.

Unfortunately, the calculated value of pressure is much too low as compared with the experimental value, and the behavior of the melting curve below the temperature at which the minimum is attained and that of the ground state have not been properly obtained within the present formulation. We feel that these difficulties can be removed if the following points are taken into account: (1) For the nearest-neighbor hard-core potential, the pressure in the disordered phase was not affected by the added kinetic term within the pair approximation. This unexpected feature would be removed even within the pair approximation if a more reasonable soft-core potential is used. (2) Since the fluid state under consideration is a superfluid, the off-diagonal long-range order parameter should be important and hence one should include this effect into the formulation. (3) The many-particle correlations higher than two particles should play an important role at lower temperatures.

It is hoped that these problems may be investigated in future publications.

References

1. J. Wilks, *The Properties of Liquid and Solid Helium*, Oxford University Press (1967).
2. J.E. Lennard-Jones and A.F. Devonshire, *Proc. Roy. Soc.* **A169**, 317 (1939); **A170**, 464 (1939).
3. T. Matsubara and H. Matsuda, *Prog. Theor. Phys.* **16**, 569 (1956).
4. J. Halow, T. Tanaka, and T. Morita, *Phys. Rev.* **175**, 680 (1968).

Impulse of a Vortex System in a Bounded Fluid

E.R. Huggins

Department of Physics, Dartmouth College
Hanover, New Hampshire

We derive, from the equation of motion of a classical constant-density fluid, the dynamic equation for the fluid impulse of a vortex system in a bounded charged or uncharged fluid. The result is an integral theorem which primarily describes the effect of nonpotential forces upon the vorticity field $\omega \equiv \mathbf{V} \times \mathbf{v}$ [or $\omega \equiv \mathbf{V} \times \mathbf{v} + (q/m)\mathbf{B}$ for charged fluids]. Although the equations treat ω as a continuous field, the main application of the theorem* has been to quantum fluids where a vortex core is usually considered to be a δ-function singularity in the field $\mathbf{V} \times \mathbf{v}$. Figure 1, however, shows how the concept of a continuous vorticity field can be applied to a quantum fluid if we redefine ω as the curl of the physical current $\mathbf{j} = \rho\mathbf{v}$.

The equation of motion of a fluid with constant density ρ can be written in the form

$$\partial\mathbf{v}/\partial t = \mathbf{v} \times (\mathbf{V} \times \mathbf{v}) - \mathbf{V}(v^2/2 + P/\rho) + \mathbf{f} \tag{1}$$

where $\mathbf{v}$ is the fluid velocity, P is the pressure, and $\mathbf{f}$ represents the total force density (per unit mass) acting on the fluid particles.

For the analysis of the dynamics of a vortex system, it is convenient to separate

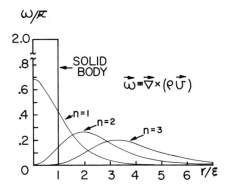

Fig. 1. Vortex core structure of the vorticity field $\omega \equiv \mathbf{V} \times \mathbf{j}$ for a Ginsburg–Landau (GL) fluid (see, e.g., Ref. 7). In the GL theory, where $v = n\bar{\kappa}/r$ ($n =$ an integer, $\bar{\kappa} = h/m$) in a vortex core, the velocity field is singular but $\rho \to 0$ at $r = 0$, so that the physical current $\mathbf{j} = \rho\mathbf{v}$ is smooth and continuous. If we define the vorticity field as the curl of $\mathbf{j}$ rather than $\mathbf{v}$, we avoid all singularities in the hydrodynamic analysis and obtain the rather appealing pictures shown here. In the $n = 1$ ground state ω is similar to that of a classical solid-body rotating core, except that the sharp edge of the classical core is rounded off. The $n = 2$ first excited state is analogous to an atomic wave function, in that it has a node at the center. Here we have a two-dimensional node running the length of the vortex giving us a "macaroni" core structure. In the $n = 3$ second excited state the node expands in much the same way as the higher angular momentum states of the hydrogen wave functions. The use of $\mathbf{V} \times \mathbf{j}$ for vortex core structure is an approach worked out in collaboration with Mr. T. Fugita.

---

* The impulse theorem[1] provides an explanation of the force exerted upon a flexible diaphragm by a stream of charged vortex rings in superfluid helium (observed by Gamota and Barmutz[2]).

$\mathbf{f}$ into a potential part $\mathbf{f}_\Omega = -\nabla\Omega$ and nonpotential part $\mathbf{g} = \mathbf{f} - \mathbf{f}_\Omega$.* Since $\nabla \times \mathbf{f}_\Omega = \nabla \times (-\nabla\Omega) = 0$, we have the formal relations for $\mathbf{f}$, $\mathbf{f}_\Omega$, and $\mathbf{g}$:

$$\mathbf{f} = \mathbf{f}_\Omega + \mathbf{g}, \qquad \mathbf{f}_\Omega = -\nabla\Omega, \qquad \nabla \times \mathbf{f} = \nabla \times \mathbf{g} \tag{2}$$

and Eq. (1) can now be written in the form

$$\partial\mathbf{v}/\partial t = -\nabla\mu + \mathbf{v} \times (\nabla \times \mathbf{v}) + \mathbf{g} \tag{3}$$

where the so-called "chemical potential" μ is

$$\mu = (P/\rho) + (v^2/2) + \Omega \tag{4}$$

Assuming that $\mathbf{f}$ is a known field, Eq. (2) uniquely specifies only $\nabla \times \mathbf{f}_\Omega$ and $\nabla \times \mathbf{g}$; thus, the separation $\mathbf{f} = \mathbf{f}_\Omega + \mathbf{g}$ is not unique and Eq. (2) is also satisfied by the fields $\mathbf{f}'_\Omega$ and $\mathbf{g}'$ given by

$$\mathbf{g}' = \mathbf{g} + \nabla\chi, \qquad \mathbf{f}'_\Omega = \mathbf{f}_\Omega - \nabla\chi \tag{5}$$

where χ is any scalar field. Because our results will be derived from Eq. (1) [or (3)] which involves only the unique combination $\mathbf{f} = \mathbf{f}_\Omega + \mathbf{g}$, all physical predictions of the theory must be invariant under the gauge transformation given by Eq. (5).

To treat the case of a constant-density fluid consisting of particles of charge q and mass m (electrons or Cooper pairs in a superconductor), it is convenient to explicitly display the electric and magnetic forces and replace $\mathbf{f}$ in Eq. (1) by $\mathbf{f} + (q/m)(\mathbf{E} + \mathbf{v} \times \mathbf{B})$. Defining the vorticity field $\boldsymbol{\omega}$ by

$$\boldsymbol{\omega} \equiv \nabla \times \mathbf{v} + (q/m)\mathbf{B} \tag{6}$$

Eq. (1) can be written in the form

$$\partial\mathbf{v}/\partial t - (q/m)\mathbf{E} = -\nabla\mu + \mathbf{v} \times \boldsymbol{\omega} + \mathbf{g} \tag{7}$$

where the Lorentz force is included in the $\mathbf{v} \times \boldsymbol{\omega}$ term. Our basic equation for the vorticity field is obtained by taking the curl of Eq. (7) and using the Maxwell equation $\nabla \times \mathbf{E} = -\partial\mathbf{B}/\partial t$. The result, which is identical in form for both charged and uncharged fluids, is

$$\partial\boldsymbol{\omega}/\partial t = \nabla \times (\mathbf{v} \times \boldsymbol{\omega} + \mathbf{g}) \tag{8}$$

Since only $\nabla \times \mathbf{g}$ appears in Eq. (8), this equation is automatically invariant under the gauge transformation of Eq. (5).

Further analysis requires the following vector identities which are not commonly found in textbooks, but which are all derived from a general form of Gauss's law† :

$$\int_V \mathbf{v} \times (\nabla \times \mathbf{v})\,d\tau = \int_S \mathbf{v}(\mathbf{v}\cdot -\hat{n})\,dS + \int_V \mathbf{v}(\nabla\cdot\mathbf{v})\,d\tau + \int_S (v^2/2)\hat{n}\,dS \tag{9}$$

$$\int_V \mathbf{v}\,d\tau = \tfrac{1}{2}\int_V \mathbf{r} \times (\nabla \times \mathbf{v})\,d\tau + \tfrac{1}{2}\int_S \mathbf{r} \times (\mathbf{v} \times \hat{n})\,dS \tag{10}$$

$$\int_V \mathbf{v}\,d\tau = -\int_V \mathbf{r}(\nabla\cdot\mathbf{v})\,d\tau + \int_S \mathbf{r}(\mathbf{v}\cdot\hat{n})\,dS \tag{11}$$

---

* This separation of $\mathbf{f}$ into potential and nonpotential forces was used to study the Magnus effect and the dissipation of energy due to vortex motion. See Huggins.[3]

† See the discussion of Gauss's law by Milne-Thomson.[4]

where $\hat{n}$ is an outward unit normal vector at S and $\mathbf{r}$ is the coordinate vector of the element $d\tau$ or dS. The only restriction is that the fixed surface S completely enclose V; the volume V may be multiply connected.

To describe the behavior of a vortex system within V, we define the "fluid impulse" $\mathbf{P}_\omega$ and "surface impulse" $\mathbf{P}_S$ by the equations

$$\mathbf{P}_\omega = (\rho/2) \int_V \mathbf{r} \times \omega \, d\tau, \qquad \mathbf{P}_S = (\rho/2) \int_S \mathbf{r} \times (\mathbf{v} \times \hat{n}) \, dS \qquad (12)$$

In an unbounded fluid, with the proper limits on $\mathbf{v}$ and ω, $\mathbf{P}_\omega$ is equal to the impulse $\mathbf{I}_\omega$ required to establish a vortex system (the Kelvin–Lamb definition[5]). The advantage of the choice of $\mathbf{P}_\omega$ in Eq. (12) is that in a bounded fluid we have a unique integration over the vorticity field. (If ω cuts through the surface S, then $\mathbf{P}_\omega$ does have an additive constant whose value depends upon the location of the origin of the coordinate system.)

The other quantity $\mathbf{P}_S$ can be interpreted as the impulse of the vortex sheets one would have if the fluid were at rest just outside the surface S. Using the second vector identity, Eq. (10), we find that $\mathbf{P}_\omega$ and $\mathbf{P}_S$ are related to the total fluid momentum $\mathbf{P} = \rho \int_V \mathbf{v} \, d\tau$ by

$$\mathbf{P} = \mathbf{P}_\omega + \mathbf{P}_S \qquad (13)$$

Differentiating $\mathbf{P}$ and $\mathbf{P}_\omega$ with respect to time, using Eq. (7) for $\partial\mathbf{v}/\partial t$ and Eq. (8) for $\partial\omega/\partial t$, and using the vector identities to convert various volume integrals to surface integrals, we get the dynamic equations

$$\partial\mathbf{P}/\partial t = \int_V \rho(\partial\mathbf{v}/\partial t) \, d\tau = \int_V \rho[\mathbf{f}_\Omega + \mathbf{g} + (q/m)(\mathbf{E} + \mathbf{v} \times \mathbf{B})] \, d\tau$$

$$+ \int_S P(-\hat{n}) \, dS + \int_V \rho\mathbf{v}(\nabla \cdot \mathbf{v}) \, d\tau + \int_S \rho\mathbf{v}(\mathbf{v} \cdot -\hat{n}) \, dS \qquad (14)$$

$$\partial\mathbf{P}_\omega/\partial t = \tfrac{1}{2}\rho \int_V \mathbf{r} \times (\partial\omega/\partial t) \, d\tau = \rho \int_V [(q/m)\mathbf{v} \times \mathbf{B} + \mathbf{g}] \, d\tau \qquad (15a)$$

$$+ \tfrac{1}{2}\rho \int_S \mathbf{r} \times \mathbf{j}_{\omega(\mathbf{v}\cdot-n)} \, dS \qquad (15b)$$

$$+ \int_V \rho\mathbf{v}(\nabla \cdot \mathbf{v}) \, d\tau + \int_S \rho\mathbf{v}(\mathbf{v} \cdot -\hat{n}) \, dS \qquad (15c)$$

$$+ \rho \int_S \tfrac{1}{2}v^2\hat{n} \, dS \qquad (15d)$$

The equation for $\partial\mathbf{P}_S/\partial t$ is easily obtained from Eq. (13). Equation (14), which is known as the momentum equation in classical hydrodynamics, is a straightforward application of Newton's second law to the fluid within V.

In Eq. (15b) we introduced the notation $j_{\omega(\mathbf{v}\cdot-n)}$ to represent the quantity $(\mathbf{v} \times \omega + \mathbf{g}) \times (-\hat{n})$. In an earlier paper[6] we showed that the flow of vorticity could be described by the conserved current tensor $j_{\omega\mathbf{v}}$, where the first index represents the orientation of ω, and the second the direction of flow. From the results of that paper, we find that the term (15b) represents the increase of $\mathbf{P}_\omega$ due to the flow of vorticity in across the boundary S.

Because $\partial\omega/\partial t$ involves $\mathbf{\nabla} \times \mathbf{g}$ rather than $\mathbf{g}$, $\partial\mathbf{P}_\omega/\partial t$ is invariant under the gauge transformation of Eq. (5). But the individual terms on the right side of Eq. (15) are not invariant. When we make the substitution $\mathbf{g} \to \mathbf{g} + \mathbf{\nabla}\chi$ there is an equal and opposite change in terms (15a) and (15b); a potential force can contribute to the vorticity current flowing in from the surface S, but the effect is cancelled out by term (15a).

All ambiguity is removed when we consider the special case that the fields ω and $\mathbf{g}$ are localized within the volume V and $\omega|_S = \mathbf{g}|_S = 0$. This would apply to the case where no vortex lines go through the surface S, and where the only non-potential forces are those acting on vortex cores.[3] If we impose the further restrictions $\mathbf{\nabla} \cdot \mathbf{v} = 0$ and $\mathbf{v} \cdot \hat{n} = 0$ (a divergence-free fluid that does not flow through S), then the equations for $\partial\mathbf{P}_\omega/\partial t$ and $\partial\mathbf{P}_S/\partial t$ become

$$\partial\mathbf{P}_\omega/\partial t = \int_V \rho\left[\mathbf{g} + (q/m)\,\mathbf{v} \times \mathbf{B}\right] d\tau + \int_S \rho(v^2/2)\,\hat{n}\,dS \qquad (16a)$$

$$\partial\mathbf{P}_S/\partial t = \int_V \rho\left[-\mathbf{\nabla}\mu + (q/m)\,\mathbf{E}\right] d\tau \qquad (16b)$$

where μ is given by Eq. (4).

The striking feature of Eqs. (16a) and (16b) is the separation of the role of potential and nonpotential forces. Potential and electrical forces* act only on the surface impulse $\mathbf{P}_S$, while nonpotential and magnetic forces act only on the fluid impulse $\mathbf{P}_\omega$. The $v^2/2$ term in Eq. (16a) plays the following role: When one blows a smoke ring against a wall, the ring expands and its fluid impulse increases even though nonpotential forces are negligible. This expansion, which is usually explained as being a consequence of an image vortex in the wall, is accounted for here by the $v^2/2$ integral.

Acknowledgments

We wish to acknowledge stimulating discussions with A. L. Fetter, L. J. Campbell, and T. Fugita, and to thank Mr. Fugita for the computer calculation of Fig. 1.

References

1. E. Huggins, *Phys. Rev. Lett.* **29**, 1067 (1972).
2. G. Gamota and M. Barmutz, *Phys. Rev. Lett.* **22**, 874 (1969).
3. E. Huggins, *Phys. Rev.* **A1**, 327, 332 (1970).
4. L.M. Milne-Thomson, *Theoretical Hydrodynamics*, MacMillan, New York (1955).
5. H. Lamb, *Hydrodynamics,* Dover, New York (1945), Section 152.
6. E. Huggins, *Phys. Rev. Lett.* **26**, 1291 (1971).
7. W.F. Vinen, in *Superconductivity,* R.D. Parks, ed., Dekker, New York (1969), Vol. ll, p. 1167.

* Electrical forces on charged impurities in a neutral fluid must be treated as nonpotential forces (see Ref. 3).

New Results on the States of the Vortex Lattice

M..Le Ray, J. P. Deroyon, M. J. Deroyon, M. Francois,* and F. Vidal†

Laboratoire d'Hydrodynamique Superfluide, Centre Universitaire
Valenciennes, France

It was found previously[1,2] that in the presence of a counterflow or a superflow normal to the axis of rotation, the vortex lattice created by the rotation is modified. These modifications are studied with second-sound attenuation measurements, and we report here some of our latest results. In particular, systematic measurements have been performed at various temperatures and with various angles β between the flow and the axis of rotation, using second-sound normal to the flow and either normal to the axis of rotation (direction Y, attenuation α'_y) or parallel to the axis (direction Z, attenuation α'_z). The main results can be summarized as follows.

Consider first the case of a flow normal to the axis of rotation. It is well known that when the second sound is propagated parallel to the vortices (direction Z) $\alpha'_z = 0$. In the presence of the normal counterflow, however, α'_z may differ from zero; this shows the presence of a vortex deformation. Below 1.8°K α'_z becomes nonzero systematically and abruptly for $K = 1$ (where $K = d/\lambda$, d is the intervortex distance, and $\lambda = h/mv_s$ is the de Broglie wavelength associated with superflow). Notice that for $K = 1$, λ becomes equal to the intervortex distance, or $\lambda/2$ is equal to the distance $d/2$ between points of zero relative velocity in pure rotation.[2] Above 1.8°K, α'_z starts only for $d/\lambda = 2/\sqrt{3}$, i.e., when λ becomes equal to the distance between vortex streets (or $\lambda/2$ equal to the distance between streets of points of zero relative velocity).

In counterflow α'_z exhibits a discontinuous behavior with sharp transitions respectively at $K = 1, 2, 3, 4$ below 1.8°K (or $2/\sqrt{3}, 4/\sqrt{3}, \ldots$, above 1.8°K) and plateaus between them. We also find unexpected values between the levels of α'_z on different plateaus (see, for example, Fig. 1).

The measurements in the Y direction show that α'_y is modified by the super-imposed flow; this behavior is also temperature dependent. The main point is that for $T < 1.5$°K the attenuation rises above this pure rotation level at values of $K < 1$, for which $\alpha'_z = 0$, i.e., for which there is no deformation of the vortices. This had been observed with superflow (see Fig. 1 of Ref. 1 and Fig. 2 of Ref. 2) and system-atically found also with counterflow in this range of temperatures.

This leads us to suggest that there are new rectilinear vortices parallel to the axis formed below $K = 1$. But in order to preserve the constancy of the total circu-lation induced by rotation, we must assume that at least part of the new vortices are of reverse sign and that the presence of these new vortices of reverse sign is compensated by either another part of the vortices which are of the same sign as

* Permanent address: Laboratoire de Mécanique des Fluides, Université de Paris, Orsay, France.
† Permanent address: Groupe de Physique des Solides, Ecole Normale Supérieure, Paris, France.

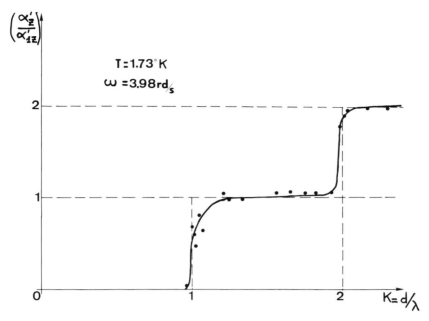

Fig. 1. Attenuation parallel to the axis with counterflow normal to this axis (the level α'_{1z} of the first plateau being taken as unity).

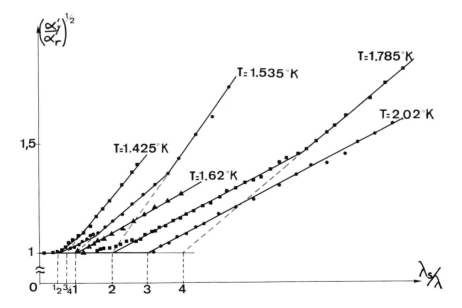

Fig. 2. $(\alpha'_y/\alpha'_r)^{1/2}$ versus λ_s/λ.

the initial vortices due to rotation only, or by the appearance on the initial vortices of several quanta.

The unexpected values of K for which singularities occur in the development of the attenuation and the simple ratios α'_y/α'_r obtained for these values of K seem to indicate that the new structures are probably located on very simple sites (the middle of a side or the center of mass of a triangular cell which, as was pointed out in Ref. 2, are points of zero relative velocity in pure rotation).

Another striking feature of the deformation of the vortices occurring in the presence of flow is revealed by the appearance of long horizontal levels with simple ratios between them (Fig. 1) and also with simple ratios with pure rotational attenuation in the Y direction[1,2] (with a ratio of one in this latter case; Fig. 1, Ref. 1; or Fig. 2, Ref. 2). This indicates the existence of characteristic states of the vortex lattice, and leads us to propose, tentatively, that integral values of K correspond to *successive eigenstates of the angular momentum* associated with each vortex, characterized by low values of the quantum numbers l and m. We believe that when the wavelength λ becomes smaller than the intervortex distance d, the vortex lattice undergoes transition from a collective state of very high quantum number l (equal to the total number of vortices present in the channel) to individual states of very low quantum number l (1 and then $2, 3, \ldots$) for each vortex. In other words, we propose to *generalize* the Onsager–Feynman condition for quantization of the circulation and to replace it by the application to superfluid helium of the quantization properties of angular momentum in quantum mechanics. Such situations would correspond to definite values of the angle $\theta = \cos^{-1}\{|m|/[l(l+1)]^{1/2}\}$ between the local elements of vortices and the axis of rotation (i.e., the second-sound direction of propagation in the case of the Z axis) and of the modulus $[l(l+1)]^{1/2}\hbar$ of the angular momentum associated with each element. According to the law of Snyder and Putney,[3] this would lead to successive definite values of the second-sound attenuation with simple ratios between them, as found in our experiments.

Some calculations based on this model allow us to associate certain critical levels with certain couples of quantum numbers. For instance, we find, using the simple principles summarized above, that for states $(l, |m| = l)$ which correspond for given l to the state of minimum deformation the ratio $\alpha'_z(l, |m| = l)/\alpha'_r = 1$, which is independent of l, and $\alpha'_y(l, |m| = l)/\alpha'_r = l + \frac{1}{2}$, where α'_r is the attenuation in pure rotation in the direction Y.

If we assume that when K increases we can have combinations of such particular states for low values of l ($l = 1, 2, 3, \ldots$), the simultaneous presence (see Fig. 1, Ref. 1 and Fig. 2, Ref. 2) of a plateau in the Z direction corresponding to $\alpha'_z/\alpha'_r = 1$ and of an increasing part of the curve α'_y versus K can then be understood. (We add as a proof that for $K = 2/\sqrt{3}$, $\alpha'_y/\alpha'_r = 3/2 \to l = 1$, and for $K = \sqrt{3}$, $\alpha'_y/\alpha'_r = 5/2 \to l = 2$).

The behavior of the vortex lattice in a flow parallel to the axis of rotation has also been studied. The study using second sound as normal to the common direction Z of the axis and the flow is plotted in Fig. 2 as $(\alpha'_y/\alpha'_r)^{1/2}$ vs. λ_s/λ. Here $\lambda_s = 2\pi s/\omega$ is the wavelength which is associated with Tkachenko's vortex waves of velocity $s = 1/2(\hbar\omega/m)^{1/2}$ which occur at the frequency of rotation ω. The experiments show that λ_s/λ, i.e., the number of de Broglie wavelengths comprised in a wavelength associated with the elastic properties of the rotating lattice, is the characteristic parameter of each curve. Note that the curves are composed of several linear parts intersecting the pure rotation level ($\alpha'_y/\alpha'_r = 1$) for $\lambda_s/\lambda = p/q$, where p and q are small integers. Thus the value $\lambda_s = 1.65d$ for the case of parallel vortices is rather analogous to the value d or $d\sqrt{3}$ in the case of the flow normal to the axis.

The effect on the vortices of the flow at various angles β to the axis of rotation has also been investigated. The role of λ_s/λ (where $\lambda = h/mv_s \cos\beta$ is associated with the projection of the superfluid velocity on the initial direction of the vortices) is also apparent on the results (Fig. 3; see also Ref. 4). Notice that $v_s \cos\beta = s$ corresponds to $\lambda_s/\lambda = 1/4$ and that $v_s \cos\beta = 4s$ corresponds to $\lambda_s/\lambda = 1$. An essential consequence of the flow *between* the vortices is the presence of a plateau or parts of decreasing slope on the curves between inflexive or linear parts corresponding to the role played by λ_s/λ, i.e., by the flow *along* the vortices. This is similar to the case of the flow normal to the axis, even for small angles like 5°.

All the available results with the flow normal to the axis show that the increase of the temperature delays the appearance of the quantized growing of the vortices; it also diminishes the ratio of attenuation for a given value of K compared with pure rotational attenuation. In the case of the flow parallel to the vortices, Fig. 2 shows that the above-mentioned intersections are also systematically delayed by the increase of T and always occur near integral values of λ_s/λ. Another striking feature is that the linear parts obtained at different temperatures are nearly parallel. The only differences between the various curves are the intersections with the increasing horizontal axis; consequently the linear parts are longer for higher temperatures. These properties suggest that very simple processes, probably analogous to that described for the case of flow with rotation, must take place. The temperature, perhaps, simply determines the distances over which they take place. What, then,

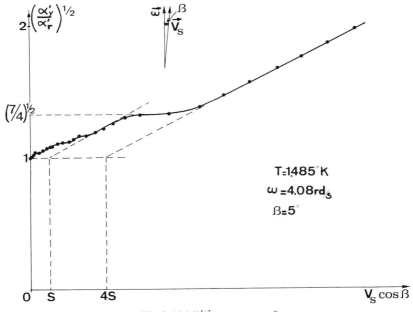

Fig. 3. $(\alpha'_y/\alpha'_r)^{1/2}$ versus $v_s \cos\beta$.

could be an explanation of this role of temperature in selecting the point of appearance of successive steps in both types of flow? One can notice that in the normal fluid the viscous penetration depth of shear waves at the frequency ω of rotation, $\delta_n = (2\eta_n/\rho_n\omega)^{1/2}$, is of the same order as the intervortex distance $d = (h/\sqrt{3}m\omega)^{1/2}$. The ratio δ_n/d can characterize the degree of coupling between vortices through both mutual friction and the shear waves in the normal fluid. This ratio, which is independent of ω, decreases quickly with the temperature ($\delta_n/d = 2$ at $T = 1.42°K$ and $\delta_n/d = 1$ at $T = 1.79°K$). Therefore, a relatively long-range order is revealed by some peculiarities occurring in the Y direction when $\lambda = 2d$ or $\lambda = d\sqrt{3}$ ($K = 1/2$ or $K = 1/\sqrt{3}$) at low temperatures and when δ_n/d is high enough. This could not be realized by the viscous waves in the normal fluid at higher temperature when δ_n/d is of order one or even less than one.

References

1. M. Le Ray, J. Bataille, M. Francois, and D. Lhuillier, *Phys. Lett.* **31A**, 249 (1970).
2. M. Le Ray, J. Bataille, M. Francois, and D. Lhuillier, in *Proc. 12th Intern. Conf. Low Temp. Phys. 1970*, Academic Press of Japan, Tokyo (1971), p. 75.
3. H.A. Snyder and Z. Putney, *Phys. Rev.* **150**, 110 (1966).
4. M. Le Ray and M. Francois, *Phys. Lett.* **34A**, 431 (1971).

Superfluid Density in Pairing Theory of Superfluidity

W.A.B. Evans and R.I.M.A. Rashid

University of Kent at Canterbury
Canterbury, Kent, England

It has long been recognized that the question of whether a system is superfluid or not may be formulated in terms of the properties of its linear response kernel, for such a kernel can tell us what the ultimate state of a system, initially in uniform motion relative to some weak scattering mechanism (walls, etc.), will be. If this final state should possess a nonvanishing current density (measured relative to the walls), then we say the system is superfluid and the magnitude of the residual current density relative to the initial current density defines the superfluid density relative to the density. More precisely, if we imagine a system at rest relative to some weak scattering mechanism, and at time $t = 0$ we impart to each atom in the system a small impulse such that its velocity is increased by a small amount v_s, then the initial current density ($t = 0+$) will be $j(0+) = \rho v_s$. At future times, if v_s is small, we have

$$j(t) = K(t)\, v_s \tag{1}$$

where $K(t)$ is the linear response kernel. After an infinite time we have formally

$$j(\infty) = K(\infty)\, v_s = \rho_s v_s \tag{2}$$

which defines the superfluid density ρ_s in terms of the response kernel in the limit $t \to \infty$. From linear response theory it is easily shown (see Evans and Rickayzen[1]) that

$$K(\infty) = \rho_s = \rho + F(i0+) \tag{3}$$

where

$$F(i\omega_p) = -\int_0^\beta d\tau\, \frac{1}{m\mathscr{V}} \sum_{k,k',\sigma} (\hat{\mathbf{v}}_s \cdot \mathbf{k}')(\hat{\mathbf{v}}_s \cdot \mathbf{k}) \left\langle \tilde{a}_{k'\sigma}^+(\tau)\, \tilde{a}_{k'\sigma}(\tau)\, \tilde{a}_{k\sigma}^+(0)\, \tilde{a}_{k\sigma}(0) \right\rangle e^{-i\omega_p \tau} \tag{4}$$

$$\omega_p = 2p\pi/\beta = 2p\pi k_B T, \qquad p = 0, \pm 1, \pm 2, \ldots \tag{5}$$

$$\tilde{a}_{k\sigma}(\tau) = e^{\mathscr{H}\tau} a_{k\sigma} e^{-\mathscr{H}\tau} \tag{6}$$

and $\hat{\mathbf{v}}_s = \mathbf{v}_s/|\mathbf{v}_s|$, $\mathscr{V}$ is the volume of the system, m is the particle mass, and $\mathscr{H}$ denotes the grand canonical Hamiltonian of the system. The subscript σ denotes spin quantum numbers (if any). $F(i0+)$ is the analytic continuation of the bounded sectionally holomorphic function $F(z)$ which is given by (4) at the special points $z = i\omega_p$. Baym and Mermin[2] have shown that such analytic continuations are unique. In the language of temperature perturbation theory, $F(i\omega_p)$ is given in terms of the sum of all linked diagrams with two current vertices at one of which a "Fourier energy" ω_p is created and at the other destroyed, Fourier energy being conserved at all internal vertices.

From (3) we see that ρ_s is given in terms of a single-body term ρ and a two-body term $(F(i0+)$. Since one expects ρ_s to vanish in the normal phase, these terms should cancel there, but if our intention is to evaluate ρ_s within some *approximation*, then such a cancellation would be no more profound than mere coincidence if our approximations for the single- and two-body Green's functions are unrelated. This kind of situation was clearly recognized by Baym and Kadanoff[3] and others who emphasized the need to ensure that qualitatively important conservation laws (e.g., number conservation, f-sum rule, momentum and energy conservation laws, etc.) still be satisfied in any useful approximation. They developed a formal procedure for generating such "conserving" approximations and here we shall apply what amounts to an equivalent technique to the superfluid density problem within the pair theory of the Bose superfluid as presented by Evans and Imry[4]. The details of our method differ slightly in that we employ temperature (as opposed to time) Green's functions. In essence we will obtain the approximate two-body propagator by the functional differentiation of the approximate single-body propagator of the pair model. Since the exact propagators obey the same relation (via functional differentiation), it is evident that our resulting approximation will satisfy all conservation laws that are expressible in terms of the single-body propagator and its functional derivative.

A convenient way of separating the various functional derivatives of the Green's function is to add a suitable "well-chosen" temperature-dependent term

$$U(\tau) = \sum_{k,\sigma} (\hat{\mathbf{v}}_s \cdot \mathbf{k}) \, a_{k\sigma}^+ a_{k\sigma} e^{-i\omega_p \tau} \tag{7}$$

to the basic Hamiltonian of the superfluid,

$$\mathscr{H} = \sum_{k\sigma} \varepsilon_k a_{k\sigma}^+ a_{k\sigma} + \tfrac{1}{2} \sum_{k,p,q,\sigma,\sigma'} V_q a_{k-q,\sigma}^+ a_{p+q,\sigma'}^+ a_{p,\sigma'} a_{k,\sigma} \tag{8}$$

and to then work with a generalized single-body propagator

$$g(k,\sigma; s_1, s_2) = \mathrm{tr}\left\{ TS(\beta) \, \tilde{a}_{k\sigma}(s_1) \, \tilde{a}_{k\sigma}^+(s_2) \right\} / \mathrm{tr}\left\{ S(\beta) \right\} \tag{9}$$

where

$$S(\beta) = P \exp\left\{ -\int_0^\beta [\mathscr{H} + U(\tau)] \, d\tau \right\} \tag{10}$$

and

$$\tilde{a}_{k\sigma}(s) = S^{-1}(s) \, a_{k\sigma} S(s) \tag{11}$$

where T and P are temperature-ordering operators that order greater temperature *arguments* and greater temperature *moduli* to the left, respectively.

It will be clear that the rules for the diagrammatic evaluation of the propagator generalized above differ from the usual prescription only in the presence of U-vertices within the diagrams at each of which a Fourier energy ω_p will be created. Thus $g(k,\sigma; s_1, s_2)$ will have the Fourier expansion

$$g(k,\sigma; s_1, s_2) = \sum_{m=0}^{\infty} \sum_{\omega_n} g_m(k,\sigma; \omega_n, \omega_n + m\omega_p) \exp\left[i\omega_n(s_1 - s_2) - im\omega_p s_2 \right] \tag{12}$$

the mth term being interpreted in terms of the sum of all diagrams containing m U-vertices only. Thus $g_0(k,\sigma; \omega_n, \omega_n)$ is the usual single-particle propagator and

$g_1(k, \sigma; \omega_n, \omega_n + \omega_p)$ corresponds to a current vertex at which a Fourier energy ω_p is created. It is consequently related to $F(i\omega_p)$ by the equation

$$F(i\omega_p) = (\varepsilon/m\mathscr{V}) \sum_{k,\sigma} (\hat{\mathbf{v}}_s \cdot \mathbf{k}) \sum_{\omega_n} g_1(k, \sigma; \omega_n, \omega_n + \omega_p) e^{i\omega_n 0^-} \tag{13}$$

where $\varepsilon = +1$ for bosons and -1 for fermions.

What has been said thus far is exact. However, we shall now evaluate $g(k, \sigma; s_1, s_2)$ within a generalized Hartree–Fock–Gor'kov decoupling approximation of its equation of motion. That this procedure does indeed lead to a conserving approximation has been shown in detail by Ambegaokar and Kadanoff[5] in the superconducting case. The resulting generalized Hartree–Fock–Gor'kov equation may be decomposed into its various m components according to (12) and these equations can be solved self-consistently starting with $m = 0$ and proceeding upward. In fact the $m = 0$ equations are nothing other than the familiar BCS–Gor'kov equations for the Fermi case and in the Bose case are the equations first proposed by Valatin and Butler[6] and whose solutions have been numerically investigated by Evans and Imry[4] who, we believe, gave a consistent treatment of the single-particle condensate for the first time. For general statistics the $m = 0$ self-consistent equations are

$$g_0(k, \sigma; \omega_n, \omega_n) = \frac{-i\omega_n + \tilde{\varepsilon}_k}{\beta[\omega_n^2 + \tilde{\varepsilon}_k^2 - \varepsilon|\Delta_0(k, \sigma)|^2]} \tag{14}$$

where

$$\tilde{\varepsilon}_k = \varepsilon_k + \sum_{k',\sigma'} (V_{k-k'}\delta_{\sigma,\sigma'} + \varepsilon V_0) \left(\frac{\tilde{\varepsilon}_{k'}}{2E_{k'}} \frac{\exp(\beta E_{k'}) + \varepsilon}{\exp(\beta E_{k'}) - \varepsilon} - \frac{1}{2} \right) \tag{15}$$

$$\Delta_0^*(k, \sigma) = - \sum_{k'} V_{k-k'} \frac{\Delta_0^*(k', \sigma)}{2E_{k'}} \frac{\exp(\beta E_{k'}) + \varepsilon}{\exp(\beta E_{k'}) - \varepsilon} \tag{16}$$

and

$$E_k = [\tilde{\varepsilon}_k^2 - \varepsilon|\Delta_0(k, \sigma)|^2]^{1/2} \tag{17}$$

The coupled self-consistent equations (15) and (16) are solved for the $\tilde{\varepsilon}_k$ and the $\Delta_0^*(k, \sigma)$, which, in turn, give the pairing single-particle propagator according to (14), from which one then obtains the thermodynamic properties of the system. Following Evans and Imry,[4] we have solved these equations for the Bose case with a pseudopotential chosen to parameterize the observed helium dispersion spectrum. The salient features of the solution are shown in Figs. 1 and 2.

We have also empirically investigated the manner in which various self-consistent quantities of the model [e.g., n_0/N, $\hbar^{-1}(\partial E_k/\partial k)_{k=0}$, etc.] vanish as we approach T_λ. In each case we find that a power law of the form $X(T) = A(1 - T/T_\lambda)^\alpha$ is accurately obeyed, where the parameters A, α, and T_λ depend on the pseudopotential used. Further analytic investigation of the nonlinear integral equations of the model in the critical region is desirable to understand this feature in a better than empirical fashion.

To obtain the superfluid density, however, we see that we must, in addition, solve the $m = 1$ component of the generalized Hartree–Fock–Gor'kov equation in order to obtain $g_1(k, \sigma; \omega_n, \omega_n + \omega_p)$, and hence $F(i\omega_p)$, using (13). Since, how-

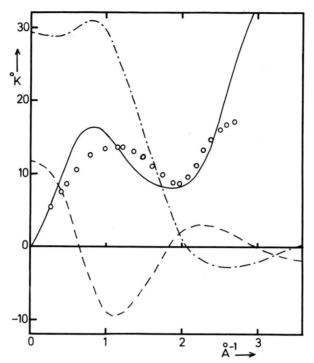

Fig. 1. Plots of $\tilde{\varepsilon}_k - \varepsilon_k - N|V_0|$ (— · — ·), $\Delta_0(k)$ (— — —), and E_k (————) at absolute zero with some experimental excitation spectrum points (after Henshaw and Woods[7]).

ever, we need $F(i0+)$, we may, provided some scattering (e.g., at walls, etc.) is present in the system (see Evans and Rickayzen[1], simply put $\omega_p = 0$, for in this case there is no discontinuity between $F(0)$ and $F(i0+)$. For this case the $m = 1$ equations lead to

$$g_1(k, \sigma; \omega_n, \omega_n) = \frac{-\left[\hat{\mathbf{v}}_s \cdot \mathbf{k} + \xi_1(\mathbf{k}, \sigma)\right]\left[(-i\omega_n + \tilde{\varepsilon}_k)^2 - \varepsilon\Delta_0^2(k, \sigma)\right]}{\beta\left[\omega_n^2 + E_k^2\right]^2} \quad (18)$$

with

$$\xi_1(\mathbf{k}, \sigma) = \sum_{k',\sigma'} (-\varepsilon V_{k-k'}\delta_{\sigma,\sigma'} - V_0)\frac{\beta\exp(\beta E_{k'})}{\left[\exp(\beta E_{k'}) - \varepsilon\right]^2}\left[\xi_1(\mathbf{k'}, \sigma') + \hat{\mathbf{v}}_s \cdot \mathbf{k'}\right] \quad (19)$$

so that $g_1(k, \sigma; \omega_n, \omega_n)$ may be obtained in terms of the self-consistent energies of the equilibrium ($m = 0$) solution and $\xi_1(\mathbf{k},\sigma)$, which is obtained by solving the *linear* equation (19), the kernel and inhomogeneous term of which depend on the equilibrium solution. From (3), (18), and (13) we find that

$$\rho_s = \rho - \frac{\beta}{m\mathcal{V}}\sum_{k,\sigma}\frac{(\hat{\mathbf{v}}_s \cdot \mathbf{k})\left[\hat{\mathbf{v}}_s \cdot \mathbf{k} + \xi_1(\mathbf{k},\sigma)\right]\exp(\beta E_k)}{\left[\exp(\beta E_k) - \varepsilon\right]^2} \quad (20)$$

which gives the superfluid density directly in terms of the solution $\xi_1(\mathbf{k},\sigma)$ to the linear equation (19). $\xi_1(\mathbf{k},\sigma)$ does, however, depend on the direction of $\mathbf{k}$ (relative

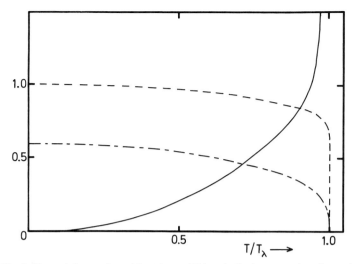

Fig. 2. Plots of the condensed fraction n_0/N (– - –), the long-wavelength quasi-particle group velocity $\hbar^{-1}(\partial E_k/\partial k)|_{k=0}/(233 \text{ m. sec}^{-1})$ (– – –), and the specific heat at constant volume $C_V/(40 \text{ J °K}^{-1} \text{ mole}^{-1})$ (——) as functions of the reduced temperature T/T_λ in our model calculation.

to $\mathbf{v}_s$) as well as on the modulus of $\mathbf{k}$. For this reason it was convenient to work with an angle-averaged $\xi_1(\mathbf{k},\sigma)$ defined by

$$\int \frac{d\Omega_k}{4\pi} \mathbf{k}\xi_1(k, \sigma) = \bar{\xi}_1(\mathbf{k},\sigma) \mathbf{v}_s \tag{21}$$

and also to define an angle-averaged pseudopotential $\bar{V}_{kk'}$ by

$$\bar{V}_{kk'}\mathbf{k}' = \int \frac{d\Omega_k}{4\pi} \mathbf{k} V_{k-k'} \tag{22}$$

whence we see that (19) leads to

$$\bar{\xi}_1(k, \sigma) = -\varepsilon \sum_{k'} \bar{V}_{kk'} \frac{\beta \exp(\beta E_{k'})}{[\exp(\beta E_{k'}) - \varepsilon]^2} \left[\bar{\xi}_1(k', \sigma) + \frac{k'^2}{3} \right] \tag{23}$$

which is essentially a one-dimensional linear integral equation for $\bar{\xi}_1(k, \sigma)$, since the latter depends on the modulus of $\mathbf{k}$ only. From (20) we see that in terms of $\bar{\xi}_1(k, \sigma)$ the superfluid density becomes

$$\rho_s = \rho - \frac{\beta}{m\mathscr{V}} \sum_{k,\sigma} \left[\bar{\xi}_1(k, \sigma) + \frac{k^2}{3} \right] \frac{\exp(\beta E_k)}{[\exp(\beta E_k) - \varepsilon]^2} \tag{24}$$

Thus, by solving the linear integral equation (23) we obtain the superfluid density according to (24).

Note that at $T = 0$ and in the presence of coherence [i.e., nonvanishing $\Delta_0(k, \sigma)$] $\rho_s = \rho$ since the second term of (24) vanishes because $\exp(\beta E_k) \to \infty$. It can be shown (see Evans and Rashid[8]) that if in the Bose case a single-particle condensate exists, this conclusion is still valid.

When coherence is absent (i.e., in the normal phase) we can show quite generally that ρ_s vanishes identically. This follows from (15) with $\tilde{\varepsilon}_k = E_k$ [since $\Delta_0(k, \sigma) \equiv 0$] since

$$\int \frac{d\Omega_k}{4\pi} (\mathbf{k} \cdot \mathbf{V}_k) \tilde{\varepsilon}_k = \frac{k^2}{m} + \int \frac{d\Omega_k}{4\pi} \sum_{k'} \varepsilon (\mathbf{k} \cdot \mathbf{V}_k) V_{k-k'} n_{k'} \qquad (25)$$

where

$$n_k = \langle a_{k\sigma}^+ a_{k\sigma} \rangle = 1/[\exp(\beta \tilde{\varepsilon}_k) - \varepsilon] \qquad (26)$$

Using (22) and

$$\mathbf{V}_k \sum_{k'} V_{k-k'} n_{k'} = \sum_{k'} V_{k-k'} \mathbf{V}_{k'} n_{k'}$$

we get

$$\int \frac{d\Omega_k}{4\pi} (\mathbf{k} \cdot \mathbf{V}_k) \tilde{\varepsilon}_k = \frac{k^2}{m} + \sum_{k'} \varepsilon \bar{V}_{k,k'} \int \frac{d\Omega_{k'}}{4\pi} \mathbf{k}' \cdot \mathbf{V}_{k'} n_{k'}$$

i.e.,

$$\overline{(\mathbf{k} \cdot \mathbf{V}_k) \tilde{\varepsilon}_k} = \frac{k^2}{m} - \sum_{k'} \varepsilon \bar{V}_{k,k'} \frac{\beta \exp(\beta \tilde{\varepsilon}_{k'})}{[\exp(\beta \tilde{\varepsilon}_{k'}) - \varepsilon]^2} \overline{(\mathbf{k}' \cdot \mathbf{V}_{k'}) \tilde{\varepsilon}_{k'}} \qquad (27)$$

where the bar denotes angular average. Comparison of (27) with (23) in the normal phase then leads to the identification

$$\bar{\xi}_1(k, \sigma) + \tfrac{1}{3} k^2 = \tfrac{1}{3} m \overline{(\mathbf{k} \cdot \mathbf{V}_k) \tilde{\varepsilon}_k} \qquad (28)$$

so that (24) may be written

$$\rho_s = \rho - \frac{1}{\mathscr{V}} \sum_{k,\sigma} \frac{\beta \exp(\beta \tilde{\varepsilon}_k)}{[\exp(\beta \tilde{\varepsilon}_k) - \varepsilon]^2} \frac{1}{3} (\mathbf{k} \cdot \mathbf{V}_k) \tilde{\varepsilon}_k = \rho + \frac{1}{\mathscr{V}} \sum_{k,\sigma} \frac{1}{3} (\mathbf{k} \cdot \mathbf{V}_k) n_k \qquad (29)$$

Integrating the second term by parts with respect to $|\mathbf{k}|$ and using $\lim_{k \to \infty} k^3 n_k = 0$, one finds

$$\rho_s = \rho - (1/\mathscr{V}) \sum_{k,\sigma} n_k = \rho - \rho = 0 \qquad (30)$$

i.e., as expected, the superfluid density vanishes identically when coherence is absent. For this result we emphasize again that the use of a conserving approximation was vital.

This cancellation still holds (see Evans and Rashid[8]) even if a single-particle condensate exists in the Hartree–Fock phase so that, from the pairing viewpoint at least, the existence of a single-particle condensate in our superfluid solution is no more than incidental and is certainly not to be regarded as a prerequisite for the existence of superfluidity.

For the same pseudopotential that we used previously we have numerically solved the above equations for the superfluid density in the Bose case. The result is shown in Fig. 3, where the curve seems to possess two inflection points. This is an unrealistic aspect of the pseudopotential we used, for, as the dashed curve in Fig. 3 illustrates, a different pseudopotential gives quite a different curve. Since we merely chose our pseudopotential to roughly parameterize the helium excitation

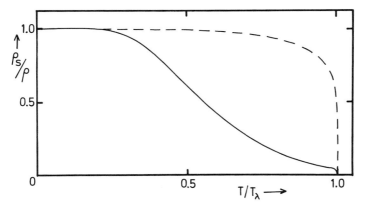

Fig. 3. The solid curve is a plot of the reduced superfluid density ρ_s/ρ for our model pseudopotential as a function of T/T_λ. For comparison purposes, the dashed curve is a similar plot for the single spherical shell pseudopotential, $V(r) = -424\delta(r-3)\,°\text{K}$.

spectrum, the choice is in no sense optimal, and we would suggest that a more suitable pseudopotential is discoverable that would improve the overall quantitative comparison with superfluid helium.

References

1. W.A.B. Evans and G. Rickayzen, *Ann. Phys.* **33**, 275 (1965).
2. G. Baym and N.D. Mermin, *J. Math. Phys.* **2**, 232 (1961).
3. G. Baym and L.P. Kadanoff, *Phys. Rev.* **124**, 287 (1961).
4. W.A.B. Evans and Y. Imry, *Nuovo Cimento* **63B**, 155 (1969).
5. V. Ambegaokar and L.P. Kadanoff, *Nuovo Cimento* **22**, 914 (1961).
6. J.G. Valatin and D. Butler, *Nuovo Cimento* **10**, 37 (1958).
7. D.G. Henshaw and A.D.B. Woods, *Phys. Rev.* **121**, 1266 (1961).
8. W.A.B. Evans and R.I.M.A. Rashid, to be published.

The ^{3}He–Roton Interaction*

H. T. Tan and M. Tan

University of Malaya†
Kuala Lumpur, Malaysia

and

C.-W. Woo ‡

Department of Physics, Northwestern University
Evanston, Illinois

For lack of more detailed knowledge concerning the interactions between elementary excitations in liquid helium, Khalatnikov's theory[1] of transport phenomena has described such interactions in terms of δ-function potentials. Recent measurements by Herzlinger and King[2] of the diffusion constant of very dilute ^{3}He–^{4}He solutions at temperatures slightly above 1°K call into question the validity of such simple models. It is indicated that the temperature variation of the kinetic coefficients may depend sensitively upon the form of the ^{3}He–roton scattering cross section. We present here, as a first leg of the journey toward obtaining the diffusion coefficient $D(T)$, a microscopic calculation of the effective ^{3}He–roton cross section.

In an earlier paper[3] (hereafter referred to as I) we established the basis for a microscopic theory of dilute ^{3}He–^{4}He solutions. Using a representation of correlated basis functions, we computed the relevant matrix elements and obtained an effective Hamiltonian in the second quantized form. The latter contained creation and annihilation operators $a_{k\sigma}^\dagger$, $a_{k\sigma}$, $c_q^\dagger$, and c_q for ^{3}He atoms and Feynman phonons, respectively. There were terms depicting various types of interactions between these entities. By carrying out a canonical transformation, the three-phonon vertex was eliminated, resulting in the replacement of $c_q^\dagger$ and c_q by $b_q^\dagger$ and b_q, which represent operators for the renormalized phonons (or rotons). From such a theory a number of physical properties were derived, including the effective mass of the ^{3}He quasiparticle, the chemical potential of ^{3}He in the solution, the volume-excess parameter α of the Bardeen–Baym–Pines theory, the effective ^{3}He–^{3}He interaction, etc. Under the conditions that (i) we stayed away from the extreme long-wavelength region and (ii) we kept the ^{3}He concentration x sufficiently small so that terms beyond linear x dependence could be neglected, the theory proved quantitatively successful. It was all the more gratifying since no adjustable parameter was allowed in the

* Work supported in part by the National Science Foundation through Grant No. GP-29130 and through the Materials Research Center of Northwestern University.
† H. T. Tan and M. Tan are with the Physics Department and the Faculty of Engineering, respectively.
‡ Alfred P. Sloan Research Fellow.

theory, the starting point of the calculation being the Lennard-Jones 6–12 potential with parameters determined by de Boer and Michels.

Our present work begins with the transformed Hamiltonian given in Eq. (56) of I. We extract from the Hamiltonian the coefficient of $a^\dagger_{l\sigma} b^\dagger_p a_{k\sigma} b_q$, i.e., the matrix element $\langle \mathbf{p}; \mathbf{l} | X' | \mathbf{q}; \mathbf{k} \rangle$. This matrix element represents in first order the scattering of a ^{3}He quasiparticle of momentum $\hbar\mathbf{k}$ by a phonon—or rather a *roton* in the case of $q \approx 2\,\text{Å}^{-1}$—of momentum $\hbar\mathbf{q}$. The momentum transfer $\Delta\mathbf{k}$ defined as $\mathbf{l} - \mathbf{k}$ equals $\mathbf{q} - \mathbf{p}$ by momentum conservation. Second-order contributions to the scattering process come from vertices depicting the emission or absorption of rotons by a ^{3}He quasiparticle. We postpone the consideration of such effects until later work.

Now, with the understanding that $\mathbf{k} + \mathbf{q} = \mathbf{p} + \mathbf{l}$, we find

$$\langle \mathbf{p}; \mathbf{l} | X' | \mathbf{q}; \mathbf{k} \rangle = \langle \mathbf{p}; \mathbf{l} | X | \mathbf{q}; \mathbf{k} \rangle \left\{ 1 - \sum_s \xi^2_{\mathbf{p}-\mathbf{s},\mathbf{s}} \right\} \left\{ 1 - \sum_t \xi^2_{\mathbf{q}-\mathbf{t},\mathbf{t}} \right\} \tag{1}$$

where

$$\langle \mathbf{p}; \mathbf{l} | X | \mathbf{q}; \mathbf{k} \rangle = \left(\left\{ -\tfrac{1}{2}[\varepsilon_0(|\mathbf{p}-\mathbf{q}|) + \omega_0(p) + \omega_0(q)] S^{44}(p) S^{44}(q) + \frac{\hbar^2 \mathbf{p}\cdot\mathbf{q}}{2m_4} \right\} S^{34}(|\mathbf{p}-\mathbf{q}|) \right.$$

$$\left. - \tfrac{1}{2}[\varepsilon_0(|\mathbf{p}-\mathbf{q}|) + \omega_0(p) + \omega_0(q)] \left(\frac{N_4}{N_3} \right)^{1/2} S^{34}(p) S^{34}(q) S^{33}(|\mathbf{p}-\mathbf{q}|) \right)$$

$$\times [N_3 N_4 S^{44}(p) S^{44}(q)]^{-1/2} \tag{2}$$

$$\xi_{\mathbf{s},\mathbf{t}} = \tfrac{1}{2} Z_{\mathbf{s},\mathbf{t}} [\omega(|\mathbf{s}+\mathbf{t}|) - \omega_0(s) - \omega_0(t)]^{-1} \tag{3}$$

and

$$Z_{\mathbf{s},\mathbf{t}} = \left\{ \frac{\hbar^2 \mathbf{s}\cdot(\mathbf{s}+\mathbf{t})}{2m_4} [S^{44}(t) - S^{44}(s)] - \frac{\hbar^2 (\mathbf{s}+\mathbf{t})^2}{2m_4} S^{44}(s) [S^{44}(t) - 1] \right\}$$

$$\times [N_4 S^{44}(p) S^{44}(q) S^{44}(|\mathbf{p}+\mathbf{q}|)]^{-1/2} \tag{4}$$

Also,

$$S^{\alpha\beta}(q) = \delta_{\alpha\beta} + [(N_\alpha N_\beta)^{1/2}/\Omega] \int [g^{\alpha\beta}(r) - 1] \exp(i\mathbf{q}\cdot\mathbf{r})\, d\mathbf{r} \tag{5}$$

and

$$\frac{(N_\alpha N_\beta)^{1/2}}{\Omega} g^{\alpha\beta}(\mathbf{r}_{i_\alpha}, \mathbf{r}_{i_\beta}) =$$

$$N_\alpha(N_\beta - \delta_{\alpha\beta}) \int \psi_0^2 \, d\mathbf{r}_1 \cdots d\mathbf{r}_{i_\alpha-1}\, d\mathbf{r}_{i_\alpha+1} \cdots d\mathbf{r}_{i_\beta-1}\, d\mathbf{r}_{i_\beta+1} \cdots d\mathbf{r}_{N_\alpha+N_\beta} \tag{6}$$

where ψ_0 denotes the ground-state eigenfunction of a mass 3–mass 4 boson solution. The excitation energies $\varepsilon_0(k) \equiv \hbar^2 k^2/2m_3$, $\omega_0(q) \equiv \hbar^2 q^2/2m_4 S^{44}(q)$, and $\omega(q)$ correspond, respectively, to the spectra of a bare ^{3}He atom, a Feynman phonon, and a renormalized phonon (roton).

This matrix element is then used in the "golden rule" to obtain the differential scattering cross section[1,2]

$$d\sigma = \frac{2\pi}{\hbar} |\langle \mathbf{p};1|X'|\mathbf{q};\mathbf{k}\rangle|^2 \delta[\omega(p) + \varepsilon(l) - \omega(q) - \varepsilon(k)] \frac{(1 - \cos\psi)\Omega}{V_{rel}} \frac{\Omega}{(2\pi)^3} d\mathbf{p} \qquad (7)$$

where $\varepsilon(k) = \hbar^2 k^2 / 2m_3^*$ is the ^{3}He-quasiparticle spectrum, ψ denotes the angle of deflection of ^{3}He, and

$$V_{rel} = |\mathbf{V}_q - \mathbf{V}_k| = \hbar \left[\frac{k^2}{m_3^{*2}} + \frac{(q - p_0)^2}{\mu^2} - \frac{2k(q - p_0)}{m_3^*} \cos\alpha \right]^{1/2} \qquad (8)$$

represents the velocity of ^{3}He relative to the roton against which it scatters. Here α is the angle between $\hat{k}$ and $\hat{q}$: the incident trajectories of the ^{3}He and the roton. Note that the entrance of p_0 and μ into Eq. (8) implies that Landau's parameterized approximation for $\omega(q)$ has been put to use:

$$\omega(q) \approx \Delta + (\hbar^2/2\mu)(q - p_0)^2 \qquad (9)$$

The differential scattering cross section $d\sigma$ is a function of $\mathbf{k}$, $\mathbf{q}$, and $\mathbf{p}$. By transformation of variables, we can express it as a function of $\mathbf{k}$, $\mathbf{q}$, and $\Delta\mathbf{k}$. The δ-function, which implies energy conservation, determines the magnitude of $\Delta\mathbf{k}$. We then integrate over the two angles which define its orientation, bearing in mind that in certain orientations there is no contribution to the cross section since energy cannot be conserved. Finally, in order to make direct comparison to Herzlinger and King's results (HK), we integrate over $\cos\alpha$ and obtain $\sigma(k,q)$. At a given temperature T, the thermal average $\langle\sigma\rangle_T$ is then taken over k and q. We find

$$\langle\sigma\rangle_{1.27K} = 0.84 \times 10^{-14} \quad cm^2 \qquad (10)$$

$$\langle\sigma\rangle_{1.69K} = 0.80 \times 10^{-14} \quad cm^2 \qquad (11)$$

in comparison to HK's 1.6×10^{-14} cm^2 and 2.3×10^{-14} cm^2, respectively. Considering that only the first-order term is included, we feel that the factor of two discrepancy is not distressing. These results are found to be in good agreement with a model calculation by Herzlinger,[4] who assumed a dipole–dipole interaction between the ^{3}He and the roton. In both calculations the strong temperature dependence of the cross section is notably absent.

Clearly, this mode of extracting $\langle\sigma\rangle_T$ from experiment is not really valid. $\sigma(k, q, \alpha)$ enters the expression for $D(T)$ through complicated integrals.[1] Unless its dependence on k, q, and α is weak, there is no way to justify factoring it out of the integrals. We offer here merely an order-of-magnitude estimate of the strength of the interaction in the spirit of Herzlinger and King.[2] Extension of this work is continuing in two directions: the application of $\sigma(k, q, \alpha)$ to directly evaluating $D(T)$, and the inclusion of second-order contributions.

We have reason to believe that such a calculation will lead to quantitatively meaningful results. The experiment by HK was carried out under conditions where diffusion is dominated by ^{3}He–roton interactions. The ^{3}He–phonon interactions were unimportant because of the relatively high temperatures ($> 1°K$), and ^{3}He–^{3}He interactions were negligible because of the very low concentration ($\sim 10^{-4}$). Our theory works best at intermediate wavelengths (q = roton momentum ~ 2 Å^{-1}) and small ^{3}He concentrations, which fit the experimental conditions perfectly.

Finally, we wish to raise an issue with theorists engaged in model calculations that require knowledge concerning interactions between various excitations in liquid and solid helium. We seriously question the validity of potentials which are typed as contact, δ-function, or separable. The ³He–roton interaction as shown here is nonlocal and noncentral. In another paper[5] we shall discuss the roton–roton interaction in pure liquid ⁴He, which appears almost equally intractable.

References

1. I.M. Khalatnikov and V.N. Zharkov, *Soviet Phys.—JETP* **5**, 905 (1957).
2. G.A. Herzlinger and J.G. King, *Phys. Lett.* **40A**, 65 (1972).
3. C.-W. Woo, H.T. Tan, and W.E. Massey, *Phys. Rev.* **185**, 287 (1969).
4. G.A. Herzlinger, private communications.
5. C.-W. Woo, this volume.

Quantum Lattice Gas Model of ^{3}He–^{4}He Mixtures*

Y.-C. Cheng and M. Schick

Department of Physics, University of Washington
Seattle, Washington

The quantum lattice gas model of Matsubara and Matsuda[1] has been applied with much success to the liquid–gas and superfluid transitions[2] of liquid ^{4}He, to the solid phase[3] of ^{4}He, and to the question of the existence of a superfluid solid phase[4] of ^{4}He. We consider here the extension of this model in order to apply it to ^{3}He–^{4}He mixtures. After defining the model we shall obtain its phase diagram within mean-field theory.

We consider N_B bosons and N_F fermions contained in a volume V which is divided into M cells of volume d^3. Let the operator which creates a boson in the cell i be denoted $\psi_i^\dagger$ and that which creates a fermion of spin s in the cell i be denoted $\phi_{is}^\dagger$. It is assumed that the interaction between all particles is characterized by a hard core which prevents multiple occupancy of any cell. Rather than obtain this restriction from the dynamics of the system, we alter the commutation relations of all field operators to ensure that no cell will be occupied by more than one particle. The following commutation relations result:

$$[\psi_i, \psi_j^\dagger] = (1 - 2\psi_i^\dagger\psi_i)\,\delta_{ij}, \qquad [\psi_i, \psi_j] = 2\psi_i\psi_j\delta_{ij}$$
$$\{\phi_{is}, \phi_{js'}^\dagger\} = \delta_{ij}\delta_{ss'} + \phi_{is'}^\dagger\phi_{is}\delta_{ij}(1 - \delta_{ss'}), \qquad \{\phi_{is}, \phi_{js'}\} = \phi_{is}\phi_{is'}\delta_{ij}(1 - \delta_{ss'})$$
$$[\phi_{is}, \psi_j^\dagger] = -\psi_j^\dagger\phi_{is}\delta_{ij}, \qquad [\psi_i, \phi_{js}^\dagger] = -\phi_{is}^\dagger\psi_i\delta_{ij}, \qquad [\psi_i, \phi_{js}] = 0$$

and the condition $\psi_i\phi_{is} = 0$. It should be noted that for i unequal to j the commutation relations are the usual fermion or boson relations. Thus the effects of fermion as well as boson statistics are explicitly included in the model. The inclusion of fermion statistics prevents the model from being mapped onto a system of spins as in other recent treatments.[5,6]

The model Hamiltonian is taken to be

$$H = \sum_{\langle ij\rangle} \{(\hbar^2/2m_Bd^2)(\psi_i^\dagger - \psi_j^\dagger)(\psi_i - \psi_j) + (\hbar^2/2m_Fd^2)\sum_s (\phi_{is}^\dagger - \phi_{js}^\dagger)(\phi_{is} - \phi_{js})$$
$$- v_{BB}\psi_i^\dagger\psi_i\psi_j^\dagger\psi_j - v_{FF}\sum_{s,s'}\phi_{is}^\dagger\phi_{is}\phi_{js'}^\dagger\phi_{js'} - v_{BF}\sum_s (\phi_{is}^\dagger\phi_{is}\psi_j^\dagger\psi_j + \psi_i^\dagger\psi_i\phi_{js}^\dagger\phi_{js})\}$$

where $\langle ij\rangle$ denotes a sum over nearest-neighbor pairs. The kinetic energy operators are simply the first finite difference approximation to the usual expression applicable to continuous fields. The above Hamiltonian and commutation relation completely define the model.

* Work supported in part by the National Science Foundation.

112

We now investigate the model within mean-field theory. Let γ denote the number of nearest neighbors, and set m_B/m_F to 4/3. The energy per cell in units of $J = \hbar^2\gamma/2m_Bd^2$ within mean-field theory is

$$e(n, x, \sigma_T) = n(1 + \tfrac{1}{3}x) - \tfrac{1}{4}\sigma_T^2 - \tfrac{1}{2}\gamma n^2\left[v_{BB} + 2x(v_{BF} - v_{BB})\right.$$
$$\left. + x^2(v_{BB} + v_{FF} - 2v_{BF})\right]$$

where n is the total density per cell, x is the fermion concentration, and $\sigma_T \equiv 2|\langle\psi_i^\dagger\rangle|$, twice the magnitude of the boson order parameter. It is to be noted that mean-field theory approximates the expectation value of the fermion tunneling terms $\phi_{is}^\dagger\phi_{is}$ by $\langle\phi_{is}^\dagger\rangle\langle\phi_{is}\rangle$, which vanishes.[7] Thus, within this approximation, the statistical properties of the fermions, contained in the model, are lost. It follows that the thermodynamic functions will be essentially the same as for a mixture of bosons and classical particles.[6]

The entropy of the system is obtained by counting the number of configurations of the system which are characterized by a density n, concentration x, and order parameter σ_T. Within mean-field theory the entropy per cell is

$$s(n, x, \sigma_T) = -nx \log nx - C \log C - D \log D$$

where

$$C(n, x, \sigma_T) \equiv \tfrac{1}{2}\left[1 - nx + B(n, x, \sigma_T)\right]$$
$$D(n, x, \sigma_T) \equiv \tfrac{1}{2}\left[1 - nx - B(n, x, \sigma_T)\right]$$
$$B(n, x, \sigma_T) \equiv \left[\sigma_T^2 + (1 - 2n + nx)^2\right]^{1/2}$$

The free energy per cell has the differential

$$df = -s\,dT + \left[\mu_B(1 - x) + \mu_F x\right] dn + n(\mu_F - \mu_B)\,dx + \mathcal{H}\,d\sigma_T$$

where μ_B and μ_F are the boson and fermion chemical potentials and $\mathcal{H}$ is the field conjugate to the superfluid order parameter. Since this field is not physically accessible, it must ultimately be set to zero. Expressions for $\mathcal{H}$, μ_B, and μ_F are obtained from partial derivatives of f and the pressure p from

$$pd^3 = \mu_B n + (\mu_F - \mu_B) nx + \mathcal{H}\sigma_T - f$$

The results are

$$\mathcal{H}(n, x, \sigma_T, T) = \tfrac{1}{2}\sigma_T\left[TB^{-1}\log(C/D) - 1\right]$$
$$\mu_B(n, x, \sigma_T, T) = 1 - n\gamma\left[v_{BB} + x(v_{BF} - v_{BB})\right] - TB^{-1}(1 - 2n + nx)\log(C/D)$$
$$\mu_F(n, x, \sigma_T, T) = \tfrac{4}{3} - n\gamma\left[v_{BF} - x(v_{BF} - v_{FF})\right] + T\left[\log nx\right.$$
$$\left. - \tfrac{1}{2}\log CD - \tfrac{1}{2}B^{-1}(1 - 2n + nx)\log(C/D)\right]$$
$$p(n, x, \sigma_T, T)\,d^3 = \tfrac{1}{4}\sigma_T^2 - \tfrac{1}{2}\gamma n^2\left[v_{BB} + 2x(v_{BF} - v_{BB}) + x^2(v_{BB} + v_{FF} - 2v_{BF})\right]$$
$$- T\left\{\tfrac{1}{2}\log CD + \tfrac{1}{2}B^{-1}\left[\sigma_T^2 + (1 - 2n + nx)\right]\log(C/D)\right\} + \sigma_T\mathcal{H}$$

The conditions for two-phase equilibria are that μ_B, μ_F, and p are equal in each phase and $\mathcal{H}$ vanishes in each phase. These conditions provide five equations relating the six quantities n, x, and σ_T of each phase, which completely determine the state of the two-phase system. Therefore, one external condition may be specified.

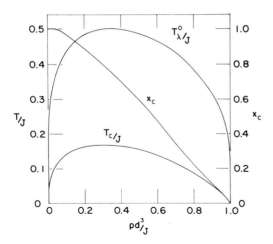

Fig. 1. Variation of pure boson lambda temperature, the consolute temperature, and the concentration with pressure.

We have solved the five equilibrium equations with all nearest-neighbor interactions set equal to zero to emphasize the dependence of phase separation upon superfluid ordering[8] and the hard-core interactions. Some features of the phase diagram are as follows: One phase of the system is normal (i.e., $\sigma_T = 0$); the other is superfluid. There is a line of second-order transitions between them, the lambda line, whose equation is

$$(2n - nx - 1)/(1 - nx) = \tanh\left[(2n - nx - 1)/2T\right]$$

This line terminates at a consolute point whose temperature is given by the simultaneous solution of the above and

$$T = \tfrac{1}{2}\left[(1 - nx) - (2n - nx - 1)^2\right]$$

The shape of the lambda line, the consolute temperature, and concentration at the consolute point x_c depend upon the external constraint. The variation of T_c, x_c, and T_λ^0, the lambda point of the pure boson phase, with pressure is shown in Fig. 1.

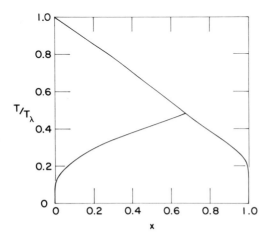

Fig. 2. Phase diagram with normal density per cell fixed at 0.93.

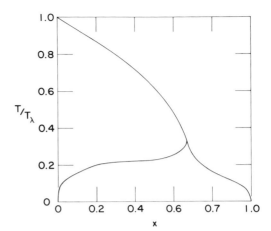

Fig. 3. Phase diagram with pressure fixed.
The value of pd^3 is 0.43.

At the absolute zero of temperature the phase separation is complete. This is undoubtedly due to the fact that mean-field theory ignores the effects of the fermion statistics as remarked earlier. Figure 2 shows the phase diagram under the constraint that the density of the normal phase is constant at 0.93, which gives an x_c in agreement with experiment. The density of the superfluid phase vanishes as T approaches zero under this constraint. Figure 3 shows the phase diagram for a constant pressure $pd^3 = 0.43$, which gives the same x_c as experiment. The density of each phase is nonzero at T equal to zero.

Clearly most of the properties of the physical system are included in the mean-field treatment of the model. It is hoped that a better approximation will also yield the incomplete phase separation.

References

1. T. Matsubara and H. Matsuda, *Prog. Theor. Phys. (Kyoto)* **16**, 569 (1956).
2. P.R. Zilsel, *Phys. Rev. Lett.* **15**, 476 (1965).
3. W.J. Mullin, *Phys. Rev. Lett.* **26**, 611 (1971); *Phys. Rev.* **A4**, 1247 (1971).
4. J.-S. Jun, Ph.D. Thesis, Case Western Reserve University, unpublished, 1969; H. Matsuda and T. Tsuneto, *Prog. Theor. Phys. (Kyoto) Suppl.* **46**, 411 (1970); K.-S. Liu and M.E. Fisher, Report # 1813, Material Sci. Center, Cornell Univ., 1972.
5. M. Blume, V.J. Emery, and R.B. Griffiths, *Phys. Rev.* **A4**, 1071 (1971).
6. S. Takagi, *Prog. Theor. Phys.* **47**, 22 (1972).
7. C.N. Yang, *Rev. Mod. Phys.* **34**, 369 (1962).
8. J.M.J. van Leeuwen and E.G.D. Cohen, *Phys. Rev.* **176**, 385 (1968).

The S = 1 Ising Model for ^{3}He–^{4}He Mixtures

W. M. Ng, J. H. Barry, and T. Tanaka

Department of Physics, Ohio University
Athens, Ohio

As a preliminary study toward understanding phase transitions in ^{3}He–^{4}He mixtures, the $S = 1$ Ising model has recently been proposed[1] by Blume, Emery, and Griffiths (BEG) and, within the Weiss molecular-field approximation, qualitatively good agreement was obtained in comparison with experiment. Previously the model as applied to magnetism had been studied in the Weiss approximation independently by Capel[2] and Blume.[3] For helium mixtures the Hamiltonian used by BEG and formulated in terms of the grand canonical ensemble is given by

$$\tilde{H} = -J \sum_{\langle i,j \rangle} S_i S_j + \Delta \sum_i S_i^2 \tag{1}$$

where the summation is taken over all distinct nearest-neighbor pairs, $S_i = 0$ corresponds to a ^{3}He atom occupying lattice site i, and $S_i = \pm 1$ corresponds similarly for a ^{4}He atom, $2J > 0$ denotes the interaction strength between neighboring ^{4}He atoms, and $\Delta \equiv \mu_3 - \mu_4$ is the difference in chemical potentials of the ^{3}He and ^{4}He atoms. There is a fixed number N of lattice sites and no vacancies are allowed. Then the thermal expectation value $\langle S_i \rangle$ simulates the superfluid order parameter and the concentration of ^{3}He atoms is given by $x \equiv 1 - \langle S_i^2 \rangle$.

To improve beyond the Weiss approximation, we examine the above model in a spin-pair approximation using the cluster-variation method.[4] We have chosen to work in a canonical ensemble in which the Hamiltonian is given by

$$H = -J \sum_{\langle i,j \rangle} S_i S_j \tag{2}$$

In applying the cluster-variation method, a trial Helmholtz free energy is formed according to

$$\mathscr{F} = \text{tr} \, \rho_t \left[H + kT \ln \rho_t \right] \tag{3}$$

where $\rho_t \equiv \rho_t^{(N)}(i_1, \ldots, i_N)$ is the N-spin trial density matrix of the system. The equilibrium free energy F is then obtained from minimization of $\mathscr{F}$ with respect to ρ_t undes the normalization condition that

$$\text{tr} \, \rho_t = 1 \tag{4}$$

To facilitate a systematic approximation scheme, reduced trial density matrices $\{ \rho_t^{(n)}(i_1, \ldots, i_n), n = 1, \ldots, N \}$ are introduced which satisfy the reducibility conditions

$$\rho_t^{(n-1)}(i_1, \ldots, i_{n-1}) = \text{tr}_{i_n} \rho_t^{(n)}(i_1, \ldots, i_{n-1}, i_n), \qquad n = 2, \ldots, N \tag{5}$$

One then introduces a cumulant decomposition of the entropy expression in (3)

116

and, retaining terms up to and including two-spin clusters (spin-pair approximation), the trial Helmholtz free energy of the problem can be written as

$$\mathscr{F} = -J \sum_{\langle i,j \rangle} \mathrm{tr}_{i,j} \rho_t^{(2)}(i,j) \, S_i S_j + kT \sum_i \mathrm{tr}_i \rho_t^{(1)}(i) \ln \rho_t^{(1)}(i)$$

$$+ kT \sum_{i>j} [\mathrm{tr}_{i,j} \rho_t^{(2)}(i,j) \ln \rho_t^{(2)}(i,j) - \mathrm{tr}_i \rho_t^{(1)}(i) \ln \rho_t^{(1)}(i)$$

$$- \mathrm{tr}_j \rho_t^{(1)}(j) \ln \rho_t^{(1)}(j)] \tag{6}$$

Following the procedure of Halow et al.,[5] the reduced trial density matrices are specified, element by element, in terms of the expectation values of the problem, while satisfying all reducibility and normalization conditions. Then the trial Helmholtz free energy can be considered as a function of the unknown expectation values of the problem and the minimization is made with respect to these quantities.

For the S = 1 Ising Hamiltonian all the reduced trial density matrices are diagonal in a representation in which all the S_i are diagonal. After some calculations these matrices are otherwise found to be

$$\rho_t^{(1)}(i) = \mathrm{diag}\,[\varepsilon_1, \varepsilon_2, \varepsilon_3] \tag{7a}$$

$$\rho_t^{(2)}(i,j) = \begin{cases} \rho_t^{(2)}(i,i') = \mathrm{diag}\,[\lambda_1, \lambda_2, ..., \lambda_9] & i, i' \text{ are nearest neighbors} \\ \rho_t^{(1)}(i)\,\rho_t^{(1)}(j), & \text{otherwise} \end{cases} \tag{7b}$$

where

$$\varepsilon_1 = \tfrac{1}{2}(a+b), \qquad \varepsilon_2 = (1-b), \qquad \varepsilon_3 = -\tfrac{1}{2}(a-b) \tag{8a}$$

$$\lambda_1 = \tfrac{1}{4}(c + 2d + e), \qquad \lambda_2 = \tfrac{1}{2}(a + b - d - e), \qquad \lambda_3 = -\tfrac{1}{4}(c - e)$$

$$\lambda_4 = \tfrac{1}{2}(a + b - d - e), \qquad \lambda_5 = (1 - 2b + e), \qquad \lambda_6 = -\tfrac{1}{2}(a - b - d + e) \tag{8b}$$

$$\lambda_7 = -\tfrac{1}{4}(c - e), \qquad \lambda_8 = -\tfrac{1}{2}(a - b - d + e), \qquad \lambda_9 = \tfrac{1}{4}(c - 2d + e)$$

having defined

$$a \equiv \langle S_i \rangle, \qquad b \equiv 1 - x \equiv \langle S_i^2 \rangle, \qquad c \equiv \langle S_i S_{i'} \rangle, \qquad d \equiv \langle S_i S_{i'}^2 \rangle, \qquad e \equiv \langle S_i^2 S_{i'}^2 \rangle \tag{8c}$$

Then the trial Helmholtz free energy can be expressed as a function of the unknown expectation values a, c, d, and e ($b \equiv 1 - x$ is prescribed in the canonical ensemble):

$$(1/N)\mathscr{F}(a,c,d,e) = -\tfrac{1}{2}zJc + kT(1-z)\sum_{k=1}^{3} \varepsilon_k \ln \varepsilon_k + \tfrac{1}{2}kTz\sum_{l=1}^{9} \lambda_l \ln \lambda_l \tag{9}$$

where z is the lattice coordination number and $\{\varepsilon_k\}$ and $\{\lambda_l\}$ are given by (8a) and (8b). Minimizing $\mathscr{F}$ with respect to a, c, d, and e, one obtains the following basic equilibrium equations:

$$(b - a)^{z-1}[(b - e) + (a - d)]^z = (b + a)^{z-1}[(b - e) - (a - d)]^z \tag{10a}$$

$$[(c + e)^2 - 4d^2]\exp(-4\beta J) = (c - e)^2 \tag{10b}$$

$$[(b - e) - (a - d)]^2(c + e + 2d) = [(b - e) + (a - d)]^2(c + e - 2d) \tag{10c}$$

$$(1 - 2b + e)^4(c - e)^2[(c + e)^2 - 4d^2] = [(b - e)^2 - (a - d)^2]^4 \tag{10d}$$

where $\beta = 1/kT$.

In a temperature T versus concentration x diagram one now calculates the λ-line (a second-order phase transition line) by using the fact that both a and d experience onset upon crossing this line, which gives from (10a), (10c), and (8c) the result that, along the λ-line $T_\lambda(x)$

$$c(T_\lambda) = (z - 1)^{-1}b = (z - 1)^{-1}(1 - x) \tag{11}$$

The equilibrium Helmholtz free energy F as a function of T and x is then calculated from

$$(1/N)F(T, x) = -\tfrac{1}{2}zJc + kT(1 - z)\sum_{k=1}^{3}\varepsilon_k \ln \varepsilon_k + \tfrac{1}{2}kTz\sum_{l=1}^{9}\lambda_l \ln \lambda_l \tag{12}$$

where $\{\varepsilon_k\}$ and $\{\lambda_l\}$ are now evaluated using the equilibrium solutions a, c, d, and e [Eq. (10)] for a given T and x. The chemical potential difference between ^{3}He and ^{4}He atoms can be obtained from

$$\Delta(T, x) \equiv \mu_3 - \mu_4 = (\partial/\partial x)[(1/N)F(T, x)] \tag{13}$$

In order to determine the phase separation curve [meeting the λ-line at the tricritical point (T^*, x^*)], the equilibrium Helmholtz free energy isotherms are plotted as a function of ^{3}He concentration $x = 1 - b$ and the appearance of a thermodynamically unstable homogeneous phase, interpreted as the occurrence of phase separation of the mixture, is found from common-tangent constructions upon the family of isotherms. An equivalent procedure in obtaining the two branches of the phase separation curve $x_0(T)$ and $x_1(T)$, respectively, is to solve the following equations:

$$\Delta(T, x_0) = \Delta(T, x_1) \tag{14a}$$

$$F(T, x_0) - Nx_0\,\Delta(T, x_0) = F(T, x_1) - Nx_1\,\Delta(T, x_1) \tag{14b}$$

The tricritical point (T^*, x^*) is located as the point at which the branches $x_0(T)$ and $x_1(T)$ meet.

The phase diagrams in the $T - x$ plane obtained from both the two-spin cluster and the Weiss approximation are compared in Fig. 1. It is seen that the curve obtained from the pair approximation lie below that found from the Weiss approximation. Also, a very slight convex curvature in the λ-line appears as a result in the two-spin approximation compared to a straight line found in the Weiss approximation. The tricritical point is found to be $kT^*/zJ = 0.271$, $x^* = 0.662$ in the pair approximation as compared to the Weiss result $kT^*/zJ = 0.333$, $x^* = 0.666$. It is also interesting to compare the phase diagrams in the $T–\Delta$ plane found from the two approximations, as shown in Fig. 2. The curve resulting from the two-spin approximation again lies below that due to the Weiss approximation, and the tricritical chemical potential difference is found as $\Delta^*/zJ = 0.4695$ in the two-spin approximation as compared to the Weiss result $\Delta^*/zJ = 0.4621$. In Figs. 1 and 2 the overall qualitative features of the phase diagrams are quite similar in both approximations, and further suggest that the locations of the tricritical quantities x^* and Δ^* are rather insensitive to the approximation used, while the quantity T^* changes more appreciably

Finally, the $T–x$ phase diagram of the BEG model in a two-spin cluster approximation is compared with experiment in Fig. 3. Aside from the fact that the model is known to be incapable of giving incomplete phase separation at low ^{3}He con-

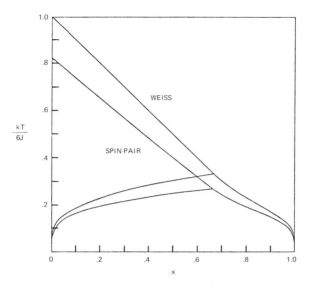

Fig. 1. Phase diagram in the T–x plane for ^{3}He–^{4}He mixtures as predicted by the BEG model in the Weiss molecular-field approximation and in the two-spin cluster approximation.

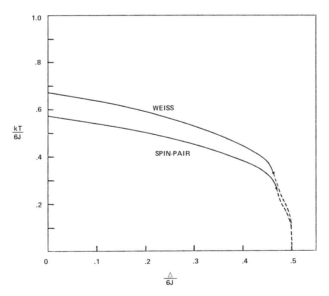

Fig. 2. Phase diagram in the T–Δ plane as predicted by the BEG model in the Weiss approximation and in the two-spin cluster approximation. The solid lines represent second-order phase transitions and dashed lines represent those of first-order.

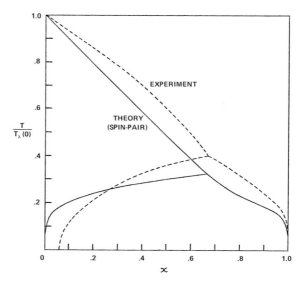

Fig. 3. Phase diagram in the T–x plane as predicted by the BEG model in the two-spin cluster approximation compared with experiment.

centration, the general qualitative features are in rather good agreement with experiment, which may lead one to speculate that the model contains some important vestige of a more complete theoretical description of helium mixtures.

Concluding Remarks

Besides using the model as one for the helium mixture problem, we have studied the $S = 1$ Ising model with additional uniaxial single-ion-type anisotropy as a model for ferromagnetism. In addition to specific heat and initial susceptibility, the equilibrium expectation values $\langle S_i \rangle$, $\langle S_i^2 \rangle$, $\langle S_i S_{i'} \rangle$, $\langle S_i S_{i'}^2 \rangle$, and $\langle S_i^2 S_{i'}^2 \rangle$ are all calculated in a two-spin cluster approximation as functions of temperature. These results, both for helium mixtures and for ferromagnetism, illustrate the value of the cluster-variation method as a controlled-approximation procedure. Aside from recent studies of rigorous high-temperature series expansions,[6] the present work offers, as far as we have found, the only approximation results beyond the Weiss approximation for the $S = 1$ Ising model.

References

1. M. Blume, V.J. Emery, and R.B. Griffiths, *Phys. Rev.* **A4**, 1071 (1971).
2. H.W. Capel, *Physica* **32**, 966 (1966).
3. M. Blume, *Phys. Rev.* **141**, 517 (1966).
4. T. Morita and T. Tanaka, *Phys. Rev.* **145**, 288 (1966).
5. J. Halow, T. Tanaka, and T. Morita, *Phys. Rev.* **175**, 680 (1968).
6. J. Oitmaa, *J. Phys. C. Sol. St. Phys.* **4**, 2466 (1971).

Low Temperature Thermodynamics of Fermi Fluids

A. Ford,* F. Mohling, and J.C. Rainwater†

Department of Physics and Astrophysics, University of Colorado
Boulder, Colorado

Introduction

In this report we shall show how the temperature coefficients in the low temperature expansions of thermodynamic quantities for a normal Fermi fluid can be determined from Landau's phenomenological theory.

Using the dilute hard-sphere Fermi gas (DHSFG) as a model, for which explicit results can be calculated from microscopic theory (as reported in the last section of this paper), we propose parameterized expansions about the Fermi surface of the basic Landau energy functions. These expansions lead to the following low temperature expansions of the heat capacity C_v, magnetic susceptibility χ, and speed of sound (squared) c^2:

$$(Nk_B)^{-1} C_v = C_1 T + C_3 T^3 + C_{3l} T^3 \ln T = 2.93T + 70.5T^3 + 57.4T^3 \ln T \qquad (1)$$

$$\chi = \chi_{T=0}(1 + M_2 T^2 + M_{2l} T^2 \ln T)$$
$$\cong (3n\gamma^2/8k_B T_F^*) \xi_0^{(0)} [1 - 0.49\xi_0^{(0)}(T/T_F^*)^2(1 + \alpha \ln T)] \qquad (2)$$

$$c^2 = \gamma_0 + \gamma_2 T^2 \cong (187.2)^2 - 12(187.2) T^2 \quad m^2 \, sec^{-2} \qquad (3)$$

where N is the number of particles in the system of volume Ω, $n = N/\Omega$, k_B is Boltzmann's constant, and it is assumed that $T < T_F^* = \hbar^2 k^2/2k_B m^*$. The quantity $\gamma/2$ is the magnitude of the single-particle magnetic moment of the (assumed) spin-1/2 fermions, whose zero-temperature effective mass is denoted by m^*.

The second equality of each of the above equations refers to measured low-pressure ($\mathscr{P} \cong 0.28$ atm) results for liquid ^{3}He for which $T_F^* = 1.68°$K, and the so-called "enhancement factor" $\xi_0^{(0)} = 3.11$. The heat capacity expression represents a least-squares fit to data below 125 m°K.[1] Analysis of measurements[2,3] of the nuclear magnetic susceptibility has yielded the cited value for $\xi_0^{(0)}$ and the T^2 coefficient in Eq. (2). The value for α has not yet been determined although experiments suggest that it is small and positive.[3,4] Finally, in Eq. (3) the value of γ_0 has been taken from the article by Wheatley[4] and γ_2 has been estimated from the experimental results of Laquer *et al.*[5]

The Landau Theory

A variety of experiments[4] have now confirmed that liquid ^{3}He conforms to Landau's Fermi liquid theory[6] at the lowest temperatures, say below $T \sim 0.05°$K.

* Based in part on Ph.D. thesis submitted to the University of Colorado by A. Ford.
† Supported in part by a Boettcher Foundation Fellowship.

The purpose of the present paper is to show how this theory can be applied consistently for temperatures up to 200 m°K.

Abrikosov and Khalatnikov[7] summarized some of the earliest applications of Landau's theory to liquid ^{3}He. Noteworthy for its extension of these results to higher temperatures for equilibrium properties is a paper by Richards,[8] and we have adopted Richards's calculational techniques to derive the results reported here. There is no inherent reason why Landau's theory cannot be applied for any conditions at which the Fermi fluid is a well-defined liquid or gas, for the notion that quasiparticle lifetimes may limit this theory to excitations above the ground state applies only to pure quantum mechanical states and not to the real statistical states which occur in nature.

It is important to realize that the quasiparticles of Landau's theory give no less than a representation of *all* the Fermi system's degrees of freedom, a representation equivalent to that of using the unperturbed set of free-particle states. Thus, there must be N quasiparticles in any normal Fermi fluid. Corresponding to each quasiparticle state k in the fluid is a quasiparticle energy ε_k, which is related to the total system energy E by functional differentiation with respect to the quasiparticle occupation number n_k:

$$\varepsilon_k = \partial E/\partial n_k \tag{4}$$

At thermodynamic equilibrium, the condition of interest in the present paper,

$$n_k = \{\exp[\beta(\varepsilon_k - g)] + 1\}^{-1} \tag{5}$$

where $\beta = (k_B T)^{-1}$ and g is the chemical potential. Based on a generalization of the exact result (28) for the DHSFG, we have taken the expansion of $\varepsilon_k^0 = \varepsilon_k(T=0)$ about the Fermi surface to be of the form

$$\varepsilon_k^0 = g^0 + (\hbar^2 k_F^2/m^*)[y + \tfrac{1}{2} a_2 y^2 + \tfrac{1}{3} y^3 (a_3 + a_4 \ln|y|) + \cdots] \tag{6}$$

where $g^0 = g(T=0)$ and the Fermi momentum is defined (for all T) by $n = k_F^3/3\pi^2$. Here $y = (k - k_F)/k_F$, and the effective mass m^* is introduced so as to agree with the conventional definition $\partial \varepsilon_k^0/\partial k|_{y=0} = \hbar^2 k_F/m^*$. The numbers a_2, a_3, and a_4 are treated as unknown parameters in the phenomenological version of Landau's theory.

The quasiparticle energy can also vary, and we shall indicate its deviation from ε_k^0 by $\delta\varepsilon_k$. Variations of this sort may be either virtual or real, but their microscopic origin is, for either case, due to quasiparticle interactions in the fluid. The interaction energy $\Omega^{-1}f(k, k')$ for quasiparticles in states k and k' is determined by the equation

$$\delta\varepsilon_k = \Omega^{-1} \sum_{k'} f(k, k') \, \delta n_{k'} \quad \text{or} \quad \Omega^{-1}f(k, k') = \frac{\partial \varepsilon_k}{\partial n_{k'}} = \frac{\partial^2 E}{\partial n_{k'} \, \partial n_k} \tag{7}$$

The unavoidable occurrence of *interacting* quasiparticles is by no means inconsistent with Eq. (5). We have for a real Fermi fluid *both* a Fermi sea, analogous to that which occurs in the ideal gas limit, as well as quasiparticle interactions.

We write $k = (\mathbf{k}, \boldsymbol{\sigma})$, in which a conventional matrix notation for the spin states is adopted, where $\boldsymbol{\sigma}$ is a vector composed of Pauli spin matrices. Then the scalar $f(k, k')$ includes both direct (s) and exchange (a) parts, according to the relation

$$f(\mathbf{k}, \boldsymbol{\sigma}; \mathbf{k}', \boldsymbol{\sigma}') = f_{kk'}^{(s)} + (\boldsymbol{\sigma} \cdot \boldsymbol{\sigma}') f_{kk'}^{(a)} \tag{8}$$

We also introduce the notation for the simple angular average of $f_{kk'}^{(s)}$

$$\bar{f}_{kk'}^{(s)} \equiv \frac{1}{2} \int_0^\pi \sin\theta \, d\theta \, f_{kk'}^{(s)} \qquad (9)$$

and similarly for $f_{kk'}^{(a)}$, where $\hat{k} \cdot \hat{k'} \equiv \cos\theta$. Then the general forms of $\bar{f}_{kk'}^{(s)}$ and $\bar{f}_{kk'}^{(a)}$, based on generalizations of the exact DHSFG results (29), are assumed to include logarithmic terms as follows*:

$$\bar{f}_{kk'}^{(s)} = \bar{f}_{kk'}^{(0)} + \bar{f}_{kk'}^{(3)}(y - y')^2 \ln|y - y'| \qquad (10)$$
$$\bar{f}_{kk'}^{(a)} = \bar{f}_{kk'}^{(4)} + \bar{f}_{kk'}^{(5)}[y^2 \ln|y| + (y')^2 \ln|y'|]$$

The logarithmic terms in Eqs. (6) and (10) are the sources of the logarithmic terms in Eqs. (1) and (2).

Liquid ^{3}He Results

The starting point of thermodynamic calculations, according to the procedure developed by Richards, is to consider the entropy formula for a Fermi fluid. After converting sums to integrals and integrating by parts, one is able to convert all expressions of interest to integrals involving the variable ε_k^0 of Eq. (6), in which the derivative of n_k occurs in the integrand. It is also convenient to use Eq. (7) in order to express this derivative in terms of ε_k^0 instead of $\varepsilon_k = \varepsilon_k^0 + \delta\varepsilon_k$.† Since the derivative of n_k is a sharp function of its argument at very low temperatures, it is then useful to introduce this argument as the integration variable

$$X \equiv (\varepsilon_k^0 - g)/k_B T \qquad (11)$$

for the quantities being calculated. The variables

$$x = \theta X = (\hbar^2 k_F^2/m^*)^{-1}(\varepsilon_k^0 - g) \qquad (12)$$

$$\xrightarrow[T \,\to\, 0]{} x_0 = (\hbar^2 k_F^2/m^*)^{-1}(\varepsilon_k^0 - g^0)$$

and

$$\theta = (\hbar^2 k_F^2/m^*)^{-1} k_B T = T/2T_F^* < 1 \qquad (13)$$

are also useful for the cited calculations. Thus, for example, inversion of Eq. (6) yields the expansion $y = x_0 - \frac{1}{2}a_2 x_0^2 + \cdots$.

In order to summarize the final results of our calculations, it is necessary to introduce some simplifying notation. We define, for $i = 0, 3, 4, 5$,

$$\bar{f}^{(i)\prime} \equiv \{\bar{f}_{kk'}^{(i)}\}_{x'=0}, \qquad D'\bar{f}^{(i)} \equiv \{(\partial/\partial x')\bar{f}_{kk'}^{(i)}\}_{x'=0}, \qquad D''\bar{f}^{(i)} \equiv \{(\partial^2/\partial x'^2)\bar{f}_{kk'}^{(i)}\}_{x'=0}$$

$$(14)$$

---

* It is nontrivial to argue that Eqs. (6) and (10) are quite general forms for a normal Fermi fluid which must apply in particular to liquid ^{3}He. The argument relies on a careful study of the general microscopic results, involving two-particle reaction matrices, from which Eqs. (28) and (29) are derived.

† Details of these calculations will be included in a more comprehensive report of the present calculations, now being prepared for publication in *Annals of Physics*.

and similarly for $D\bar{f}^{(i)}$, $DD'\bar{f}^{(i)}$, etc. We also define

$$F_0^{(i)} \equiv 2\rho_1 \bar{f}^{(i)'}_{x=T=0} \tag{15}$$

where $\rho_1 \equiv m^* k_F / 2\pi^2 \hbar^2$ is the density of quasiparticle states in the Fermi fluid at the Fermi surface. For the particular case $i = 0$, $F_0^{(0)} = F_0$ is the familiar angular-averaged Landau interaction energy evaluated at the Fermi surface. A further quantity of interest is the also familiar angular average

$$F_1 \equiv 6\rho_1 \int_{-1}^{+1} \cos\theta \; d(\cos\theta) f_{k_F k_F}^{(s)} \tag{16}$$

The parameters of Eqs. (6) and (14)–(16) appear in the final thermodynamic expressions which we have derived. Thus, the coefficients of Eqs. (1)–(3) can be deduced from our results, given to lowest orders in T, as follows.

Heat capacity at constant volume:

$$(Nk_B)^{-1} C_v = \frac{\pi^2}{2}\left\{ \frac{T}{T_F^*} - \frac{\pi^2}{4}\left(\frac{T}{T_F^*}\right)^3\left[\left(C_3' - \frac{1}{3}C_{3l}'\right) - C_{3l}' \ln\frac{T}{T_F^*}\right]\right\}$$

$$C_3' = \rho_1 DD'\bar{f}^{(0)} - 1.08 F_0^{(3)} + \tfrac{1}{5}[3 + 11a_2 - 8a_2^2 + 7a_3 + 5.90a_4] \tag{17}$$

$$C_{3l}' = F_0^{(3)} - 1.4a_4$$

Nuclear magnetic susceptibility:

$$\chi = \frac{3}{8}\frac{n\gamma^2}{k_B T_F^*}\xi_0^{(0)}\left[1 - \frac{\pi^2}{12}\xi_0^{(0)}\left(\frac{T}{T_F^*}\right)^2\left(M_2' + M_{2l}' \ln\frac{T}{T_F^*}\right)\right]$$

$$\xi_0^{(0)} = (1 + F_0^{(4)})^{-1}$$

$$M_2' = (1 + a_2 - a_2^2 + a_3 + \tfrac{1}{3}a_4) + J + \rho_1 DD'\bar{f}^{(0)} - 1.68 F_0^{(3)}$$
$$\quad + 1.046(a_4 + 2F_0^{(5)}) - M_{2l}' \ln 2 \tag{18}$$

$$M_{2l}' = a_4 + 2F_0^{(5)} - F_0^{(3)}$$

$$J = 2\rho_1 D^2 \bar{f}^{(4)'} - 2\rho_1 D\bar{f}^{(4)'}(2\rho_1 D\bar{f}^{(4)'} - 2 + a_2)$$

Speed of sound, squared:

$$c^2 = \frac{\gamma'\hbar^2 k_F^2}{3mm^*}\left[\gamma_0' + \frac{\pi^2}{12}\gamma_2'\left(\frac{T}{T_F^*}\right)^2\right]$$

$$\gamma_0' = (1 + F_0^{(0)}) \tag{19}$$

$$\gamma_2' = \left(2 - \frac{k_F}{m^*}\frac{dm^*}{dk_F}\right)\left(1 - \frac{1}{2}a_2\right) + \frac{1}{2}k_F\frac{da_2}{dk_F}$$

$$= \rho_1 k_F \frac{d}{dk_F}(D'\bar{f}^{(0)})_{x=0} - \left(1 - \frac{2k_F}{m^*}\frac{dm^*}{dk_F}\right)(\rho_1 D'\bar{f}^{(0)})_{x=0}$$

where $\gamma' = C_p/C_v$ is very close to unity for liquid ^{3}He.

We have also deduced three identities relating the Landau parameters of Eqs. (6) and (14)–(16). There is the familiar result for the effective mass, derived originally by Landau,[6,7]

$$m^*/m = 1 + \tfrac{1}{3}F_1 \tag{20}$$

There is a result which is closely related to Eq. (19) at $T = 0$,

$$\frac{dg^0}{dk_F} = \frac{\hbar^2 k_F}{m^*}(1 + F_0^{(0)}) \tag{21}$$

and there is an identity

$$\frac{k_F}{m^*}\frac{dm^*}{dk_F} + 2\rho_1(D'\bar{f}^{(0)})_{x=0} + a_2 = 1 \tag{22}$$

Comparison of Eqs. (1)–(3) with Eqs. (17)–(21) yields a set of values for various combinations of the Landau parameters for liquid ^{3}He. Thus, the coefficient $C_1 = 2.93^\circ\text{K}^{-1}$ gives $T_F^* = 1.68^\circ\text{K}$, which implies that $m^*/m = 2.97$ and that $F_1 = 5.90$. The coefficient $\gamma_0 = (187.2\ \text{m sec}^{-1})^2$ then gives $F_0 = F_0^{(0)} = 10.32$ and $k_B^{-1}k_F dg^0/dk_F = 38.1^\circ\text{K}$. The value $\xi_0^{(0)} = 3.11$ implies that $F_0^{(4)} = -0.68$. From these numbers we may conclude that on the Fermi surface the average direct quasiparticle interaction is strongly repulsive whereas the average exchange interaction is weakly attractive.

Next we consider the logarithmic coefficients C_{3l} and M_{2l}. Examination of the former yields the large value

$$F_0^{(3)} = 22.5 + 1.4a_4 \tag{23}$$

while for the magnetic susceptibility coefficient we can only conclude that

$$(0.49)\,\alpha = (\pi^2/12)(a_4 + 2F_0^{(5)} - F_0^{(3)}) \tag{24}$$

Thus if $|a_4|$ and α are both small numbers ($\lesssim 1$), as might be expected, then $F_0^{(5)} \sim 10$. In this case, the logarithmic terms of Eqs. (10) are all large and important.

Finally, one can show from consideration of the coefficients C_3, M_2, and γ_2 in Eqs. (1)–(3) that

$$\rho_1 DD'\bar{f}^{(0)} = -8.83 + (a_2 - 1) + 1.6(a_2 - 1)^2 - 1.4a_3 + 0.327a_4 \tag{25}$$

$$J = 18.9 + 0.34F_0^{(5)} - 0.6(a_2 - 1)^2 + 0.4a_3 + 0.16a_4 \tag{26}$$

$$[2\rho_1 D'\bar{f}^{(0)}]_{x=0}^2 + [2\rho_1 D'\bar{f}^{(0)}]_{x=0} \cong 7 - 2(a_2 - 1)(2\rho_1 D'\bar{f}^{(0)})_{x=0}$$
$$+ k_F\frac{d}{dk_F}(2\rho_1 D'\bar{f}^{(0)})_{x=0} - (a_2 - 1) - (a_2 - 1)^2 + k_F\frac{da_2}{dk_F} \tag{27}$$

The most significant of these last results is the second, for it shows, assuming $F_0^{(5)} \sim 10$, that $J \sim 20$, which is a large value. Upon referring to Eq. (18) for J, we then conclude, assuming $a_2 \sim 1$, that $J \sim 2\rho_1 D^2\bar{f}^{(4)'} \sim 20$, which means that the attractive exchange interaction $\bar{f}_{kk'}^{(4)}$ must have a sharp minimum near the Fermi surface. This fact is undoubtedly closely related to the success of spin-resonance theories used to treat magnetic phenomena in liquid ^{3}He.

In passing, we note that both of the results (17) and (18) agree with those derived by Richards[8] for the special case $a_2 - 1 = a_3 = a_4 = f_{kk'}^{(3)} = f_{kk'}^{(5)} = 0$.

Dilute Hard-Sphere Fermi Gas

We now present the results of an explicit calculation* for the thermodynamic properties of a model system, the dilute hard-sphere Fermi gas (DHSFG). The hard-sphere diameter is denoted by a_S. These results are valid also for a dilute gas of arbitrary intermolecular potential if a_S is the two-particle scattering length, to second order in perturbation theory. It is assumed $k_F |a_S| \ll 1$.

A microscopic calculation based on reaction-matrix perturbation theory yields the following expansions about the Fermi surface, to $O(k_F a_S)^2$:

$$\varepsilon_k^0 = \frac{\hbar^2 k_F^2}{2m} + \frac{2}{3\pi} \frac{\hbar^2 k_F^2}{m} k_F a_S + \frac{\hbar^2 k_F^2}{m} y(1 + \tfrac{1}{2} y)$$

$$+ \frac{2}{15\pi^2} \frac{\hbar^2 k_F^2}{m} (k_F a_S)^2 \{11 - 2 \ln 2 + 4y(1 - 7 \ln 2) - 4y^2(1 + \tfrac{1}{2} \ln 2) \quad (28)$$

$$+ \tfrac{1}{3} y^3 (67 - 54 \ln 2) - 10 y^3 \ln y + \cdots\}$$

where $\quad y \equiv (k - k_F)/k_F \ll 1$,

$$\bar{f}_{kk'}^{(s)} = \frac{2\pi \hbar^2 a_S}{m} + \frac{8\hbar^2 k_F a_S^2}{3m} [1 + \tfrac{1}{2}(y_1 + y_2) + \tfrac{5}{8}(y_1^2 + y_2^2)$$

$$+ \tfrac{7}{8} y_1 y_2 - (1 + 2y_1 + 2y_2 + \tfrac{1}{4} y_1^2 + \tfrac{1}{4} y_2^2 - 2y_1 y_2) \ln 2$$

$$- \tfrac{3}{4}(y_1^2 \ln |y_1| + y_2^2 \ln |y_2|) + O(y^3 \ln |y|)]$$

$$\bar{f}_{kk'}^{(a)} = - \frac{2\pi \hbar^2 a_S}{m} - \frac{16\hbar^2 k_F a_S^2}{3m} \{1 + \tfrac{1}{2}(y_1 + y_2) - \tfrac{13}{8}(y_1^2 + y_2^2) + \tfrac{51}{16} y_1 y_2$$

$$+ [\tfrac{1}{2} - 2(y_1 + y_2) + \tfrac{5}{4}(y_1^2 + y_2^2) + \tfrac{1}{2} y_1 y_2] \ln 2$$

$$+ \tfrac{3}{4}(y_1 - y_2)^2 \ln |y_1 - y_2| + O(y^3 \ln |y|)\} \quad (29)$$

where $y_i = (k_i - k_F)/k_F$, $i = 1, 2$. The $\ln |y|$ terms in ε_k^0 and f give rise to the $\ln T$ terms in the thermodynamic functions.

We have calculated the entropy from ε_k^0 and f, as explained in the previous section. The result is, to $O(k_F a_S)^2$,

$$S = Nk_B \left\{ \frac{T}{T_F} \left[\frac{\pi^2}{2} + \frac{4}{15} (7 \ln 2 - 1) (k_F a_S)^2 \right] \right.$$

$$+ \left(\frac{T}{T_F} \right)^3 \left[-\frac{\pi^4}{20} - 5.22(k_F a_S)^2 + \frac{2\pi^2}{5} (k_F a_S)^2 \ln \frac{T}{T_F} \right] + O\left(\frac{T}{T_F} \right)^4 \right\} \quad (30)$$

It is quite straightforward to calculate all other thermodynamic quantities. Thus,

---

* Details will be given in a forthcoming publication.

we obtain

$$C_v = T \left(\frac{\partial S}{\partial T} \right)_{k_F} = N k_B \left\{ \frac{T}{T_F} \left[\frac{\pi^2}{2} + \frac{4}{15} (7 \ln 2 - 1) (k_F a_S)^2 \right] \right.$$

$$+ \left(\frac{T}{T_F} \right)^3 \left[-\frac{3\pi^4}{20} - 11.71 (k_F a_S)^2 + \frac{6\pi^2}{5} (k_F a_S)^2 \ln \frac{T}{T_F} \right]$$

$$\left. + O\left(\frac{T}{T_F} \right)^4 \right\} \tag{31}$$

$$E = E_{T=0} + \int_0^T C_v \, dT$$

$$= N \varepsilon_F \left\{ \frac{3}{5} + \frac{2}{3\pi} k_F a_S + \frac{4}{35\pi^2} (11 - 2 \ln 2) (k_F a_S)^2 \right.$$

$$+ \left(\frac{T}{T_F} \right)^2 \left[\frac{\pi^2}{4} + \frac{2}{15} (7 \ln 2 - 1) (k_F a_S)^2 \right]$$

$$\left. + \left(\frac{T}{T_F} \right)^4 \left[-\frac{3\pi^4}{80} - 3.67 (k_F a_S)^2 + \frac{3\pi^2}{10} (k_F a_S)^2 \ln \frac{T}{T_F} \right] + O\left(\frac{T}{T_F} \right)^5 \right\} \tag{32}$$

where $\varepsilon_F \equiv \hbar^2 k_F^2 / 2m = k_B T_F$. The zero-temperature energy has been given by several authors.[9] Upon constructing $F = E - TS$, we may calculate the pressure and chemical potential:

$$\mathcal{P} = -\left(\frac{\partial F}{\partial \Omega} \right)_{T,N} = \frac{k_F}{3\Omega} \left(\frac{\partial F}{\partial k_F} \right)_{T,N}$$

$$= n \varepsilon_F \left\{ \frac{2}{5} + \frac{2}{3\pi} k_F a_S + \frac{16}{105\pi^2} (11 - 2 \ln 2) (k_F a_S)^2 + \left(\frac{T}{T_F} \right)^2 \frac{\pi^2}{6} \right.$$

$$+ \left(\frac{T}{T_F} \right)^4 \left[-\frac{\pi^4}{40} - 1.41 (k_F a_S)^2 + \frac{2\pi^2}{15} (k_F a_S)^2 \ln \frac{T}{T_F} \right]$$

$$\left. + O\left(\frac{T}{T_F} \right)^5 \right\} \tag{33}$$

$$g = \left(\frac{\partial F}{\partial N} \right)_{T,\Omega} = \frac{k_F}{3N} \left(\frac{\partial F}{\partial k_F} \right)_{T,\Omega}$$

$$= \varepsilon_F \left\{ 1 + \frac{4}{3\pi} k_F a_S + \frac{4}{15\pi^2} (11 - 2 \ln 2) (k_F a_S)^2 \right.$$

$$+ \left(\frac{T}{T_F} \right)^2 \left[-\frac{\pi^2}{12} - \frac{2}{15} (7 \ln 2 - 1) (k_F a_S)^2 \right]$$

$$\left. + \left(\frac{T}{T_F} \right)^4 \left[-\frac{\pi^4}{80} + 0.141 (k_F a_S)^2 + \frac{\pi^2}{30} (k_F a_S)^2 \ln \frac{T}{T_F} \right] + O\left(\frac{T}{T_F} \right)^5 \right\} \tag{34}$$

The above results all reduce to the proper ideal gas expressions when $a_S = 0$, in which case all $\ln T$ terms vanish. Equation (34) for g in the limit $T \to 0$ agrees with Eq. (28) for ε_k^0 when $y = 0$ $(k = k_F)$. We have also calculated the $O(T^2)$ part of g by an entirely different method, and the result, which is in agreement with Eq. (34), yields identity (22).

It may be shown from thermodynamics that

$$c^2 = -\gamma' \frac{\Omega^2}{mN} \left(\frac{\partial \mathscr{P}}{\partial \Omega} \right)_{T,N} = \frac{\gamma' k_F}{3mn} \left(\frac{\partial \mathscr{P}}{\partial k_F} \right)_T \quad \text{and} \quad \gamma' \equiv \frac{C_p}{C_v} = 1 - \frac{T}{C_v} \left(\frac{\partial S}{\partial \Omega} \right)_T^2 \left(\frac{\partial \mathscr{P}}{\partial \Omega} \right)_T^{-1}$$

Upon calculating the necessary partial derivatives from Eqs. (30) and (33) and multiplying out the expansions, one obtains

$$
\begin{aligned}
c^2 = \frac{2}{3} \frac{\varepsilon_F}{m} \Bigg\{ & 1 + \frac{2}{\pi} k_F a_S + \frac{8}{15\pi^2}(11 - 2\ln 2)(k_F a_S)^2 \\
& + \left(\frac{T}{T_F} \right)^2 \left[\frac{5\pi^2}{12} - \frac{8}{45}(7\ln 2 - 1)(k_F a_S)^2 \right] \\
& + \left(\frac{T}{T_F} \right)^4 \left[-\frac{\pi^4}{16} - 1.46(k_F a_S)^2 + \frac{\pi^2}{5}(k_F a_S)^2 \ln \frac{T}{T_F} \right] + O\left(\frac{T}{T_F} \right)^5 \Bigg\}
\end{aligned}
\tag{35}
$$

Therefore, we predict a $T^4 \ln T$ term in the speed of sound for both the DHSFG and liquid ^{3}He.

We have also calculated the magnetic susceptibility of the DHSFG. The result is

$$
\begin{aligned}
\chi = \frac{3}{8} \frac{n\gamma^2}{k_B T_F^*} \zeta_0^{(0)} \Bigg\{ & 1 + \frac{8}{15\pi^2}(7\ln 2 - 1)(k_F a_S)^2 - \zeta_0^{(0)} \left(\frac{T}{T_F} \right)^2 \\
& \times \left[\frac{\pi^2}{12} + 0.935(k_F a_S)^2 - \frac{1}{3}(k_F a_S)^2 \ln \frac{T}{T_F} \right] \Bigg\}
\end{aligned}
\tag{36}
$$

where $\frac{1}{2}\gamma$ is the magnitude of the single-particle magnetic moment and the enhancement factor $\zeta_0^{(0)}$ is given by

$$\zeta_0^{(0)} = [1 - (2/\pi)k_F a_S - (8/3\pi^2)(1 - \ln 2)(k_F a_S)^2]^{-1} \tag{37}$$

Not surprisingly, the above DHSFG expressions do not agree particularly well with experiment when an appropriate value of $k_F a_S$ for liquid ^{3}He is chosen. However, the DHSFG model is quite instructive, since it tells us the mathematical form for the low-temperature dependence of a thermodynamic variable, and such forms do agree with experiment. As shown in the previous section, the $\ln T$ terms observed experimentally are not inconsistent with straightforward application of the Landau theory. Furthermore, because it is an exactly soluble model, the DHSFG serves as a useful means for checking the internal consistency of the theory. One may show, for example, that upon extracting the parameters of Eqs. (6) and (14)–(16) from Eqs. (28) and (29), identities (20)–(22) are satisfied for the DHSFG.

References

1. A.C. Mota, R.P. Platzek, R. Rapp, and J.C. Wheatley, *Phys. Rev.* **177**, 266 (1969).
2. J.C. Wheatley, *Phys. Rev.* **165**, 304 (1968).
3. H. Ramm, P. Pedroni, J.R. Thompson, and H. Meyer, *J. Low Temp. Phys.* **2**, 539 (1970).
4. J.C. Wheatley, in *Quantum Fluids,* D.F. Brewer, ed., North-Holland, Amsterdam (1966), p. 183.
5. H.L. Laquer, S.G. Sydoriak, and T.R. Roberts, *Phys. Rev.* **113**, 417 (1959).
6. L.D. Landau, *Zh. Eksperim. i Teor. Fiz.* **30**, 1058 (1956) [*Soviet Phys.—JETP* **3**, 920 (1957)].
7. A.A. Abrikosov and I.M. Khalatnikov, *Rep. Prog. Phys.* (Phys. Soc., London) **22**, 329 (1959).
8. P.M. Richards, Phys. Rev. **132**, 1867 (1963).
9. T.D. Lee and C.N. Yang, *Phys. Rev.* **105**, 1119 (1957); V.M. Galitskii, *Zh. Eksperim. i Theor. Fiz.* **34**, 151 (1958) [*Soviet Phys.—JETP* **7**, 279 (1958)].

Density and Phase Variables in the Theory of Interacting Bose Systems*

P. Berdahl

Department of Physics, Stanford University
Stanford, California

The introduction of canonical density and phase variables for Bose systems may be accomplished in a variety of ways. It is the purpose of this paper to give a unified presentation of the introduction of these variables, within the context of field theory.

Whereas it is customary to use the canonically conjugate pair $\psi(\mathbf{x})$ and $\psi^\dagger(\mathbf{x})$ as field operators, it is possible to replace them by two operators $\psi_1(\mathbf{x})$ and $\psi_2(\mathbf{x})$, respectively. By this replacement one thus demands that they satisfy the commutation rules

$$[\psi_1(\mathbf{x}), \psi_2(\mathbf{y})] = \delta(\mathbf{x} - \mathbf{y}) \tag{1a}$$

$$[\psi_1(\mathbf{x}), \psi_1(\mathbf{y})] = [\psi_2(\mathbf{x}), \psi_2(\mathbf{y})] = 0 \tag{1b}$$

and that the Hamiltonian is given by

$$H = (\hbar^2/2m) \int d^3\mathbf{x} \, \nabla\psi_2(\mathbf{x}) \cdot \nabla\psi_1(\mathbf{x})$$

$$+ \frac{1}{2} \int\int d^3\mathbf{x} \, d^3\mathbf{y} \, \psi_2(\mathbf{x}) \psi_2(\mathbf{y}) V(\mathbf{x} - \mathbf{y}) \psi_1(\mathbf{y}) \psi_1(\mathbf{x}) \tag{2}$$

It is not necessary, however, that ψ_1 and ψ_2 be the adjoints of one another, nor, therefore, that H be Hermitian.

The introduction of the Hermitian operators $\rho(\mathbf{x})$ for the particle density and $\sigma(\mathbf{x})$ for the conjugate phase can be achieved by expressing ψ_1 and ψ_2 in terms of these two operators. For this purpose it is necessary to require that $\psi_2(\mathbf{x})\psi_1(\mathbf{x}) = \rho(\mathbf{x})$, and that the commutation rules for ρ and σ,

$$[\sigma(\mathbf{x}), \rho(\mathbf{y})] = -i\delta(\mathbf{x} - \mathbf{y}) \tag{3a}$$

$$[\sigma(\mathbf{x}), \sigma(\mathbf{y})] = [\rho(\mathbf{x}), \rho(\mathbf{y})] = 0 \tag{3b}$$

are compatible with the rules (1a)–(1b) for ψ_1 and ψ_2. This can be done in a variety

* Research supported by the Office of Naval Research under Contract No. NR024-011, N00014-67-A-0112-0047.

of ways, but we are here particularly interested in the expressions

$$\psi_1^{(\gamma)}(\mathbf{x}) = \{\exp[i\sigma(\mathbf{x})]\}\,\rho^\gamma(\mathbf{x}) \tag{4a}$$

and

$$\psi_2^{(\gamma)}(\mathbf{x}) = \rho^{1-\gamma}(\mathbf{x})\exp[-i\sigma(\mathbf{x})] \tag{4b}$$

The c-number γ introduced here is arbitrary at this point; it has been affixed as a superscript to ψ_1 and ψ_2 to emphasize the dependence upon γ.

The substitution of Eqs. (4a) and (4b) into Eq. (2) now permits an examination of the Hamiltonian in terms of density and phase variables (the argument $\mathbf{x}$ of ρ and σ will often be omitted for brevity):

$$H^{(\gamma)} = (\hbar^2/2m)\int d^3\mathbf{x}\,[\rho(\nabla\sigma)^2 - i\gamma\,\nabla\sigma\cdot\nabla\rho + i(1-\gamma)\,\nabla\rho\cdot\nabla\sigma + \gamma(1-\gamma)(\nabla\rho)^2\,\rho^{-1}]$$

$$+ \frac{1}{2}\iint d^3\mathbf{x}\,d^3\mathbf{y}\,V(\mathbf{x}-\mathbf{y})\,\rho(\mathbf{x})\,[\rho(\mathbf{y}) - \delta(\mathbf{x}-\mathbf{y})] \tag{5}$$

In the derivation of this expression the commutability of $\nabla\sigma(\mathbf{x})$ with any power of $\rho(\mathbf{x})$ has been used, a fact which follows from the commutation rule (3a) if the δ-function is considered as the limit of a differentiable even function. It may be noted that a simple quadratic dependence on the field variables appears in the kinetic energy of the form (2) for the Hamiltonian whereas such a dependence is found in the potential energy of the form (5).

The particular choice $\gamma = 0$ in Eq. (5) yields a non-Hermitian form of the Hamiltonian closely related to that previously derived by Bogoliubov and Zubarev while introducing a density representation.[1] They obtained this Hamiltonian by going from the coordinates $\mathbf{x}_i$ of the particles as variables over to the Fourier components $\rho(\mathbf{k})$ of the density by means of the relation $\rho(\mathbf{k}) = \sum_i^N \exp(-i\mathbf{k}\cdot\mathbf{x}_i)$. The present approach is quite different, but the Hamiltonian of Bogoliubov and Zubarev may be obtained with the following procedure. Fourier transforms for $\rho(\mathbf{x})$, $\sigma(\mathbf{x})$, and $V(\mathbf{x})$ are defined by

$$\rho(\mathbf{x}) = \Omega^{-1}\sum_\mathbf{k}[\exp(-i\mathbf{k}\cdot\mathbf{x})]\,\rho(\mathbf{k}) \tag{6a}$$

$$\sigma(\mathbf{x}) = \sum_\mathbf{k}[\exp(i\mathbf{k}\cdot\mathbf{x})]\,\sigma(\mathbf{k}) \tag{6b}$$

$$V(\mathbf{x}) = \Omega^{-1}\sum_\mathbf{k}[\exp(i\mathbf{k}\cdot\mathbf{x})]\,v(\mathbf{k}) \tag{6c}$$

where Ω is the volume. These expressions are substituted into Eq. (5) (with $\gamma = 0$). The choice of the $\rho(\mathbf{k})$-representation in which $\sigma(\mathbf{k}) = -i\partial/\partial\rho(\mathbf{k})$ then yields the Hamiltonian of Bogoliubov and Zubarev.

Another special case which is of importance is characterized by $\gamma = \frac{1}{2}$. In this case the kinetic energy part of Eq. (5) has the form

$$T^{(1/2)} = (\hbar^2/2m)\int d^3\mathbf{x}\,\{\rho(\nabla\sigma)^2 + (\nabla\rho)^2/4\rho + (i/2)[\nabla\rho,\nabla\sigma]\} \tag{7}$$

so that the Hamiltonian becomes a Hermitian operator. In an extension of the theory of Bogoliubov and Zubarev, Chan and Valatin[2] have likewise derived a

Hermitian form of the Hamiltonian. It is related to the form (5) for the case $\gamma = \frac{1}{2}$ in the same way as the Hamiltonian of Bogoliubov and Zubarev is related to the form (5) for the case $\gamma = 0$.

It is now of interest that there is a canonical nonunitary transformation connecting the field operators of Eqs. (4) for $\gamma = 0$ to the same operators for $\gamma \neq 0$. For this purpose one must find a transformation operator S such that

$$\psi_1^{(\gamma)}(\mathbf{x}) = S\psi_1^{(0)}(\mathbf{x})\,S^{-1} \tag{8a}$$

and

$$\psi_2^{(\gamma)}(\mathbf{x}) = S\psi_2^{(0)}(\mathbf{x})\,S^{-1} \tag{8b}$$

It is shown elsewhere[3,5] that the solution to Eqs. (8a) and (8b) is

$$S = e^w, \qquad w = -\gamma \int d^3\mathbf{x}\rho(\mathbf{x})\ln\rho(\mathbf{x}) \tag{9}$$

The operator w is unique up to an additive c-number constant. Equations (2), (4), (5), and (8) imply that

$$H^{(\gamma)} = SH^{(0)}S^{-1} \tag{10}$$

which shows that the eigenvalues of the Hamiltonian are independent of γ. Since $H^{(1/2)}$ is a Hermitian operator, the eigenvalues of the various non-Hermitian Hamiltonians are all real. Using the ρ-representation, Chan and Valatin[2] earlier derived the transformation operator of Eq. (9) for the special case $\gamma = \frac{1}{2}$.

The Hermitian Hamiltonian is particularly convenient for the perturbation treatment of the weakly interacting Bose gas at low temperatures. Under the assumption of translational invariance the average density $n = N/\Omega$ must be independent of position. In the perturbation method used here it is assumed that the density fluctuation operator

$$\rho'(\mathbf{x}) \equiv \rho(\mathbf{x}) - n \tag{11}$$

is small (formally) compared to n. The Hamiltonian can be written in the form

$$H = H_0 + H_1 + H_2 + \cdots \tag{12a}$$

where

$$
\begin{aligned}
H_0 = (\hbar^2/2m) \int d^3\mathbf{x}\,[(\nabla\rho')^2/4n + n(\nabla\sigma)^2] - \tfrac{1}{2}\sum_{\mathbf{k}} \hbar^2 k^2/2m \\
+ \frac{1}{2}\int\!\!\int d^3\mathbf{x}\,d^3\mathbf{y}\,V(\mathbf{x}-\mathbf{y})\,[n + \rho'(\mathbf{x})]\,[n + \rho'(\mathbf{y}) - \delta(\mathbf{x}-\mathbf{y})]
\end{aligned} \tag{12b}
$$

$$H_1 = (\hbar^2/2m)\int d^3\mathbf{x}\,[-\rho'(\nabla\rho')^2/4n^2 + \rho'(\nabla\sigma)^2] \tag{12c}$$

and

$$H_j = (\hbar^2/8mn)\int d^3\mathbf{x}(-\rho'/n)^j\,(\nabla\rho')^2, \qquad j \geq 2 \tag{12d}$$

The terms in the Hamiltonian have been classified according to the powers of ρ' and σ that they contain. Since H_0 contains at most quadratic terms in ρ' and σ, it may be immediately diagonalized with a linear transformation. Fourier variables are introduced with Eqs. (6a)–(6c), and canonical variables $\pi_{\mathbf{k}}$ and $\pi_{\mathbf{k}}^{\dagger}$ are defined (for $\mathbf{k} \neq 0$) by

$$\rho(\mathbf{k}) = -iN^{1/2}S_0^{1/2}(k)\left[\pi_{\mathbf{k}}^{\dagger} - \pi_{-\mathbf{k}}\right] \tag{13a}$$

and

$$\sigma(\mathbf{k}) = \tfrac{1}{2}N^{-1/2}S_0^{-1/2}(k)\left[\pi_{\mathbf{k}} + \pi_{-\mathbf{k}}^{\dagger}\right] \tag{13b}$$

where

$$S_0(\mathbf{k}) \equiv \left[1 + 4mnv(k)/\hbar^2k^2\right]^{-1/2} \tag{14}$$

In terms of $\pi_{\mathbf{k}}$ and $\pi_{\mathbf{k}}^{\dagger}$ the approximate Hamiltonian H_0 has the form

$$H_0 = E_0 + \sum_{\mathbf{k} \neq 0} \eta_0(k)\,\pi_{\mathbf{k}}^{\dagger}\pi_{\mathbf{k}} \tag{15}$$

where

$$E_0 = \tfrac{1}{2}v(0)\,n^2\,\Omega - \tfrac{1}{2}\sum_{\mathbf{k} \neq 0}\left[\hbar^2k^2/2m - \eta_0(k) + nv(k)\right] \tag{16}$$

is the approximate ground-state energy and

$$\eta_0(k) \equiv \hbar^2k^2/2mS_0(k) \tag{17}$$

The approximate ground state $|0\rangle$ is defined by $\pi_{\mathbf{k}}|0\rangle = 0$ for all $\mathbf{k} \neq 0$. The excited states of the system are generated by the repeated operation of $\pi_{\mathbf{k}}^{\dagger}$ upon $|0\rangle$. They thus represent a system of noninteracting phonons with the energies given by Eq. (17). According to Landau's criterion,[4] superfluidity simply demands that $\eta_0(k)/k$ is a nonvanishing positive quantity. If the interaction potential is not too strong, this criterion is satisfied.

In order to illustrate the use of perturbation theory in this formalism, the leading correction to the ground-state energy has been obtained. The lowest-order correction E_1 is just the expectation value of H_1 of Eq. (12c) in the ground state $|0\rangle$. This contribution vanishes, however, due to the fact that when H_1 is expressed in terms of π and $\pi^{\dagger}$ the result is a trilinear form. The first nonvanishing contribution to the ground-state energy is E_2, the sum of the values obtained by taking H_2 of Eq. (12d) to first order and H_1 of Eq. (12c) to second order. The calculation is straightforward with the result that

$$E_2 = \frac{\hbar^2}{8mN}\sum_{\mathbf{q}_1\mathbf{q}_2}q_1^2S_1S_2 - \frac{1}{96N}\left(\frac{\hbar^2}{m}\right)^2\sum_{\mathbf{q}_1\mathbf{q}_2}\frac{1}{(\Delta E)S_1S_2S_3}\left[\mathbf{q}_1\cdot\mathbf{q}_2S_3(1 + S_1S_2) + \text{c.p.}\right] \tag{18}$$

where $\Delta E \equiv \eta_0(q_1) + \eta_0(q_2) + \eta_0(q_3)$, $S_i \equiv S_0(\mathbf{q}_i)$, $\mathbf{q}_3 \equiv -\mathbf{q}_1 - \mathbf{q}_2$, and c.p. means + cyclic permutations of the indices 1, 2, and 3. Each of the double summations above diverges strongly for large values of the wave vectors $\mathbf{q}_1$ and $\mathbf{q}_2$, in a manner reminiscent of divergences in quantum hydrodynamics, but the total is finite and even vanishes if $v(k)$ is set equal to zero. The value for E_2 given here is in analytical agreement with several other theories.[5] In particular, it agrees with the expression given by Lee.[6]

Finally, we consider the calculation of the structure factor for the ground state,

$$S(k) \equiv N^{-1}\langle\rho(\mathbf{k})\rho(-\mathbf{k})\rangle$$

With Eq. (13a) one easily shows that the value of $S(k)$ in the approximate ground state $|0\rangle$ is $S_0(k)$, thus justifying the notation of Eq. (14). To calculate $S(k)$ to higher order, one may use perturbation theory to calculate corrections to the approximate ground state $|0\rangle$ and take the expectation value of $\rho(\mathbf{k})\rho(-\mathbf{k})$, but this is not necessary. One can show that for $\mathbf{k} \neq 0$

$$S(\mathbf{k}) = 1 + (2/n)\,\partial E/\partial v(\mathbf{k}) \tag{19}$$

provided E is an eigenstate of the Hamiltonian. Applying this relation to the ground state, one obtains

$$S(k) \approx S_0(k) \tag{20}$$

from Eq. (16), and using the value we have obtained for E_2 in the form given by Lee,[6] one obtains the leading correction to the structure factor:

$$
\begin{aligned}
S_2(q_1) = \frac{1}{16N}\sum_{q_2}\Bigg\{ & 4S_1^3\left(1 + \frac{q_2^2}{q_1^2} + \frac{q_3^2}{q_1^2}\right)(1 - S_2)(1 - S_3) \\
& + \frac{1}{(\Delta E)\,S_2 S_3}\left(\frac{1}{\Delta E} - \frac{2mS_1}{\hbar^2 q_1^2}\right)\left[\frac{\hbar^2}{m}\mathbf{q}_1\cdot\mathbf{q}_2 S_3(1 - S_1)(1 - S_2) + \text{c.p.}\right]^2 \\
& - \frac{4mS_1^2}{\hbar^2 q_1^2}\frac{1}{(\Delta E)\,S_2 S_3}\left[\frac{\hbar^2}{m}\mathbf{q}_1\cdot\mathbf{q}_2 S_3(1 - S_1)(1 - S_2) + \text{c.p.}\right] \\
& \times \left[\frac{\hbar^2}{m}\mathbf{q}_1\cdot\mathbf{q}_2 S_3(1 - S_2) + \frac{\hbar^2}{m}\mathbf{q}_1\cdot\mathbf{q}_3 S_2(1 - S_3) - \frac{\hbar^2}{m}\mathbf{q}_2\cdot\mathbf{q}_3 S_2(1 - S_2)(1 - S_3)\right]\Bigg\}
\end{aligned}
$$

Acknowledgment

The author is indebted to Professor Felix Bloch for numerous stimulating discussions, without which this paper could not have been written.

References

1. N.N. Bogoliubov and D.N. Zubarev, *Soviet Phys.—JETP* **1**, 83 (1955).
2. Hong-mo Chan and J.G. Valatin, *Nuovo Cimento* **19**, 1, 118 (1961).
3. P. Berdahl, to be published.
4. L. Landau, *J. Phys. USSR* **5**, 71 (1941).
5. P. Berdahl, Ph.D. Thesis, Stanford University, 1972, unpublished.
6. D.K. Lee, *Phys. Lett.* **37**, A49 (1971).

Subcore Vortex Rings in a Ginsburg-Landau Fluid*

E.R. Huggins

Department of Physics, Dartmouth College
Hanover, New Hampshire

We will discuss the hydrodynamic limit of the Schrödinger equation, Maxwell's equations applied to a laser beam, and the Abrahams–Tsuneto[1] $T = 0$ time-dependent Ginsburg–Landau (GL) theory of superconductivity. The main aim will be to illustrate the close relationship between the three theories and to discuss some features of the vortex solutions. Of particular interest will be the subcore vortex rings in the time-dependent GL theory.

The Schrödinger Fluid

To illustrate the approach, we will begin with the Schrödinger fluid.[2] Starting with the Schrödinger equation,

$$-\frac{\hbar}{i}\frac{\partial \psi}{\partial t} = \frac{1}{2m}\left(\frac{\hbar}{i}\mathbf{V} - q\mathbf{A}\right)^2 \psi + q\phi\psi \tag{1}$$

we make the usual substitutions

$$\psi = R\exp(-im\phi/\hbar); \qquad \rho = R^2; \qquad \mathbf{v} \equiv -\mathbf{V}\phi - (q/m)\mathbf{A} \tag{2}$$

where $R(\mathbf{x}, t)$ and $\phi(\mathbf{x}, t)$ are real functions. Separating the real and imaginary terms gives the two equations

$$\frac{\partial \rho}{\partial t} + \mathbf{V}\cdot(\rho\mathbf{v}) = 0 \tag{3}$$

$$\frac{\partial \phi}{\partial t} = \frac{v^2}{2} - \frac{1}{2}\left(\frac{\hbar}{m}\right)^2\frac{\nabla^2 R}{R} + (q/m)\phi \tag{4}$$

Equation (3) is the continuity equation for the current $\mathbf{j} \equiv \rho\mathbf{v}$. If we associate $-(1/2)(\hbar/m)^2\nabla^2 R/R$ with an effective quantum mechanical pressure P/ρ, then Eq. (4) is Bernoulli's equation.

The Maxwell Fluid

The idea is to rewrite Maxwell's equations, giving a hydrodynamic description of the behavior of a laser beam. Limiting our discussion to a medium of constant

* Work supported by the National Science Foundation through the U.S.–Japan Cooperative Science Program.

135

index of refraction, assuming no free charges or currents, and choosing a gauge where $\mathbf{V} \cdot \mathbf{A} = 0$ and the scalar potential is zero, Maxwell's equations reduce to the wave equation

$$- \frac{1}{c^2} \frac{\partial^2 \mathbf{A}}{\partial t^2} + \mathbf{V}^2 \mathbf{A} = 0 \tag{5}$$

where $\mathbf{E} = -\partial \mathbf{A}/\partial t$, $\mathbf{B} = \mathbf{V} \times \mathbf{A}$. To find solutions applicable to a laser beam of frequency ω_0, we make the substitution

$$\mathbf{A}(\mathbf{x}, t) = \tilde{\mathbf{A}}(\mathbf{x}, t) \exp(-i\omega_0 t) \tag{6}$$

Equation (6) is not strictly a Fourier expansion because we are allowing for slow "hydrodynamic" time variations of the amplitude $\tilde{\mathbf{A}}(\mathbf{x}, t)$. In the use of Eq. (6), we will explicitly assume that

$$\partial \tilde{\mathbf{A}}(\mathbf{x}, t)/\partial t \ll -i\omega_0 \tilde{\mathbf{A}}(x, t) \tag{7}$$

Substituting Eq. (6) in Eq. (5) and dropping the small $\partial^2 \mathbf{A}/\partial t^2$ term, the wave equation is reduced to the following Schrödinger equation for a Maxwell fluid:

$$- \frac{\hbar}{i} \frac{\partial \tilde{\mathbf{A}}}{\partial t} = \frac{1}{2m} \left(\frac{\hbar}{i} \mathbf{V} \right)^2 \tilde{\mathbf{A}} - \frac{\hbar \omega_0}{2} \tilde{\mathbf{A}} \tag{8}$$

where $mc^2 \equiv \hbar \omega_0$ is the energy of the photons in the laser beam. Going from Eq. (5) to Eq. (8) is formally equivalent to taking the nonrelativistic limit of the Klein–Gordon equation. Here it is the energy of the photons, which is so precisely defined in a laser beam, that plays the role of rest mass.

The subsidiary condition $\mathbf{V} \cdot \tilde{\mathbf{A}} = 0$ can be handled by substituting $\tilde{\mathbf{A}} = \mathbf{V} \times \mathbf{W}$ and working with the $\mathbf{W}$ field. But for the present discussion we will make the simplifying (and restrictive) assumption that $\tilde{\mathbf{A}} = \hat{z}\psi$ is a z-directed field. Then $\mathbf{V} \cdot \tilde{\mathbf{A}} = 0$ implies $\psi(r, \theta)$ has no z dependence. Making the substitution of Eq. (2) in Eq. (8) (with $q = 0$), we again get a continuity equation for $\mathbf{j} = \rho \mathbf{v}$, and the Bernoulli equation

$$\frac{\partial \phi}{\partial t} = \frac{v^2}{2} - \frac{1}{2} \left(\frac{\hbar}{m} \right)^2 \frac{\mathbf{V}^2 R}{R} - \frac{c^2}{2} \tag{9}$$

Note that when $\partial \phi/\partial t = 0$, $\rho = $ const, we get the expected solution $v^2 = c^2$.

The Ginsburg—Landau Fluid

Here we start with the Abrahams–Tsuneto[1] $T = 0$ time-dependent Ginsburg–Landau theory, as given by the Lagrangian or free energy written by Werthamer[3] (with $q = 2e$). We make the following definitions of the quantities λ_D, γ, ξ, λ_L, and ψ:

$$\lambda_D^2 \equiv [4/N(0)](1/4\pi q^2); \qquad \gamma c^2 \equiv n_s(\hbar/2\tau_{tr}\Delta_0) v_f^2/3$$

$$\xi^2 \equiv \hbar^2 c^2 \gamma/2\Delta_0^2; \qquad \lambda_L^2 \equiv \lambda_D^2/2\gamma; \qquad \psi \equiv \Delta/\Delta_0 \tag{10}$$

Then the Lagrangian equation $\delta \mathscr{L}/\delta \psi^* = 0$ becomes

$$\left(\frac{\hbar}{i} \frac{\partial}{\partial t} + q\phi \right)^2 \psi = \gamma c^2 \left\{ \left(\frac{\hbar}{i} \mathbf{V} - q \frac{\mathbf{A}}{c} \right)^2 \psi + \frac{\hbar^2}{\xi^2} (\psi^* \psi \psi - \psi) \right\} \tag{11}$$

To obtain the "nonrelativistic limit" of this second-order wave equation, we make the substitution

$$\psi = \tilde{\psi}(\mathbf{x}, t) \exp\left[-i(m\gamma c^2/\hbar)\, t\right] \tag{12}$$

where $m\gamma c^2$ plays the role of an effective rest energy. (With $m =$ the mass of a Cooper pair, $m\gamma c^2 \approx 0.1$ eV, which is large compared to the energies involved in the binding of pairs.) Neglecting small terms of the order $1/\gamma m c^2$, we obtain a Schrödinger equation of the form

$$-\frac{\hbar}{i}\frac{\partial\tilde{\psi}}{\partial t} - q\phi\tilde{\psi} + \frac{\gamma m c^2}{2}\tilde{\psi} = \frac{1}{2m}\left(\frac{\hbar}{i}\mathbf{V} - q\frac{\mathbf{A}}{c}\right)^2\tilde{\psi} + \frac{\hbar^2}{2m\xi^2}(\tilde{\psi}^*\tilde{\psi}\tilde{\psi} - \tilde{\psi}) \tag{13}$$

To obtain the hydrodynamic equations, we make the substitutions of Eq. (2), again obtaining a continuity equation for $\mathbf{j} = \rho\mathbf{v}$, and the following Bernoulli equation:

$$\frac{\partial\phi}{\partial t} = \frac{v^2}{2} - \frac{1}{2}\left(\frac{\hbar}{m}\right)^2\frac{\nabla^2 R}{R} + \frac{1}{2}\left(\frac{\hbar}{m}\right)^2\frac{1}{\xi^2}(\rho - 1) + \left(\frac{q}{m}\right)\phi - \frac{\gamma c^2}{2} \tag{14}$$

Linear Vortex Cores

In order to compare the vortex solutions of the three quantum fluids, we will limit our discussion to uncharged fluids ($q = 0$), introduce the notation $\bar{\kappa} = \hbar/m$, and neglect the $-\gamma c^2/2$ term in Eq. (14) (a constant term which may be spurious). The three Bernoulli equations may now be written in the compact form (with $\mathbf{v} = -\nabla\phi$ and $\rho = R^2$):

$$\frac{\partial\phi}{\partial t} = \frac{v^2}{2} - \frac{\bar{\kappa}^2}{2}\frac{\nabla^2 R}{R} + \left[-\frac{c^2}{2}\right]_{\text{Maxwell}} + \left[\frac{\bar{\kappa}^2}{2\xi^3}(\rho - 1)\right]_{\text{GL}} \tag{15}$$

where the square brackets are added for the corresponding fluids.

To obtain the core structure of a straight vortex, we make the usual choice $\phi = -n\bar{\kappa}\theta$ [with n as an integer to keep $\psi = R\exp(-i\phi/\bar{\kappa})$ single valued], which gives $\partial\phi/\partial t = 0$, $\mathbf{v} = \hat{\theta}n\bar{\kappa}/r$, and $v^2/2 = n^2\bar{\kappa}^2/2r^2$. Because of the $1/r^2$ singularity, we can neglect nonsingular terms near $r = 0$, and all three equations reduce to the Schrödinger fluid equation which has the exact solution $R \sim r^n$, $\rho \sim r^{2n}$. As a result, all three fluids have the same inner core structure, as illustrated in Fig. 1, where we have graphed the $n = 1$ solutions for ρ.

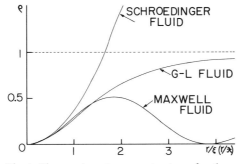

Fig. 1. The $n = 1$ vortex core structures for three quantum fluids.

For the GL fluid ρ is the well-known Ginsburg–Pitaevskii[4] solution. The exact solution for the Maxwell fluid is $\rho_n = [J_n(r/\hat\lambda)]^2$, where J_n is the nth-order Bessel function and $\hat\lambda = \lambda/2\pi$ is the wavelength of the light. The corresponding vector potential is $\mathbf{A}_n = \hat{z}J_n(r/\hat\lambda)\exp(in\theta - i\omega_0 t)$, which is the set of transverse magnetic modes for a circular waveguide at the cutoff frequency. (Our restriction $\mathbf{A} = \hat{z}\psi$ eliminated other waveguide modes.) Since solutions of Maxwell's equations are well known, the Maxwell–Bernoulli equation should serve as an important guide for the study of the hydrodynamics of higher-spin GL theories.

Vortex Rings

Because vortex rings do not remain at rest, time-dependent theories are required for their analysis. If we consider circular rings, then we may assume that the ring moves forward at a uniform speed $\mathbf{V}_v$, so that $\phi = \phi(\mathbf{x} - \mathbf{V}_v t)$ and $\partial\phi/\partial t = -\mathbf{V}_v \cdot \nabla\phi = \mathbf{V}_v \cdot \mathbf{v}$. Limiting our discussion to the GL fluid, the equation we have to solve is

$$-\mathbf{V}_v \cdot \mathbf{v} + \tfrac{1}{2}v^2 - \tfrac{1}{2}\bar\kappa^2\{(\nabla^2 R/R) - [(R^2 - 1)/\xi^2]\} = 0 \qquad (16)$$

where $\mathbf{v}$, the velocity field of a singular cored ring, is identical to the magnetic field of a current loop of infinitesimal cross section.

In one special case we obtain a fairly accurate qualitative picture without a computer solution of Eq. (16). Starting with the picture that ρ must be zero at the singular core of the ring and return to one in a distance of the order of the healing length ξ, we see that a ring of radius $R_r \ll \xi$ (a "subcore" ring) will produce an essentially spherical bubble in ρ. The bubble will be of radius ξ and thus obscure the ring structure of the velocity field $\mathbf{v}$. All that survives in the physical current $\mathbf{j} = \rho\mathbf{v}$ is the long-range dipole backflow illustrated in Fig. 2. The strength or (magnetic) moment of this backflow is proportional to the circulation κ and area A of the ring.

Because of the continuity equation for $\mathbf{j}$, we may treat the bubble as a sphere moving through a conserved fluid. Such a sphere has a dipole backflow so that the fluid can get out of the path of the sphere. The strength or moment of the backflow is proportional to the volume of the sphere and its speed. We can thus find the speed of the ring by equating dipole backflows; the result is

$$V_v = \alpha(\rho\kappa A)/2\pi\xi^3\rho \equiv P_\omega/m \qquad (17)$$

where $\alpha \approx 1$, $P_\omega = \rho\kappa A$ is the "fluid impulse" of the ring, and m is the effective mass of the fluid displaced by the bubble. Unlike a classical singular cored ring, whose velocity goes to infinity as its radius goes to zero, the speed of rings in a GL fluid reaches a maximum when $R_r \approx \xi$ and then goes to zero as the ring disappears.

The subcore ring has several features in common with the roton in liquid helium. It has a minimum energy, namely that required to create the bubble. (As long as $R_r \ll \xi$, the shape of the bubble and therefore the energy of the ring will be insensitive to R_r.) And both the subcore ring and the roton wave function have long-range dipole backflows.[5] Comparison of a detailed computer solution of the subcore ring with the well-known experimental data for rotons should provide an

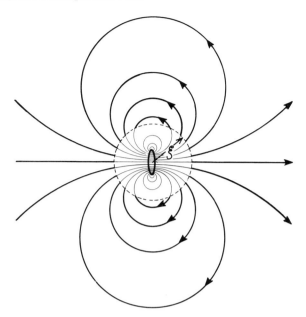

Fig. 2. Subcore ring in a Ginsburg–Landau fluid. When the radius of the ring is smaller than the healing length ξ the bubble in ρ covers the ring structures in $\mathbf{v}$ and only the longrange backflow survives in the physical current $\mathbf{j} = \rho\mathbf{v}$.

interesting check of the applicability of the Ginsburg-Landau theory to superfluid helium. A very special feature of the subcore ring is that it is part of a continuous series of solutions that evolve into the macroscopic quantized vortex ring.

Acknowledgments

This work has benefited considerably from discussions with T. Fujita and M. Matsuoka. We wish to acknowledge the support of the University of Kyoto and express our gratitude for the hospitality of T. Tsuneto. The author also thanks A. Campbell for computer calculations in the vortex ring.

References

1. E. Abrahams and T. Tsuneto, *Phys. Rev.* **152**, 416 (1966).
2. R.P. Feynman, in *The Feynman Lectures*, Addison-Wesley, Reading, Mass. (1963), Vol. III, Lecture 21.
3. N.R. Werthamer, in *Superconductivity*, R.D. Parks, ed., Dekker, New York (1969), Vol. 1, p. 366.
4. V.L. Ginsburg and L.P. Pitaevskii, *Soviet Phys.—JETP* **7**, 1240 (1958).
5. R.P. Feynman and M. Cohen, *Phys. Rev.* **102**, 1189 (1956).

3
Static Films

Multilayer Helium Films on Graphite*

Michael Bretz†

Department of Physics, University of Washington
Seattle, Washington

The λ-point specific heat anomaly for very thin films of helium is known to round, shift to lower temperatures, and finally disappear altogether at a few monolayers. The superfluid onset shifts in temperature also, but more rapidly with decreasing coverage than the heat capacity peak[1-4].

Our recent experiments on monolayer helium films adsorbed on an exfoliated graphite substrate demonstrate the crucial importance of substrate homogeneity and cleanliness for monolayer film properties.[5] Since all multilayer helium film experiments were performed on presumably heterogeneous substrates, it seemed necessary to test whether a graphite substrate would influence multilayer film behavior to any appreciable extent.

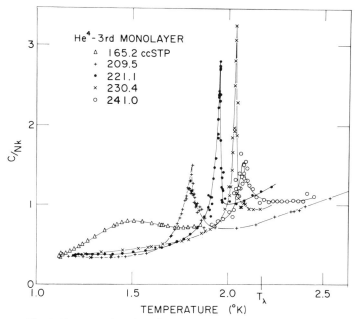

Fig. 1. Heat capacity of the ^{4}He film through third layer coverage.

* Research supported by the National Science Foundation.
† Present address: University of Michigan, Ann Arbor, Michigan.

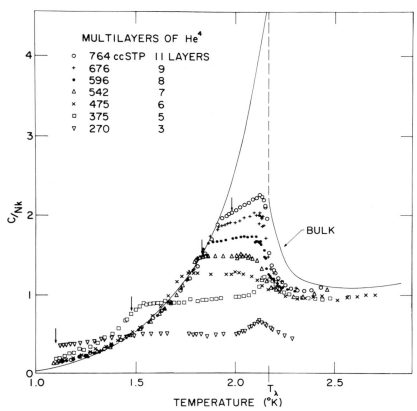

Fig. 2. Heat capacity of the multilayer ^{4}He film. The solid line is the heat capacity of bulk helium and the arrows indicate cell relaxation onsets for coverages of 270, 375, 542, and 764 STP cc.

The first complete monolayer of helium on graphite forms a 2D solid with a θ_{Debye} of approximately 60°.[6] Its melting transition has been seen to the third layer, showing that this solid remains distinct and stable when multilayers are present. The second monolayer (96–165 cm^3 STP) seems gaslike with no striking heat capacity structure, but the third monolayer (165–230 cm^3 STP) displays a broad, low bump which apparently grows into a sharp λ-like peak at 2.1°K with completion of the layer.* (See Fig. 1.) It appears that this is the melting transition of the second layer which has been compressed somewhat by the presence of the third layer. A 5–10% compression makes these melting temperatures and densities consistent

---

* The data presented (with the exception of the 165.2 cm^3 STP run) were corrected for the desorption contribution to the heat capacity (see Dash et al.[7]) and the amount and heat capacity of the imperfect gas in the dead volume. The contribution of the first solid monolayer and empty cell were subtracted out. The corrections were quite large for the thicker films and high temperatures, amounting to about 50% of the total signal at 11 layers. Since gas desorbs from the cell with increased temperature, the coverage values given in the figure are for 0°K. Thus, the true coverage of the 764 cm^3 STP run at the lower truncation point, or elbow, is 737 cm^3 STP; at the upper truncation point it is 720 cm^3 STP; and at 2.45°K it has dropped to 685 cm^3 STP. On the other hand, at 2.5°K the 209.5 cm^3 STP run depletion is only 5 cm^3.

with those of first layer solidification. Moreover, the low-temperature constant heat capacity can be accounted for by a third-layer 2D gas contribution. For thicker films the solid dissolves, evidently due to increasing communication between layers, so the entire film except for the solid monolayer tends toward liquid behavior.

Continuing to increase the coverage, we find only a remnant of the 2.1°K peak. A series of plateaus build to high heat capacity values which are recognized as truncations of the bulk λ peak[8,9] (See Fig. 2.) These plateaus are somewhat narrower than the bulk, with a small shift to lower temperatures, consistent with previous film experiments. Truncation of peaks is not uncommon for other systems but is usually quite close to the critical temperature.[10,11] The truncation in these systems might be caused by impurities within the crystallites which limit the coherence length. For helium the coherence length appears limited by film thickness rather than by impurities. Theoretical peak rounding of an Ising system with a finite-sized square lattice[12] has been applied to helium films[13] with a prediction that $C^{max} = [4.5(J/g)/v \log_{10}(D/D_0)$, where $v = 2/3$, D is the film thickness, and D_0 is some constant. The low-temperature truncating points are in excellent agreement with $[5.35(J/g)/(2/3)] \log_{10}(N/217)$, where N is the coverage in cm³ STP. At 11 layers there is a distinct deviation from this fit which could mark interference from adjacent films or partial filling of the exfoliated graphite pores. Ferrell and co-workers[14] have predicted that when the coherence length of helium becomes comparable to the film thickness the long-wavelength modes of second sound normal to the surface will become clamped, stopping a further increase in the specific heat as $T \to T_\lambda$. They, also expect C^{max} to be logarithmic with film thickness.

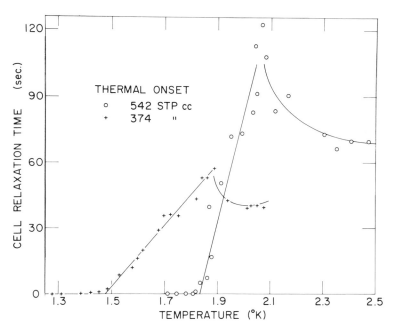

Fig. 3. Heat capacity cell relaxation times. Onsets are taken to be where the straight lines intersect the temperature axis.

The heat capacity cell relaxation time provides additional experimental data. The heater and thermometer are on the outside of a copper capsule which contains the graphite and helium gas. During heating the capsule will not, in general, be equilibrated with the graphite, so a subsequent post-heat thermometer cooling will occur before cell and contents acquire thermal equilibrium. Figure 3 shows the measured cell relaxation times for the 374 and 542 cm^3 STP coverages. The abrupt "onset" in cell relaxation time with temperature is undoubtedly related to the thermal conductivity of the film. These onsets are shown in Fig. 2 as arrows. It is evident that they correspond closely to the heat capacity truncation points, which seems to indicate that superfluidity disappears whenever there is a restriction in the growth of the correlation length in liquid helium.

References

1. H. P. R. Frederikse, *Physica* **15**, 860 (1949).
2. D. F. Brewer, A. J. Symonds, and A. L. Thomson, *Phys. Rev. Lett.* **15**, 182 (1965).
3. E. A. Long and L. Meyer, *Phys. Rev.* **85**, 1030 (1952).
4. K. Fokkens, K. W. Taconis, and R. de Bruyn Ouboter, *Physica* **32**, 2129 (1966).
5. M. Bretz and J. G. Dash, *Phys. Rev. Lett.* **26**, 963 (1971).
6. M. Bretz, G. B. Huff, and J. G. Dash, *Phys. Rev. Lett.* **28**, 729 (1972).
7. J. C. Dash, R. E. Peierls, and G. A. Stewart, *Phys. Rev.* **2A**, 932 (1970).
8. H. C. Kramers, J. D. Wasscher, and C. J. Gorter, *Physica* **18**, 329 (1952).
9. R. W. Hill and O. V. Lounasmaa, *Phil. Mag.* **2**, 1143 (1957).
10. M. Rayl, O. E. Vilches, and J. C. Wheatley, *Phys. Rev.* **165**, 698 (1968).
11. M. B. Salamon, *Phys. Rev.* **2B**, 214 (1970).
12. M. E. Fischer and A. E. Ferdinand, *Phys. Rev.* **185**, 832 (1969).
13. M. A. Moore, *Phys. Lett.* **37A**, 345 (1971).
14. R. A. Ferrell, N. Menyhand, H. Schmidt, F. Schwabl, and P. Szepfalusy, *Phys. Lett.* **24A**, 493 (1967).

Submonolayer Isotopic Mixtures of Helium
Adsorbed on Grafoil*

S.V. Hering, D.C. Hickernell,† E.O. McLean,‡ and O.E. Vilches

Department of Physics, University of Washington
Seattle, Washington

Monolayers or submonolayers of helium adsorbed on different substrates have been studied for several years. Heat capacity signatures for He on rouge,[1] Vycor,[2] and bare and rare-gas-plated copper[3-5] resemble those of two-dimensional (2D) Debye systems. Only recently measurements done on a commercially available variety of graphite, Grafoil,§ have given results that can be compared to 2D quantum and classical gases.[6] Several prominent features have been reported in the range between 1/4 and 2/3 monolayer, mainly: (a) classical k/atom specific heats at 4°K; (b) peaks around 1.2°K appearing only for bosons (^{4}He) at the lower coverages, assumed due to condensation induced by weak substrate lateral fields[6,7]; (c) peaks around 3°K occurring for coverages at which a well-defined fraction of the Grafoil lattice hexagons x_G is occupied by He atoms ($x_G = 1/4$ and $x_G = 1/3$); (d) spin-ordering effects appearing for ^{3}He at very low temperatures.[8] Comparisons have been made with the properties of liquid and solid ^{3}He and ^{4}He, and it appears that 2D films on Grafoil have properties corresponding to those of the three-dimensional (3D) systems (see, e.g., a recent report on the melting line).[9]

The third quantum system available in three dimensions is ^{3}He–^{4}He mixtures. As is well known,# liquid mixtures show ideal 3D Fermi gas behavior at very low temperature for $x_3 = N_3/(N_3 + N_4) < 6.4\%$. Above this concentration mixtures separate into two phases with the temperature of separation T_{ph} depending on x_3. Addition of ^{3}He to liquid ^{4}He lowers the λ temperature until for $x_3 \cong 0.674$, $T_\lambda = T_{ph} \cong 0.871°K$.[12] Solid mixtures separate into two phases for all concentrations at temperatures somewhat lower than for the liquid mixtures.[13] Submonolayer films of mixtures of ^{3}He and ^{4}He adsorbed on Vycor show above 1°K the behavior of 2D Debye systems characteristic of the pure isotopes on the same substrate, with a constant addition to the specific heat interpreted as a classical contribution of the ^{3}He atoms in the mixture.[2,14] Multilayer films show the effect of preferential ^{4}He adsorption by the substrate, producing a phase separation or concentration gradient normal to the wall of the pores. The excess ^{3}He atoms on the free surface give a

* Work supported by the National Science Foundation.
† Present address: Cryogenic Center, Stevens Institute of Technology, Hoboken, N.J.
‡ Present address: Department of Physics, UCLA, Los Angeles, Calif.
§ Grafoil is a trademark of the Union Carbide Corp.
# For a review of the properties of ^{3}He–^{4}He solutions see Keller[10] and Anderson et al.[11]

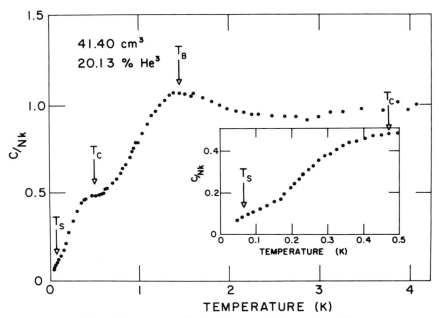

Fig. 1. Measured specific heat for $x = 0.25$ and $x_3 = 0.20$.

contribution to the specific heat[15] and to the surface tension[16] that can be compared with submonolayer specific heat measurements of ^{3}He films on Grafoil.

The present specific heat measurements have been made on submonolayer mixture films adsorbed on Grafoil. Submonolayers were chosen both to study the evolution of effects obtained with the pure isotopes and to avoid the "vertical" concentration gradient. Four coverages x were chosen based on the results with the pure isotopes*: $x = 0.25$ and $x = 0.38$ had the characteristics of quantum gases, while $x = 0.44$ and $x = 0.60$ closely corresponded to $x_G = 1/4$ and $x_G = 1/3$, respectively. Measurements were done for $0.04°K < T < 4°K$ and for eight ^{3}He concentrations. A partial description of the experimental setup has been given in Ref. 8.

Figure 1 shows specific heat results for one of the mixtures. Similar curves have been obtained for each coverage and ^{3}He concentration measured. Indicated in Fig. 1 are the temperatures at which anomalies are observed. The behavior of the film features is as follows.

For the two lower coverages ($x = 0.25$ and $x = 0.38$) the maximum attributed to lateral condensation[7] (T_B) is observed to decrease in magnitude and in temperature as the mole fraction of ^{3}He is increased. The low-temperature spin anomaly[8] at T_s evolves fairly regularly from pure ^{4}He to ^{3}He. At about 0.4°K and low ^{3}He mole fractions a shoulder is seen for the $x = 0.38$ coverage at T_c, Fig. 1. The temperature of this anomaly is perhaps only slightly dependent upon the concentration of the mixture until, for the highest concentrations, the shoulder is no longer evident.

---

* Measurements were done at approximately constant number of adsorbed atoms. One monolayer of ^{3}He corresponds to 108.6 cm^3 STP of gas, while one monolayer of ^{4}He corresponds to 113.1 cm^3 STP. Monolayer coverages x given in text are fractions of the ^{3}He monolayer number.

The $x = 0.44$ coverage shows a rather broad peak at 2°K for $0 < x_3 < 0.5$. This is different from results of Ref. 6, where a sharp maximum was observed for ^{4}He and a small cusp for ^{3}He at 3°K. The present study has a peak diminishing in magnitude but constant in temperature as the ^{3}He concentration is increased, but the maximum disappears for $0.5 < x_3 < 1$. This coverage also shows an anomaly at 0.4°K for low ^{3}He concentrations similar to that of the $x = 0.38$ coverage, and a low temperature rise that evolves into the spin peak of pure ^{3}He.

The last coverage ($x = 0.6$) is slightly larger (3%) than the coverage for which ^{3}He has its highest peak near 3°K.[17] The heights and temperatures of the 3°K peaks decrease as the ^{4}He mole fraction is increased. This is consistent with the observation that the number of atoms needed to obtain the maximum peak height is about 1.5% larger for the lighter isotope.

Entropies at 4°K for all x and x_3 have been calculated by linearly extrapolating to $T = 0$°K the C vs T curves from the lowest measured specific heat points and numerically integrating

$$\int_{T=0}^{T=4\,^\circ\mathrm{K}} (C/T)\,dT$$

Results are plotted in Fig. 2. For comparison, also plotted in the same figure is the entropy at 4°K of a mixture of two ideal 2D gases with the atomic masses of ^{3}He

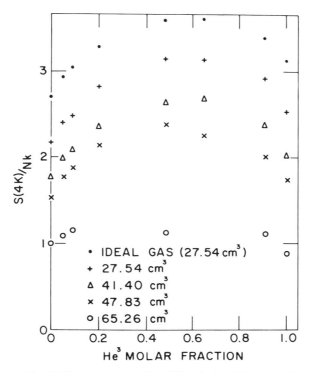

Fig. 2. Film entropies at $T = 4$°K calculated from specific heat data and for mixture of ideal ^{3}He and ^{4}He point particles at $x = 0.25$.

and ^{4}He and of the same density as the $x = 0.25$ coverage. It is evident that for the three lowest coverages the adsorbed films are mixed at $4°$K; the rounded maximum in the graph is due to the classical entropy of mixing $x_3 \ln x_3 + (1 - x_3)\ln(1 - x_3)$. The $x = 0.6$ coverage does not follow the pattern of the other coverages. It could be that mixing occurs below our temperature range.

A more detailed analysis of the mixing process can be made by comparing the measured specific heat of a mixture at certain coverage with that of a film formed by the separated isotopes in the same proportion and at the same coverage. One can then calculate by subtraction the excess specific heat and from it the excess entropy. Results show that most of the excess entropy appears below T_c and that all the films are mixed at $1°$K. This is reinforced by the decrease of T_B with increased x_3 in the manner expected from Ref. 7 if the ^{3}He is just acting as a spacer of the ^{4}He atoms. If the very low-temperature anomaly at T_s is due to ^{3}He–^{3}He spin interactions, then the films should be isotopically separated at that temperature since the magnitude of the anomaly corresponds to that of the fraction of ^{3}He in the mixture, and T_s remains essentially unchanged.

Graphs of the temperatures of specific heat anomalies vs. x_3 (Fig. 3) for the three lower coverages bear some resemblance to the phase separation–superfluid diagram for bulk liquid mixtures of helium. A full description of the data and its analysis[18] will be published elsewhere.

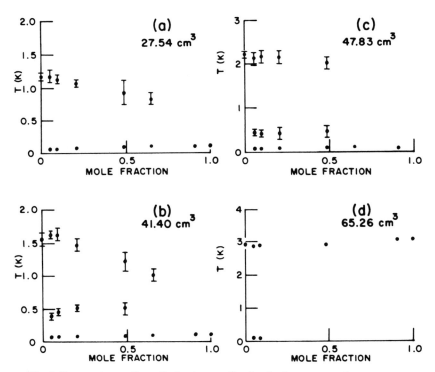

Fig. 3. Temperatures of specific heat anomalies for the four measured coverages as a function of x_3.

References

1. H. P. R. Frederikse, *Physica* **15**, 860 (1949).
2. D. F. Brewer, *J. Low Temp. Phys.* **3**, 205 (1970).
3. D. Princehouse, Ph. D. Thesis, Univ. of Washington, 1971, unpublished.
4. W. D. McCormick, D. L. Goodstein, and J. G. Dash, *Phys. Rev.* **168**, 249 (1968).
5. G. A. Stewart and J. G. Dash, *Phys. Rev.* **A2**, 918 (1970).
6. M. Bretz and J. G. Dash, *Phys. Rev. Lett.* **26**, 963 (1971); **27**, 647 (1971).
7. C. E. Campbell, J. G. Dash, and M. Schick, *Phys. Rev Lett.* **26**, 966 (1971).
8. D. C. Hickernell, E. O. McLean, and O. E. Vilches, *Phys. Rev. Lett.* **28**, 789 (1972).
9. M. Bretz, G. B. Huff, and J. G. Dash, *Phys. Rev. Lett.* **28**, 729 (1972).
10. W. E. Keller, *Helium-3 and Helium-4*, Plenum, New York, (1969).
11. A. C. Anderson, W. R. Roach, R. J. Sarwinski, and J. C. Wheatley, *Phys. Rev. Lett.* **16**, 263 (1966).
12. T. A. Alvesalo, P. M. Berglund, S. T. Islander, G. R. Pickett, and W. Zimmerman, *Phys. Rev.* **A4**, 2354 (1971).
13. D. O. Edwards, A. S. McWilliams, and J. G. Daunt, *Phys. Rev. Lett.* **9**, 195 (1962).
14. A. E. Evenson, Thesis, Univ. of Sussex, 1968, unpublished.
15. D. F. Brewer, A. Evenson, and A. L. Thomson, *Phys. Lett.* **35A**, 307 (1971).
16. H. M. Guo, D. O. Edwards, R. E. Sarwinski, and J. T. Tough, *Phys. Rev. Lett.* **27**, 1259 (1971).
17. E. O. McLean, Ph.D. Thesis, Univ. of Washington, 1972, unpublished.
18. D. C. Hickernell, Ph.D. Thesis, Univ. of Washington, 1972, unpublished.

Spatial Ordering Transitions in Quantum Lattice Gases*

R.L. Siddon and M. Schick

Department of Physics, University of Washington
Seattle, Washington

We wish to investigate the effects of statistics and quantum mechanical tunneling on the spatial ordering transitions of adsorbed gases. To do so, we consider a system of atoms adsorbed on a substrate which is divided into M cells of area d^2 corresponding to adsorption sites. Restricting the atoms to these cells results in a quantum lattice gas governed by the Hamiltonian

$$H - \mu N = (\hbar^2/2md^2) \sum_{\langle ij \rangle} (\psi_i^\dagger - \psi_j^\dagger)(\psi_i - \psi_j) + v \sum_{\langle ij \rangle} n_i n_j - \mu \sum_i n_i$$

where ψ_i is the annihilation operator for an atom on the ith site and $\langle ij \rangle$ denotes a sum over nearest-neighbor pairs. The first term is the first finite difference approximation to the usual kinetic energy operator and enables the atoms to tunnel from site to site. The parameter v is the strength of the interaction between nearest neighbors and is taken to be repulsive. In principle, v can be calculated from the interparticle potential and localized wave functions. It is therefore a functional of the potential and particle mass. The effect of the hard-core interaction is to prevent multiple occupation of a site. For bosons, we follow Matsubara and Matsuda[1] and incorporate this effect by altering the commutation relation of the boson field operators to read

$$[\psi_i, \psi_j^\dagger] = (1 - 2\psi_i^\dagger \psi_i)\delta_{ij}, \qquad [\psi_i, \psi_j] = -2\psi_j \psi_i \delta_{ij}$$

If we restrict ourselves to spinless fermions, which is sufficient for our purposes, the usual anticommutation relations

$$\{\psi_i, \psi_j^\dagger\} = \delta_{ij}, \qquad \{\psi_i, \psi_j\} = 0$$

prevent multiple occupancy and need not be changed.

The simplest approximation that can be applied to the above Hamiltonian is that of mean field theory in which a cluster of one site is considered to interact with its neighbors via mean fields. This yields a cluster Hamiltonian

$$(H - \mu N)_{cl} = -z\xi(\psi_i^\dagger \langle \psi_j \rangle + \langle \psi_j^\dagger \rangle \psi_i) + (z\xi + zv \langle n_j \rangle - \mu) n_i$$

where $\xi \equiv \hbar^2/2md^2$, z is the number of nearest neighbors, and the angular brackets denote ensemble average. Upon imposing the condition that $\langle \psi_i \rangle$ vanishes for a fermion system and for a nonsuperfluid boson system[2] (which we assume to be the case), we see that in mean field theory the only difference between a quantum and classical system is a shift in chemical potential. Otherwise, the phase diagrams are

* Supported in part by the National Science Foundation.

the same. It is therefore necessary to employ a higher-order cluster approximation such as the Bethe–Peierls approximation in which the interactions within a cluster of one site and its z nearest neighbors are treated exactly. The cluster interacts with the rest of the system by mean fields. The resulting cluster Hamiltonian is

$$(H - \mu N)_{cl} = (z\xi - \mu) n_c - \xi \sum_\delta (\psi_c^\dagger \psi_\delta + \psi_\delta^\dagger \psi_c) + [vn_c - (\phi - \xi + \mu)] \sum_\delta n_\delta$$

where $\sum_\delta$ is a sum over the z edge sites, the subscript c denotes the center site of the cluster, and ϕ is the mean field. In this case, the vanishing of $\langle \psi_i \rangle$ enables us to ignore any mean field coupling to $\psi_\delta^\dagger$ or ψ_δ. Since each site can be occupied or empty, there are 2^{z+1} basis states for the cluster. In the occupation number representation, the the diagonal elements of $(H - \mu N)_{cl}$ are, apart from a shift in the chemical potential, simply the classical eigenvalues. The effect of tunneling and statistics is reflected in the presence of off-diagonal elements, $-\xi$ for bosons and $\pm\xi$ for fermions, which connect states of the cluster which differ only in the arrangement of the atoms. The eigenvalues of the cluster Hamiltonian and the cluster partition function $\mathscr{Q}$ are obtained straightforwardly.

For simplicity, we consider a square lattice for which there are 32 basis states. Since the repulsive interaction tends to cause the atoms to form an ordered array in which alternating sites are occupied, we further subdivide the substrate into two interpenetrating sublattices and introduce two cluster Hamiltonians and two coupling fields in the usual way.[3] The coupling fields are determined from the self-consistency condition that the average occupation of an A site, n_A, be the same regardless of whether it is viewed as the center site of an A cluster or the edge site of a B cluster and similarly for the occupation n_B of a B site. Let the density and order parameter of the system be denoted n and η, so that $n = (1/2)(n_A + n_B)$ and $\eta = n_A - n_B$. The form of the cluster Hamiltonian then allows the self-consistency conditions to be written as

$$-\partial(\log \mathscr{Q}_A)/\partial\beta\varepsilon = n + \tfrac{1}{2}\eta, \qquad -\partial(\log \mathscr{Q}_B)/\partial\beta\varepsilon = n - \tfrac{1}{2}\eta$$

$$\tfrac{1}{4}\partial(\log \mathscr{Q}_A)/\partial\beta V_B = n - \tfrac{1}{2}\eta, \qquad \tfrac{1}{4}\partial(\log \mathscr{Q}_B)/\partial\beta V_A = n + \tfrac{1}{2}\eta$$

where $\beta = 1/kT$, $\varepsilon \equiv 4\xi - \mu$, and $V_{A,B} = \phi_{A,B} - \xi + \mu$.

The resulting phase diagram is symmetric in the $T - n$ plane about the density $1/2$. Its form depends on the value of $\gamma \equiv \xi/v$ which measures the relative strength of the tunneling term. For $\gamma = 0$, the system is classical and can exist in either ordered or disordered phases which are separated by a line of second-order transitions. The maximum transition temperature occurs at $n = 1/2$. For γ nonzero but less than about 0.33, the transition temperature is reduced due to the disordering effect of tunneling. To order γ^2, the transition temperature at $n = 1/2$ has the same form for bosons and fermions and is given by

$$kT/v = (kT/v)_{cl} - 0.99\gamma^2 \tag{1}$$

where $(kT/v)_{cl} = 1/(2 \log 2)$.

The effect of statistics on the transition temperatures is of fourth order in γ. As the temperature is lowered still further, the system undergoes a second-order transition to the disordered state. This behavior is an artifact of the Bethe–Peierls approximation and has been noted previously.[3,4] As the value of γ increases, the region of the phase diagram which corresponds to an ordered state shrinks and vanishes for $\gamma \approx 0.33$. The phase diagrams for a few values of γ are shown in Fig. 1. The effects of statistics are ignorable on the scale shown. The magnitude of the specific heat per site just above and below the transition temperature at $n = 1/2$ is shown as a function of γ and $T_c(\gamma)$ in Fig. 2.

To summarize, the maximum temperature of transitions to the ordered phase depends, in decreasing order of importance, on the interaction strength v (order γ^0), the particle mass (order γ^2), the statistics (order γ^4). Since no transition takes place for $\gamma \gtrsim 0.33$, the effect of statistics can be ignored.

The above may be applied to helium monolayers adsorbed on graphite[5] if it is assumed that the behavior of the maximum ordering temperature, which, for a triangular lattice,[6] occurs at $n = 1/3$, is similar to that which occurs at $n = 1/2$ for the square array. For helium on graphite, $d \approx 2.46$ Å and $\gamma \approx 0.1$. From Eq. 1, one obtains

$$T_3 - T_4 = \frac{v_3 - v_4}{2k\log 2} - 0.99\left[\frac{\hbar^2}{2m_3 d^2 k}\right]^2 \frac{k}{v_3}\left[1 - \frac{v_3}{v_4}\left(\frac{m_3}{m_4}\right)^2\right]$$

where the subscripts refer to the isotope. If it were assumed that v_3 were equal to v_4, then T_3 is less than T_4, indicating that the heavier isotope orders more easily, as expected. However, experiments[7] indicate that $T_3 - T_4 = 0.1°$K. The above ex-

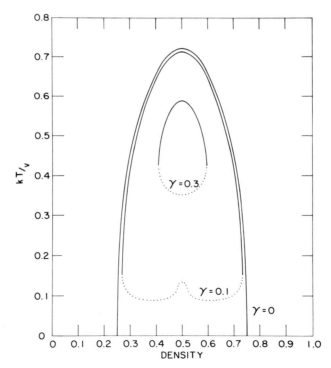

Fig. 1. Phase diagram in the T–n plane for three values of γ.
The unphysical transitions are shown by dotted lines.

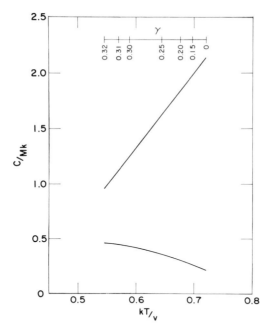

Fig. 2. Specific heat per site just above and below the transition temperature at density one-half, as a function of γ.

pression indicates, then, that $v_3 - v_4 \approx 0.2°\text{K}$. The larger value of v_3 is to be expected if it is recalled that v can be derived from a matrix element and that the ^{3}He wave functions are somewhat less localized on a site than those of ^{4}He. This spreading leads to the increase in v.

References

1. T. Matsubara and H. Matsuda, *Prog. Theor. Phys. (Kyoto)* **16**, 569 (1956).
2. C. N. Yang, *Rev. Mod. Phys.* **34**, 369 (1962).
3. Y. Y. Li, *Phys. Rev.* **84**, 721 (1951).
4. P. N. Anderson, *Phys. Rev.* **80**, 922 (1950).
5. M. Bretz and J. G. Dash, *Phys. Rev. Lett.* **27**, 647 (1971).
6. C. E. Campbell and M. Schick, *Phys. Rev.* **A4**, 1919 (1972).
7. E. O. McLean, Ph.D. Thesis, Univ. of Washington, 1972, unpublished.

A Model of the ^{4}He Monolayer on Graphite

S. Nakajima*

Department of Physics, University of Illinois
Urbana, Illinois

The experiments of Dash and co-workers show that the He monolayer adsorbed on graphite is a new, 2D quantum system.[1,9] The simplest model is the tight binding approximation, in which the zeroth-order wave function of the adsorbed atom is localized around the center of the hexagonal unit cell of graphite. Including the interaction between localized atoms as well as the tunneling of the atom from one center to another, we obtain the Hamiltonian for the ^{4}He monolayer

$$H = \frac{1}{2} \sum J_{jl} n_j n_l - \mu \sum n_j - \frac{1}{2} \sum K_{jl} a_j^* a_l \tag{1}$$

Here a_j and a_j^* are destruction and creation operators of the ^{4}He atom localized at the jth center, $n_j = a_j^* a_j$ is the occupation number, and μ is the chemical potential. For a given μ, the atomic concentration is given by

$$x_g = N^{-1} \sum \langle n_j \rangle \tag{2}$$

or vice versa, where N is the total number of adsorption sites.

The Hamiltonian for the ^{3}He monolayer in the same tight binding approximation is very much similar to the Hubbard Hamiltonian of electrons in the theory of the metal–insulator transition. In our case, the interaction J_{jj} is virtually infinite because of the hard-core repulsion between He atoms. Unfortunately, there is no simple mathematical way of eliminating those states in which two ^{3}He atoms with opposite spins occupy the same site. In the case of ^{4}He, on the other hand, Matsubara and Matsuda[2] (MM) eliminated J_{jj} by replacing Bose operators with Pauli spin operators as

$$n_j \to \frac{1}{2} - S_{jz}, \qquad a_j \to S_{jx} + iS_{jy} \tag{3}$$

In this way, our problem of the ^{4}He monolayer becomes equivalent to the 2D spin problem defined by the Hamiltonian

$$H = C - \eta \sum S_{jz} + \frac{1}{2} \sum J_{jl} S_{jz} S_{lz} - \frac{1}{2} \sum K_{jl} (S_{jx} S_{lx} + S_{jy} S_{ly}) \tag{4}$$

Here C is a constant and

$$\eta = \frac{1}{2} \sum_j J_{jl} - \mu \tag{5}$$

plays the role of an external magnetic field.

---

* On leave from Institute for Solid State Physics, University of Tokyo, Japan.

The Hamiltonian Eq. (4) was proposed originally by Matsubara and Matsuda as a quantum lattice model for bulk liquid ^{4}He. However, the basic lattice there is a mere mathematical convention to simplify the continuous configuration space of the atom, whereas in our problem it is defined by the substrate and Eq. (4) is almost equivalent to the tight binding approximation Eq. (1). Thus the MM model is much more appropriate and realistic for the ^{4}He monolayer.

Qualitative features of the model may be seen by applying the molecular field approximation. For simplicity, we retain positive J and K only for nearest neighbors ($J > K$). Then this approximation predicts three possible phases, which we call I, II, and III, respectively. They are characterized by three different modes of ordering of equivalent spins in Eq. (4).

Phase I appears for large η (small x_g) and is characterized by the ferromagnetic spin ordering in the z direction. Thus the vacant substrate is represented by the state Φ in which all spins are parallel to the z axis, and the spin reversal means the adsorption of the He atom. The reversed spin actually travels as a magnon, which is identified with the adsorbed atom tunneling over the substrate. Thus, in phase I, the monolayer behaves like a Bose gas.

Phase II is characterized by the ferromagnetic spin ordering in a direction which makes a finite angle θ with the z axis. Energetically, this phase is favorable when η is small. In the molecular field approximation, this phase appears when η becomes lower than $3 (J + K)$. Operators $S_{j\pm} = S_{jx} \pm i S_{jy}$ or equivalently a_j have non-vanishing expectation values (broken gauge symmetry), so that this is a superfluid phase. The state Ψ in which all spins are parallel to the θ direction is related to the vacuum Φ as

$$\Psi = \prod_{j=1}^{N} \{\cos(\theta/2) + [\sin(\theta/2)] S_{j-}\} \Phi \tag{6}$$

Expanding the right-hand side, we see that Ψ is asymptotically equivalent to the state of Bose condensation, in which a macroscopic number of adsorbed atoms occupy the zero-wave number state.

The quasiparticle spectrum may be obtained by applying the spin wave theory to the fluctuation from Ψ. It is essentially Bogoliubov's spectrum; the excitation energy $E(k)$ is proportional to k when the wave number k is small. The amplitude of the fluctuation associated with this Goldstone mode is divergent as $k \to 0$. In the case of 3D systems, the divergence is suppressed by the phase volume proportional to k^2 when we integrate over the wave vector. In our case of the 2D system at finite temperatures there remains a logarithmic divergence, which means the instability of the assumed superfluid long-range order in accordance with Hohenberg's general statement.[3]

Thus, if we take into consideration the fluctuation, phase II disappears. The phase boundary predicted by the molecular field approximation between phases I and II merely indicates the region in which the short-range order may bring about some anomalies such as the broad heat capacity peak observed experimentally around 1°K. This interpretation is in contrast with the one given by Campbell et al.[4] Mathematically the effect of the short-range order may be included by extending the Bogoliubov–Tyablikov method of two-time Green's functions one step further, as has been done by Richards[5] and also by Kondo and Yamaji[6] for 1D Heisenberg models. The detail of this part of our theory will be published elsewhere.

Now, when η is lowered further ($\eta < 3J$ in the molecular field approximation) and if the nearest-neighbor repulsion J is strong enough, as it is on the graphite surface, there appears phase III. The triangular lattice of the graphite surface may be regarded as consisting of three sublattices; each lattice point of each sublattice is surrounded by six nearest neighbors which belong to the other two sublattices. Phase III is characterized by the spin ordering, in which spins are antiparallel to the z axis on one of three sublattices and parallel on the other two. For this state of perfect positional ordering, $x_g = 1/3$. It has lower translational symmetry than the basic graphite lattice and therefore may be called a 2D solid, as is usually done in the lattice model of bulk ^{4}He.[7]

The heat capacity peak experimentally observed around $x_g = 1/3$ seems to indicate the transition to phase III. Such a transition has already been studied by Campbell and Schick[8] with use of the Ising model ($K = 0$) and the Bethe–Peierls approximation. When the tunneling term is included and the fluctuation from the perfect order is taken into consideration, say, by the spin wave theory, spins are not quite parallel or antiparallel to the z axis even at the absolute zero of temperature. This zero-point precession means that there exist some defects in our 2D solid even at $T = 0$, so that x_g can differ from $1/3$. However, it is an open question whether phase III including the zero-point defects may be identified with the 2D solid ^{4}He observed by Dash et al. for $x_g > 1/3$.

The quasiparticle spectrum obtained by the spin wave theory for phase III has the form const $+ \alpha k.$[2] This quasiparticle is the defect tunneling through our 2D solid and may be called a defecton. The 2D Debye phonon observed by Dash et al. for $x_g > 1/3$ is definitely outside the scope of our tight binding approximation, which only includes the lowest (Einstein) vibrational state of the atom.

Acknowledgment

The author wishes to thank Professor John Bardeen for his kind hospitality and illuminating discussion.

References

1. M. Bretz and J. C. Dash, *Phys. Rev. Lett.* **26**, 963 (1971); **27**, 647 (1971).
2. T. Matsubara and H. Matsuda, *Prog. Theor. Phys.* **16**, 569 (1956).
3. P. C. Hohenberg, *Phys. Rev.* **158**, 383 (1967).
4. C. E. Campbell, J. C. Dash, and M. Schick, *Phys. Rev. Lett.* **26**, 966 (1971).
5. P. M. Richards, *Phys. Rev. Lett.* **27**, 1800 (1971).
6. J. Kondo and K. Yamaji, *Prog. Theor. Phys.* **47**, 807 (1972).
7. H. Matsuda and T. Tsuneto, *Prog. Theor. Phys. (Suppl.)* **46**, 861 (1970); W. Mullin, *Phys. Rev.* **A4**, 1247 (1971).
8. C. E. Campbell and M. Schick, *Phys. Rev.* **A5**, 1919 (1972).
9. J. C. Dash, this volume.

Thermal Properties of the Second Layer of Adsorbed ³He*

A.L. Thomson, D.F. Brewer and J.Stanford

Physics Laboratory, University of Sussex
Falmer, Brighton, Sussex, England

Measurements were made of the specific heats of ³He films with thicknesses between one and two atomic layers. The adsorbent was Vycor porous glass on which previous experiments[1] have shown that the first adsorbed layer possesses a specific heat proportional to T^2, which is indicative of a two-dimensional solid. The measurements presented here were obtained with five different fractions of the second layer ranging from one-fourth up to the completed layer; the results are shown in Fig. 1. No evaporation corrections have been made on these data. As can be seen, the specific heats of successively higher coverages attain the same values at progressively lower

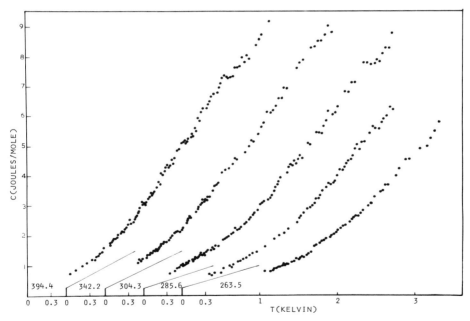

Fig. 1. The specific heat C of various coverages (in cm³ STP) between one and two layers of ³He films as a function of temperature.

* The work was carried out with the support of the Science Research Council (Grant Nos. 3/9/1832 and B/SR/4261, and a Research Studentship), and the U.S. Air Force Office of Scientific Research through the European Office of Aerospace Research under Grant EOOAR 68-0023.

temperatures, and this most probably is due to the second layer possessing a larger specific heat than the first.

We assume that the specific heats of the two layers can be considered separately since the tightly bound first layer is probably unaffected by the presence of a helium layer above it. The total specific heat will therefore be analyzed by the simple addition of the thermal capacities of two separate layers. Previous determinations [1] have shown that the specific heat of the first layer is given by $(0.3 \pm 0.02) T^2$ J mole^{-1} deg^{-1}, and hence the specific heat of the second layer can be determined. The results of this procedure are shown in Fig. 2. The temperature at which an evaporation correction becomes necessary, i.e., when a measurable vapor pressure over the adsorbed sample occurs, is indicated by arrows in Fig. 2. For all of the coverages the specific heat appears to vary linearly with temperature, and a best straight line has been drawn through the points.

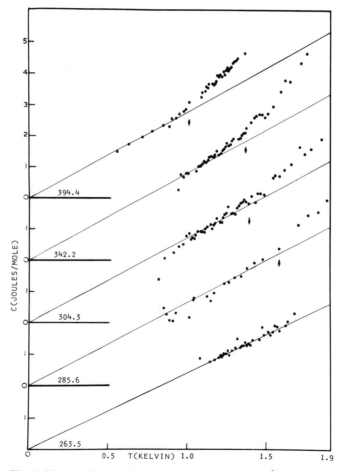

Fig. 2. The specific heat C_2 of various coverages (in cm^3 STP) of the atoms in the second layers in ^{3}He films as a function of temperature.

A linear specific heat would be expected for a degenerate two-dimensional gas, and the slope of our lines can be used to determine m^*, the effective mass of the atoms in the second layer, using the following equation for the specific heat:

$$C/R = (4\pi^3 m^* k/3h^2)\, T\alpha$$

For the complete layer the area per atom[4] has been found to be $\alpha = 13.3\text{Å}^2$. Four of the lines in Fig. 2 are consistent with a value for C of $(2.8 \pm 0.2)\, TJ\, \text{mole}^{-1}\, \text{deg}^{-2}$, while the remaining slope falls outside this range of values by only 0.2. The value obtained for four of the coverages would thus indicate an effective mass $m^* = (0.4 \pm 0.02)m_3$, where m_3 is the mass of a bare ³He atom. This may be compared with the effective mass of ³He atoms in bulk liquid, which falls in the range of $2.8m_3$ to $4.5m_3$ and is dependent upon the applied pressure.[2] Furthermore, a recent determination[3] of the effective mass of a layer of ³He adsorbed on bulk liquid ⁴He gives a value of $(2.07 \pm 0.3)m_3$. The constancy of our specific heat coefficient suggests that the second-layer atoms are condensed into a two-dimensional liquid rather than forming a gas. This is so since one would expect a change in coverage in a gas layer to cause a significant change in m^*.

An alternative way to view our results is to describe the degenerate behavior of the layer in terms of a Fermi temperature. To do this, one must know the density of the atoms in the second layer n_s since the equation for the specific heat in terms of the Fermi temperature T_F is

$$C = \pi^2 N_s k T / 3 T_F$$

and thus a value for T_F of $\sim 10°\text{K}$ is obtained. If the atoms in the second layer condense, both α and T_F would remain approximately constant with coverage change. Alternatively, if the atoms behave as a gas, T_F would be proportional to coverage.

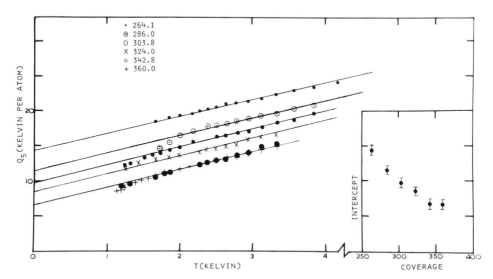

Fig. 3. The isosteric heat of adsorption Q_s as a function of temperature for various coverages (in cm³ STP). The continuous lines have a slope of 5/2. The inset diagram is the extrapolated Q_s value at absolute zero as a function of coverage.

Heats of adsorption were determined at coverages approximately equal to those of the specific heats runs, and this was accomplished by monitoring the sample pressure as a function of temperature. Corrections for evaporation were made to the results in order to obtain values for true isosteric conditions and then the isosteric heat of adsorption was found from

$$Q_s = R \, d(\ln p)/d(1/T)$$

where R is the gas constant. The results are shown in Fig. 3. The straight lines drawn through the points all have a slope of $(5/2) \, T$ since this seemed a reasonable fit to the data.

References

1. D. F. Brewer, A. Evenson, and A. L. Thomson, *J. Low Temp. Phys.* **3,** 603 (1970).
2. A. C. Anderson, G. L. Salinger, W. A. Steyert, and J. C. Wheatley, *Phys. Rev. Lett.* **6,** 331 (1961), and **7,** 295 (1961); D. F. Brewer, J. G. Daunt, and A. K. Sreedhar, *Phys. Rev.* **115,** 836 (1959).
3. H. M. Guo, D. O. Edwards, R. E. Sarwinski, and J. T. Tough, *Phys. Rev. Lett.* **27,** 1259 (1971).
4. A. J. Symonds, thesis, Univ. of Sussex, England.

Nuclear Magnetic Resonance Investigation of ^{3}He Surface States in Adsorbed ^{3}He–^{4}He Films*

D.F. Brewer, D.J. Creswell, and A.L. Thomson

Physics Laboratory, University of Sussex
Falmer, Brighton, Sussex, England

We have made nuclear susceptibility measurements of ^{3}He atoms adsorbed on the free surface of a ^{4}He film. Apart from providing a unique opportunity for studying the magnetic properties of a two-dimensional Fermi system, this experiment is of interest in relation to recent theoretical and experimental investigations of "surface states" in ^{3}He–^{4}He mixtures.[1-3] Specific heat measurements[2] carried out with a Vycor substrate similar to that used in this work have shown that there is a strong dependence on whether or not a free surface is present. With a free surface the specific heat rises almost linearly as expected for a degenerate Fermi gas, while at full pores the limiting low-temperature specific heat is independent of temperature, indicating that the ^{3}He atoms are behaving as a nondegenerate gas. The latest surface tension data[3] have confirmed Andreev's predictions[1] that ^{3}He atoms are bound on the free ^{4}He surface with a binding energy ε_0 equal to 1.8°K and that if the surface concentration is sufficiently dilute, they constitute a two-dimensional gas of quasiparticles with effective mass $m^* = 1.7m_3$, where m_3 is the bare ^{3}He mass. For higher surface densities, however, there appears to be an attractive interaction V_0 between the surface quasiparticles, and Guo et al.[3] estimate that this alters the chemical potential by approximately -0.7°K for a complete layer.

In our experiment an isotopic mixture was adsorbed in porous Vycor glass, the pores of which are on average 60 Å in diameter. The convoluted internal surface of this substrate provides a total surface area of about 100 m^2 within a sample volume of about $\frac{1}{2}$ cm^3, thereby allowing investigation of a macroscopic number of surface atoms. Continuous-wave NMR techniques have been employed, the susceptibility χ being obtained by electronically integrating the area under the absorption line using phase-sensitive detection. A conventional ^{3}He cryostat enabled temperatures within the range 0.3–1.5°K to be conveniently maintained.

In Fig. 1 the data for a mixture containing 9% ^{3}He are plotted as χ/C, C being the Curie constant, as a function of the inverse of temperature T for three coverages ranging from 88 cm^3 STP ($\sim$ 3 layers) up to 150.3 cm^3 STP, which completely fills the Vycor pores. Within the scatter, all the data superimpose remarkably onto a unique curve resembling the degeneracy curve of a Fermi gas[4] or of liquid ^{3}He. In

* The work was carried out with the support of the Science Research Council (Grant Nos. 3/9/1832 and B/SR/4261, and a Research Studentship), and the U.S. Air Force Office of Scientific Research through the European Office of Aerospace Research under Grant EOOAR 68-0023.

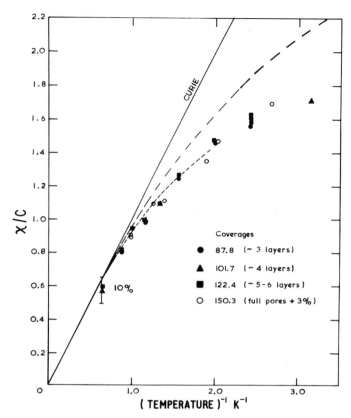

Fig. 1. The nuclear magnetic susceptibility χ of various coverages (in cm³ STP) of a 9 % ³He isotopic mixture plotted versus inverse temperature. The line (— — —) represents the behavior of pure ³He liquid and the line (------) represents the expected value for a bulk mixture.

order to interpret these data, it is necessary to consider the structure of helium films adsorbed on Vycor.[5] It is known that to a first-order approximation the film consists of a tightly bound monolayer (44 cm³, ⁴He) of density equivalent to solid helium at 400 atm; a second layer which is more mobile and has the density of liquid helium at 30 atm; and the remaining fluid with essentially the density of bulk liquid helium; so that the effect of the wall has not become small until the third layer. Furthermore, as a result of its higher atomic mass, a ⁴He atom is more tightly bound to the wall than is a ³He atom and thus in a mixture containing only 9 % ³He, ⁴He atoms are preferentially adsorbed into the first layer certainly and possibly into the second as well, to the exclusion of ³He. The effect has been measured via vapor pressures to be temperature dependent and in a sample with pores full the ³He concentration in the center of the pores decreases as the temperature is raised. In the following analysis we have allowed for this phenomenon by calculating the number of pure ⁴He layers that must be preferentially adsorbed at each temperature to give the measured effective concentration. Although strictly only applicable to the full-pore case, it is felt that this procedure should be sufficiently precise for the lower coverages.

At full pores where there is no free surface the susceptibility appears to obey the behavior expected for a bulk mixture.[7] The lower dashed curve in Fig. 1 was calculated from the expression for an imperfect Fermi gas,

$$C/T_F\chi = C/T_F\chi_{\text{ideal}} - 4k_F a/3\pi$$

where C is the Curie constant, T_F the degeneracy temperature, k_F the Fermi momentum, and χ_{ideal} the susceptibility of an ideal Fermi gas of the appropriate concentration. The scattering length a was taken from the work of Husa et al.[7] The curve shown in Fig. 1 has not been extended into the region where a bulk phase separation occurs but there is satisfactory agreement at high temperatures.

For the three coverages where a free surface exists the fraction of ³He atoms in the surface layer was calculated using the parameters of Guo et al.[3] It was assumed that above the adsorbed pure ⁴He layers the mixture is of uniform density and thus the total free surface area S and the "volume" V accessible to the ³He atoms can be calculated from the known adsorption parameters of the substrate.[6] It is then straightforward to calculate the number of atoms at the surface N_s and the number in the "volume" region N_v by equating the chemical potentials of each phase. The chemical potential μ_s of the surface phase is given by

$$\mu_s = kT\ln\left[\exp(T_F^*/T) - 1\right] - \varepsilon_0 - \tfrac{1}{2}n_s V_0$$

where $n_s = N_s/S$ and the degeneracy temperature T_F^* is

$$T_F^* = h^2 n_s/4\pi m_s^* k$$

and the value of m_s^* is taken to be $1.7m_3$ in accordance with the surface tension data. The volume phase was treated as a three-dimensional gas of quasiparticles of effective mass $2.32m_3$ and the chemical potential μ_v calculated from Stoner's tables.[4] The result of this analysis is that the majority of ³He atoms ($\gtrsim 90\%$) are found to reside on the free surface throughout the experimental temperature range. Since it is only at the highest temperatures that the "volume" region becomes appreciably populated, and because here both phases have an almost Curie susceptibility, a negligible error is incurred in ascribing the measured magnetic degeneracy at any temperature to the surface atoms alone. Using the following expression for the susceptibility of a two-dimensional Fermi gas,

$$\chi = (N_s\mu^2/kT_F^{**})\left[1 - \exp(-T_F^{**}/T)\right]$$

an estimate can be made of the magnetic degeneracy temperature T_F^{**}, the latter being related to T_F^* by $T_F^{**} = T_F^* - X$, where X is a ferromagnetic exchange parameter. From the limiting low-temperature behavior of the data in Fig. 1, T_F^{**} is found to be $0.5 \pm 0.1°$K.

In view of the significant difference in average atomic spacing for the various coverages shown in Fig. 1, it is surprising that the value $T_F^{**} = 0.5°$K seems to be appropriate to all the data. Two interpretations of the constant T_F^{**} have been considered. Either the ³He condenses into puddles of two-dimensional liquid or it spreads uniformly over the surface in a gaslike way. In the former case, the constancy of T_F^{**} is to be expected since the interatomic spacing and therefore the spin-dependent forces will presumably not change. Assuming the spacing to be that of bulk liquid ³He under its saturated vapor pressure the degeneracy temperature

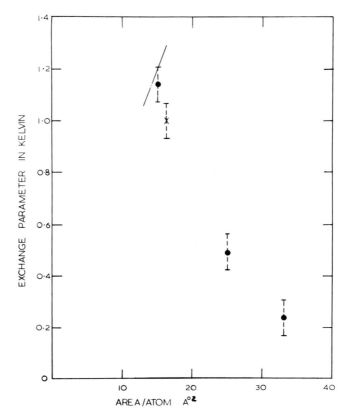

Fig. 2. The exchange parameter X, defined as $T_F^* - T_F^{**}$, as a function of areal density of ^{3}He atoms. The solid line represents the values for pure ^{3}He liquid.

T_F^* (ignoring exchange forces) is found to be 1.6°K, requiring an exchange interaction of 1.1°K, which is very close to the value in bulk liquid. On the other hand, if the surface ^{3}He atoms behave in a gaslike manner, then T_F^* will increase as the amount of surface ^{3}He increases, and the exchange interaction must also vary strongly if T_F^{**} is to be constant. This variation of exchange is shown in Fig. 2 and is compared with the behavior of bulk liquid ^{3}He.

References

1. A. F. Andréev, *Soviet Phys.—JETP* **23**, 939, (1966).
2. D. F. Brewer, A. Evenson, and A. L. Thomson, *Phys. Lett.* **35A**, 307 (1971).
3. H. M. Guo, D. O. Edwards, R. E. Sarwinski, and J. T. Tough, *Phys. Rev. Lett.* **27**, 1259, (1971).
4. J. McDougall and E. C. Stoner, *Phil. Trans. Roy. Soc.* **237**, 67, (1938); E. C. Stoner, *Phil. Mag.* **28**, 257, (1939).
5. D. F. Brewer, *J. Low Temp. Phys.* **3**, 205, (1970).
6. P. J. Reed, Ph.D. Thesis, unpublished.
7. D. L. Husa, D. O. Edwards, and J. R. Gaines, *Phys. Lett.* **21**, 28, (1966).

Nuclear Magnetic Relaxation of Liquid ^{3}He in a Constrained Geometry*

J.F. Kelly and R.C. Richardson

Laboratory of Atomic and Solid State Physics, Cornell University
Ithaca, New York

We have measured the nuclear magnetic relaxation properties of liquid ^{3}He contained within a vessel filled with fine carbon powder. The purpose of the experiments was to gain a better understanding of the surface relaxation mechanisms for ^{3}He atoms at low temperatures in order to: (1) interpret the NMR relaxation rates in thin films, and (2) look for more general manifestations of the apparent magnetic thermal coupling between ^{3}He and paramagnetic salts observed at low temperatures.[1]

The apparatus for the experiment consisted of a pulsed NMR spectrometer[2] and a ^{3}He–^{4}He dilution refrigerator. The specimens were contained within a cylindrical epoxy chamber 3 cm long and 1 cm in diameter. The carbon powder used in the experiments, Carbolac I,† consisted of graphitelike particles with an average diameter of 9×10^{-7} cm. The powder was packed within the coil form so that it occupied approximately 10% of the volume within the cylinder. The relaxation measurements were made over the temperature interval 0.035–4.2°K and in magnetic fields corresponding to ^{3}He Larmor frequencies between 850 kHz and 37 MHz. T_1 was measured using a standard sequence of two 90° pulses, and T_2 using the technique of spin echoes with multiple 180° pulses.

The results of the T_1 measurements using liquid ^{3}He with an isotopic impurity concentration of 0.2% ^{4}He and at a pressure of 0.3 atm are shown in Figs. 1 and 2. Figure 1 shows the temperature dependence of T_1 for ^{3}He contained in the carbon particles using a Larmor frequency of 3 MHz. The relaxation time passes through a maximum near 1°K and decreases linearly with temperature below about 0.3°K. Figure 2 shows the quite striking linear dependence of T_1 upon the Larmor frequency, measured at 0.04°K. The measurements were made using two different specimens of Carbolac I: one with the powder as furnished by the manufacturer, and the second with a specimen of the powder which had been heated in an H_2 oven to 950°C prior to being filled into the sample chamber. Direct measurements of the magnetic susceptibility of the powder show that the powder has a large susceptibility, obeying Curie's Law down to at least 16 m°K. The magnitude of the susceptibility corresponds to about 40 paramagnetic electrons per carbon particle for the material in the form furnished by the manufacturer.[3] ESR measurements on similar powders have shown

---

* Work supported in part by the National Science Foundation through Grant No. GP-29682 and also under Grant No. GH-33637 through the Cornell Materials Science Center, Report No. 1850.

† Source: Cabot Corp., Boston, Mass.

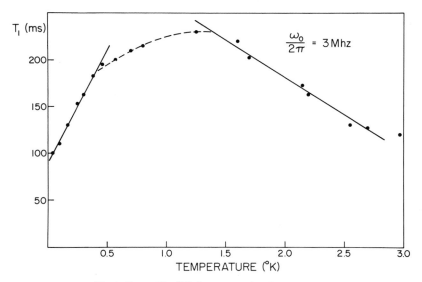

Fig. 1. T_1 vs. T for ^{3}He in untreated carbon sample.

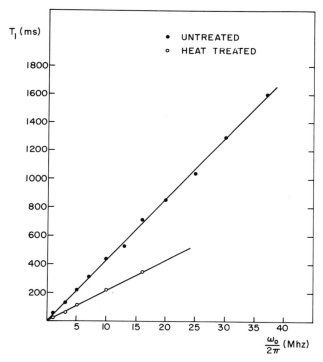

Fig. 2. T_1 vs. ^{3}He Larmor frequency at $T = 0.04°$K.

that the effect of the vacuum heat treatment is to liberate oxygen that is bonded to essentially free electrons of the carbon located at lattice imperfections,[4] thereby increasing the number of electrons which may interact with the ^{3}He atoms at the surface. The increase in T_1 with increasing field has been measured throughout the temperature interval of these experiments, but has been most systematically studied at 0.04°K. The effect of the heat treatment of the powder on T_1 was to change the slope of the field dependence by approximately a factor of two, decreasing the relaxation time in large fields. The fact that the ^{3}He relaxes more rapidly when the number of electronic relaxation sites is increased makes it clear that the magnetic dipolar interaction between the ^{3}He and electrons in the carbon is the dominant mechanism for the recovery of the ^{3}He to thermal equilibrium.

The diffusion coefficient of bulk liquid ^{3}He is sufficiently large (between 10^{-3} and 10^{-5} cm^2 sec^{-1} in this temperature interval) that the maximum time between random collisions of a bulk ^{3}He atom with the surface is less than 10^{-9} sec. The observed relaxation times are more than 10^6 greater than this so that the ^{3}He atoms have time to make a great many collisions with the paramagnetic electrons of the carbon before thermal equilibrium is established. The limiting time in the relaxation process is quite likely that required for the electrons to come to thermal equilibrium with each other, since the energy stored in the nuclear Zeeman levels of the ^{3}He is much greater than the exchange heat capacity of the electrons. In this case the observed relaxation time would be given by $T_1 = \tau_e(k_3 + k_e)/k_e$, where τ_e is the intrinsic time for the "electrons" to equilibrate and k_3 and k_e are the heat capacities of the ^{3}He and "electron" energy states. In the limit $k_3/k_e \gg 1$, and $k_3, k_e \propto (H/T)$,[2] we conclude from Figs. 1 and 2 that $\tau_e \propto HT$ (using this model). The constant k_e depends upon the number of electronic relaxation sites, so that the decrease in T_1 with the liberation of additional sites in the heat treatment process is also consistent with this model.

Measurements of T_2 were made with the same specimens of ^{3}He and Carbolac I, and in addition measurements were made with a 15% solution of ^{4}He in ^{3}He. The ^{4}He is more strongly attracted to walls than ^{3}He and forms a film over the surface. It has been observed in the thermal resistance measurements with paramagnetic salts that the boundary resistance increases sharply with the addition of sufficient ^{4}He to make a complete film over the surface.[1] In this case we used sufficient ^{4}He to give several layers of ^{4}He film. The relaxation rate T_2^{-1} is usually given by the product of the mean square width of the spectral line due to the interaction of the nuclei with the effective "local" field $\langle \gamma H_L \rangle^2$, where γ is the gyromagnetic ratio, and the characteristic time between motional fluctuations of the nuclei τ_n, so that $1/T_2 = \langle \gamma H_L \rangle^2 \tau_n$. In this case the local field is provided by the magnetic dipolar interaction with the "electrons" and depends upon the polarization of the electrons and their relaxation rate. The time τ_n is the average time the ^{3}He atom spends in the environment of the "electron" and should be independent of the applied magnetic field. Because of the short range of the magnetic dipolar interaction, the coating of the carbon particles with ^{4}He had the effect of increasing T_2 by a factor of 30, measured in experiments performed at 5 MHz. Figure 3 shows the measurements of T_2 with various Larmor frequencies as a function of temperature for the untreated specimen of Carbolac I with the 15%

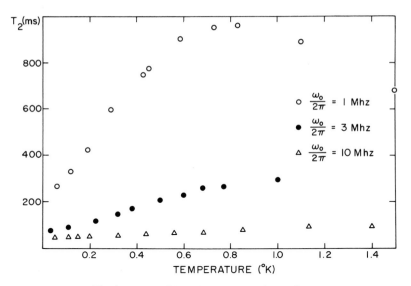

Fig. 3. T_2 vs. T with a mixture of 15% ^{4}He in ^{3}He.

^{4}He solution. The data for the two specimens are, however, qualitatively the same and show several important features: (1) At the higher temperatures, near 0.5°K, T_2 has a very large dependence on magnetic field; and (2) In the larger magnetic fields, T_2 is essentially independent of temperature. In terms of our crude model we can use the high-field data to conclude that τ_n is also independent of temperature and that the film thickness does not change with temperature. The large change in T_2 as we go to lower fields, including a large temperature dependence in lower fields, is a measure of the variation of $\langle \gamma H_L \rangle^2$ with applied field and temperature. The data may be convincingly represented by a single function if the values of T_2 are plotted versus f_0^2/T, where f_0 is the Larmor frequency. Apparently $\langle \gamma H_L \rangle^2$ approaches a saturation value at sufficiently low temperatures or in sufficiently large fields. We have no theory to predict this functional behavior; however, both the polarization of the electrons and the relaxation rate of the electrons depend upon the applied field and temperature.

In order to develop a consistent theoretical model to account for these results, the work should be extended to lower temperatures and to other surface systems. Preliminary measurements on ^{3}He within the pores of Vycor glass[5] and measurements in zeolite[6] have qualitative features, such as the linear dependence of T_1 upon applied magnetic field, that are the same as are found in these studies. We conclude, then, that for a large class of experiments measuring properties of ^{3}He in such restricted geometries and in thin films the interaction with paramagnetic electronic sites is the dominant relaxation mechanism. In order to determine the characteristic frequency of the ^{3}He motion in these environments, a detailed theory of the mechanism is necessary.

References

1. W. C. Black, A. C. Mota, J. C. Wheatley, J. H. Bishop, and P. M. Brewster, *J. Low Temp. Phys.* **4**, 391 (1971).
2. A. S. Greenberg, Thesis, Cornell University, 1972, unpublished.
3. R. A. Buhrman, J. F. Kelly, W. P. Halperin, and R. C. Richardson, to be published.
4. H. Harker, C. Jackson, and W. F. K. Wynne-Jones, *Proc. Roy. Soc. London* **A262**, 328 (1961).
5. A. Zabell, Cornell University, private communication.
6. P. Monod and J. A. Cowen, Service de Physique du Solide et de Resonance Magnétique, Centre d'Etudes Nucléaires de Saclay, Technical Report, 1967, unpublished.

Low-Temperature Specific Heat of ^{4}He Films in Restricted Geometries*

R.H. Tait,† R.O. Pohl, and J.D. Reppy

Laboratory of Atomic and Solid State Physics, Cornell University
Ithaca, New York

Introduction

Confinement of superfluid helium to a system of random small (~ 100Å) pores markedly affects its properties, as was demonstrated by Pobell *et al.*[1]; they observed the oscillations of superfluid helium in a U tube filled with tightly packed lampblack or porous Vycor glass and noted a large increase in the normal fluid density over that of the bulk. At low temperatures they saw a linear dependence of ρ_n on T in contrast to the bulk T^4 dependence derived from other experiments. Padmore[2] proposed a model for long-wavelength phonons in a "zero-dimensional geometry" that gave a linear temperature dependence and a constant heat capacity at low temperatures given by

$$C = (V/d^3)\, k_{\mathrm{B}}$$

where V is the total volume of the He II in the porous system, d is the average pore diameter, and k_{B} is Boltzmann's constant. Heat capacity measurements on helium in disordered, restricted geometries have been made by other workers, but they have only been made at temperatures greater than $0.5°$K, which is too high to test Padmore's prediction. These measurements have also been done on partially filled pores and have pointed out interesting properties of helium films under these conditions.

The present heat capacity work was carried out both to evaluate the validity of Padmore's model and to extend to lower temperatures ($0.06–0.6°$K) the ^{4}He film studies.

Experimental Method

Porous Vycor glass was chosen as a substrate because of its small pore size and because of the large body of work[3] on the same substrate at higher temperatures. The surface area and open volume of our Vycor were determined from a BET nitrogen adsorption isotherm and found to be 120 m^2 g^{-1} and 0.19 cm^3 g^{-1}, res-

---
* Work supported in part by the National Science Foundation through Grant No. GP-27736 and also under Grant No. GH-33637 through the Cornell Materials Science Center, Report No. 1861. This work also received support from the U.S. Atomic Energy Commission under contract AT (11-1)-3151, Technical Report No. COO-3151-10.
† NSF Graduate Fellow.

pectively. If one assumes a highly idealized model—namely that the open volume in the glass is arranged in uniform, nonintersecting cylindrical pores—then the radius of the pores is 32 Å.

The 4.79 g of Vycor used in this experiment was contained in a thin-walled copper vessel and attached to it at one end both thermally and mechanically by Stycast 2850 GT epoxy. The copper container was supported by nylon pins and thermally grounded to the dilution refrigerator by 8 cm of two-mil copper wire. The ^{4}He was introduced into the sample chamber by a thin Cu–Ni capillary.

The total heat capacity was measured by passing a current pulse through a 1000-ohm nichrome wire heater wound around the copper vessel and observing the resulting change in the sample temperature with a calibrated carbon resistance thermometer attached to the copper chamber.

Results

The heat capacity of each of five coverages from one-half layer to filled pores (where layer capacities were determined using the statistical layer model of the Sussex group[3]) was measured from 0.06 to 1.0°K. The results are given in Fig. 1, where it can be seen that the agreement between the present work and that of the Sussex group is good, with the exception of the filled-pores measurement, for which it is possible that some helium was lost due to a leak in the sample chamber. The present data agree more closely with the Sussex group's data for 8/10-filled pores.

The first layer of adsorbed ^{4}He behaved like a two-dimensional solid in that its specific heat varied as T^2. In the temperature range 0.06–0.6°K the heat capacity data for one-half layer and one layer can be characterized by the two-dimensional Debye temperatures 20 and 17°K, respectively. Note, however, that in the one-layer data a deviation from the T^2 dependence is observed at 0.6°K. This is not understood at present.

The most striking feature of our data is the large increase in the heat capacity upon increasing the coverage from one layer to $1\frac{1}{2}$ layers and the subsequent large decrease in heat capacity upon completion of the second layer.

If we assume that the heat capacity of the first layer remained unaffected by the addition of more ^{4}He, then the heat capacity of the helium above the first layer for this $1\frac{1}{2}$-layer coverage had a roughly linear temperature dependence down to 0.06°K. This is shown in Fig. 2, where we have plotted the specific heat C_v. We have $C_v = 1.4T$ J mole^{-1} °K^{-1}. A similar temperature dependence has been observed by Stewart and Dash[4] and McCormick et al.[5] for helium on argon-plated copper for coverages of less than one monolayer although these measurements were made only above 0.5°K.

The large heat capacity of the incomplete second layer indicates that these atoms were significantly more mobile than those atoms in the first layer. A detailed picture of this behavior, however, does not exist at present.

By adding sufficient helium to give a coverage of about $2\frac{1}{2}$ layers, we arrived at half-filled pores. For this case the heat capacity was found to be lower than the $1\frac{1}{2}$-layer result. For ^{4}He coverages corresponding to more than half-filled pores, persistent currents with onset temperatures greater than 0.6°K have been observed,[6] indicating that a superfluid transition has taken place and that the ^{4}He above the

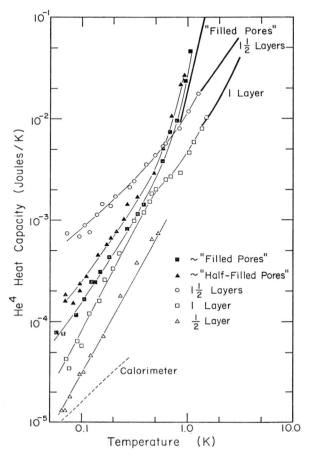

Fig. 1. Heat capacity data for the five coverages studied. The
heavy lines are the results obtained by the Sussex group.

first layer is in an ordered superfluid state. The drop in entropy in going to a more
ordered state explains the sharp drop in heat capacity upon completion of the second
layer.

The temperature dependence (roughly $T^{3/2}$) and the magnitude of the heat
capacity of the half-filled pores corrected for the first layer heat capacity is in fair
agreement with that expected for surface contributions due to the temperature
variation of surface tension and surface "ripplon" modes.[7] The fact that adding
additional helium to fill the pores further lowered the heat capacity strengthens the
argument for a free surface contribution.

The heat capacity of the filled pores exceeded that of bulk He II, and even after
subtraction of the first layer heat capacity the filled-pore specific heat (plotted in
Fig. 3) exceeded by far that of bulk superfluid at the same temperature. The magnitude
of the observed heat capacity is comparable to that expected from the Padmore

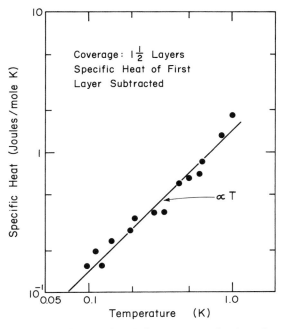

Fig. 2. Specific heat of the helium above the first layer for a total helium coverage of $1\frac{1}{2}$ layers.

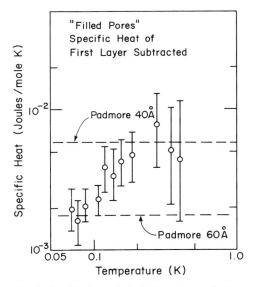

Fig. 3. Specific heat of the helium above the first layer for "filled pores." The Padmore model predictions for two pore sizes are shown as dashed lines.

model. However, the uncertainty in the amount of ^{4}He in the Vycor for this coverage, the disagreement at higher temperatures with earlier work, the large scatter in the data, and the fact that the result is the small difference between two large numbers, all combine to make any other firm conclusions premature. Further work is planned to correct these deficiencies.

References

1. F. D. Pobell, H. W. Chan, L. R. Corruccini, R. P. Henkel, S. W. Schwenterly, and J. D. Reppy, *Phys. Rev. Lett.* **28**, 542 (1972).
2. T. C. Padmore, *Phys. Rev. Lett.* **28**, 1512 (1972).
3. A. J. Symonds, unpublished thesis, Univ. of Sussex, 1965; A. E. Evenson, unpublished thesis, Univ. of Sussex, 1968; D. F. Brewer, *J. Low Temp. Phys.* **3**, 205 (1970).
4. G. A. Stewart and J. G. Dash, *Phys. Rev.* **A2**, 918 (1970).
5. W. D. McCormick, D. L. Goodstein, and J. G. Dash, *Phys. Rev.* **168**, 249 (1968).
6. H. W. Chan, A. W. Yanof, F. D. M. Pobell, and J. D. Reppy, this volume.
7. D. F. Brewer, in *Proc. 10th Intern. Conf. Low Temp. Phys., 1966*, VINITI, Moscow, (1967), Vol. 1, p. 145.

Mean Free Path Effects in ^{3}He Quasiparticles: Measurement of the Spin Diffusion Coefficient in the Collisionless Regime by a Pulsed Gradient NMR Technique*

D.F. Brewer and J.S. Rolt

School of Mathematical and Physical Sciences, University of Sussex
Falmer, Brighton, Sussex, England

The low-temperature properties of bulk liquid ^{3}He are broadly consistent with Landau quasiparticle theory although they do not, in a precise sense, verify its correctness. The temperature dependence in the limit $T \to 0$ is that predicted for any Fermi system, and the magnitude of the coefficients cannot be calculated explicitly. Moreover, accurate comparison of the temperature variations below 50 m°K, where it is most significant, is made doubtful by uncertainties in the temperature scale. We report here the results of an experiment of a different sort, namely the direct observation of mean free path effects in the quasiparticles. The present measurements are of the spin diffusion coefficient of ^{3}He confined to narrow channels at temperatures low enough for the quasiparticles to be in the collisionless regime. Our results are in agreement with modified Landau theory which does not involve the usual temperature-dependent mean free path.[1] In these experiments the bulk and the mean-free-path-limited diffusion coefficients can be measured virtually coextensively and simultaneously, which greatly facilitates their direct comparison. A similar experiment on heat conduction has already been reported.[2]

The experiments were performed in a dilution refrigerator. The sample chamber was made of Epibond 100 A and contained ten Vycor disks (mean pore diameter 71 Å) of 7 mm diameter and 1.2 mm thickness. The steady resonance field was 500 G and a crossed coil technique was used, the receiver coil being embedded in the sample chamber and the transmitter coil being located outside the glass vacuum jacket.[3]

The bulk liquid ^{3}He, outside the Vycor pores but inside the experimental cell, provided a convenient method for obtaining the temperature. The NMR technique described below allowed the diffusion coefficient of this ^{3}He to be measured separately and in a short time so that the root mean square diffusion distance $(\overline{x^2})^{1/2} = (2Dt)^{1/2}$ remained less than the expected gap between the Vycor disks even at the lowest temperatures. The temperature was then obtained from the published values[4] of D. The bulk diffusion coefficients were confirmed by measurements made independently in another bulk sample chamber containing no Vycor disks located below the main cell and thermally connected to it by a column of ^{3}He.[3]

* This work was supported by the Science Research Council and the U.S. Office of Aerospace Research (European Office, Grant EOOAR 68-0023).

A simple $90°-\tau-180°$ pulse sequence was used to determine the characteristic spin–spin relaxation times for the system, with τ varying from 1 to 40 msec. The resulting curve of echo amplitude versus time was found to be the sum of two exponential terms with characteristic decay times T_2 of about 2 and 80 msec. The amplitude of the latter component was about 5% of the total maximum signal and was associated with the liquid ^{3}He which surrounded the Vycor disks (Fig. 1).

The diffusion coefficient was measured at fixed τ using the pulsed gradient technique. The gradient G was variable between 12 and 250 G cm^{-1} with a constant gradient pulse rise and fall time over the entire range. In addition, a continuous gradient of about 2 G cm^{-1} was needed to narrow the echoes and to relax to some extent the stringent requirements on gradient pulse equality.

The full pulse sequence for diffusion coefficient measurements is shown in Fig. 2, which also serves to define the parameters. The echo height is the sum of the signals from inside and outside the Vycor pores, i.e.,

$$A_T(t) = A_V(0)\exp(-t/T_{2V})\exp(-D'f) + A_B(0)\exp(-t/T_{2B})\exp(-Df)$$
$$= A_V(t)\exp(-D'f) + A_B(t)\exp(-Df)$$

where A_T is the amplitude of the total signal, A_V and A_B are the amplitudes of the signals from the ^{3}He inside and outside the pores, T_{2V} and T_{2B} the corresponding transverse relaxation times, D is the bulk diffusion coefficient of ^{3}He, and D' is its measured mean-free-path-limited value, and f is a complicated function of the quantities G, G_0, δ, and Δ, which are shown in Fig. 2. Initially D, and hence the temperature, was measured by setting τ to about 5 msec so that most of the signal

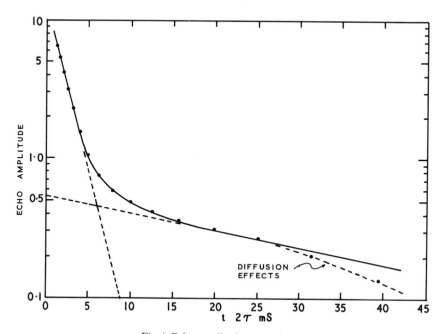

Fig. 1. Echo amplitude versus time.

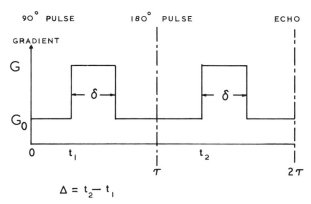

Fig. 2. Pulse sequence for the pulsed gradient experiment.

originated from the bulk ^{3}He. The small correction ($\sim 5\%$) necessary to take into account the first term in Eq. (1) was almost constant because $D' \ll D$. The value of τ used to measure D' needs to be chosen with some care; a suitable compromise between various factors leads to $\tau \sim 2$ msec and at the time of the echo $A_V \approx A_B$.

The results are shown in Fig. 3, where the quantity D'' ($\propto D'$) has been plotted against $D^{-1/2}$. The temperatures shown were calculated from D using the bulk diffusion coefficients of Anderson et al.[4]

Before discussing the results in terms of Fermi liquid theory, it is necessary to consider the geometric structure of the system. Ideally the collisionless regime should be examined using a system of uniform cylindrical pores in which no barriers to transport exist along the axes. Porous Vycor glass certainly does not fulfill this requirement but it is a suitable system to study because of its high porosity (~ 0.3) and hence large signal-to-noise ratio. Gas diffusion experiments such as those in Ref. 6 have shown that mass transport in Vycor is partly restricted because of tortuous paths and blind pores.

However, it is possible to show from our results by use of a detailed argument from the theory of diffusion under bounded conditions[7] that any insurmountable barriers that exist in Vycor must be separated by distances well in excess of 10^{-4} cm, i.e., considerably greater than the diffusion distances. We do, however, expect, because of the tortuous paths, the diffusion coefficient will be reduced by a "tortuosity factor" X, and we write, as usual, the measured diffusion coefficient $D' = X^2 D''$, where D'' is the corrected diffusion coefficient. We can obtain a measure of X by noting that at the higher temperatures diffusion should not be mean free path limited, and hence the corrected diffusion coefficient D'' should approach the bulk value. Hence we obtain $X \approx 0.3$. Possible errors in X do not effect our conclusions.

The data in Fig. 3 clearly show that D'' does not increase rapidly as the temperature is decreased, but reaches the constant value $24 \pm 1 \times 10^{-5}$ cm^2 sec^{-1}, in strong contrast to the bulk diffusion coefficient. The broken line represents the theoretical curve obtained from Rice's theory[1] for channels of diameter 70 Å: agreement is fairly good, although the measured diffusion coefficient levels off at a higher temperature than predicted. According to Rice, the spin diffusion coefficient

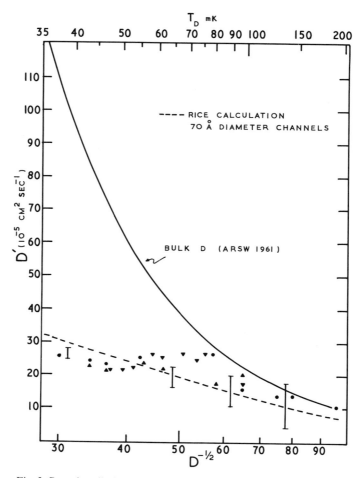

Fig. 3. Boundary-limited and bulk diffusion coefficients, compared with
Rice's theory.[1]

in the quasiparticle collisionless regime is given by

$$D'_{qp} = \tfrac{1}{3}(1 + \tfrac{1}{4}Z_0)v_0\lambda_B\left(\frac{1+v}{1-v}\right) \tag{2}$$

where λ_B is an effective mean free path, Z_0 is the usual Landau parameter, v_0 is the mean quasiparticle velocity, and v, the specular reflection coefficient, is the fraction of quasiparticles incident on the wall that is scattered specularly. We initially make the reasonable assumption of complete accommodation at the wall, i.e., $v = 0$, and this seems to be supported by the excellent agreement between pure ^{3}He viscosities as measured with an oscillating quartz crystal[8] and by the attenuation of sound. We make the further approximation that $\lambda_B = d$, the mean channel diameter, which is true for a cylindrical channel. Hence

$$D'_{qp} = \tfrac{1}{3}(1 + \tfrac{1}{4}Z_0)p_F\,d/m^* \tag{3}$$

where we have put $v_0 = p_F/m^*$, in which the Fermi momentum p_F is independent of the effective mass m^* of the diffusive carriers, and is a function only of number density. Using experimental bulk values[10] for Z_0 and m^*, we find $D' = 4.2 \times 10^{-4}$ cm^2 sec^{-1}, in good agreement with experiment. Note that for *bare* particle diffusion, according to simple kinetic theory,

$$D'_b = \tfrac{1}{3} p_F \, d/m_3 \approx 3.8 \times 10^{-3} \quad \text{cm}^2 \, \text{sec}^{-1} \tag{4}$$

(where m_3 is the bare ^{3}He atomic mass), which is an order of magnitude larger. We consider this to be well outside the accumulated experimental uncertainties, and thus to constitute strong evidence for the validity of the Landau quasiparticle concept. Any value of v greater than zero would increase the discrepancy between experiment for both bare and quasiparticle diffusion, and these measurements therefore support both the quasiparticle picture and a specular reflection coefficient of zero.

In order to discuss these results more fully, it is necessary to consider the effect of the first tightly bound adsorbed ^{3}He layer. A simple noninteracting layer model is known to describe the susceptibility down to 0.4K,[11] but not to lower temperatures.[3] The interlayer interactions are expected to be complex, and further analysis must await a better theoretical model for adsorbed ^{3}He. It is not likely, however, to affect the conclusions drawn above.

References

1. M. J. Rice, *Phys. Rev.* **165**, 288 (1968).
2. D. S. Betts, D. F. Brewer, and R. S. Hamilton, in *Proc. 12th Intern. Conf. Low Temp. Phys.*, 1970, Academic Press of Japan (1971), p. 155.
3. D. F. Brewer and J. S. Rolt, *Phys. Rev. Lett.*: submitted also this volume.
4. A. C. Anderson, W. Reese, R. J. Sarwinski, and J. C. Wheatley, *Phys. Rev. Lett.* **7**, 220 (1961).
5. E. O. Stejskal and J. E. Tanner, *J. Chem. Phys.* **42**, 288 (1964); E. O. Stejskal, *J. Chem. Phys.* **43**, 3597 (1965).
6. D. Basmadjian and K. P. Chu, *Can. J. Chem.* **42**, 946 (1964).
7. B. Robertson, *Phys. Rev.* **151**, 273 (1966); R. C. Wayne and R. M. Cotts, *Phys. Rev.* **151**, 264 (1966); J. E. Tanner and E. O. Stejskal, *J. Chem. Phys.* **49**, 1768 (1968).
8. M. P. Bertinat, D. S. Betts, D. F. Brewer, and G. J. Butterworth, *Phys. Rev. Lett.* **28**, 472 (1972).
9. W. R. Abel, A. C. Anderson, and J. C. Wheatley, *Phys. Rev. Lett.* **17**, 74 (1966); B. M. Abraham, D. Chung, Y. Eckstein, J. B. Ketterson, and P. R. Roach, *J. Low Temp. Phys.* **6**, 521 (1972).
10. J. C. Wheatley, *Quantum Fluids*, D. F. Brewer, ed., North-Holland, Amsterdam (1966), p. 206.
11. D.F. Brewer, D.J. Creswell, and A.L. Thomson, in *Proc. 12th Intern. Conf. Low Temp. Phys.*, 1970, Academic Press of Japan, Tokyo (1971), p. 157.

Adsorption of ^{4}He on Bare and on Argon-Coated Exfoliated Graphite at Low Temperatures*

E. Lerner and J.G. Daunt

Cryogenics Center, Stevens Institute of Technology
Hoboken, New Jersey

Measurements have been made of adsorption isotherms of ^{4}He on bare, exfoliated graphite (Grafoil[†1]) and on the same substrate coated with a completed monolayer of argon. The isotherms were measured at 4.2, 6.18, 7.90, 9.65, 11.60, 13.50, 15.08, and 18.55°K in the pressure range 0.25–25 Torr using an apparatus which has been described previously. Thermomolecular pressure corrections were less than 3% even at the lowest temperatures and the accuracy of pressure measurement was within 0.5%. Details of the method of temperature and pressure measurement and of the corrections used in the calculations of the mass adsorbed have been given previously.[2,3]

The Grafoil adsorbent was in the form of a ribbon 0.013 cm thick, 1.9 cm wide, mass 33.81 g wound around a copper post and enclosed in a cylindrical copper vessel. The surface area of the adsorbent was determined from adsorption isotherm measurements of both argon and nitrogen at 77.3°K and by following computational procedures already outlined.[4] The argon data at 77.3°K yielded a surface area of 19.9 m^2/g^{-1}, or a total surface area for the adsorbent of 672 m^2.

The isotherms were found to be very reproducible and reversible. A typical family of isotherms for ^{4}He on the Grafoil coated with a monolayer of argon is given in Fig. 1. The general shape of this family of isotherms,[‡] particularly those at 7.90, 6.18, and 4.2°K, suggests that at pressures below about 0.5 Torr the system becomes a two-phase system with a critical temperature between 6.18 and 7.9°K.

From these data and from the isotherms on the bare Grafoil we computed the isosteric heats of adsorption Q_{st} as a function of the coverage. Figure 2 shows the results of this computation.

It will be seen from Fig. 2 that ^{4}He adsorption on the argon-coated Grafoil yields Q_{st} values much lower than the values on the bare Grafoil at all coverages. Moreover, the relative independence of Q_{st} of coverage at low ^{4}He coverages on the argon-coated Grafoil indicates a relatively high degree of homogeneity of this surface. On the bare Grafoil the variations of Q_{st} with coverage at low coverage

* Supported by contracts with the Office of Naval Research and Department of Defense (Themis Program) and by a grant from The National Science Foundation.
† Manufactured by Union Carbide Products Division, New York.
‡ See for comparison the isotherms for the two-phase submonolayer adsorbed system of krypton on NaBr reported by Fisher and McMillan.[5]

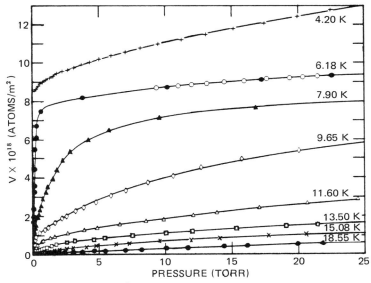

Fig. 1. Adsorption isotherms of ⁴He on monolayer of argon on Grafoil at temperatures as marked. The open points on the 6.18°K curve were taken for decreasing pressures.

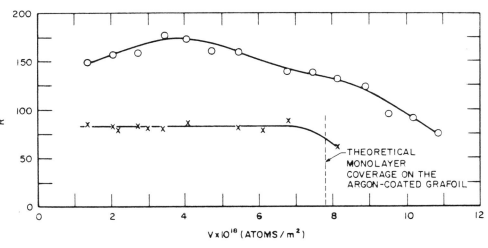

Fig. 2. The isosteric heat of adsorption Q_{st}/R as a function of coverage. (Circles) ⁴He on bare Grafoil; (crosses) ⁴He on monolayer of argon on Grafoil.

are quite marked and, unfortunately, do not allow an unambiguous extrapolation of Q_{st} to zero coverage.

In Fig. 2 the completed monolayer coverage 7.9×10^{18} atoms m⁻² on argon-coated Grafoil is shown by the vertical broken line and it appears that at this value of coverage Q_{st} is decreasing in value significantly. It is of interest to compare these data with those obtained previously, either theoretically or experimentally. Hagen *et al.*[6] theoretically estimated the binding energy E of ^{4}He on bare graphite at zero coverage to be 189°K. Allowing for the differences between E and Q_{st}, this value is somewhat larger than our results would indicate but not excessively so. Novaco and Milford[7] calculated E for ^{4}He on graphite coated with a monolayer of argon at zero coverage and found $E = 71$°K. This, as Novaco and Milford[7] pointed out, is essentially the same as their previous calculations[8] for ^{4}He on copper coated with a monolayer of argon, indicating that the material below the argon coating plays only a minor role. Experimentally, we found[4] that Q_{st} (for $T \simeq 10$°K) extrapolated to zero coverage for ^{4}He on argon-coated copper was 76°K, which is to be compared with the value $Q_{st} \simeq 82$°K for ^{4}He on the argon-coated Grafoil found in this paper. Clearly all these results are consistent with each other and lead to a firmer understanding of our knowledge of ^{4}He adsorption at low temperatures.

Acknowledgment

We acknowledge gratefully the help given by Mr. W.H.P. Van Iperen in this work.

References

1. M. Bretz and J. G. Dash, *Phys. Rev. Lett.* **26**, 963 (1971).
2. J. G. Daunt and C. Z. Rosen, *J. Low Temp. Phys.* **3**, 89 (1970).
3. E. Lerner and J. G. Daunt, *J. Low Temp. Phys.* **6**, 241 (1972).
4. J. G. Daunt and E. Lerner, *J. Low Temp. Phys.* (in press).
5. B. B. Fisher and W. G. McMillan, *J. Chem. Phys.* **28**, 549 (1958).
6. D. E. Hagen, A. D. Novaco, and F. J. Milford, in *Proc. II Intern. Symp. on Adsorption-Desorption Phenomena, Florence,* Academic, New York, (1971).
7. A. D. Novaco and F. J. Milford, *Phys. Rev.* **A5**, 783 (1972).
8. A. D. Novaco and F. J. Milford, *Phys. Rev.* **A4**, 1136 (1971).

Ellipsometric Measurements of the Saturated Helium Film[*]

C.C. Matheson and J.L. Horn

School of Mathematics and Physics, University of East Anglia
Norwich, England

The Ellipsometric Technique

The ellipsometric technique was first used to measure helium film thicknesses by Burge and Jackson[1] in 1951, and ellipsometric measurements of the film have been reported sporadically over the past two decades.

The method is based on measuring the change in state of polarization of light reflected from a substrate on which the helium film forms. In general, the state of polarization of a polarized beam is completely specified by two parameters, the phase difference between the p and s components, and their relative amplitudes. Reflection results in a change in these two parameters. This change may be expressed in terms of the ratio of the appropriate Fresnel reflection coefficients. For a clean, film-free surface this ratio may be written

$$r^p/r^s = \tan \bar{\psi} \exp(i\bar{\Delta}) \tag{1}$$

where $\bar{\Delta}$ is the phase change and $\bar{\psi}$ is the relative amplitude change produced by reflection. If, however, the surface is covered by a film of thickness d, the ratio becomes

$$R^p/R^s = \tan \psi \exp(i\Delta) \tag{2}$$

with

$$R^{p,s} = (r_F^{p,s} + r_M^{p,s} e^{-2i\delta})/(1 + r_F^{p,s} r_M^{p,s} e^{-2i\delta})$$

where $r_F^{p,s}$ and $r_M^{p,s}$ are the reflection coefficients for the film and substrate interfaces, respectively, and $\delta = (2\pi/\lambda)\,d(n_F^2 - n_V^2 \sin^2\phi)^{1/2}$ with n_F and n_V the refractive indices of the film and the vapor, respectively, and ϕ the angle of incidence of the light of wavelength λ.

It is clear that measurements of $(\bar{\Delta}, \bar{\psi})$ and (Δ, ψ) yield information about the optical properties of both bare and film-covered surfaces. An ellipsometer measures these parameters.

Previous ellipsometric measurements of helium films have measured changes in only one of the parameters Δ or ψ (usually Δ) because of the difficulty in making sufficiently rapidly the necessary manual manipulations required to measure both parameters simultaneously. In fact, the values of *both* Δ and ψ are necessary to solve Eq. (2) completely.

A fully automatic ellipsometer has been developed[2] which will track continuously and rapidly changes in both Δ and ψ. Thus we can readily observe formation and

[*] This work was supported by the Science Research Council.

destruction of the helium film. This paper reports some preliminary measurements of the film obtained with this instrument and discusses means of analyzing the data.

The basic optical system of the automatic ellipsometer is conventional. The compensator is fixed. However, the polarizer and analyzer prisms are rotated by servomotors. The polarization states of the incident and reflected beams are separately modulated. The doubly modulated beam incident on a photomultiplier is analyzed by phase-sensitive detection techniques, and the servomotors are driven to obtain minimum intensity. The polarizer and analyzer prisms are mounted in the centers of Moiré fringe circles, the angular positions of which are measured by Moiré fringe counters. The azimuths of the polarizer and analyzer prisms are recorded digitally to a precision of $0.001°$. These azimuths are related by simple geometry to the parameters Δ and ψ that characterize the reflecting surface. A sensitivity of 1 mdeg in Δ and ψ corresponds to a sensitivity to a helium film thickness change of between 0.3 and 4 Å, depending on the thickness. In practice, for a static helium film, measured values of Δ and ψ are constant to within a few millidegrees.

Experimental Details

We have measured the variation in helium film thickness d with height h for h between 2 and 40 mm at different temperatures in the range 1.5–2.05°K.

The experiments were carried out using a metal cryostat with tails designed for optical work. Windows in the helium can and outer vacuum jacket were of strain-birefringence-free glass.

The substrate on which the film was observed was a glass optical flat on which gold had been evaporated. This was mounted vertically in a bath of helium. The height at which the film thickness was measured was altered by raising or lowering the bath level by means of fountain pumps. This height was monitored by means of a capacitance depth gauge.[3]

To obtain the "bare substrate" values, two small lightbulbs were mounted in front of the substrate in order to allow the film to be "burnt off."

Temperature was measured with a germanium resistance thermometer and was maintained constant to within 10 $\mu°$K.

Methods of Analysis

Ellipsometric data on the helium film can be analyzed in a number of ways. Given measured values of $\bar{\Delta}$ and $\bar{\psi}$ appropriate to the bare surface and the values of Δ and ψ obtained when the surface is covered with a film, it is possible[4] to solve Eq. (2) for the film thickness d by writing it in the form

$$Az^2 + Bz + C = 0, \qquad \text{where } z = e^{-2i\delta} \tag{3}$$

A, B, C are dependent on Δ and ψ and on n_V, n_F, ϕ, λ, and the refractive index of the substrate, which is complex. To solve Eq. (3) *exactly* for d, it is necessary to know the refractive index of the film n_F. If a value for n_F is assumed, two values of d are obtained (corresponding to the two solutions of the quadratic equation). These two values will, in general, be complex. Since d must be real, however, any imaginary part is due either to experimental error or an incorrect value of n_F. The value of d

with the smallest imaginary part is taken to be the correct solution. Clearly the magnitude of the imaginary part of d is an indication of the reliability of the calculated value of d.

If the value of n_F is unknown, it is possible to solve Eq. (2) *numerically*[4] to obtain values of both d and n_F. For the case of very thin films, however, there is a relatively large uncertainty in the value of n_F obtained, and hence in the value of d.

If the films are very thin, linear relationships hold between d and Δ and between d and ψ. These relationships are of the form

$$\bar{\Delta} - \Delta = \alpha d \tag{4}$$

$$\psi - \bar{\psi} = \beta d \tag{5}$$

where α and β are constants which depend on n_V, n_F, ϕ, λ, and the refractive index of the substrate. The value of d obtained from Eq. (4) is a good approximation only for films of the order of 100 Å or less. (When d is 100 Å the error in d introduced by taking the linear approximation is 6%. When $d = 400$ Å the error has risen to 25%.) On the other hand, Eq. (5) is a valid approximation for helium films up to about 500 Å, at which thickness the error is 3%. Even at $d = 1000$ Å the error is only 12%. It is, however, possible to amend values of d obtained from Eq. (5) over the thickness range 450–1500 Å by applying a polynomial correction of order three.

Analysis of Results

We have analyzed our results using the different methods outlined in the previous section. Throughout the following, values of n_F have been assumed.[5] Further work is in progress in an attempt to fit both d and n_F to the experimental data.

Figure 1 is a typical plot of the variation of helium film thickness with height.

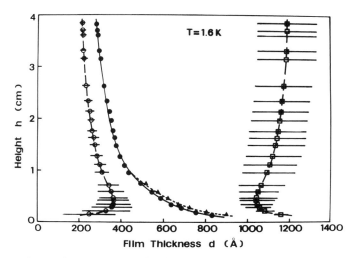

Fig. 1. The dependence of helium film thickness on height. Open circles and squares: the two solutions for d obtained from Eq. (3). Full circles: the solution for d obtained from Eq. (5). Triangles: the polynomial correction.

The variation of d obtained from the two solutions of Eq. (3) is shown. The error bars on these curves represent the magnitudes of the computed imaginary parts of d. The full curve shows the variation of d obtained from Eq. (5). The polynomial correction is also indicated.

It is obvious that the exact solution of Eq. (2) does not yield physically sensible variations of d with height. This may be because we are taking $\bar{\Delta}$ and $\bar{\psi}$ values to be those for a clean substrate when in fact they are probably the values appropriate to a very thin contaminant film overlaying the gold. A theoretical investigation into the effect of such a contaminant film leads to the conclusion that the apparent d value will have its greatest error when d is in the region around 600 Å. A further investigation of the effect of a contaminant layer on the full solution of the ellipsometric equations is being carried out. Meanwhile experimental data have been analyzed with the corrected linear approximation based on Eq. (5), since the effect of a contaminant film on this solution is not as serious.

The variation of helium film thickness with height is generally taken to be of the form

$$d = Kh^{-1/\alpha} \qquad (6)$$

where K is the film thickness at $h = 1$ cm. Fitting the data from the curve in Fig. 1 based on the corrected linear approximation to Eq. (6) gives $K = 445 \pm 5$ and $\alpha = 2.58 \pm 0.01$, where the errors quoted are standard deviations on a least squares fit to the experimental data points.

A similar least squares fit to the data points shown by open circles in Fig. 1 in which the data points were weighted in inverse proportion to the computed imaginary part of d leads to the value of $K = 307 \pm 2$ and $\alpha = 3.81 \pm 0.01$. An apparent increase in α with increasing film thickness might be deduced from this analysis if the data for very small heights which is evidently in error were not included.

Over a series of ten experimental runs at different temperatures values of K varying between 300 and 700 Å have been found. The corresponding values of α vary from 1.5 to 3.9. No systematic variation of K or α with temperature was observed. Values of α vs. K are plotted in Fig. 2.

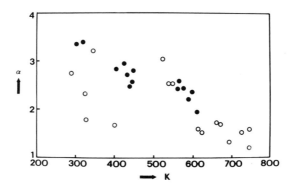

Fig. 2. The variation of α with K. Full circles denote data from a single run. Open circles denote data from eight other runs.

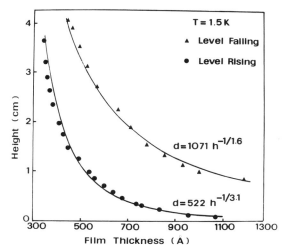

Fig. 3. The variation of film thickness with height before (triangles) and after (circles) destroying the film.

An interesting preliminary result was obtained during one run when d was measured while the bath level was lowered at a rate of about 6 mm min^{-1}. The film was then burnt off and allowed to reform, the bath level then being raised immediately. The observed variation of d with h in these two cases is shown in Fig. 3.

Acknowledgments

One of us (J.L.H.) wishes to acknowledge the award of a University of East Anglia Research Grant. The automatic ellipsometer was built in collaboration with Bellingham and Stanley Ltd.

References

1. E. J. Burge and L. C. Jackson, *Proc. Roy. Soc. A.* **205**, 270 (1951).
2. C. C. Matheson, J. G. Wright, H. Norris, and R. Gundermann, to be published.
3. D. G. Blair and C. C. Matheson, *Cryogenics* **10**, 513 (1970).
4. F. L. McCrackin, NBS Tech. Note 479 (1969).
5. M. H. Edwards, *Can. J. Phys.* **36**, 884 (1958).

Momentum and Energy Transfer Between Helium Vapor and the Film*

D.G. Blair and C.C. Matheson

School of Mathematics and Physics, University of East Anglia
Norwich, England

The torsional oscillator helium film detector described in an accompanying paper[1] gives a number of anomalous results. First, all helium film thicknesses observed were between three and ten times that expected. Second, the contribution of the vapor to the period of oscillation was much less than calculated from published values of helium viscosity. Third, the film formation curve (thickness vs. time) for film flow driven by van der Waals forces showed a sharp peak at a film thickness about 10% of the maximum film thickness.

However, these results are entirely explained if the momentum transfer ratio between helium vapor and the film depends on the film thickness as follows.

The vapor contribution $\Delta\tau$ to the period of oscillation of a conical torsional oscillator of surface area A, oscillating in helium vapor of density ρ, is given by

$$\Delta\tau = \frac{1}{2}\frac{A\rho\tau\delta}{m}\left[\frac{f^2(4\lambda^2 + 4\delta\lambda + \delta^2) - f(16\lambda^2 + 8\lambda\delta) + 16\lambda^2}{f^2(2\lambda^2 + 2\lambda\delta + \delta^2) - f(8\lambda^2 + 4\lambda\delta) + 8\lambda^2}\right]$$

where τ is the period of oscillation, m is the mass of the oscillator, δ is the penetration depth for viscous waves in the vapor, given by $(\eta\tau/\pi\rho)^{1/2}$, f is the momentum transfer ratio between the vapor and the oscillating substrate, and λ is the mean free path for atoms in the vapor. The derivation of this formula follows from a hydrodynamic approach corrected for viscous slip, and involves the valid approximation that the change in period of oscillation $\Delta\tau$ is small.

Looking at the dependence of $\Delta\tau$ on f shown in Fig. 1, we find that for a typical oscillator (frequency 10 Hz) $\Delta\tau$ is roughly independent of f until f falls below 10^{-2}. That is, for a 10-Hz oscillator mean free path effects are negligible until $f \approx 10^{-2}$. For $f < 10^{-2}$, $\Delta\tau$ becomes very strongly dependent on f.

In practice, however, our oscillator is covered by a helium film. In this situation, the value of f is determined by the interaction between the vapor and the helium *film* rather than the conical *substrate. In itself* a film of thickness $d = 300$ Å is expected to contribute a period change $\Delta\tau_{\text{film}}$ of about 70 μsec for a typical oscillator at 2°K. In addition there will be a film-thickness-dependent vapor contribution $\Delta\tau_{\text{vap}}$ provided that (a) $f \approx 10^{-3}$ and (b) f is an increasing function of film thickness. For example, if f were to increase by a factor of two as the film thickness increased from

* The work was supported by the Science Research Council and the University of East Anglia.

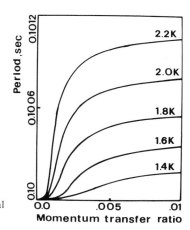

Fig. 1. The dependence of the period of oscillation of a torsional oscillator on the momentum transfer ratio f.

0 to 300 Å, the *total* film contribution to the period of oscillation $\Delta\tau_{tot}$ would be dependent on temperature, varying between 200 and 700 μsec in the temperature range above 1.4°K. On this basis $\Delta\tau_{tot} = \Delta\tau_{film} + \Delta\tau_{vap}$, and $\Delta\tau_{vap}$ may be considered as an additive amplification factor.

The temperature dependence of the film thickness amplified in this way will not vary as ρ_n/ρ as expected for $\Delta\tau_{film}$. However, it will still be a monotonically decreasing function of temperature.

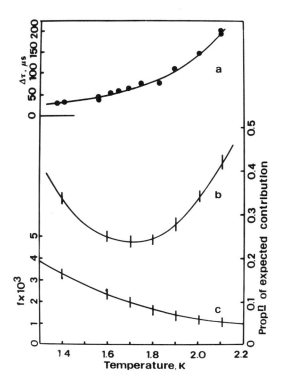

Fig. 2. (a) The observed variation of the period of oscillation $\Delta\tau$ with temperature in the absence of a helium film. (b) The temperature variation of the proportion of the expected vapor contribution calculated from the smoothed data in (a). (c) The temperature variation of f calculated from (b).

By measuring the temperature dependence of $\Delta\tau$ (i.e., the period of oscillation in the absence of a film), it is possible to obtain further information about the momentum transfer ratio between the vapor and the substrate. Infrared radiation (~ 1 mW) from an incandescent lamp was used to "burn off" the helium film. In Fig. 2 (a) $\Delta\tau$ is plotted against temperature. The contribution of the vapor is much smaller than would be expected if f were unity. The proportion of the expected contribution is plotted in Fig. 2 (b) and from it the calculated temperature dependence of f is shown in Fig. 2 (c).

Finally, there remains to be explained the sharp peak in the period of oscillation as the film thickness increases from zero to the equilibrium thickness. Two typical examples are shown in Fig. 3 (a). The peak in each curve could represent a resonance in the momentum transfer ratio. It does not represent an effect associated with film flow since the flow rates required would be too high. Coinciding with each peak there is a sharp increase in damping of the oscillator as might be expected for an increase in f.

Thus the anomalous results obtained with the torsional oscillator can be interpreted in terms of very low values of f.

To obtain further information about the film–vapor interaction, we have examined the behavior of the helium film on the *levitated* substrate using an automatic ellipsometer as described in the accompanying papers.[1,2] In contrast to the dynamic measurements so far described in this paper, measurements made in this way can only yield information about the accommodation coefficient a and cannot be directly related to the momentum transfer ratio f. However, a is always less than or equal to f.

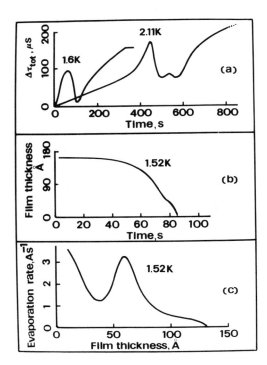

Fig. 3. (a) Film formation at different temperatures on the oscillating substrate showing possible resonant behavior in f. (b) Evaporation of a helium film from an isolated substrate. (c) The dependence of evaporation rate on helium film thickness calculated from (b), showing possible resonant behavior in a.

We have made two types of measurement with this apparatus. One is to observe the behavior of an equilibrium film when the bath level is lowered away from the substrate and flow fiber, leaving the film isolated from the bath. The other is to look at the formation of the film from the vapor on the initially bare substrate. In the first case we find that the film is not stable when it is isolated from the bath. Most of the film evaporates, leaving a small residual layer. Typical evaporation behavior is shown in Fig. 3 (b). There is a delay of from 2 to 4 min before evaporation is observed. This delay is consistent with the presence of a droplet held by surface tension on the end of the flow fiber.

The cause of the film evaporation is probably lack of thermal equilibrium. A significant quantity of ambient radiation enters the cryostat through the windows— we estimate that the substrate absorbs about 6 μW of radiation. The temperature of the substrate is then determined by the rate at which the heat can be conducted away by the vapor. For large values of a, the temperature difference between the substrate and the bath is $\sim 20 \ \mu°K$, but if $a < 10^{-3}$, temperature differences higher than 20 m°K may be reached.

Unexpectedly, the rate of evaporation implied in Fig. 3 (b) appears to be a decreasing function of film thickness. The opposite might be expected, since as the film becomes thinner the van der Waals attraction to the substrate becomes stronger and the latent heat of evaporation increases. However, since $a \lesssim f$, the results from the torsional oscillator lead us to expect that a is an increasing function of film thickness. Now the dynamic equilibrium between the levitated substrate and the bath depends on a in an inverse manner to the dependence of the period of the torsional oscillator on f. The sustained temperature difference ΔT between the substrate and the bath (for constant incident radiation) will increase as a decreases, provided a is sufficiently small for free-molecular effects to be important, and that dynamic equilibrium is attained in a time short compared with the duration of the evaporation. The latter proviso holds provided a is not $\ll 10^{-3}$. Then during evaporation, as a decreases ΔT increases, leading to an increased evaporation rate.

Looking at the evaporation rate as a function of the thickness in detail [Fig. 3(c)], we find there is a "resonance" at $d \approx 60$ Å.

It appears, therefore, that both the torsional oscillator data and the data from the levitator may be interpreted by postulating similar behavior for a and f; both a and f are small, they are increasing functions of d, and both possess a "resonance" at about 60 Å. It is difficult to imagine a microscopic process to account for resonance at this thickness.

Although we are unaware of any other measurements of f, Mate and Sawyer[3] quote values of a near unity.

The film does not completely evaporate after the substrate is removed from the bath. A residual film remains, about 10 Å thick. If this residual film is "burnt off", it reforms in about 10 sec. From adsorption isotherms it is possible to determine the temperature difference required for the residual film to be maintained. A 10-Å film at 1.6°K requires the vapor pressure for the film to be less than 10% of the saturated vapor pressure. This corresponds to temperature differences much larger than 1°K. However, the isotherms for ^{4}He approach the pressure axis at film thickness of ~ 8 Å, implying that the last two layers of helium are very strongly bound, and should not be destroyed by radiation at all. This being so, all the residual thicknesses measured should be increased by 8 Å. Then we are in the region on the isotherms for 40–60% saturation, implying temperature differences not greater than 3°K.

Either of these results, however, is quite inconsistent with the calculated temperature differences. Again, however, a low value of a may be interpreted as a correction to the percentage saturation, bringing the required temperature differences down by a factor of the order of a.

In an attempt to clarify our results, we are observing the formation of unsaturated films and measuring the temperature of the substrate.

References

1. D. G. Blair and C. C. Matheson, this volume.
2. C. C. Matheson and J. L. Horn, this volume.
3. C. F. Mate and S. P. Sawyer, *Phys. Rev. Lett.* **20**, 834 (1968).

Nuclear Magnetic Resonance Study of the Formation and Structure of an Adsorbed ^{3}He Monolayer*

D.J. Creswell, D.F. Brewer, and A.L. Thomson

Physics Laboratory, University of Sussex
Falmer, Brighton, Sussex, England

For several years there has been speculation concerning the nature of an adsorbed monolayer of helium atoms, with special interest as to whether the state of this layer is like a liquid, a solid, or a collection of atoms isolated by the potentials of the adsorbent sites.†[1] Although the monolayer density corresponds to that of a high-pressure solid, the observed temperature dependences of the specific heat[1-3] ($C \propto T^2$) and nuclear magnetic susceptibility[4] ($\chi \propto T^{-1}$) are consistent with either solid-like or liquid-like behavior. We present here measurements of the spin–lattice (T_1) and spin–spin (T_2) relaxation times for ^{3}He atoms adsorbed on Vycor glass; from these results it is possible to estimate both the correlation time τ_c, which is related to mobility, and the second moment of the nuclear magnetic resonance line M_2, which depends on the spatial arrangement of the atoms. We find that the configuration of the atoms is strongly dependent on the fractional coverage θ, in particular indicating a solid arrangement for values of $\theta \gtrsim 1/2$.

A continuous-wave magnetic resonance technique was used at resonant frequency $\omega_0/2\pi = 6.24$ MHz. The time T_1 was determined by saturating the signal and observing the time constant of the recovery when the saturation condition was removed. Two methods were used for measuring T_2, depending on the width of the resonance line, which was quite broad (about 2/3 G) at the lowest coverages but became progressively narrower as the coverage increased. The resonances from a monolayer upward all had the same linewidth (~ 0.16 G), which was therefore taken to be due to the inhomogeneity of the applied magnetic field. Thus for each broad line T_2 was taken as the reciprocal of the difference between the observed width of the line at half-height and the width due to the magnet inhomogeneity. For the complete monolayer T_2 was obtained from measurements of T_1 and the saturation factor Z given by

$$Z = 1/(1 + \gamma^2 H_1^2 T_1 T_2) \tag{1}$$

where γ is the gyromagnetic ratio of the ^{3}He nuclei and H_1 is the amplitude of the radiofrequency magnetic field.

* The work was carried out with the support of the Science Research Council (Grant Nos. 3/9/1832 and B/SR/4261, and a Research Studentship), and the U.S. Air Force Office of Scientific Research through the European Office of Aerospace Research under Grant EOOAR 68-0023.

† The literature on this subject is too extensive to be quoted in detail in this paper. Manchester[1] reviewed some of the earlier work with ^{4}He and questioned strongly whether the assumption of solidlike configuration was justified. Dash and co-workers have also discussed this question extensively: see, for example, Bretz *et al.*[2]

Values of T_1 and T_2 are plotted versus fractional monolayer coverage θ in Fig. 1. The very strong coverage dependence of T_1 and T_2 makes it immediately clear that the state of the adsorbate changes drastically during the formation of the monolayer. However, T_1 is always much larger than T_2, so it is most unlikely that at any coverage it is in a gaseous or liquid state, where T_1 and T_2 would be approximately equal. Moreover, the large T_1/T_2 ratio indicates that the measurements were made in the regime $\omega_0 \tau_c > 1$. In this region the expressions for T_1 and T_2 derived from the three most commonly used correlation functions (Gaussian, Lorentzian, and exponential)[5] take on particularly simple forms so that τ_c can be estimated from the ratio T_1/T_2 and M_2 from T_2 itself. The results are shown in Figs. 2 and 3, where τ_c and M_2 are plotted against θ.

The first thing to be noted is that all three correlation functions give values of τ_c which decrease as the monolayer is built up. Thus the reduction of linewidth with increasing coverage is due at least partly to motional narrowing. Second, in order

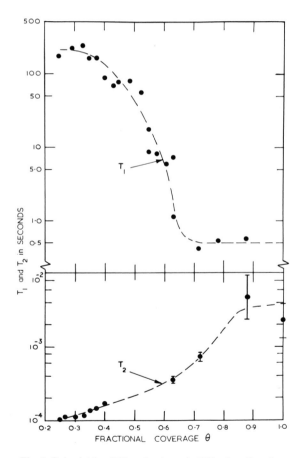

Fig. 1. Spin–lattice (T_1) and spin–spin (T_2) relaxation times plotted against fractional monolayer coverage. Full monolayer coverage corresponds to 38.6 cm³ STP.

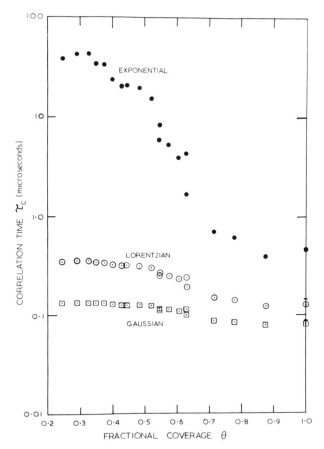

Fig. 2. The correlation time τ_c plotted against fractional monolayer coverage θ for the three correlation functions.

to choose which correlation function is appropriate to the adsorbed film, we must
inspect the magnitude and coverage dependence predicted for M_2. If its magnitude
is determined by the local fields due to neighboring ³He nuclei, then it is possible
to estimate its value for a complete layer from the known atomic spacing. Assuming
triangular packing, we expect $M_2 \sim 4 \times 10^8$ (rad sec^{-1})^2. This is consistent with
any of the correlation functions for a complete layer, but not for the low coverages,
where the Lorentzian and Gaussian predictions are one or two orders of magnitude
too large. Such a large M_2 could be caused by paramagnetic impurities (even though
great care was taken to exclude them) which would contribute a local field approximately 10^3 times larger than that of the helium atoms, resulting in an increase of
order 10^6 in M_2 for those atoms next to the impurities. The following argument,
however, suggests that this is unlikely.

The coverage dependence of an impurity-dominated M_2 will be determined
by the distribution of the ³He atoms with respect to the impurities. If the distribution
were random, then M_2 would be approximately constant. On the other hand, if

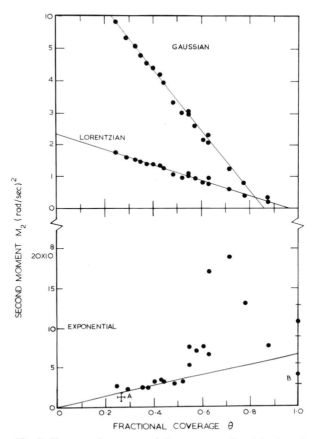

Fig. 3. The second moment of the resonance line M_2 plotted against fractional monolayer coverage θ for the three correlation functions. The point A is a direct measurement of M_2. Point B is the value of M_2 for a complete layer.

they were preferentially adsorbed near the impurities, M_2 would be expected to vary as $1/\theta$. The actual variation predicted by the Gaussian and Lorentzian functions (Fig. 3) seems to decrease linearly with θ. It is difficult to envisage an atomic distribution which would produce such a variation, and consequently we conclude that the Gaussian and Lorentzian correlation functions are inappropriate for our system and that the Vycor surface is essentially free from paramagnetic impurities. This conclusion is considerably reinforced by a single direct measurement of M_2 which could be made for the broadest line, which is shown in Fig. 3. Although the accuracy is not high, the result is quite consistent with the exponential predictions and incompatible with the other two.

Not only the magnitude of M_2, but also its variation with coverage as predicted by the exponential correlation function, can be understood reasonably well in terms of the local field produced by the ^{3}He atoms themselves. At the lower coverages it is plausible to expect that the atoms would be adsorbed in a random distribution, giving a linear increase in M_2 with coverage as shown in Fig. 3. Above $\theta \gtrsim 1/2$, however, there is a sudden increase in M_2 indicating that at these coverages the atoms prefer to condense into a tightly packed phase, presumably consisting of islands on the Vycor surface. As more and more atoms are adsorbed into the spaces between the islands to complete the monolayer, M_2 seems to decrease again, suggesting perhaps that these last atoms attract other helium atoms from the surrounding islands to bring about an increase in the average interatomic spacing.

In the interpretation of our values of τ_c it is important to note that we find no significant dependence of T_1 and T_2 on temperature, indicating that the mobility associated with τ_c is due to some kind of tunneling. Up to $\theta \sim 1/2$ the adsorption of atoms in a random manner into sites on the Vycor surface would imply that τ_c is associated with the time taken for atoms to jump from one site to another and is therefore determined by the time taken to tunnel through the potential barriers presented by the Vycor. The variation of τ_c with coverage shows the existence of a heterogeneous surface with the deepest potential wells being filled first and atoms subsequently adsorbed experiencing lower barrier heights. In the region $\theta \gtrsim 1/2$, where we have deduced that condensation is occurring, the interpretation of τ_c is quite different. Here the mobility is probably caused by exchange tunneling of particles as found in bulk solid ^{3}He.[5,6] If τ_c is the characteristic time for an atom to exchange places with any of z nearest neighbors, then J, the exchange frequency for a single pair of atoms, will be $\sim 1/z\tau_c$. For our complete monolayer we have obtained a value of τ_c of 0.5×10^{-6} sec and therefore, assuming triangular close packing of the condensed phase, where the number of nearest neighbors would be six, we find that $J \sim 0.4$ MHz. This should be compared to bulk ^{3}He solid where, for a similar interparticle spacing, the measured value of J is of order 10^{-2} MHz.[7] Apparently the two-dimensional solid has a larger exchange frequency, and in view of the absence of atoms above it this is perhaps not surprising.

References

1. F. D. Manchester, *Rev. Mod. Phys.* **39**, 383 (1967).
2. M. Bretz, G. B. Huff, and J. G. Dash, *Phys. Rev. Lett.* **28**, 729 (1972), and references therein.
3. D. F. Brewer, A. Evenson, and A. L. Thomson, *J. Low Temp. Phys.* **3**, 603 (1970).
4. G. Careri, M. Santini, and G. Signorelli, in *Proc. 9th Intern. Conf. Low Temp. Phys., 1964*, Plenum, New York (1965), p. 364; A. L. Thomson, D. F. Brewer, and P. J. Reed, in *Proc. 10th Intern. Conf. Low Temp. Phys., 1966*, VINITI, Moscow (1967), Vol. 1, p. 338; D. F. Brewer, D. J. Creswell, and A. L. Thomson, in *Proc. 12th Intern. Conf. Low Temp. Phys., 1970*, Academic Press of Japan, Tokyo (1971), p. 157.
5. M. G. Richards, in *Advances in Magnetic Resonance*, J. S. Waugh, ed., Academic, New York (1971), Vol. 5.
6. R. A. Guyer and L. I. Zane, *Phys. Rev.* **188**, 445 (1969).
7. M. G. Richards, J. Hatton, and R. P. Gifford, *Phys. Rev.* **139**, A91 (1965); L. H. Nosanow and W. J. Mullin, *Phys. Rev. Lett.* **14**, 133 (1965).

Pulsed Nuclear Resonance Investigation of the Susceptibility and Magnetic Interaction in Degenerate ^{3}He Films*

D.F. Brewer and J.S. Rolt

Physics Laboratory, University of Sussex
Falmer, Brighton, Sussex, England

Continuous-wave nuclear resonance measurements have been made recently of the susceptibility and spin–lattice relaxation times on films[1,2] of pure ^{3}He and ^{3}He–^{4}He mixtures in the temperature range between 1.5 and 0.4°K for various film thicknesses. In the region investigated the films either were classical, in the sense that their susceptibility followed a Curie law, or showed only small (up to $\sim 26\%$) Fermi degeneracy, depending on coverage and temperature. It was shown that the susceptibility was consistent with a statistical layer model in which the first two layers could be considered as magnetically independent of each other and of the rest of the film, and each layer region of the film exhibited the same susceptibility as that of bulk ^{3}He at the same interatomic spacing. Such an interpretation is not necessarily unique, and it is expected to be an approximation. In the present report we show that there are large deviations from this model at lower temperatures down to 50 m°K, in such a direction as to indicate a lower Fermi temperature in the real film than in the model, or a larger ferromagnetically inclined exchange interaction.

The adsorbed phase was formed on a substrate of Vycor porous glass consisting of ten disks of 7 mm diameter and 1.2 mm thickness cut from sheet. The total pore volume was 0.142 cm^3, the average pore diameter was 71 Å, and the total surface area was 79.7 m^2. A schematic diagram of the apparatus is shown in Fig. 1 and is described more fully elsewhere in this volume.

Mixing chamber temperatures could be measured with a powdered cerium magnesium nitrate pill immersed in the fluid, using a mutual inductance bridge method. Speer carbon resistance thermometers (220 ohms, $\frac{1}{2}$ W) were placed in the mixing chamber and in the sample ^{3}He above and below the Vycor chamber. In addition, we could measure the spin diffusion coefficient D of the bulk liquid ^{3}He surrounding the Vycor disks and also of the bulk liquid in the small chamber below the experimental cell.[3] These diffusion coefficient measurements could be used as an additional thermometric parameter by comparing D with previous measurements.[4] A CMN thermometer could not be incorporated into the sample chamber for fear of contaminating the Vycor substrate with adsorbed water. Temperatures quoted below are probably accurate to within about 3 m°K at 50 m°K.

* This work was supported by the Science Research Council and by the U.S. Air Force Office of Scientific Research through the European Office of Aerospace Research (Grant EOOAR 68-0023).

200

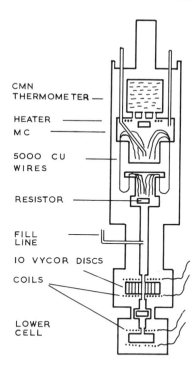

CMN
THERMOMETER —

HEATER

M C

5000 CU
WIRES

RESISTOR

FILL
LINE

10 VYCOR DISCS

COILS

LOWER
CELL

Fig. 1. Low-temperature part of apparatus.

The NMR receiver coil was embedded in the epoxy and surrounded the Vycor disks. The transmitter coil was a saddle Helmholz pair located around the glass vacuum jacket and the experiments were performed in a field of about 500 G (resonance frequency ~ 1.6 MHz) provided by a copper-wound, liquid nitrogen-cooled sixth-order solenoid located in the nitrogen bath. This magnet provided a field homogeneity of $3:10^5$ over the sample volume of about 0.5 cm³.

The susceptibility χ of the adsorbed ³He was obtained by measuring the amplitude $A(t)$ of the free induction tail following a single 90° pulse. This amplitude is given by

$$A(t) = A(\Delta t) e^{-t/T_2} \tag{1}$$

where T_2 is a measure of the total spin dephasing time, which is determined in our case largely by the field inhomogeneity, and Δt is the sum of the receiver recovery time and the length of the 90° pulse itself. If the pulse is narrow enough for the dephasing of the nuclear moments during the time Δt to be negligible, then $A(\Delta t) = A(0) \propto M_0 = \chi H_0$, where M_0 is the magnetic moment in the steady field H_0. In our experiments the width of the 90° pulse was typically 50 μsec and the receiver recovery time was about 15 μsec, whereas T_2 was about 3 msec; consequently a simple extrapolation to the zero of time to obtain $A(0)$ could be made with negligible error. As usual, the circuit parameters are not sufficiently accurately known for absolute evaluation of χ and a normalization procedure must be used. In our case we have normalized our results to the Curie value in the region 1.5–2.0°K, where we

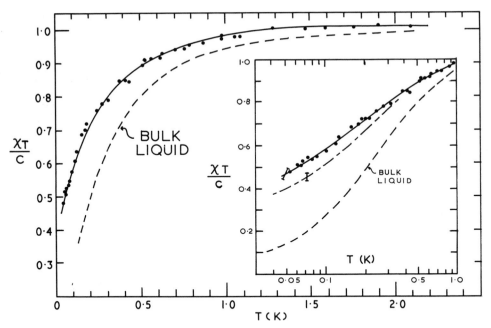

Fig. 2. Main diagram: linear plot of $\chi T/C$ (see text) vs. T; the full line through the points represents the experimental results; the broken curve is for bulk liquid (Ref. 6). Inset: the same data plotted semilogarithmically to bring out the low-temperature part. The lower broken curve is for bulk liquid, the upper broken curve represents the prediction of the statistical layer model, and the points are the experimental results.

find $\chi T = C$ = const and where previous measurements* in this laboratory of χ as a function of coverage and temperature give reason to suppose that the susceptibility has its free spin value.

Our values of $\chi T/C$ vs. T are shown in Fig. 2 together with the most recent bulk liquid measurements[6] at the saturated vapor pressure. These results were obtained from four experimental runs, and the scatter indicates that the reproducibility is of the same order as the precision. It is clear from the linear plot of Fig. 2 that in this system Curie's law is followed down to an appreciably lower temperature than in bulk liquid ^{3}He and that the total susceptibility is always significantly higher. Furthermore, the curve through the experimental points does not extrapolate readily to the absolute zero.

We now attempt to interpret these results in terms of the statistical layer model already mentioned. The assumptions of this model are that (i) the first layer has an interparticle spacing corresponding to that of solid ^{3}He at a molar volume of 15.5 cm^3; (ii) the second layer corresponds to bulk liquid at a molar volume of about 26 cm^3; (iii) higher layers have the density of bulk liquid at the saturated vapor pressure; and (iv) the first two layers behave independently of each other and of the rest of the film, and these three regions have the properties of the bulk phase with the same

---

* For a review of these experiments and further references see Brewer.[5] Also see Brewer et al.[1]

interparticle spacing. Although crude, the model is useful for the interpretation of a number of properties of helium. In the present case we write, for a film containing n_t atoms having an average magnetic moment M per atom,

$$n_t M = n_1 M_1 + n_2 M_2 + \sum_{i>2} n_i M_i(l) \tag{2}$$

where i labels the layer with magnetic moment M_i per atom containing n_i atoms and $M(l)$ is the magnetic moment per atom of bulk liquid at the saturated vapor pressure. At temperatures below 0.4°K our results show an increasing divergence from the predictions of this model, the observed susceptibility being considerably higher than the predictions (the divergence occurs only below 0.15°K if the second layer is taken to be Curie). This is shown most clearly in the insert to Fig. 2 in which the data are plotted logarithmically. The quantities n_i in Eq. (2) were obtained from the adsorption characteristics of the Vycor glass as deduced from the usual adsorption isotherms. The error bars shown in Fig. 2 (insert) indicate the uncertainties associated with these measurements.

If we retain the basic ideas of the statistical layer model, the second layer susceptibility can be calculated from Eq. (2) and is shown in Fig. 3 together with the experimental susceptibility of bulk ^{3}He at the same atomic spacing and with calculated values of the Curie law susceptibility. This layer might be expected, in the model, to behave like a two-dimensional Fermi system with an unknown degeneracy temperature. Also in the same figure we show for comparison the susceptibility of a two-dimensional Fermi gas of appropriate surface number density, calculated from the equation

$$\frac{\chi}{(\chi T)_{T \to \infty}} = \frac{\chi}{C} = \frac{1}{T_D^{**}} \left(1 - \exp - \frac{T_D^{**}}{T} \right) \tag{3}$$

where T_D^{**}, the two-dimensional magnetic degeneracy temperature, replaces T_D, the corresponding ideal gas degeneracy temperature, and is taken to be 0.05°K for the best fit. The fit, though good in general temperature dependence, may be fortuitous, although it is clear that there are deviations by a factor of three or more from bulk liquid and from a Curie law at 50 m°K.

A true comparison of this idea with the results can only be made at temperatures well below T_D^{**} where the leveling off of χ to its constant low-temperature limit becomes apparent and a proper determination of T_D^{**} can be made. The application of such an expression was suggested by Goldstein's phenomenological approach[7] to the problem of the bulk liquid, where a similar substitution is quite successful, particularly in the temperature region well below the bulk magnetic degeneracy temperature T_F^{**}. It is doubtful whether this analysis can profitably be pursued any further at present, but it is interesting to note that for bulk ^{3}He at the density of the second layer

$$T_F^{**} = T_F^* \left(1 + \frac{G_0}{4} \right) = T_F \left(\frac{m}{m^*} \right) \left(1 + \frac{G_0}{4} \right) \approx \frac{T_F}{21} \approx 0.3°K$$

where T_F is the ideal Fermi temperature, T_F^* the experimental Fermi temperature deduced from thermal data and G_0 the usual Landau parameter. In the case of the film, by contrast, $T_D^{**} \approx T_D/80$; consequently either the spin-independent in-

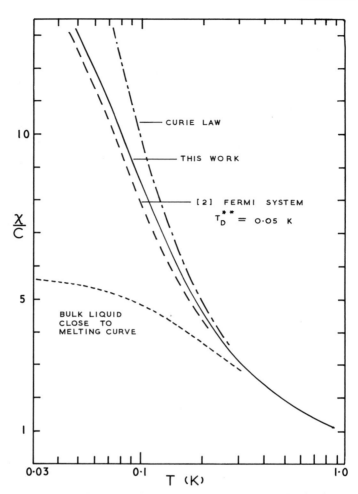

Fig. 3. A plot of χ/C vs. T for the second layer (see text), showing large deviations from bulk liquid and from Curie behavior at the lower temperatures and approximate agreement with a two-dimensional Fermi gas.

teractions (responsible for m^*) or the spin-dependent interactions [responsible for $(1 + \frac{1}{4}G_0)$], or both, are considerably larger in the film than in bulk, tending to make it more nearly ferromagnetic. To confirm these ideas, independent estimates of m^* and G_0 for a film must be made and our experiments continued to lower temperatures. We have assumed above that the first layer obeys Curie's law and that the second layer behaves in an unknown way. If we assume that the layers above the first obey the predictions of the layer model, we can attribute the deviations to the first layer and then calculate the effective susceptibility of the first layer and in particular look for a Curie–Weiss behavior. The results show that the first layer susceptibility follows the equation $1/\chi = T/1.257C$ to high accuracy and indicates that about half of the second layer appears to follow Curie's law. In view of the uncertain structure of the Vycor surface and the possibility of second-layer localization, this is indeed a possibility.

References

1. D. F. Brewer, D. J. Creswell, and A. L. Thomson, in *Proc. 12th Intern. Conf. Low Temp. Phys., 1970,* Academic Press of Japan, Tokyo (1971), p. 157; and to be published.
2. D. F. Brewer, A. J. Symonds, and A. L. Thomson, *Phys. Lett.* **13**, 298 (1964).
3. D. F. Brewer and J. S. Rolt, this volume.
4. W. R. Abel, A. C. Anderson, W. C. Black, and J. C. Wheatley, *Physics* **1**, 6, 337 (1965); A. Tyler, *J. Phys. C. Sol. St. Phys.* **4**, 1479 (1971).
5. D. F. Brewer, *J. Low Temp. Phys.* **3**, 205 (1970).
6. J.R. Thompson, H. Ramm, J.F. Jarvis, and H. Meyer, *J. Low Temp. Phys.* **2**, 521 (1970); H. Ramm, P. Pedroni, J. R. Thompson, and H. Meyer, *J. Low Temp. Phys.* **2**, 539 (1970).
7. L. Goldstein, *Phys. Rev.* **113**, A52 (1964).

Measurements and Calculations of the Helium Film Thickness on Alkaline Earth Fluoride Crystals

E.S. Sabisky and C.H. Anderson

RCA Laboratories
Princeton, New Jersey

Accurate measurements have been made of the thickness of helium films adsorbed on alkaline earth fluoride crystals as a function of the van der Waals potential. The helium film thickness, as we have shown,[1] is measured through the detection of simple standing wave patterns of a given frequency set up between the substrate which supports the film and the liquid–gas interface, the film thickness being determined by counting the number of wavelengths across the film and calculating the magnitude of the wavelength from the existing frequency. The van der Waals potential which is solely responsible for the adsorbed film is obtained by measuring the height of the crystal above the free liquid surface in the case of saturated films and by measuring the helium gas pressure at the position of the resonant peaks for unsaturated films. The substrates are cleaved single crystals of CaF_2, SrF_2, and BaF_2 doped with 0.02 mole % of the paramagnetic ion divalent thulium.

The experimental values of the film thickness are obtained from the resonant peaks in the intensity of the thulium spin temperature as the helium gas pressure, for example, is varied for unsaturated films. Typical experimental curves for frequencies between 18 and 60 GHz are shown in a previous publication.[1] To first order, the film thickness at the resonant peaks is given by the expression

$$d_A = N\lambda = NC_0/v; \qquad N = \tfrac{1}{4}, \tfrac{3}{4}, \dots \tag{1}$$

where d_A is the simple acoustic thickness, N is the number of wavelengths, and λ is the acoustic wavelength calculated from the velocity of sound in bulk liquid helium C_0 and the microwave frequency v. Two small corrections have to be added to the above equation, the dispersion of the velocity of sound and a frequency-dependent phase shift correction introduced at the boundaries. The "true" film thickness is therefore given by the modified expression

$$d = (C_0 N/v\,[1 + \delta(v)] + d_0(v) \tag{2}$$

where $\delta(v)$ is the dispersion correction and $d_0(v)$ is the phase shift correction. Experimental values for these two corrections have been obtained and will soon be published.

Figure 1 shows the standing wave pattern for a particular set of conditions. It should be emphasized that this curve is the end result of the study and was obtained using an expression relating the film thickness to the experimentally measured values of the helium gas pressure. This type of plot permits a study of the detailed shape

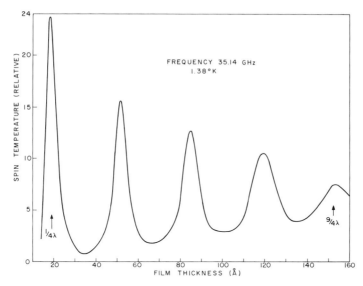

Fig. 1. A typical curve of the relative spin temperature where the film thickness
was calculated from the helium gas pressure readings.

of these resonances, which should provide insight into the attenuation of sound at
these very high frequencies. Curves with the shape shown in Fig. 1 can be obtained
from a simple acoustic theory approach to the problem. The theory contains two
unknown parameters, the attenuation of sound in the film and the loss rate from the
crystal to the film. The attenuation coefficient for phonon frequencies near 35 GHz
is found to be 8×10^{-3} Å^{-1} using this simple approach. However, before anything
can be said conclusively about the loss rate and the attenuation, a much more
extensive theory must be developed and in addition more frequency- and temperature-dependent data are needed.

The experimental data given in this paper were taken on SrF_2 at 1.38°K for
both saturated and unsaturated films and are given in Fig. 2. In the figure the van
der Waals potential per helium atom is plotted as a function of the "true" film thickness. Each data point represents the average of two or more measurements. Typical
reproducibility of the potential is $\pm 2\%$ over the entire range. The major uncertainty
in the film thickness of 1% comes from the value used for C_0, 236.6 m sec^{-1}. The
phase shift correction is known from experiment to about ± 0.5 Å. The uncertainty
due to the dispersion is less than 0.5%. The inverse cube law is only applicable for
very thin films, while the inverse fourth-power law is not reached for a thickness of
240 Å.

Richmond and Ninham[2] have calculated the van der Waals potential for helium
films on alkaline earth fluoride substrates using the full Lifshitz formula.[3] They
obtained good agreement with our earlier, uncorrected data. We have refined their
calculation and compared the results to the more precise data given in this paper.

In order to use the exact Lifshitz formula, models of the complex dielectric
susceptibility of the materials must be developed. Following Parsegian and Ninham,[4]
we represent the dielectric functions as a sum of Lorentzian terms with the small

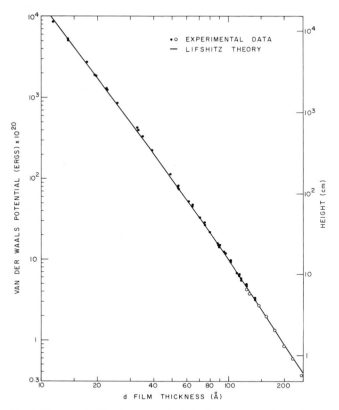

Fig. 2. The van der Waals potential of a helium atom on the surface
of the helium film which is adsorbed on a SrF₂ cleaved surface.
The data represented by closed and open points are for unsaturated
and saturated films, respectively.

corrections due to damping left out. That is, the response function is given by

$$\varepsilon(i\zeta) = 1 + \sum_n \frac{A_n}{1 + (\zeta/w_n)^2} \tag{3}$$

where the w_n are a set of resonant frequencies and the constants A_n represent a measure
of the oscillator strength associated with each resonance. The dielectric function is
written in this way because it is essential that the optical dielectric constant be
reproduced accurately, the results being less sensitive to the number and location
of the resonant frequencies.

The dielectric function for liquid helium can be represented by the three major
transitions of atomic helium centered at the 1^1S-2^1P transition (21.2 eV), the 1^1S
ionization limit (24.6 eV), and the double ionization transition (79 eV). The response
function used for helium is

$$\varepsilon(i\zeta) = 1 + \frac{0.016}{1 + (\zeta/w_1)^2} + \frac{0.036}{1 + (\zeta/w_2)^2} + \frac{0.0047}{1 + (\zeta/w_3)^2} \tag{4}$$

This function gives the correct dielectric constant of 1.057 and also produces the effective number of electrons per helium atom of two.

The optical and dielectric properties of the alkaline earth fluorides have been studied up to 50 eV. At first we used a ten-band model in the UV region but later found that the dielectric function was just as well represented by a simpler three-band model with frequencies centered at the three main groups of transitions observed in these materials. The response function used in the calculation for SrF_2 including the infrared band is

$$\varepsilon(i\zeta) = 1 + \frac{4.03}{1 + (\zeta/w_1)^2} + \frac{0.757}{1 + (\zeta/w_2)^2} + \frac{0.065}{1 + (\zeta/w_3)^2} + \frac{0.25}{1 + (\zeta/w_4)^2} \tag{5}$$

where the w's are 0.028, 13.5, 23.1, and 29 eV, respectively. These models have no physical meaning; they are merely used to provide a reasonable facsimile to the dielectric function.

The results of the calculation using the full Lifshitz formula and the given response function are represented by the solid line in Figs. 2 and 3. In order to better exhibit the deviations from the inverse cube law, the function Vd^3 (van der Waals potential times the cube of the thickness) is plotted as a function of the thickness in Fig. 3. There are two scales in the figure, shifted with respect to one another so that the comparison can be made with and without the dispersion and phase shift corrections. Since these small corrections have been obtained independent of the calculation, the marked improvement in the agreement when they are added provides an independent check on them. The first few points near 10 Å are probably off because the perturbation scheme implicit in the way the corrections are made is beginning to fail in thin films (10 Å is about three statistical-atomic layers of helium). Probably

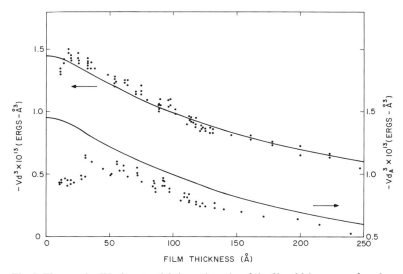

Fig. 3. The van der Waals potential times the cube of the film thickness as a function of the film thickness. The solid line is theory. The data points on the lower half of the graph are for the uncorrected film thickness, while those on the upper half are after corrections of dispersion and phase shift have been made.

the most exceptional and convincing feature of Fig. 3 is the way the experimental points and the theoretical calculation both exhibit the same subtle deviation from the simple inverse cube law due to retardation.

The calculations for CaF_2, SrF_2, and BaF_2 all give the same results to within 2%. There is no simple reason for this but it does agree with the measurements. The response function of the substrate was varied and it was found that rather drastic changes in it do not significantly change the result. The results in the thin film region can be changed by something like a 3% change in the potential for realistic changes in the helium response function. Thus, the agreement between the measurements and calculations given in Figs. 2 and 3 provides a real verification of the Lifshitz theory. In addition, calculations for other materials, including glass substrate, have been done and from these calculations we conclude that the thickness scale used in the past for glass substrates must be about 30–40% too large.

Accurate and reliable measurements of the film thickness on cleaved surfaces of SrF_2 crystals at 1.38°K have been presented. The films were measured between 10 and 250 Å using an acoustic interferometry technique. The exceptionally good agreement found between the film thickness measurements and those calculated using the Lifshitz formula leaves very little uncertainty about the thickness on these crystals. It must also be concluded that for dielectric substrates and clean metal surfaces the film thickness, when greater than the surface roughness, is given by the Lifshitz formula about as accurately as can be obtained from present measurements.

References

1. C. H. Anderson and E. S. Sabisky, *Phys. Rev. Lett.* **24**, 1049 (1970).
2. P. Richmond and B. W. Ninham, *J. Low Temp. Phys.* **5**, 177 (1971).
3. I. E. Dzyaloshinskii, E. M. Lifshitz, and L. P. Pitaevskii, *Adv. Phys.* **10**, 165 (1961).
4. V. A. Parsegian and B. W. Ninham, *J. Chem. Phys.* **52**, 4578 (1970).

The Normal Fluid Fraction in the Adsorbed Helium Film*

L.C. Yang

Physics Department, University of California
Los Angeles, California
and
Jet Propulsion Laboratory, California Institute of Technology
Pasadena, California

Marvin Chester

Physics Department, University of California
Los Angeles, California

and

J.B. Stephens

Jet Propulsion Laboratory, California Institute of Technology
Pasadena, California

The resonant frequency of a thickness-shear mode oscillating quartz crystal is lowered by virtue of the mass loading of any film laid down on its surfaces.[1,2] For a uniformly deposited, thin solid film with thickness less than 1% or so of the crystal thickness this change Δf, is given by the formula

$$- \Delta f = 2(2f^2/c\rho_q)\,\sigma \tag{1}$$

Here σ represents that part of the mass per unit area which remains rigidly coupled to the substrate motion, f is the resonant frequency, c is the shear wave velocity in the crystal, and ρ_q is the density of the quartz.

Similarly, the mass loading behavior of an adsorbed liquid film uniformly deposited on the crystal can be shown to be given by the formula

$$- \Delta f = 2(2f^2/c\rho_q)\,\beta \tag{2}$$

where

$$\beta = (\eta\alpha/2\pi f)\,\mathrm{Im}\{(1 + i)\tanh\left[(1 + i)\,\alpha D\right]\} \tag{3}$$

and

$$\alpha = (\pi\rho f/\eta)^{1/2} \tag{4}$$

D represents the film thickness, ρ its density, and η its viscosity, and $1/\alpha$ represents the "viscous wavelength" in the liquid film. The extra factor of two is included to allow for loading on both of the two faces of the crystal.

* This work was supported by a grant from the California Institute of Technology President's Fund.

For the limiting case of thin films one finds

$$\beta = \rho D = \sigma, \qquad \alpha D \ll 1 \qquad (5)$$

For the regimes of interest in our experiments, in which 24-MHz crystals were used, we find, employing the normal fluid viscosity and density in Eq. (4), that the viscous wavelength always far exceeds the thickness of the films with which we work except perhaps at relative saturations well in excess of 99%. Hence we recover Eq. (1) as applying to our experiments with the understanding that $\sigma \to \sigma_n = D\rho_n$, since no viscosity is associated with ρ_s.

By virtue of this relationship, we have measured the mass adsorbed as a function of the helium gas pressure P in our experimental chamber up to the saturation vapor pressure $P_0(T)$, for a series of temperatures. The resulting adsorption isotherms $[\sigma$ vs. $P/P_0(T)]$ for ^{4}He below the λ point show a characteristic point of departure from those for a nonsuperfluid film such as ^{3}He at sub-λ-point temperatures, ^{4}He at temperatures above the λ point, and O_2, N_2, and Ar at liquid oxygen and nitrogen temperatures obtained by the same experimental apparatus. The latter is characterized by a continuous, monotonic increasing function when the film thickness is beyond the first two film layers. We have attributed these evident points of departure to the onset and presence of superfluid fraction in the ^{4}He film. This fraction remains uncoupled to the substrate and hence does not load the quartz crystal as does a normal film. Several other possible explanations such as the critical velocity, relaxation of elementary excitations, and thermal conduction increase associated with the commencement of superfluid have been considered and proven to be unfeasible.[3]

Figure 1 exhibits several representative isotherms obtained by employing a 24-MHz, 39°49′, Y-cut quartz crystal. The results were essentially the same for both Al and Au substrates (i.e., the electrode surfaces of the crystals) and exhibited no adsorption hysteresis.[3] If we take as the onset of superfluidity those points on this figure at which the adsorption isotherm curve begins to turn away

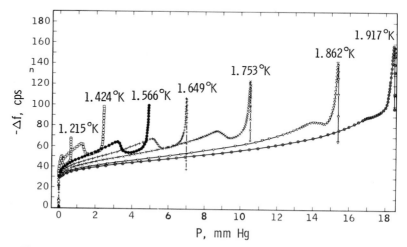

Fig. 1. Representative experimental data on the frequency changes observed for different helium gas pressures over the range of temperatures explored.

from a monotonic increase, our onset points fall as shown on Fig. 2. The dashed curves in Fig. 2 represent the limits between which past reported results* lie, and the full curve represents some mean of these results. The fact that our points generally fall slightly lower than this curve suggests that we are seeing earlier into the commencement of superfluidity than heretofore reported.

Related experiments[5] have been reported in the literature where a superfluid onset did not make itself manifest in the measurements. We attribute our observations of this effect to (a) the increased sensitivity and frequency stability gained by employing the high frequencies and liquid-helium-immersed oscillators that we have used, and (b) the success of our efforts to reduce the power dissipation in our crystals down to the order of $2 \mu W$ or less. A careful survey of the effects of power dissipation upon our observations indicates that this is a crucial factor. The superfluid regime tends to be masked completely at the milliwatt power levels conventionally employed with this technique.

Figure 3 represents a plot of an observed adsorption signal $-\Delta f$ against $[\log P_0(T)/P]^{-1/3}$ at a temperature $1.649°K$. On the basis of the Frenkel–Halsey–Hill theory[4] this should reflect the behavior of the nonsuperfluid mass adsorbed as a function of the thickness D of the film in the following manner:

$$D/D_0 = \{\theta/[T\ln P_0(T)/P]\}^{1/3} \qquad (6)$$

Here D/D_0 represents the "number of film layers," where D_0 is currently taken as 3.6 Å. The characteristic temperature θ measures the strength of the van der Waals force between the wall and the helium atoms.

As expected, for relative saturations $P/P_0(T)$, between 0.1 and 0.72 a linear

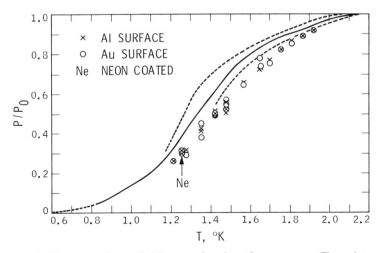

Fig. 2. The onset of superfluidity as a function of temperature. The points shown represent our results for three different substrates. (The point marked "Ne" was obtained with neon substrates over the Au and Al electrodes.) The dashed curve represents the limits between which past reported results lie. (See Fig. 2 of Ref. 4.)

* See Fig. 2 of Ref. 4 and the references cited in Ref. 4.

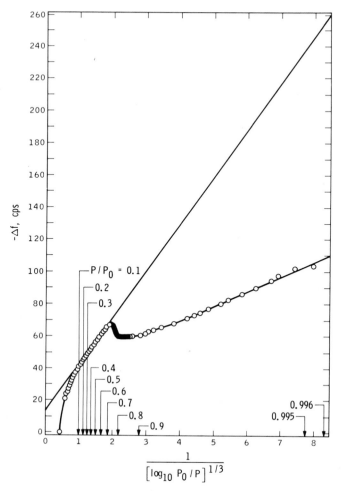

Fig. 3. Our experimental adsorption signal at $T = 1.649°K$ plotted in Frenkel–Halsey–Hill fashion as a function of $(\log P_0/P)^{-1/3}$ to illustrate the method of obtaining the van der Waals strength θ and the normal fluid fraction as a function of film thickness. The abscissa is proportional to the film thickness.

regime is observed before the onset of superfluidity. The slope of the extrapolated line drawn through this regime should reflect the bulk total helium density ρ because at these levels of saturation one expects

$$d\sigma/dD = \rho_{bulk} = 0.146 \quad g\,cm^{-3}, \qquad 0.1 < P/P_0 < 0.7, \qquad T = 1.649°K \quad (7)$$

Employing Eq. (6), and the sensitivity of our instrument, which, via Eq. (1), is $d\sigma/d(-\Delta f) = 3.9 \times 10^{-10}$ g cm^{-2} Hz^{-1}, plus the values in Eq. (7), one finds $\theta = 39 \pm 6°K$.

Our experimental data points in Fig. 3 fall below the extrapolated total fluid density line discussed above for relative saturations in excess of 72%. As mentioned, we attribute this to the elimination of the superfluid fraction from the film mass-loading effect. Hence we expect our data points to extrapolate to a straight line of lower slope $d\sigma_n/dD = \rho_n$ (bulk). As one can see from the figure, this is in fact the case. Using this notion, one calculates a residual superfluid fraction of about $75 \pm 3\%$, whereas the tabulated vapor pressure bulk value is 80%. The plots of the data at other temperatures which we have explored yield similar results, namely the same value of θ and normal fluid fractions slightly in excess of bulk values (about 15% higher). Our current activity emphasizes the adsorption experiments using a 28-MHz X-cut quartz crystal microbalance. For this case we expect to observe the full helium adsorption isotherm since both normal and superfluid fractions are coupled to the longitudinal mode crystal oscillation and contribute to the Δf.

References

1. G. Sauerbrey, *Z. Physik* **155**, 206 (1959).
2. C. D. Stockbridge, in *Vacuum Microbalance Techniques*, K. H. Behrndt, ed., Plenum, New York (1966), Vol. 5, p. 147–205.
3. M. Chester, L. C. Yang, and J. B. Stephens, *Phys. Rev. Lett.* **29**, 211 (1972).
4. R. S. Kagiwada, J. C. Fraser, I. Rudnick, and D. Bergman, *Phys. Rev. Lett.* **22**, 338 (1969).
5. J. B. Brown and M. G. Tong, in *Proc. 12th Intern. Conf. Low Temp. Phys., 1970,* Academic Press of Japan, Tokyo (1971), p. 113.

4
Flowing Films

Helium II Film Transfer Rates Into and Out of Solid Argon Beakers*

T.O. Milbrodt and G.L. Pollack

Department of Physics, Michigan State University
East Lansing, Michigan

The purpose of this experiment is to study He II film flow over a surface whose interaction with the He is well understood. To accomplish this, we are conducting beaker-filling and-emptying experiments in which the beakers are made of solid Ar. There is a van der Waals attractive potential between the Ar and He atoms, the magnitude of which may be calculated from experimentally derived He–He and Ar–Ar potentials. This potential in turn determines the thickness of the He film on the beaker walls.

The most difficult part of the experiment is making the Ar beakers. We make them by freezing liquid Ar in a mold. The mold is in two parts. Its base is made of Teflon and is attached to a brass block on which an electric heater and a platinum resistance thermometer are mounted. These are used for controlling and measuring the temperature of the lower part of the mold. The upper part of the mold shapes the main body of the beaker and consists of two concentric glass tubes. The inner tube is sealed at the bottom and contains a heater. The bottom of the outer tube extends beyond the inner tube and fits into a seat in the Teflon to make a gas-tight seal. Both tubes are attached at their tops to a brass collar that holds another heater and thermometer. A stainless steel tube soldered to this collar is used to admit Ar gas to the mold and also to move the top part of the mold up and down from outside the cryostat. The mold and Ar beaker are contained in a glass chamber in a conventional cryostat.

To make a beaker, we first condense a sufficient amount of liquid Ar in the mold by setting the mold temperature $0.5°K$ above the Ar triple-point temperature of $83.8°K$ and admitting Ar gas to the mold. We then slowly freeze the liquid from bottom to top by gradually reducing the mold temperature. This takes about 12 hr. After all of the liquid has frozen we free the solid from the mold and also cool it somewhat by gently pumping on it. The glass upper part of the mold is then pulled from its seat in the Teflon and lifted upward to leave the beaker standing free. We then transfer liquid He into the cryostat to further cool the Ar to $4.2°K$.

Our beakers are typically 3.5 cm in height, with a bore length of 2.5 cm, an i.d. of 0.30 cm, and an o.d. of 0.55 cm. The Ar is transparent, but usually has a few visible defects. Quite often one of the defects will penetrate the wall of a beaker, allowing it to leak He I and rendering it useless for transfer experiments. We have stored beakers

* This research has been supported by the U.S. Atomic Energy Commission.

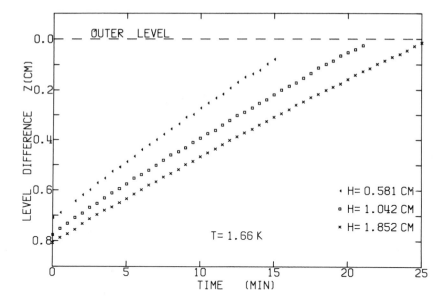

Fig. 1. Level difference as a function of time for three beaker fillings at three different heights H from the outer level to the beaker rim.

for over a week at 4.2°K. During this time the film transfer rates are reproducible to within the accuracy of our measurement.

The temperature of the He bath is controlled below T_λ with a pressure regulator and pump. A beaker-filling run is initiated by admitting He to the chamber from the bath through a needle valve until the liquid level in the chamber is about 1 cm higher than it is inside the beaker. A beaker-emptying run is initiated by using a He II fountain in the chamber to fill the beaker. We observe the liquid level in the beaker with a cathetometer and record the level height every 30 sec.

Figure 1 is a plot of the liquid level inside the beaker as a function of time for three beaker-filling runs. The ordinate Z is the difference between the inner and outer levels. The transfer rate σ (cm^3 sec^{-1} cm^{-1} of circumference) is proportional to the slope of the $Z(t)$ curve. Differentiating the raw data directly by taking the difference of adjacent points results in a great deal of scatter because of random errors in the height measurements, so it is necessary to use a smoothing procedure.[1] The transfer rates corresponding to the data shown in Fig. 1, after smoothing, are plotted in Fig. 2.

We may represent σ in the usual way by the relation

$$\sigma = (\rho_s/\rho) V_s d \qquad (1)$$

where d is the thickness of the adsorbed He film at the lip of the beaker, V_s is the maximum superfluid velocity in the film, and ρ_s/ρ is the superfluid fraction.* The film thickness d may be calculated, using the treatment of Frenkel,[3] by applying the condition that the potential energy of a He atom on the surface of the film must be equal to the potential energy of an atom on the surface of the bulk liquid. This results

* See, e.g., Keller.[2]

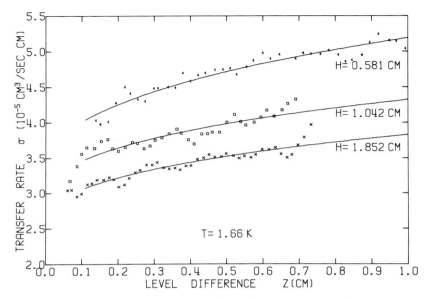

Fig. 2. Transfer rate as a function of level difference for the three beaker fillings of Fig. 1. The solid curves are of the form $\sigma = 1/(A - B \ln Z)$.

in the relation

$$d = kH^{-1/3} \tag{2}$$

where H is the height above the bulk liquid surface. We have used published van der Waals potential parameters for helium[2] and argon[4] from which we calculate that for an argon substrate $k = 2.92 \times 10^{-6}$ cm$^{4/3}$.

For He superfluid flow in channels of width $< 10^{-3}$ cm many experiments have yielded results for the critical velocity that are represented well by the Leiden formula[5]

$$V_{s,c} \approx d^{-1/4} \text{ (cgs units)} \tag{3}$$

We use this relation for $V_{s,c}$ to determine V_s in Eq. (1).

Combining Eqs. (1)–(3) with our calculated k, we arrive at

$$\sigma = 7.1 \, (\rho_s/\rho) \, H^{-1/4} \times 10^{-5} \quad \text{cm}^3 \text{ sec}^{-1} \text{ cm}^{-1} \tag{4}$$

If one takes H to be the distance from the higher He level to the top of the beaker, then Eq. (4) predicts that σ should be constant for a given filling run.

It is clear from Fig. 2, however, that σ is dependent on level difference. This dependence of superfluid flow rate on driving chemical potential difference has been observed by others[1,6] and treated theoretically.[7,8] In accord with the theoretical treatment, we fit our beaker-filling transfer rates to the form

$$\sigma = 1/(A - B \ln Z) \tag{5}$$

The solid lines in Fig. 2 are fits to Eq. (5).

In order to compare our results for the beaker fillings with Eq. (4), we fit them to the form

$$\sigma = \sigma_0 H^{-n} \tag{6}$$

Table I. Comparison between Theory and Experiment for the He II Film Flow Rates

| Temperature, °K | Beaker fillings | | | | Beaker emptyings | | Eq. (4) |
| | $Z = 1.0$ cm | | $Z = 0.1$ cm | | $Z = 1.0$ cm | $Z = 0.1$ cm | |
	$\sigma_0 \times 10^5$	n	$\sigma_0 \times 10^5$	n	$\sigma_0 \times 10^5$	$\sigma_0 \times 10^5$	$\sigma_0 \times 10^5$
1.77	4.35	0.236	3.16	0.263	3.94	2.52	4.95
1.66	4.43	0.214	3.50	0.192	4.16	2.98	5.57
1.47	4.56	0.212	3.52	0.195	4.71	3.53	6.35

where, because of the pressure head dependence, σ must here be taken for a given constant pressure head. We have calculated $\sigma(Z = 1$ cm$)$ and $\sigma(Z = 0.1$ cm$)$ from the parameters A and B derived from each filling run and have made a fit to Eq. (6) for both values of Z. The results from the data from one beaker are displayed in Table I. Also in Table I for comparison are corresponding values of σ_0 obtained from Eq. (4).

For beaker emptyings the higher He level is the inner level, which changes continuously while the beaker empties. We fit these transfer rates to

$$\sigma = H_I^{-1/4}/(A - B \ln Z) \qquad (7)$$

where H_I is the distance from the inner level to the beaker lip. We also make an additive correction of 0.07 cm to Z for both emptying and filling fits to compensate for the observed capillary rise of the He in the beaker. Transfer rates from three beaker emptyings as well as the corresponding fitted curves to Eq. (7) are shown in Fig. 3.

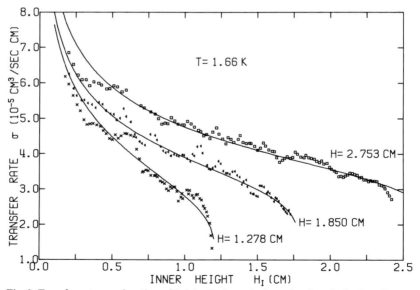

Fig. 3. Transfer rate as a function of height H_I from the inner level to the beaker rim at three different heights H from the outer level to the beaker rim. The solid curves are of the form $\sigma = H_I^{-1/4}/(A - B \ln Z)$.

Equation (7) already includes the H dependence, so we calculate σ_0 ($Z = 1$ cm) and $\sigma_0(Z = 0.1$ cm) directly from the fit parameters A and B. Mean values for σ_0 from our data are displayed in Table I. These emptying data were obtained using the same beaker as used for the filling data.

Since one expects V_s to approach $V_{s,c}$ for small pressure heads, the transfer rates we have observed are lower than predicted by Eq. (4). We have numerical agreement with Eq. (4) for pressure heads of about 1 cm. Also, our observed temperature dependence seems to be weaker than is predicted by Eq. (4).

In summary, then, we have made solid Ar beakers and studied emptying and filling of the beakers by superfluid film transfer. We have studied the dependence of the transfer rate σ on level difference, film height, and temperature. We found that the observed film height dependence and observed dependence on driving force were generally in good agreement with theory. Our flow rates were 20–30% lower than expected from theory. The results of beaker emptyings and fillings usually agree with each other to 10%.

References

1. C. J. Duthler and G. L. Pollack, *Phys. Rev.* **A3**, 191 (1971).
2. W. E. Keller, *Helium-3 and Helium-4*, Plenum, New York (1969).
3. J. Frenkel, *J. Phys. USSR* **2**, 365 (1940).
4. G. L. Pollack, *Rev. Mod. Phys.* **36**, 748 (1964).
5. W. M. Alphen, G. J. Van Haasteren, R. De Bruyn Ouboter, and K. W. Taconis, *Phys. Lett.* **20**, 474 (1966).
6. D. R. Williams and M. Chester, *Phys. Rev.* **A4**, 707 (1971).
7. J. S. Langer and J. D. Reppy, *Prog. in Low Temp. Phys.*, C. J. Gorter, ed., North-Holland, Amsterdam (1970), Vol. VI, p. 1.
8. M. Chester and R. Ziff, *J. Low Temp. Phys.* **5**, 285 (1971).

Preferred Flow Rates in the Helium II Film*

R.F. Harris-Lowe and R.R. Turkington

Royal Military College
Kingston, Ontario, Canada

We wish to report on our most recent results in an extended series of experiments designed to investigate the reproducibility of helium film flow rates out of a Pyrex beaker at a temperature of 1.35°K. The experimental apparatus and procedures are described in detail elsewhere.[1] The basic technique involved the use of a coaxial capacitance depth gauge and associated bridge circuit to monitor changes in the helium level in the beaker. The method allowed an accurate determination of the flow rate every 10 sec. A fountain pump in the experiment space was used to change the level in the beaker by controlled amounts.

Our method resulted in an improvement in the resolution by a factor of about 40 over any previous measurements. Figure 1 displays the results obtained during an experiment in which the beaker was filled to the rim and allowed to empty via

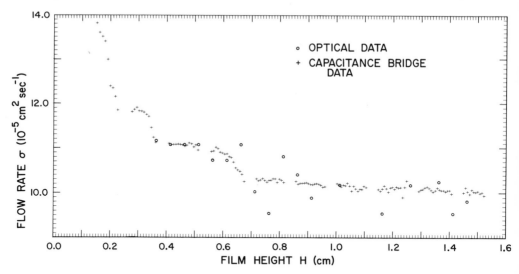

Fig. 1. Experimental values for the helium II film flow rate σ versus the film height H, where σ is the volume flow rate in $cm^3 sec^{-1} cm^{-1}$ of beaker periphery and the film height H is the distance between the helium surface in the beaker and the beaker rim. The optical data were calculated from timed cathetometer measurements of the position of the helium surface within the beaker.

* The research for this paper was supported by the Defence Research Board of Canada, Grant number 9510-23.

the film. The optical data presented, which compare favorably with any previous measurements, were obtained at the same time as the capacitance bridge data. The improvement in resolution is obvious, and the latter data allow us to observe a structure in the flow rate versus film height data which is normally present but which is completely obscured in the results obtained optically. It is clear that many of the discrepancies between the results of various workers who studied the film height dependence[2] were more apparent than real, and were in fact a consequence of the low resolution available optically.

One major difficulty in our efforts to study the reproducible features of the above structure has been the effect of long-term drift in the bridge circuit electronics. It takes approximately 1hr to obtain even a moderate amount of flow rate data, and in this time a change in the overall system gain of greater than 0.2 % can result in serious difficulties when the data are analyzed.

Figure 2 displays the results of an experiment which was carried out in a manner designed to obtain a great deal of data and at the same time compensate for the effects of long-term drift. Briefly, the experiment consisted in seven successive emptyings from the rim. In each, once the helium level reached the lower limit of a selected narrow range of values (0.56 cm $\leq H \leq$ 0.66 cm) the fountain pump was employed to raise the level to the upper limit. Observations were continued and the procedure was repeated until we had obtained data on the flow rate in this interval through ten successive passes. A correction was then applied to the results of each emptying to compensate for changes in the overall system gain, which was monitored at regular intervals throughout the experiment. Figure 2 (a) shows the results from one of the seven emptyings and Fig. 2 (b) shows the combined results for all seven. The features displayed on this figure are characteristic of all the data obtained to date. First, we note the sequence of sharp transitions in flow rate between stable values which show only a small height dependence. Second, there is an obvious tendency for these transitions to occur at the same film heights, in this case with approximate values of 0.19 and 0.36 cm. Another prominent feature is the layered appearance of the results obtained over the film height interval 0.56–0.66 cm. There are approximately 830 data points plotted in this interval which were obtained during 62 separate passes, and clearly they form four distinct and two indistinct levels.

The statistical distribution of these measurements can be seen more clearly in Fig. 3, where they have been presented in histogram form. Figure 3 (a) compiles all the results obtained over a film height interval of 0.5 mm and Fig. 3 (b) all the results over a larger interval of 1.0 mm. Here the layers of Fig. 2 show up as a set of distinct peaks with an approximate even spacing of 0.26×10^{-5} cm^2 sec^{-1}. This magnitude of spacing between preferred levels is considerably less than that reported from early optical data which first suggested the possibility of preferred rates.[3] However, the difference can be accounted for as a direct result of the poor resolution of the optical data, which made necessary the combination of data taken over a film height range of 30 mm. The effect of increasing the film height interval even to 1 mm can be seen by examining Figs. 3 (a) and 3 (b). The increase in interval both broadens the peaks and completely obliterates the evidence for one indistinct peak indicated by an arrow.

The features displayed in these results allow us to draw certain conclusions about the present status of flow rate measurement and theory. First, the existence of a

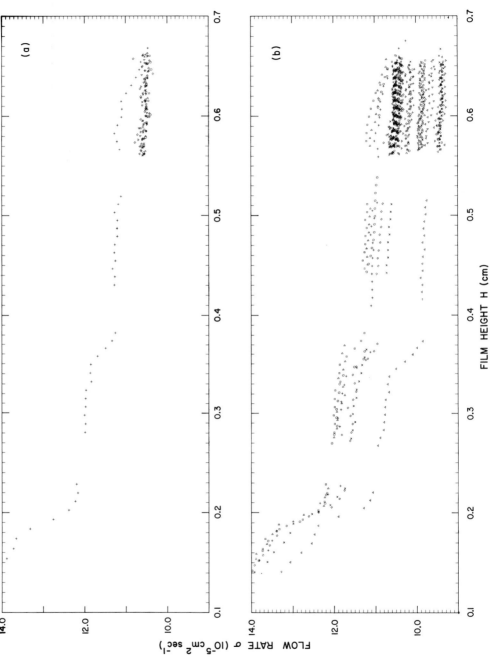

Fig. 2. Experimental values for the helium II film flow rate σ versus the film height H. (a) The results of one of the seven successive emptyings from

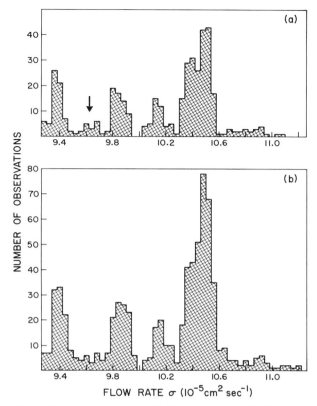

Fig. 3. Flow rate histograms. (a) All the results obtained over a film height interval of 0.5 mm; (b) all the results over a larger interval of 1 mm.

stepped flow rate versus film height profile means that earlier attempts to describe the flow-rate height dependence by a simple functional relationship based on low-resolution data have to be reexamined. Second, it is clear that the flow rate cannot be characterized by a single well-defined value, even when the substrate and other experimental parameters are fixed. This fact, which has not been generally recognized, means that we should reexamine attempts to relate changes in flow rate to changes in the various experimental parameters such as temperature or substrate. Until we can understand the nature of the observed spontaneous changes in flow rate it is impossible to distinguish between these changes and those which are the result of the variation of some experimental parameter. Finally, these results exhibit a variety of features which present critical velocity theories for narrow channels do not even start to describe. This fact casts considerable doubt on the employment of such theories to describe flow in more complicated geometries when they are obviously inadequate when dealing with the relatively simple geometry of film experiments. Hopefully, the phenomena displayed here will provide at least some clues toward the development of an adequate theory of critical velocities both in the film and in more complicated channels. In particular, the existence of preferred flow rate values, at least on a short-term basis, and the clear tendency for transitions to occur at specific values of the film height will have to be accounted for by a successful theory.

References

1. R. R. Turkington and R. F. Harris-Lowe, *J. Low Temp. Phys.* **10** (to be published).
2. L. C. Jackson and L. G. Grimes, *Phil. Mag.* **7**, 454 (1958).
3. R. F. Harris-Lowe, C. F. Mate, K. L. McCloud, and J. G. Daunt, *Phys. Lett.* **20**, 126 (1966).

Superfluidity of Thin Helium Films[*]

H. W. Chan, A. W. Yanof, F. D. M. Pobell[†], and J. D. Reppy

Laboratory of Atomic and Solid State Physics, Cornell University
Ithaca, New York

The question of the existence of superfluidity in unsaturated ^{4}He films has been investigated by a number of experimental techniques. The techniques used include mass and thermal transport,[1-5] propagation of third sound,[6,7] and the presence of long-lived persistent current flow states.[8] The onset temperature for superfluidity is found to decrease as the film thickness is reduced.

The present experiment attempts to examine the question of superfluidity in the thinnest possible films, one or two atomic layers adsorbed on a solid substrate. We take the existence of stable, nondissipative flow states as the criterion for superfluidity. This avoids the possible ambiguities which appear in measurements relying on the mobility of the helium film. Indeed, Herb and Dash[9] recently reported a precursor of mass transport above T_λ as well as two distinct onset temperatures of enhanced mobility below T_λ. Third-sound measurements, while useful in work on somewhat thicker films, may not be possible in the very thin films of one or two atomic layers.

The experimental method is similar to that used by Henkel *et al.*,[8] who observed persistent currents in unsaturated films with onset temperatures above 1°K. In the present experiment the cryostat is equipped with a recirculating ^{3}He refrigerator so that persistent currents can be created and observed at temperatures as low as 0.5°K. A substrate of porous Vycor glass is used to provide a large surface area for the film. The thermal properties of helium in Vycor have been extensively studied by the Sussex group.[10]

The helium film is formed by introducing a measured quantity of gas into a sample volume containing the Vycor substrate. The surface area and open volume of our Vycor sample are determined by a BET analysis of a nitrogen adsorption isotherm. We have used a molecular area of 16.2Å^2 for the nitrogen. This yields an estimated area of 8030 m^2 and an open volume of 13.0 cm^3 for the 66.84 g of Vycor contained in the sample chamber. A volume of 12.1 liters STP of ^{4}He gas is required to completely fill the pores of this Vycor with liquid. The average density of the liquid in the pores is therefore 0.168 g cm^{-3}. The average full-pore density of helium is reduced to 0.158 g cm^{-3} when a monolayer (3.78 liters STP) of neon is preplated on the substrate. (The preplated neon is assumed to have the density of bulk, solid neon.) These average helium densities are larger than 0.145 g cm^{-3}, the value for bulk helium at 1°K.

* Work supported in part by the National Science Foundation through Grant No. GP-27736 and also under Grant No. GH-33637 through the Cornell Materials Science Center, Report No. 1859.
† Present address: KFA, Jülich, West Germany.

In the neon preplated run 3.45 liters STP of helium were first introduced into the sample chamber; but the attempt to create and observe persistent currents at this coverage was unsuccessful. This means that superfluidity does not exist in this temperature-coverage domain, above 0.5°K and below 3.45 liters STP of helium. When 4.16 liters STP of helium was adsorbed onto the substrate a signal proportional to the angular momentum was barely discernible above the noise; but an additional 0.17 liter STP of helium on the substrate caused a thirty-fold increase in the signal size. Temperature was slowly increased until the current completely decayed. The above procedure was repeated for successively higher coverages of helium until the Vycor pores were full. The conditions of formation of persistent currents at all coverages were standardized as follows: The sample chamber was warmed above the lambda point, rotated at 175 rpm, and cooled to 0.45°K under rotation; rotation at 0.45°K was continued for 1 hr to ensure thermal relaxation. The angular momentum measured at 0.5°K prior to decay is plotted in Fig. 1 as a function of helium coverage.

Figure 2 traces out L_p, the angular momentum, as a function of temperature for four different coverages. There is a reversible portion in the low-temperature range of each curve. When the temperature is brought up to a particular point instability appears in the superflow and the signal decreases irreversibly. At low coverages the decay is sharp in temperature and rapid in time; in the case of the 4.48 liters STP film at 0.59°K the signal decreased by 60% of the undecayed value for each decade of time of observation. We have chosen to define the onset temperature as the temperature of maximum instability of flow. Note that this definition is different from that of transport and wave propagation experiments. Onset temperatures of He film on bare Vycor were obtained following the same procedure.

In Fig. 3 onset temperature as a function of ^{4}He coverage is shown for both the bare Vycor and the neon-preplated substrate. Onset temperature increases linearly

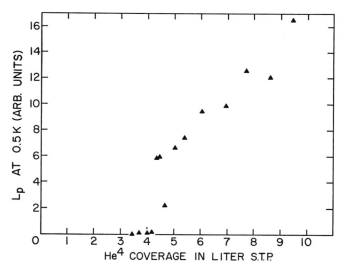

Fig. 1. Initial angular momentum at 0.5°K prior to decay. Angular momentum L_p in arbitrary units is plotted as a function of ^{4}He coverage.

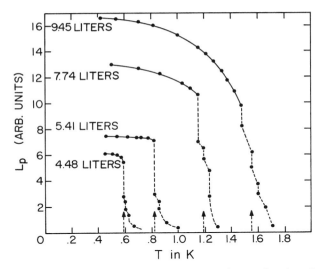

Fig. 2. Angular momentum L_p in arbitrary units as a function of temperature for four different coverages. The onset temperature at each coverage is indicated by a dashed arrow.

with the quantity of helium adsorbed. Note these two interesting features: First, for a given onset temperature the quantity of helium needed on the neon-preplated substrate is about 2.5 liters STP less than that on bare Vycor; second, the onset temperature is the same for full pores with or without neon preplating. A possible

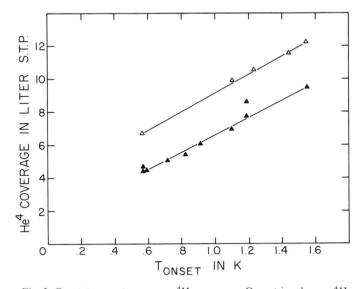

Fig. 3. Onset temperature versus ^{4}He coverage. Open triangles are ^{4}He film on bare Vycor substrate and closed triangles are ^{4}He film on Vycor preplated with one layer of neon. The two highest coverage points represent full pores.

inference is that the first layer of helium on bare Vycor does not participate in super-flow and can be replaced by neon. Translating the ^{4}He coverage into film thickness, it is found that a ^{4}He film of approximately 10 Å on bare Vycor and a film of about 7 Å on neon-preplated Vycor have the same onset temperature of 0.57°K.

The observation of stable superflow in a film of two atomic layers at 0.5°K encourages us to extend this work to lower temperatures where one hopes to study superfluidity in still thinner films.

References

1. E. Long and L. Meyer, *Phys. Rev.* **79**, 1031 (1950).
2. R. Bowers, D. F. Brewer, and K. Mendelssohn, *Phil. Mag.* **42**, 1445 (1951).
3. E. Long and L. Meyer, *Phys. Rev.* **98**, 1616 (1955).
4. D. F. Brewer and K. Mendelssohn, *Proc. Roy. Soc.* **A260**, 1 (1960).
5. K. Fokkens, K. W. Taconis, and R. DeBruyn Ouboter, *Physica* **32**, 2129 (1966).
6. K. R. Atkins, *Phys. Rev.* **113**, 962 (1959).
7. I. Rudnick, E. Guyon, K. A. Shapiro, and S. A. Scott, *Phys. Rev. Lett.* **19**, 488 (1967).
8. R. P. Henkel, G. Kukich, and J. D. Reppy, in *Proc. 11th Intern. Conf. Low Temp. Phys., 1968,* St. Andrews University Press, Scotland (1969).
9. J. A. Herb and J. G. Dash, this volume.
10. D. F. Brewer, *J. Low Temp. Phys.* **3**, 205 (1970); A. J. Symonds, Ph.D. Thesis, Univ. of Sussex, unpublished.

Superfluidity in Thin ^{3}He–^{4}He Films*

B. Ratnam and J. Mochel†

Department of Physics and Materials Research Laboratory, University of Illinois
Urbana, Illinois

Third-sound resonance experiments[1,2] have shown that the superfluid helium film can be studied with remarkable precision. In a series of experiments to study the effect of ^{3}He on a ^{4}He film of constant thickness we have observed several qualitative changes in the propagation of third sound.

The resonance cell used was an improved cell of the type described in Ref. 1. Slips of clear, fused quartz were edge flame sealed to make a thin, rectangular cell whose interferometrically measured average inner separation was 4.26 μm, which is an order of magnitude smaller than for the cell of Ref. 1. The inner cell length was 1.79 cm, which corresponds to a half wavelength at resonance. An atmosphere of argon sealed within the cell provided a well-characterized substrate‡ at low temperatures. Controlled amounts of ^{3}He (P_3) and ^{4}He (P_4) were diffused for several hours *in situ* into the cell, and the cell was rapidly cooled to liquid nitrogen temperatures. The room-temperature helium diffusion time constant was ~ 1 hr. Runs weeks apart were found to reproduce to within several percent.

The cell resonant frequency f_0 corresponding to a third-sound velocity $C_3 = 3.58 f_0$ cm sec^{-1} and the resonance $Q = f_0/\Delta f_0$, where Δf_0 is the resonance width, were simultaneously measured for several ^{3}He–^{4}He films. Figure 1 shows the resonant frequency as a function of temperature for a 3.87 atomic layer ^{4}He film (an additional layer of ^{4}He is considered frozen to the substrate) with successive nominal additions of 0, 0.5, 1, 2, and 4 atomic layers (at $T = 0°$K) of ^{3}He. The corresponding 300°K fill pressures P_4 and P_3 were, respectively, 552 and 0; 551 and 50; 554 and 100; 555 and 199; and 557 and 400 mm Hg. The densities of the ^{3}He and ^{4}He liquid layers were assumed to have their bulk values.

The pure ^{4}He film was approximately of constant thickness over most of the temperature range of interest. Film thickness calculations for the ^{3}He–^{4}He films are difficult since the two-component thermodynamics is a function of film thickness. Recent measurements however, confirm the existence of ^{3}He surface states with 2°K lower energy than bulk ^{3}He states in superfluid ^{4}He.[4,5] Furthermore, the large-surface-area heat capacity experiments of Brewer[6-8] with ^{3}He and ^{4}He are best understood by assuming the ^{3}He goes to the ^{4}He surface. Finally, the larger zero-point volume of the ^{3}He atom in a rapidly changing van der Waals potential

* This research was supported in part by the Advanced Research Projects Agency under Contract HC 15-67-C-0221 and National Science Foundation Grant GH 33634.
† A.P. Sloan Fellow.
‡ Anderson and Sabisky[3] have calculated the van der Waals potential for helium films on an argon substrate.

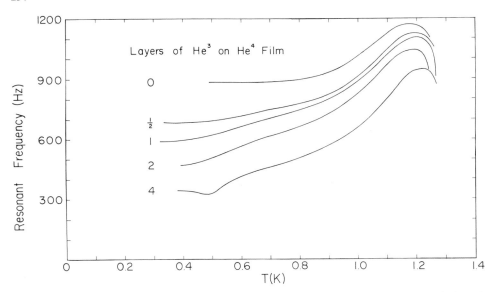

Fig. 1. Curves of third-sound cell resonant frequency f_0 versus temperature for a 3.87 atomic layer (at $T = 0°$K) liquid ^{4}He film with (top to bottom) 0, 0.53, 1.1, 2.1, and 4.3 atomic layers (at $T = 0°$K) of ^{3}He. The third-sound velocity C_3 is given by $C_3 = 3.58 f_0$ cm sec^{-1}. The transition temperatures for third-sound propagation are temperatures above which no third sound is seen ($Q \lesssim 1$).

would tend to push up the ^{3}He to the surface of ^{4}He. Thus as the cell is cooled the condensing ^{3}He progressively coats the ^{4}He film.

The behavior of the resonant frequencies exhibits three especially noticeable features: (1) The transition temperature for third-sound propagation is unaffected by the presence of ^{3}He; (2) the resonant frequency is smaller for the ^{3}He–^{4}He films; and (3) the resonant frequency temperature dependence changes as one goes from zero to four atomic layers of ^{3}He.

Several effects must combine to give an unchanged onset temperature, the transition temperature for third-sound propagation. The ^{3}He in surface states[4,5] near the onset temperature of the richest ^{3}He film is approximately 0.5 atomic layer, where we have used the Andréev–Kompaneets[9] expression

$$n_s \approx (c/T^{1/2}) e^{\varepsilon_0/T}$$

where n_s is the density of occupied ^{3}He surface states, c is the ^{3}He concentration in the film, estimated at 0.1, and $\varepsilon_0 \approx 2°$K.[5] However, with the exception of the two-atomic-layer ^{3}He experiments, all onset temperatures fall within a 10 m°K temperature interval. We would have expected the addition of ^{3}He to shift the onset temperature through three effects: (i) The healing length[10,11] is decreased; (ii) the film thickness is increased; and (iii) the superfluid density is decreased through a lower ^{3}He–^{4}He bulk superfluid transition temperature. The first two effects would increase the onset temperature and the third would reduce it.

The simplest interpretation of the reduction in the velocity of third sound is that F', the van der Waals force per unit mass for the ^{3}He–^{4}He film, is "geometrically" different from F, the corresponding force for the pure ^{4}He film. If we assume that

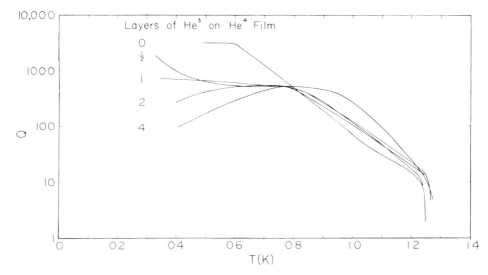

Fig. 2. Curves of third-sound cell $Q = f_0/\Delta f_0$, where f_0 is the resonant frequency and Δf_0 the resonance width, versus temperature for a 3.87 atomic layer (at $T = 0$ K) liquid ^{4}He film with (a) 0, (b) 0.53. (c) 1.1, (d) 2.1, and (c) 4.3 atomic layers (at $T = 0$ K) of ^{3}He. The transition temperatures for third-sound propagation are the temperatures for which $Q \lesssim 1$. Note the anomalous behavior of the one-half atomic layer ^{3}He on ^{4}He film.

ρ_s, the average superfluid density of a ^{4}He film, is unaffected by the addition of ^{3}He, the ratio of the third-sound velocity in the ^{3}He–^{4}He film to the third-sound velocity in the ^{4}He film is*

$$\frac{C_3'}{C_3} = \left(\frac{F'}{F}\right)^{1/2} = \left[\left(1 - \frac{\rho_3}{\rho_4}\frac{4}{3}\right) + \frac{\rho_3}{\rho_4}\frac{4}{3}\left(\frac{d_4}{d_4 + d_3}\right)^4\right]^{1/2} \tag{1}$$

where ρ is the density, d is the liquid film thickness, the prime refers to the ^{3}He–^{4}He film, and the subscripts 3 and 4 refer to ^{3}He and ^{4}He, respectively. The final expression is a sum of two terms. The first is an interfacial (liquid ^{3}He to liquid ^{4}He) term due to the different atomic volumes of ^{3}He and ^{4}He and the second is the "geometric" term due to the thickening of the film. The calculated ρ_3/ρ_4 values from Eq. (1) of 0.752, 0.659, 0.654, and 0.667 for 0.5, 1, 2, and 4 atomic layers of ^{3}He on 4 atomic layers of ^{4}He are, with the exception of the first, surprisingly close and can be compared with the bulk ratio of 0.565. This also means that the volume per atom due to zero-point motion for ^{3}He is only 12% larger than for ^{4}He in the film.

The resonance Q for the various ^{3}He–^{4}He films is displayed as a function of temperature in Fig. 2. The Q measurements were taken in one of two ways: by manual adjustment of the oscillator frequency to give a $\pi/4$ phase shift on either side of the resonance signal, or by observing the logarithmic decrement of the resonance signal. At the highest Q's measured the resonance signal decayed with a typical

* We use the third-sound equation $C_3^2 = \bar{\rho}_s/\rho F d$.

time constant of several seconds. The pure ^{4}He curve saturates at the lowest temperatures. In our preliminary cells we have observed the Q to increase as the cell separation was decreased, indicating that our dominant third-sound attenuation has been vapor-phase damping.[12] The Q has further been observed to depend on film thickness. No detailed study of these dependences has yet been made.

The Q curves display three noticeable features. (1) The resonance Q's span four decades; (2) at high temperatures increasing amounts of ^{3}He raise the Q, which increases with decreasing temperature; and (3) at low temperatures the Q is generally depressed either with increasing amounts of ^{3}He or with decreasing temperature.

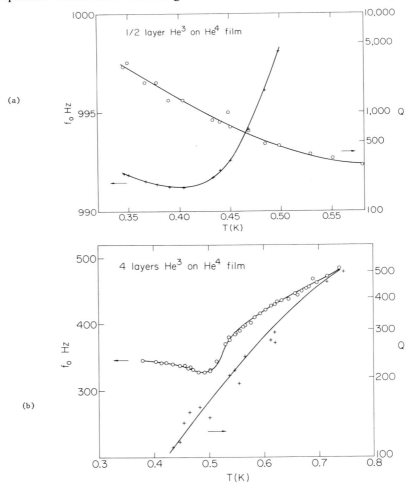

Fig. 3. An exploded low-temperature plot of the resonant frequency and Q vs. temperature for (a) 0.53 atomic layers (at $T = 0°$K) of ^{3}He on 3.94 atomic layers (at $T = 0°$K) of liquid ^{4}He and (b) 4.3 atomic layers (at $T = 0°$K) of ^{3}He on 4.09 atomic layers (at $T = 0°$K) of liquid ^{4}He Fig. 3(b). In (a) we see the surface states filling with ^{3}He and thus reducing the ^{3}He concentration in the ^{4}He film. In (b) we see the ^{3}He–^{4}He phase separation at an unexpectedly low temperature.

In preliminary pure ^{4}He runs we observed Q's $\approx 3 \times 10^4$ for films thinner than the one reported here. This unusually wide range of Q values reinforces our belief that the resonance cells provide a well-isolated thermodynamic state of the superfluid film.

At high temperatures the vapor above the film determines the resonance Q. We argue that the ^{3}He reduces the strength of the capillary waves at the surface of the ^{4}He film.[13] These capillary waves could provide the means for heat transfer to the vapor phase. The ^{3}He could also suppress direct evaporation of ^{4}He atoms from the superfluid state.

In our 4.26-μm separation cell at low temperatures the vapor phase does not produce significant dissipation. The ^{3}He present then dominates the Q. The low-temperature measurements on a one-half atomic layer ^{3}He on a four atomic layer ^{4}He film are shown in Fig. 3(a). Note that the velocity begins to rise as the Q is increasing. This behavior can be understood in the following way: With the vapor phase depleted the ^{3}He concentration within the ^{4}He film begins to decrease and the density of ^{3}He surface states begins to increase. Thus, it is the ^{3}He atoms in solution which reduce the Q at low temperatures. Additional ^{3}He beyond the first atomic layer also goes into solution and further reduces the Q.

Figure 3(b) shows the low-temperature tails of the velocity and Q curves for the four atomic layer ^{3}He on a four atomic layer ^{4}He film. Anomalously large heater powers were required for these resonances. Furthermore, the power dependence of the resonant frequency became large and changed sign, going through the minimum. The plotted resonant frequencies have been corrected to zero power. Bulk thermodynamics suggests a phase separation temperature of $\sim 0.7°$K for these [Fig. 3(b)] ^{3}He concentrations, which is considerably higher than the temperature of the minimum.

The dispersion of third sound is measured by the harmonics of the resonance cell. A temperature-independent deviation from $C_3 = \omega/k$ of about 7% was observed. We tend to ascribe this to geometric effects, which at present we do not fully understand. Generally all measurements were made where f_0 and Q were insensitive to further reductions in power.

We have shown that small quantities of ^{3}He in an adsorbed superfluid ^{4}He film leave the critical onset temperature unchanged but profoundly affect the third-sound velocity and attenuation. These experiments, while presently producing more questions than answers, can be used to study ^{3}He in proximity to a superfluid. This is perhaps the next best thing to studying ^{3}He in a superfluid state.

Acknowledgments

We would like to thank T.E. Washburn for assistance in taking some of the experimental data.

References

1. B. Ratnam and J. Mochel, *Phys. Rev. Lett.* **25**, 711 (1970).
2. B. Ratnam and J. Mochel, *J. Low Temp. Phys.* **3**, 239 (1970).
3. C. M. Anderson and E. S. Sabisky (private communication).
4. K. N. Zinov'eva and S. T. Bolarev, *Zh. Eksperim. i Teor. Fiz.* **56**, 1089 (1969) [*Soviet Phys.—JETP* **29**, 585 (1969)].

5. M. M. Gvo, D. O. Edwards, R. E. Saruwinski, and J. T. Tough, *Bull. Am. Phys. Soc.* **16**, 87 (1971).
6. D. F. Brewer, *J. Low Temp. Phys.* **3**, 205 (1970).
7. D. F. Brewer, A. Evenson, and A. L. Thomson, *J. Low Temp. Phys.* **3**, 603 (1970).
8. D. F. Brewer, A. Evenson, and A. L. Thomson, *Phys. Lett.* **35A**, 307 (1971).
9. A. F. Andréev and D. A. Kompaneets, *Zh. Eksperim. i Teor. Fiz.* **61**, 2459 (1971).
10. V. L. Ginzburg and L. P. Pitaevskii, *Zh. Eksperim. i Teor. Fiz.* **34**, 1240 (1958) [*Soviet Phys.—JETP* **7**, 858 (1958)].
11. Yu. G. Mamaladze, *Zh. Eksperim. i Teor. Fiz.* **52**, 729 (1967) [*Soviet Phys.—JETP* **25**, 479 (1967)].
12. D. Bergman, *Phys. Rev.* **188**, 370 (1969).
13. K. R. Atkins, *Can. J. Phys.* **31**, 1165 (1953).

On the Absence of Moderate-Velocity Persistent Currents in He II Films Adsorbed on Large Cylinders

T. Wang* and I. Rudnick

Physics Department, University of California
Los Angeles, California

Consider a glass cylinder approximately 1 in. in diameter as shown in Fig. 1, and suppose there is an adsorbed film of helium on the outer surface. Suppose this film is rotating about the axis of the cylinder. Then the superfluid velocity v_s must obey the Onsager–Feynman relation

$$\int v_s \cdot dl = nh/m \qquad (1)$$

where h is Planck's constant, n is an integer, and m is the mass of the helium atom. Moreover, if the speed is not great, it is energetically unfavorable to have vortex lines (parallel to the axis of the cylinder), and the circulation given by Eq. (1) is the same along any path lying in the film and encircling the cylinder. The helium is in a unique quantum state and one which may be characterized by a very high quantum number (if $v_s \simeq 10$ cm sec^{-1} then $n \simeq 10^5$). The velocity v_s must vary as r^{-1}, where r is the radius, if the circulation is to be a constant, but since the films we are concerned with are always significantly thinner than 10^2 Å, the velocity should be sensibly constant throughout the film.

Persistent currents of high velocity have been observed under conditions where helium fills the voids in a finely porous substance.[1,2] What is more to the point is that they have been observed by Reppy and co-workers[3-5] in the adsorbed films which coat such porous substances with average speeds up to about 50 cm sec^{-1}.[3] Thus we are here investigating the possible existence of a *particular* type of persistent current in a film—the existence of persistent currents has already been demonstrated.

In Fig. 1 the three third-sound strips, which are thin evaporated films of aluminum, are arranged axially at equal intervals around the cylinder. If two are operated as pulsed third-sound sources, then the difference in arrival times at the third strip can be used to find the velocity of the superfluid relative to the cylinder.

In one such experiment the glass cylinder was 1 in. in diameter and 2 in. long. Through the use of slip rings and brushes, measurements could be made while the cylinder was rotating. The cylinder was contained in a can, which in turn was immersed in a temperature-stabilized liquid helium bath. Helium gas was admitted into the can at a pressure p less than the saturated vapor pressure leading to an

---

* Present address: Physics Section, Jet Propulsion Laboratory, Pasadena, California.

239

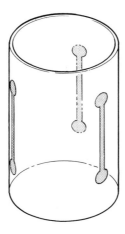

Fig. 1. Persistent current cylinder.

adsorbed film on the glass cylinder. The attenuation of third sound is high for very thin and very thick adsorbed helium films.[6] The pressure p was chosen so that the film was of intermediate thickness. The time of flight of third-sound pulses with and against the velocity of the rotating surface was determined at several speeds of rotation with increasing and decreasing rotational speeds. Some data are shown in Fig. 2, where we plot the speed of rotation of the superfluid component v_s against the surface velocity of the glass cylinder v_{sub}. The thickness d of the film can be calculated from the relation $d^3 = (T_v/T) \ln (p_0/p)$, where T_v is the "van der Waals temperature," T is the temperature, p_0 is the saturated vapor pressure, and p is the pressure of the vapor in equilibrium with the film. We have used $T_v = 87$ (layers)^3 °K consistently in the past for glass substrates. With this value the film thicknesses at 1.21, 1.41, 1.54, and 1.68°K are respectively 5.3, 9, 12.5, and 16.7 atomic layers. There is increasing evidence that T_v is much smaller. If we use the value of 30 calculated by Sabisky and Anderson[7] from the macroscopic theory of van der Waals forces by Lifshitz, the corresponding thicknesses are 3.7, 6.3, 8.8, and 11.7 atomic layers. The line in the figure is at an angle of 45° and it is clear that the superfluid moves with the substrate and that there is no evidence of a persistent current.

A defect in the measurement stemmed from the fact that the times of flight in the two directions were independently determined. This required at least 2 min and if there were a short-lived persistent current, it might not be detected. Toward the end of discovering such persistent currents we alternately triggered the two third-sound sources. The oscilloscope was gated to receive the alternate signals, respectively, on its A and B channels, the triggering and gating frequency being between 20 and 60 Hz. In this way we could examine the two signals for any small difference in time of flight. Within an approximate error of 1 cm sec^{-1} we were unable to detect evidence of a persistent current during any phase of the rotation.

The role of the helium gas in the can deserves attention. In the steady state it rotates with the cylinder if there were a persistent current, there would be a relative velocity between the superfluid and the gas. One might expect that as a result of

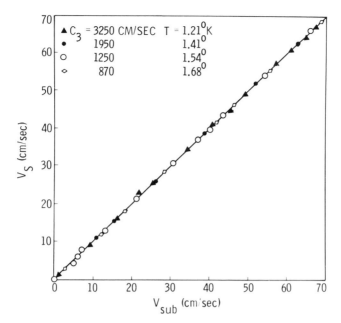

Fig. 2. The velocity of superfluid vs. rotation velocity.

the continuous evaporation and condensation processes taking place there would be a transfer of momentum from the gas to the superfluid which would eliminate this relative velocity. Several points can be made which show this line of reasoning to be incorrect. (1) Osborne[8] has shown that in a steady heat current parallel to the surface of He II the vapor above that surface has the velocity of the normal component in the counterflow. (2) Putterman,[9] and subsequently Goodstein and Saffman,[10] showed that it is a consequence of two-fluid hydrodynamics that momentum balance occurs between the vapor and the normal component of the helium. (3) The presence of vapor did not destroy the persistent current in the experiments cited earlier.[3-5] Thus the absence of a persistent current cannot be blamed on the presence of the gas.

A novel type of resonator was also used. It consists of two closely fitted glass cylinders one inside the other. They were cut from a medical syringe, the inner cylinder coming from the plunger and the outer one from the body of the syringe. The annular space was approximately 1 in. in diameter and 2 in. in length. The cylinders were polished until they were optically transparent. The outer cylinder was glued at its edges to the inner cylinder after the third-sound strips were deposited on the inner cylinder. It is probable that the cylinders were in contact but this contact was at points rather than lines, since we observed third-sound signals propagating completely around the cylinder. The signal-to-noise ratio was slightly better than that obtained with the single cylinder. Again, we never observed any relative motion between the substrate and superfluid.

We also addressed ourselves to the question of whether the method of generating and detecting the third sound destroyed the persistent current. The dc current in the detecting strip is about 100 μA, which is two orders of magnitude smaller than the power needed to raise the strip to its normal conducting state. Thus this strip is at the temperature of the helium film. Moreover, we have observed third-sound pulses which travel several times around a cylinder. However, the strip used as a source is known to be hotter than the film since it is driven to its normal conducting state by the input pulse. Using the arrangement of Fig. 1, we made the measurements with a single pulse only to each of the two sources, comparing travel times on a storage oscilloscope. This pulse had a duration of 1 m sec, and even if the film at the source was not superfluid for this length of time, it is difficult to see how this by itself could absorb all the momentum of a persistent current. Again, we never observed a persistent current.

We have described experiments where the rotations are initiated while $T < T_\lambda$. We have also initiated rotations at $T > T_\lambda$ with the rotation continuing as T is reduced below T_λ. No relative velocity was observed when the rotation was then gradually reduced to zero.

Finally, we would remark that this is an experiment with a negative result— we set out to observe a certain phenomenon and failed in its observation. The significance of such a failure depends crucially on the quality of the effort. We believe the effort was a good one and have ourselves accepted the tentative conclusion that persistent currents of velocity greater than about 1 cm sec^{-1} on cylinders greater than about 1 cm are not observable. Persistent currents are known to be metastable and as such are characterized by a decay time. Perhaps our results suggest that the decay time for a metastable superfluid quantum state, characterized entirely by a quantum number n which is very different from the equilibrium value is very short in our case and very long for films adsorbed on a porous system. After all, in the latter case there are many ways in which the elementary circulations can change their shape and location without changing the total circulation, and there is no way in our case that this can be done.

References

1. J.S. Langer and J.D. Reppy in *Progress in Low Temperature Physics*, C.J. Gorter, ed. North-Holland, Amsterdam (1970), Vol. VI.
2. H. Kojima, W. Veith, S.J. Putterman, E. Guyon, and I. Rudnick, *Phys. Rev. Lett.* **27**, 11, 714 (1971).
3. R.P. Henkel, G. Kukich, and J.D. Reppy, in *Proc. 11th Intern. Conf. Low Temp. Phys. 1968*, St. Andrews University Press, Scotland (1969), p. 178.
4. R.P. Henkel, E. Smith, and J.D. Reppy, *Phys. Rev. Lett.* **23**, 1276 (1969).
5. H.W. Chan, A.W. Yanof, F.D.M. Pobell, and J.D. Reppy, this volume.
6. T.G. Wang and I. Rudnick, *J. Low Temp. Phys.* (to be published).
7. E.H. Sabisky and C.H. Anderson, preprint; and this volume.
8. D.V. Osborne, *Proc. Phys. Soc.* **80**, 1343 (1962).
9. S.J. Putterman, Doctoral Thesis, Rockefeller Univ., New York (1970).
10. D.L. Goodstein and P.G. Saffman, *Proc. Roy. Soc. London* **A325**, 447 (1971).

Thermodynamics of Superflow in the Helium Film*

D.L. Goodstein and P.G. Saffman

California Institute of Technology
Pasadena, California

The saturated superfluid film is capable of acting as a siphon, allowing the superfluid to drain out of any containing vessel to find the lowest available gravitational level. As in any siphon, the pressure in the film is lower when it is flowing than when it is at rest, owing to the Bernoulli effect. Since the film has a free surface which presumably cannot support a pressure discontinuity, it was expected that the flowing film would collapse on itself, like a siphon with flimsy walls. In particular, at typical film flow velocities a reduction in thickness of about one-quarter was expected.[1]

On the other hand, a different kind of reasoning leads to an opposite prediction. For a stationary film in equilibrium the film–vapor interface is basically isobaric at the saturated vapor pressure P_{sv}. This is because, with respect to the bulk–vapor interface, the gas pressure is reduced gravitationally and increased by the van der Waals forces from the wall, both by the same amount. At the interface for a film thickness Z and height y above the bath the balance of gravitational and van der Waals potentials gives the familiar approximate formula

$$\chi(Z, y) = gy - \beta/Z^3 = 0 \tag{1}$$

where χ is the potential with respect to the bulk interface, β a constant, and z a coordinate normal to the wall. Now suppose the film is set into motion and the thickness is reduced. The gas far away is undisturbed, so that the gas pressure at each point retains the same hydrostatic distribution,

$$P_g(z, y) = P_{sv} \exp[- \chi(z, y)/kT] \tag{2}$$

At the new interface, at any fixed y, the gas is closer to the wall, senses a larger van der Waals force, and consequently has an equilibrium pressure higher than P_{sv}. This is unstable, causing some gas to condense and restore the original film thickness.

We thus have contradictory predictions about whether the thickness will change. It has been found experimentally that the second argument gives the correct result—no change in thickness is observed.[2] We are left with a dilemma, however, since the free surface must somehow be supporting a pressure discontinuity. According to the two-fluid model, if we take the film to be incompressible, we find

$$\frac{P(z)}{\rho} + \frac{1}{2}\frac{\rho_s}{\rho} u_s^2 + \chi(z, y) = \text{const} \tag{3}$$

* Work supported in part by the Research Corporation.

where ρ is the density, ρ_s the superfluid density, and u_s the superfluid velocity. At any point inside of the film $z \leq Z$, since χ is fixed, P must decrease when u_s rises from zero; hence the pressure discontinuity at the interface, where $P_g = P_{sv}$, remains unchanged.

In order to investigate the boundary conditions at the interface, we have examined the conditions for thermodynamic equilibrium between a film obeying the two-fluid model and its vapor.[3]* Choosing as a subsystem a cylinder of unit height and cross-sectional area with its base in the wall and its top in the gas, the condition for equilibrium is that

$$\delta(F - \mu N) = 0$$

where F is the free energy, μ the chemical potential, and N the number of helium atoms in the subsystem. The subsystem is completely determined if we specify, T, Z, and u_s^2, so that at constant temperature $F - \mu N$ must be stationary with respect to independent variations in Z and u_s^2. The free energy F is computed by integrating the free energy density (from the two-fluid model) across the film from zero to Z and across the gas from Z to one. This leaves an error in free energy owing to the transition region between film and gas, which is equal to γ, the surface tension. γ is a function of the same variables as the film–gas system, $\gamma = \gamma(T, Z, u_s)$. Adding all contributions, we find

$$\delta(F - \mu N) = \left(P_g - P + \frac{\partial \gamma}{\partial Z} \right) \delta Z + \left(\frac{1}{2} \rho_s Z + \frac{\partial \gamma}{\partial u_s^2} \right) \delta u_s^2 \qquad (4)$$

where P_g and P are the gas and film pressures evaluated at the interface. The requirement that the coefficients of δZ and δu_s^2 vanish independently give us two boundary conditions,

$$\tfrac{1}{2} \rho_s Z + (\partial \gamma / \partial u_s^2)_{T,Z} = 0$$

which may be integrated immediately,

$$\gamma + \tfrac{1}{2} \rho_s Z u_s^2 = \gamma_0(T, Z) \qquad (5)$$

where γ_0 is the surface tension at rest, and

$$P = P_g + (\partial \gamma / \partial Z)_{T,u_s^2} \qquad (6)$$

Equation (6) shows how it is possible to have a pressure discontinuity at a plane interface. The pressure P is the work required per unit change of the volume of the film. If γ depends on where the interface is, then extra work is required to change γ, in addition to ordinary compression. Thus $\partial \gamma / \partial Z$ is an effective pressure discontinuity.

Equation (5) is a sort of surface Bernoulli effect. We note in passing that it has the same form as the well-known surface tension of a type II superconductor in an applied magnetic field, except that Z must be replaced by λ, the penetration depth. Substituting in Eq. (6) we find that

$$P = P_g - \tfrac{1}{2} \rho_s u_s^2 \qquad (7)$$

* The analysis has been made more rigorous in the present paper, but does not differ in any of the results.

This is exactly the discontinuity required to keep Z constant when u_s changes, according to Eq. (3). The experimental result is thus predicted on thermodynamic grounds, and the dilemma is resolved.

The two-fluid model upon which our analysis is based is valid only below some unspecified critical velocity u_{sc}, beyond which there can be no equilibrium nondissipative flow. Since it is known that $u_{sc} \to 0$ as $Z \to \infty$, it follows that our analysis is meaningful only for small Z. The behavior of u_{sc} at very large Z is not well established, but if, for example, there is an upper bound on Zu_s, then as $Z \to \infty$, $\rho_s Zu_s^2 \to 0$, so that the correction to the surface tension becomes negligible as the film thickens. This is why it is reasonable that Z plays a role similar to the penetration depth of a superconductor, as mentioned above; equilibrium reversible superflow has only a finite penetration from its boundary.

The analysis we have made may be tested by means of an unexpected prediction it makes for the speed of third sound in a moving film. Third sound is a sloshing motion in the film, restored by the van der Waals forces. Its speed is fixed basically by the conservation of mass, together with an equation from the two-fluid model,

$$\partial u_s/\partial t + \nabla\bar\mu(T, Z, u_s^2) = 0 \qquad (8)$$

where $\bar\mu = \mu + \frac{1}{2}u_s^2$. Equation (8) reduces to Eq. (3) when $\partial u_s/\partial t = 0$. When the film is in motion the third-sound waves are carried along with the motion, but part of the convection effect arises from a term in Eq. (8) whose coefficient is $(\partial\mu/\partial u_s^2)_{T,Z}$. According to our analysis,

$$(\partial\bar\mu/\partial u_s^2)_{T,Z} = 0 \qquad (9)$$

Equation (9) is a consequence of the boundary conditions, Eqs. (5) and (6). When this result is used in the film equations of motion the result for C_3, the speed of third sound, is

$$C_3 = \frac{1}{2}\frac{\rho_s}{\rho}u_s \pm C_3^\circ\left[1 + \frac{1}{4}\left(\frac{\rho_s}{\rho}\frac{u_s}{C_3^\circ}\right)^2\right]^{1/2} \qquad (10)$$

where C_3° is the speed at zero superfluid velocity and the two solutions are for upstream and downstream propagation. If $C_3^\circ \gg u_s$, we have

$$C_3 \approx \frac{1}{2}(\rho_s/\rho)u_s \pm C_3^\circ \qquad (11)$$

Pickar and Atkins[4] have observed third sound in a moving film and analyzed their data using

$$C_3^{PA} = u_s \pm C_3^\circ \qquad (12)$$

Although Eq. (10), with its unexpected factor of one-half in the convection term, fits the experimental results better than does Eq. (12), an independent check is still needed. The reader is referred to Ref. 3 for further details.

We wish to discuss one further point concerning the attenuation of third sound. Although fundamentally simple, third sound is very complicated in detail, involving coupled disturbances in the film, the gas, and even the substrate. Bergman[5]* has studied the linearized hydrodynamic problem in detail but the result was a predicted

---

* Dr. Bergman has pointed out to us that, contrary to our belief when writing Ref. 3, the mechanism discussed here was included in his analysis.

attenuation some two or three orders of magnitude smaller than that observed in thick films. This has prompted speculation that the attenuation has other than hydrodynamic origin and requires postulation of a new quantum mechanical principle.[6] We wish to disagree with this point of view.

Our analysis has prompted us to isolate one of the many attenuation mechanisms and to compute its effect alone. Since the inverse frequency of third-sound experiments is very long ($\sim 10^{-3}$ sec) compared to all other relaxation times in the system, we suppose the gas to have the same hydrostatic distribution that led us to expect a constant film thickness under flow as described in the second paragraph of this paper. For the same reasons given there the gas tends to condense in the troughs (and evaporate at the peaks) of the wave, opposing the wave and leading to attenuation. The magnitude of this attenuation is easily calculated[3] and the result is very nearly equal to that observed in thick films (~ 0.5 cm^{-1}). This one effect alone, isolated from Bergman's analysis, yields an attenuation some two to three orders of magnitude larger than his final result.

Thus it is incorrect to believe that Bergman has shown that all the attenuation mechanisms combined cannot account for the observed damping of the wave. Instead, his analysis has given rise to a subtle and nearly exact cancellation of an attenuation of the correct magnitude. Needless to say, given a coupled set of seven linearized differential equations to deal with, there are many points at which boundary conditions, linearization procedures, and other approximations need to be tested for their effect on this delicate balance. The point to be made here is that a hydrodynamic mechanism capable of producing the observed result is indeed present in the film, so that one cannot at this point rule out hydrodynamic origin for the attenuation.

References

1. V.M. Kontorovich, *Zh. Eksperim, i. Teor. Fiz.* **30**, 805 (1956) [*Soviet Phys.—JETP* **3**, 770].
2. W.E. Keller, *Phys. Rev. Lett.* **24**, 569 (1970).
3. D.L. Goodstein and P.G. Saffman, *Proc. Roy. Soc. A* **325**, 447 (1971).
4. K.A. Pickar and K.R. Atkins, *Phys. Rev.* **178**, 389 (1969).
5. D.J. Bergman, *Phys. Rev.* **A3**, 2058 (1971).
6. S.J. Putterman, R. Finkelstein, and I. Rudnick, *Phys. Rev. Lett.* **27**, 1967 (1971).

Mass Transport of ^{4}He Films Adsorbed on Graphite*

J.A. Herb and J.G. Dash

Department of Physics, University of Washington
Seattle, Washington

Thin ^{4}He films adsorbed on various substrates have superfluid onset temperatures T_0 which vary with film thickness (or relative pressure P/P_0) in a regular manner, so that results of a large number of different studies fall on a common curve of T_0 vs. P/P_0.[1-3] Because of this it has been felt that the understanding of onset phenomena need not include details of the substrate interaction (or at least that all substrates can be considered identical.[4] Recently some experiments[5-8] using different substrates have yielded slightly higher T_0 values, suggesting that superflow onset is substrate influenced in multilayer films. This conjecture is supported by our current studies of flow on graphite, which, according to recent heat capacity data,[9-11] is a manifestly more uniform surface. We report here a complex of "onset" phenomena which have not been seen in films on typical substrates.

Our study is of mass transport of ^{4}He multilayers on Grafoil† along directions parallel to the basal plane graphite surfaces. The structure of Grafoil inferred from manufacturer's information, X-ray diffraction, optical and electron microscopy, gas adsorption, and calorimetry consists of laminae of exfoliated natural graphite oriented and interleaved as in a loose stack of cards. Typical dimensions are 5 μm laterally between visible line defects and a few hundred angstroms in thickness. The crystallites are oriented with basal planes parallel to the Grafoil sheet surface, with mean deviation of 7°. Through the thickness of a 0.01–in.–thick sheet there are $\sim$ 6000 basal planes exposed for gas adsorption. The material has very high chemical purity and can be heated to high temperatures to remove adsorbed gases. These features are important, we believe, in understanding the distinctions between the flow properties of ^{4}He in the Grafoil flow channel and the flow of films on chemically and crystallographically less uniform substrates.

The experimental cell consists of two chambers connected through a Grafoil gasket, permitting surface transport along the basal plane direction. The supply chamber contained a quantity of ballast Grafoil used to stabilize the vapor pressure P. The exit chamber was continually pumped to a low vacuum with a mass-spectrometer leak detector which measures the flow rate R (except that the high rates below onset B were estimated from dynamic pressure rise in the pumping line). Except for the highest flow rates, the depletion of adsorbate from the supply chamber was negligible over the course of an experimental day.

* Research supported by the National Science Foundation.
† Union Carbide Corp., Carbon Products Division, New York.

Sample results of over 20 coverages are shown in Figs. 1 and 2. We distinguish four distinct regimes as follows:

(a) At lowest P/P_0, $R \propto e^{-q/kT}$, with $q \simeq$ the heat of adsorption. The exponential law is consistent with "normal" transport of film and/or vapor driven by a pressure difference equal to the vapor pressure. Pollack *et al.*[12] found similiar T dependence for arrival times of low-density He gas pulses traveling along tubes at low temperatures and they show such behavior consistent with migration times of adsorbing–desorbing gas.

(b) At relative pressure $P/P_0 \gtrsim 0.02$, R begins to rise above the exponential law at $T \simeq 2.4°K$. The rise is smooth and increases monotonically as T falls. This regime is termed the "precursor."

(c) The precursor is followed by a relatively abrupt increase ("onset A") at $T_A \simeq T_\lambda$. Although T_A is effectively constant, the initial rate of increase and dependence at lower T change markedly and nonmonotonically with P/P_0.

(d) There is a second, more dramatic rise ("onset B") at lower T. The rates first increase gradually and then rise more steeply: increases are typically greater than 10^3 on cooling $0.1°K$ below T_B. The temperature T_B increases monotonically with P/P_0.

Of these observations it is the trend of T_B with P/P_0 that is most comparable to previous work, resembling the common curve of "superfluid onset" temperatures T_0. In Fig. 3 we compare our T_B values with the results of experiments on various

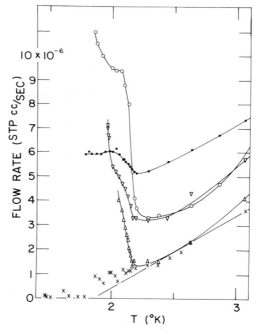

Fig. 1. Flow rates of ^{4}He across Grafoil basal planes for various partial pressures $P/P_0(T_\lambda)$: ($\times$) 4.3×10^{-4}; ($\triangle$) 0.0179; ($\bigcirc$) 0.0294; (∇) 0.0129; ($\bigcirc$) 0.1500.

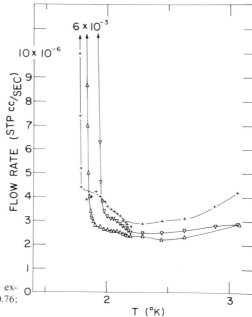

Fig. 2. Flow rates of ⁴He for higher coverages exhibiting onset B [measured as $P/P_0(T_B)$] : (+) 0.76; (Δ) 0.80; (∇) 0.87.

substrates using different techniques.[1-3] Although the T_B and T_0 curves are of the same general shape, the T_B values are shifted to considerably higher temperatures. We feel that the similarities in the curves suggest that the same basic mechanism is at work in all the onset experiments but modified by the substrate. Studies[13,14] have shown graphite surfaces to be more uniform than typical substrates and hence we feel T_B is more indicative of an "ideal" film.

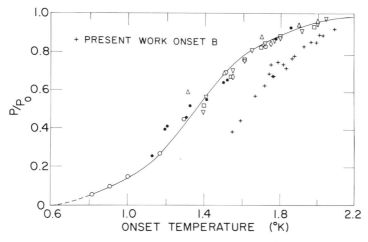

Fig. 3. Onset curve obtained from previous experiments [1-3] with present work added.

The occurrence of onset A, while startling, is not completely novel. Long and Meyer[15] reported two sets of flow rate transitions, one lying near the onset curve (Fig. 3) and the other occurring at T_λ for all coverages. The transitions at T_λ appeared only with small pressure differences across the flow path, whereas the transitions near the T_0 curve occurred with large pressure differences. Long and Meyer could not explain the high-temperature transitions but did get similar results with a cooling experiment.[15] However, no transitions at T_λ have been reported by other workers. Our type A transitions may be the same as the Long and Meyer transitions but we find onset A occurring to $P/P_0 \simeq 0.02$ while Long and Meyer found their transitions only for $P/P_0 > 0.15$. Long and Meyer used flow surfaces of stainless steel and Pyrex, substrates probably not as uniform as Grafoil.

The occurrence of two distinct flow regimes is novel, although liquid ^{4}He in porous Vycor shows a "superfluid" transition at $T < T_\lambda$ and yet a density maximum at T_λ.[4] The precursor has not been seen previously in He film flow but it seems analogous to the rise in electrical conductivity at $T > T_C$ of thin-film superconductors,[16-18] ascribed to the greater role of fluctuations in lower-dimensional systems.[19]

References

1. K. Fokkens, K.W. Taconis, and R. De Bruyn Ouboter, *Physica* **32**, 2129 (1966).
2. R.P. Henkel, G. Kukich, and J.D. Reppy, in *Proc. 11th Intern. Conf. Low Temp. Phys., 1968*, St. Andrews Univ. Press, Scotland (1969).
3. K.R. Atkins and I. Rudnick, *Progress in Low Temperature Physics*, C.J. Gorter, ed. North-Holland, Amsterdam (1970), Vol. 6.
4. D.F. Brewer, *J. Low Temp. Phys.* **3**, 205 (1970).
5. C.H. Anderson and E.S. Sabisky, *Phys. Rev. Lett.* **24**, 1049 (1970).
6. B.L. Blackford, *Phys. Rev. Lett.* **28**, 414 (1972).
7. M. Chester, L.C. Yang, and J.B. Stephens, *Phys. Rev. Lett.* **29**, 211 (1972).
8. H.W. Chan, A.W. Yanof, F.D.M. Pobell, and J.D. Reppy, this volume.
9. M. Bretz and J.G. Dash, *Phys. Rev. Lett.* **26**, 963; **27**, 647 (1971).
10. M. Bretz, G.B. Huff, and J.G. Dash, *Phys. Rev. Lett.* **28**, 729 (1972).
11. D.C. Hickernell, E.O. McLean, and O.E. Vilches, *Phys. Rev. Lett.* **28**, 789 (1972).
12. F. Pollack, H. Logan, J. Hobgood, and J.G. Daunt, *Phys. Rev. Lett.* **28**, 346 (1972).
13. A. Thomy and X. Duval, *J. Chim. Phys.* (*Paris*) **66**, 1966 (1969); **67**, 286, 1101 (1970).
14. N.N. Roy and G.D. Halsey, Jr., *J. Low Temp. Phys.* **4**, 231 (1971).
15. E. Long and L. Meyer, *Phys. Rev.* **85**, 1030 (1952).
16. J.E. Crow, R.S. Thompson, M.A. Klenig, and A.K. Bhatnagar, *Phys. Rev. Lett.* **24**, 371 (1970).
17. T.H. Geballe, A. Menth, F.J. Di Salvo, and F.R. Gamble, *Phys. Rev. Lett.* **27**, 314 (1971).
18. R.F. Frindt, *Phys. Rev. Lett.* **28**, 299 (1972).
19. P.A. Lee and M.G. Payne, *Phys. Rev. Lett.* **26**, 1537 (1971).

Helium Film Flow Dissipation with a Restrictive Geometry*

D.H. Liebenberg

Los Alamos Scientific Laboratory, University of California
Los Alamos, New Mexico

We are continuing measurements of the dissipation in the thermally driven superfluid helium film. Previously results[1] have been obtained using a glass post of either constant or stepped cross section to determine the applicability of the fluctuation dissipation model[2] to the development of dissipation in film flow. Recently further support for the model was obtained from observations of the saturation of the fluctuations at large dissipation levels.[3] The fundamental relation of the fluctuation-dissipation model as developed from film flow[1] is

$$\nabla\mu = -A_f\, vK \exp(-E_0/kT) \qquad (1)$$

where A_f is the film cross section normal to the flow, v the attempt frequency, K the quantum of circulation, and E_0 the energy of a critical fluctuation. The gradient of the chemical potential for thermally driven flow was taken as $\nabla\mu = -s\,\nabla T$. Earlier measurements with capillary potential probes[4] were extended to examine the spread of dissipation similar to measurements by Keller[5] for gravitationally driven film flow. These results indicate the dissipation can spread out over a centimeter of flow path. Thus it seemed useful to determine what effect the limitation of dissipation along the flow path would have on the character and growth of dissipation.

The apparatus is similar to that previously described.[1] However, the glass post has a constant diameter of 1 cm except for a notch located between the heater element and the bath level, as shown in Fig. 1. The notch is about 0.025 cm wide at the surface and extends ~ 0.15 cm deep into the glass. The circumference of the post and the film cross section are reduced by 30% and thus the flow path is restricted at this position on the post. This is more than sufficient to assure that the dissipation will be confined to this region. However, the gradient of the dissipation will increase by a significant factor, 40 in the above sample.

A thin-film superconducting probe[6] was evaporated onto the bottom of this notch to permit dissipation measurements in the region of maximum dissipation. The apparatus was contained in a sealed chamber surrounded by a temperature-controlled bath. Temperature stability was $\pm 5\ \mu°\mathrm{K}$.

The observed growth of dissipation was similar to that observed for the constant-diameter posts. At low superfluid velocities along the post the dissipation is characterized by the fluctuation-dissipation model. The onset of saturation is observed at larger flow velocities.[3]

* Work performed under the auspices of the United States Atomic Energy Commission.

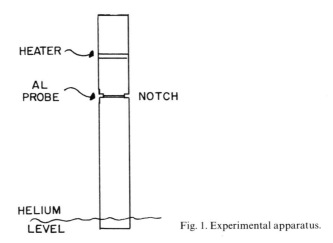

Fig. 1. Experimental apparatus.

As saturation is approached an instability is observed in the measured dissipation. The instability has been determined to occur above the dissipation level where possible film desaturation effects are observed as described by Campbell and Liebenberg.[3] Such an instability is not seen in flow measurements along a constant-diameter post. There is no apparent hysteresis of this instability within the $\sim$ 1-sec time constant of the apparatus. We suggest that the instability is related to the boundary conditions imposed on the dissipation that increases the dissipation gradient in this flow process. The system appears to be driven permaturely into saturation. This would be expected if in Eq. (1) the gradient and not the averaged difference in chemical potential is important.

References

1. D.H. Liebenberg, *Phys. Rev. Lett.* **26**, 744 (1971).
2. J.S. Langer and M.E. Fisher, *Phys. Rev. Lett.* **19**, 560 (1967); S.V. Iordanskii, *Soviet Phys.—JETP* **21**, 467 (1965).
3. L.J. Campbell and D.H. Liebenberg, *Phys. Rev. Lett.* **29**, 1065 (1972).
4. D.H. Liebenberg, *J. Low Temp. Phys.* **5**, 267 (1971).
5. W.E. Keller, *Helium-3 and Helium-4*, Plenum, New York (1969), p. 308; W.E. Keller, in *Proc. 12th Intern. Conf. Low Temp. Phys. 1970,* Academic Press of Japan, Tokyo (1971), p. 125.
6. D.H. Liebenberg and L.D.F. Allen, *J. Appl. Phys.* **41**, 4050 (1970).

Dissipation in the Flowing Saturated Superfluid Film *

J.K. Hoffer, J.C. Fraser, E.F. Hammel, L.J. Campbell
W.E. Keller, and R.H. Sherman

Los Alamos Scientific Laboratory, University of California
Los Alamos, New Mexico

The flow of the saturated helium II film through an inverted U-tube connecting two reservoirs of bulk liquid has been measured with significantly improved precision over previously reported work[1] to provide a direct experimental determination of the intrinsic dissipation in the film as a function of velocity and temperature.

The apparatus is shown schematically in the insert in Fig. 2. The U-tube is made of glass tubing with 2.0-cm-i.d. arms joined at the top by a 1.0-cm-long constriction of 0.45 cm i.d. The lower ends of the U-tube are connected to concentric cylindrical capacitors, one of which, C_1, is sealed off from the experimental helium reservoir R and serves as the tank capacitor in a back-diode oscillator[2] running at 20 MHz. The measuring electronics are sufficiently sensitive to detect changes of 200 Å in the level of the bulk helium in the 0.01 cm gap between the plates of C_1. The other capacitor, C_2, open to R, serves to balance the capillarity in C_1 and to buffer disturbances in the liquid level in R. Film flow was initiated by raising a displacer out of R. Subsequent changes x of the liquid level in the gap (a long outflow followed by damped oscillations) were monitored and recorded at intervals of 1.25 sec.

Robinson[3] was the first to give a detailed theoretical analysis of the isothermal flow of superfluid helium between a small reservoir of cross-sectional area a and an infinite reservoir, connected by a fine capillary or slit. This same theory, when applied to film flow in a U-tube of radius r, indicates that the system represents a harmonic oscillator of frequency ω_i given by

$$\omega_i^2 = 2\pi \frac{\rho_s g}{\rho a} \left[\int_s \frac{ds}{r\,\delta(H)} \right]^{-1} \tag{1}$$

where g is the gravitational constant and ρ_s/ρ is the superfluid fraction. Here it is necessary to integrate the film thickness $\delta(H)$ over the entire path length s as a function of the height H above the bulk fluid level.[4]

If the flow is not perfectly isothermal, then the fountain effect will also play a role in determining ω_i, and for small displacements Robinson has shown that the system is described by two coupled differential equations. For the present nearly isothermal system these can be reduced to one equation:

$$\ddot{x} + (\omega_i^2 \rho T S^2 A_s/g\kappa)\,\dot{x} + \omega_i^2 x = 0 \tag{2}$$

* Work performed under the auspices of the United States Atomic Energy Commission.

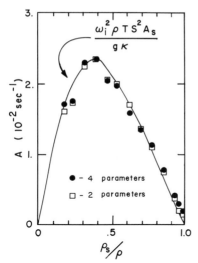

Fig. 1. The damping coefficient as a function of the bulk superfluid density.

where κ represents the total heat conductivity between C_1 and R and A_s is the effective surface area across which this heat conduction occurs. The damping coefficient

$$A \equiv \omega_i^2 \, \rho T S^2 A_s / g \kappa \tag{3}$$

can be calculated from known properties of ^{4}He and the present geometry.

The oscillation portions of the experimental data, x vs. t, were fitted to the damped oscillator equation

$$x = x_0 e^{-At/2} \sin \omega t \tag{4}$$

which is a solution of Eq. (2), where $\omega_i^2 = \omega^2 + (A/2)^2$. The values of A so determined are plotted in Fig. 1 as open squares, along with the line calculated from Eq. (3). The agreement between the calculated and observed values of A indicates that the observed damping is entirely due to a small (measured to be $< 10 \, \mu^\circ$K) departure from isothermality.

The values of ω_i^2 as obtained from the fits of Eq. (2) and scaled by ρ_s/ρ are plotted as open squares in Fig. 2. According to Eq. (1), these values should be constant. However, the deviations which occur at the highest temperatures are not due to departures of ρ_s/ρ in these films from the bulk liquid values but rather to the fact that the Robinson theory[3] does not include the effects of intrinsic dissipation, which according to the fluctuation model[5] results in a decay of the superfluid velocity v_s:

$$dv_s/dt_{\text{fluctuation}} = -K \mathscr{A} v \, \exp(-b/v_s) \tag{5}$$

Here K is the unit of circulation $\sim 10^{-3}$ cm^2 sec^{-1}; $\mathscr{A}$ is the cross-sectional area of the superflow; v is the "attempt frequency" density in cm^{-3} sec^{-1}, and b is a "barrier velocity."

The experiment measures $\dot{x}$, which can be related to the velocity of the superfluid in the 0.45-cm constriction by the continuity equation

$$\dot{x} a = v_s p \delta \rho_s / \rho \tag{6}$$

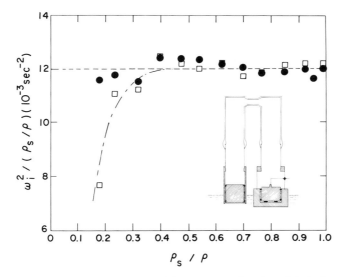

Fig. 2. The undamped isothermal frequency as a function of the bulk superfluid density. Open squares represent the results of computer fits to Eq. (4). Solid points represent fits to Eq. (8).

where a is the area of the gap between the capacitor plates, p is the internal perimeter of the constriction, and δ is the thickness of the film at the constriction. It is assumed that only the superfluid fraction of the film flows and that intrinsic dissipation is important only in the constriction. Hence $\mathscr{A} = p\delta$; substituting for v_s in Eq. (5), one has

$$\ddot{x}\big|_{\text{fluctuation}} = -\frac{\rho_s(p\delta)^2 K v}{\rho a}\exp\left(-\frac{\rho_s p\delta b}{\rho a|\dot{x}|}\right) \qquad (7)$$

$$\equiv -C\exp(-B/|\dot{x}|)$$

This term must be added to Eq. (2) yielding

$$\ddot{x} + A\dot{x} + (\dot{x}/|\dot{x}|)C\exp(-B/|\dot{x}|) + \omega_i^2 x = 0 \qquad (8)$$

Since this equation does not admit a solution in closed form, the coefficients A, B, C, and ω_i^2 were fitted to the experimental x vs. t data by a numerical integration.

The values of ω_i^2 and A obtained from these four-parameter fits are plotted in Figs. 1 and 2 as solid symbols. The values of A are in good agreement with those obtained from Eq. (2). However, since the new values of ω_i^2 are directly proportional to ρ_s/ρ as measured in bulk helium, there is no justification for assuming that ρ_s/ρ in the saturated films studied here are different from bulk values. At higher temperatures the increasing magnitude of the intrinsic dissipation limits the velocity and thus the amplitude of the oscillations. Also, the period is quite long and the damping *per second* A is still quite high. Hence the oscillations above 2.1°K are effectively damped out before a full period has been executed. Thus, it is not surprising that the two-parameter fits give erroneous values for ω_i^2 while measuring A quite accurately.

Since the raw data are so precise, $\dot{x}$ and $\ddot{x}$ can be determined by numerical

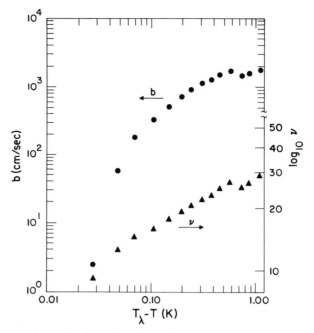

Fig. 3. The barrier velocity b and the attempt frequency density ν as a function of T_λ(bulk) $-$ T. Note that $\log_{10} \nu$ is plotted, and this in turn is plotted on a log scale.

differentiation. Plotting $\ln f(\dot{x})$ vs. $1/\dot{x}$, where

$$f(\dot{x}) \equiv -\ddot{x} - A\dot{x} - \omega_i^2 x = C \exp(-B/\dot{x}) \qquad (9)$$

one expects a straight line with slope $-B$ and intercept $\ln C$. In fact, the data exhibit a slight curvature, suggesting that higher-order terms in v_s should be included in Eq. (5). Although this curvature appears in the deviations from the fits to Eq. (8) as a general pattern, the B and C values represent the data to within a few percent.

The values of ν and b, calculated from the C and B values, respectively, are shown in Fig. 3. It has been assumed that $\delta = 200$ Å in the constriction. It is of importance to note that all of the data presented here were obtained in the same 8-hr period, with the exception of the data at 1.28°K, which were obtained on the previous day.

According to the fluctuation model,[5] the values of ν should be roughly constant or increase only slightly with temperature. The data show on the contrary that ν decreases approximately 20 orders of magnitude in going from 1.0 to 2.14°K. Such a strong temperature dependence in any physical quantity is quite unusual (especially in one which is predicted to be constant!), and remains a challenge to be explained. The fluctuation model also predicts that $bT\rho/\rho_s = $ const. However, the b values in Fig. 3 do not exhibit such behavior, being too low at higher temperatures.

Although the temperature dependence of these coefficients is in serious disagreement with the predictions of the fluctuation model, the form of the intrinsic dissipation in the saturated films studied here does follow closely that predicted for the flow of bulk helium [Eq. (7)]. The form of intrinsic dissipation predicted for the flow of thin helium films[5] has not been observed, nor, indeed, should it be expected for the relatively thick saturated film.

References

1. E.F. Hammel, W.E. Keller, and R.H. Sherman, *Phys. Rev. Lett.* **24**, 712 (1970).
2. D.H. Lester (to be published).
3. J.E. Robinson, *Phys. Rev.* **82**, 440 (1951).
4. K.R. Atkins, *Proc. Roy. Soc.* (*London*) **A203**, 119, 240 (1950).
5. J.S. Langer and J.D. Reppy, in *Progress in Low Temperature Physics*, C.J. Gorter, ed. North-Holland, Amsterdam (1970), Vol. 6, pp. 1–35.

Dissipation in Superfluid Helium Film Flow *

J.F. Allen, J.G.M. Armitage, and B.L. Saunders

Department of Physics, University of St. Andrews
North Haugh, St. Andrews, Fife, Scotland

The dissipation processes which accompany superfluid flow in saturated helium films have been studied as a function of flow rate, temperature, and film height or thickness. The film flow into or out of a glass beaker was driven at a chosen rate by expanding or compressing bellows which were attached to the bottom of the beaker. The bellows were driven remotely via a hydraulic link to minimize vibration. The cryostat was essentially vibration-free, including elimination of boiling nitrogen from contact with the helium dewar.

The bath and beaker levels were separately monitored by capacitance sensors to an accuracy of $\pm 0.5 \mu$m.

At low rates of bellows drive, film flow was initiated and the beaker level exhibited damped oscillations before settling down to a steady position different from that of the bath. On ceasing the drive the oscillation recommenced about a beaker level coincident with the bath. This behavior was confined to drive rates giving level differences not greater than 15 μm. At higher drive and flow rates dissipation became much greater and no oscillations occurred.

Two regimes of dissipation could therefore be distinguished. The regime in which flow rates great enough to yield steady level differences of more than 15 μm has been interpreted in terms of a Langer–Fisher[1] type of dissipation, similar to that observed by Liebenberg.[2] This will be the subject of a separate publication.

The other regime, involving level differences of less than 15 μm, has been attributed to a different mechanism. Both the damping of the inertial oscillations and the associated steady level differences showed satisfactory agreement with the thermal dissipation mechanism originally suggested by Robinson.[3]

In the "Robinson" region the dissipation is linear in the velocity, and the motion can be described by a damped oscillator equation

$$\ddot{x} + 2k\dot{x} + \omega_0^2 x = 0 \qquad (1)$$

with the solution

$$x = x_0 \sin \omega t \, e^{-kt} - (2k/\omega_0^2) \langle \dot{x} \rangle \qquad (2)$$

in which k is the damping constant and ω_0 the undamped frequency. This represents damped oscillations about a steady level difference x_s which is given by

$$x_s = (2k/\omega_0^2) \langle \dot{x} \rangle \qquad (3)$$

* Supported by the Science Research Council through a Research Grant and a Research Studentship.

There was satisfactory correlation between the damping coefficient k measured from the decay of oscillations and the steady level differences x_s set up by the low flow rates, which confirmed that the same mechanism governed both processes.

There have been, in fact, two proposals made to account for a damping proportional to velocity. While Robinson proposed that energy was lost from the flow because of irreversible heat exchange between the beaker and reservoir, Allen suggested that if film flow were associated with vorticity in the film, then roton–vortex interaction should cause energy loss. By observing the damping as a function of both frequency and temperature it has been possible to decide which of these mechanisms is operative.

The Robinson model gives the damping constant k_R as

$$k_R = \omega_0^2 s^2 T \rho A / 2g\kappa \tag{4}$$

where s is the entropy per unit mass, ρ is the total density, A is the cross-sectional area of the beaker, and κ is a conductance defined by

$$\dot{Q} = \kappa \, \Delta T \tag{5}$$

$\dot{Q}$ is the heat flow between the beaker and the reservoir and ΔT is the temperature difference between them. Previous workers[5,6] have made estimates of κ in order to test expression (4) but have not found good agreement, the observed damping always being larger than that predicted.

If one assumes a mechanism of the Allen type such that there is a frictional force per unit volume proportional to the flow velocity, i.e., $f_A = \gamma v_s$, then the damping constant becomes $k_A = \gamma/2$. It does not seem likely that the quantity γ will contain the frequency of oscillation, so that it is possible to distinguish between the two mechanisms by measuring k at constant temperature for various film lengths. The frequency of oscillation depends upon the length, so the ω^2 dependence predicted by expression (4) can be tested. If the damping at two frequencies ω_1 and ω_2 is measured, then from Eq. (4)

$$(\omega_1/\omega_2)^2 = k_1/k_2$$

The results obtained are shown in Table I.

These results show that the damping does depend on the frequency and that the ω^2 relation describes the dependence reasonably well.

The temperature dependence of k provides a further means of deciding between the Robinson and Allen mechanisms. Presumably, a roton-type interaction will become more effective as the temperature and thus the roton density increases. The damping due to this mechanism should therefore be expected to increase exponen-

Table I

T, °K	Ratio of film lengths	(Ratio of frequencies)^2	Ratio of damping constants
1.880	1.53	2.9 ± 0.2	3.40 ± 0.30
	1.22	2.1 ± 0.2	2.40 ± 0.25
	1.25	1.4 ± 0.1	1.40 ± 0.15
1.425	2.0	3.0 ± 0.2	3.80 ± 0.50

tially toward T_λ. In contrast, the damping described by (4) should show a maximum at about $2.05°K$ because the ω^2 term includes the quantity ρ_s, which decreases very rapidly near T_λ.

Before applying (4), however, we need a measure of κ and its temperature dependence.

Previously, when κ has been estimated the final agreement has not been good. We have therefore measured κ directly. The method was to apply a steady, known amount of heat to the beaker while monitoring both the bath and beaker levels. The fountain effect produced a steady level difference which was measured and used to calculate the temperature difference from the fountain effect equation. Then κ was calculated from Eq. (5). The values of κ were found to agree well with a suggestion by Atkins[7] that distillation between the beaker and the reservoir can provide an efficient heat transfer mechanism. The mass evaporated per second is[8]

$$\frac{dm}{dt} = A'\left(\frac{M}{2\pi RT}\right)^{1/2}\frac{dp}{dT}\Delta T$$

where A' is the area of the liquid surface in the beaker and p is the saturated vapor pressure. Since the heat flow is $Q = L\,dm/dt$, where L is the latent heat, we can define a conductance κ_e due to evaporation

$$\kappa_e = LA'\left(\frac{M}{2\pi RT}\right)^{1/2}\frac{dp}{dT} \tag{6}$$

A' must be calculated from the profile of the meniscus in the beaker. The calculated values of κ_e compare well with the experimental values for κ, as can be seen from Table II.

The experimental error in κ was about 10% because the measurements were restricted to very low heat inputs and very small temperature differences.

The temperature dependence of the damping coefficient k is shown in Fig. 1. The experimental points were determined from the pen recorder traces of the oscillations. The solid line was determined from Eq. (4) using κ_e from Eq. (6) and not the experimental values of κ. No fitted parameters were employed.

The agreement between theory and experiment is gratifyingly good. The occurrence of the maximum in the damping coefficient at the expected temperature, together with the observed ω^2 dependence, points strongly to the Robinson mechanism as the cause of dissipation in the low-flow-rate region.

Although the results do not support the Allen mechanism, they are not in conflict with his observations. If the damping had been due to a roton–vortex interaction, then the relation

$$k \propto \rho_n \propto \exp(-\Delta/k_B T)$$

Table II

T, $°K$	κ_e, ergs $°K^{-1}$ sec^{-1}	κ(exp), ergs $°K^{-1}$ sec^{-1}
1.814	9.06×10^8	11.0×10^8
1.778	7.90×10^8	7.2×10^8
1.678	6.12×10^8	5.9×10^8

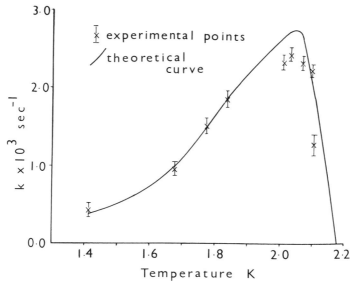

Fig. 1. Temperature dependence of oscillation damping constant.

should hold. Allen measured the temperature dependence of k up to his visual observational limit of 1.9°K and plotted ln k against $(1/T)$. He found a straight-line dependence with a gradient $\Delta/k_B \simeq 8°$K, which agreed well with the roton energy; this was taken as evidence supporting the roton explanation. If the same graph is plotted using the present values of k given by Eq. (4) and (6) for temperatures between 1.0 and 2.0°K, a straight line is once again found, this time with a slope of 9.2°K, in fair agreement with the Allen result.

Martin[6] repeated Allen's measurements using capacitance and not visual methods, but he obtained a value of the slope of 4°K, which does not give a satisfactory fit with either mechanism.

It is not easy to determine comparable values for k from the results of other workers since the precise conditions are not often described. Atkins's[7] value of k from one temperature only appears to be higher than ours by a factor of more than two. We are generally in agreement with Picus's[5] observations. The results of Hammel et al.[9] are difficult to compare since they reported a varying damping rate in a single run.

Martin[6] discussed damping, finding the damping to flatten below 1°K with an indication of a rise below 0.5°K. There is considerable scatter in his results and he does not place much credence in them. If one follows Eq. (4) and (6), the damping would be expected to flatten below 1°K due to the fall in dp/dT and rise again since s^2 is diminishing by T^6 while κ is decreasing exponentially.

We believe that our results are not in disagreement with experiments reported verbally by de Bruyn-Ouboter at the April 1972 EPS meeting in Freudenstadt (the proceedings are not to be published). The latter authors described the damping of inertial oscillations in immensely long ($\sim$ 100 m) films with correspondingly long periods ($\sim$ 1 hr).

References

1. J.S. Langer and M.E. Fisher, *Phys. Rev. Lett.* **19**, 560 (1967).
2. D.H. Liebenberg, *Phys. Rev. Lett.* **26**, 744 (1971); *J. Low Temp. Phys.* **5**, 267 (1971).
3. J.E. Robinson, *Phys. Rev.* **82**, 440 (1951).
4. J.F. Allen, *Nature* **185**, 831 (1960).
5. G.S. Picus, *Phys. Rev.* **94**, 1459 (1954).
6. D.J. Martin, PhD Thesis, 1970, Oxford, unpublished.
7. K.R. Atkins, *Proc. Roy. Soc.* **A203**, 240 (1950).
8. K.R. Atkins, D. Rosenbaum, and H. Seki, *Phys. Rev.* **113**, 751 (1959).
9. E.F. Hammel, W.E. Keller, and R.H. Sherman, *Phys. Rev. Lett.* **24**, 712 (1970).

Direct Measurement of the Dissipation Function of the Flowing Saturated He II Film *

W.E. Keller and E.F. Hammel

Los Alamos Scientific Laboratory, University of California
Los Alamos, New Mexico

We describe here experiments to measure the profile of the chemical potential μ of the saturated liquid He II film as it flows into or out from a cylindrical beaker and show how the results provide a direct measurement of the dissipation function of the system in quantitative agreement with predictions of the thermal fluctuation model of intrinsic dissipation.[†]

A scale drawing of the glass beaker used in these experiments is shown at the left in Fig. 1. The beaker is fitted with eight potential probes[2] opening at points $p(p = 1\text{–}8)$ on the inside wall, beginning as close to the inside rim as possible and with a vertical spacing of ~ 0.4 cm. We obtain the chemical potential per gram $\bar{\mu}_p$ at p relative to that of the inside bulk liquid level (at height h_i and potential $\bar{\mu}_i$) through the relation

$$\bar{\mu}_p - \bar{\mu}_i = \Delta\bar{\mu} = (h_p - h_i)\,\Delta\bar{\mu}_t/\Delta h = g(h_p - h_i) \tag{1}$$

where h_p is the liquid level height in the probe tube p and $\Delta\bar{\mu}_t$ is the total potential equal to $g(h_o - h_i) = g\,\Delta h$ (h_o is the height of the outside bulk liquid level and g is the gravitational constant).

A glass cylinder with copper top and bottom plates surrounds the beaker. The bottom plate has a cup for supporting the beaker and a small hole for admitting the bath, so that h_o is always at the bath level and always falling. The top plate is attached to a thin-wall tube which passes through a sliding seal at the cryostat top, allowing the beaker to be raised or lowered with respect to h_o in order to initiate a flow process. The various liquid level heights are read with a cathetometer as a function of time; and since the beaker is 1.68 cm i.d., the level changes are sufficiently slow that all heights can be registered in a time short relative to these changes.

On the right side in Fig. 1 is a typical plot of the level heights h_o, h_i, and h_p in mm as a function of time in min for a filling (inflow) process at $T = 1.585°$K. These are raw data uncorrected for capillarity (~ 1.2 mm) in the probe tubes. When these corrections are made and the height observations for each probe inserted into Eq. (1) the upper set of curves in Fig. 2 is generated. Here each curve represents for a fixed time (constant $\Delta\bar{\mu}_t$) the chemical potential profile of the flow process along the flow path S (referenced to the top of the beaker) on the inner beaker wall. Here $|\Delta\bar{\mu}|$

* Work performed under the auspices of the U.S. Atomic Energy Commission.
† See, e.g., Langer and Reppy.[1]

rather than $\Delta\bar{\mu}$ is plotted against S so that the inflow results may be readily compared with the corresponding data for an emptying process (outflow) shown in the lower section. There are two striking features in this comparison:

(1) The extent of dissipation along S (i.e., where $|\nabla\bar{\mu}| = d|\Delta\bar{\mu}|/dS \neq 0$) differs considerably in the two cases. Usually in outflows the dissipation was found confined to the region above probe 1, and the spreading shown in the figure was rarely seen, whereas for inflows the pattern was consistently broad and often extended nearly to the bulk liquid level as shown.

(2) We note that in both cases $\nabla\bar{\mu}$ is largest near the beaker top, indicating that most of the dissipation occurs here; however, if we let S_0 be that point on S at which the discontinuity between the maximum value of $\nabla\bar{\mu}$ and $\nabla\bar{\mu} = 0$ appears, we see that for outflow S_0 is consistently at $S = 0.3$ cm, just when the inside rim begins to flare (this would be expected if it is assumed that $\nabla\mu$ should be a maximum where the cross-sectional area of the film is a minimum and further if this area is determined from simple geometric considerations). On the other hand, for inflow, S_0 occurs at larger values of S, i.e., farther down the inside wall; and although as $\Delta\mu_t$ decreases S_0 moves up the wall, S_0 never rises above probe 1. This behavior suggests that in these inflows the thickness of the film at least part way down the *inner* wall—i.e., when $S_0 > 0.3$ cm—is governed by the *outside* level height, an effect substantiated by direct measurements of the film thickness in a different apparatus.[3]

It is possible to derive from the data in Figs. 1 and 2 the functional dependence of $|\nabla\bar{\mu}|$ upon v_s, the superfluid velocity. $|\nabla\bar{\mu}|$ has been obtained by measuring the slope of the curves of Fig. 2 at each measured point and at the various S_0's using a tangimeter. Although this method is somewhat subjective, depending upon how the curves are drawn through the experimental points, the consistency of the results for several similar experiments lends considerable credibility to it. A value of $v_s(p)$

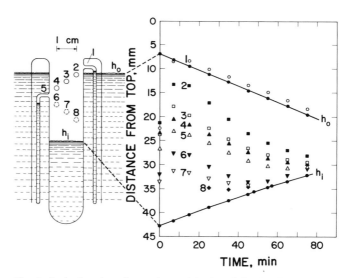

Fig. 1. Scale drawing of experimental beaker (left) showing potential probe tube insertion points ($p = 1-8$) on inner wall; and height vs. time measurements (right) for a typical film inflow experiment.

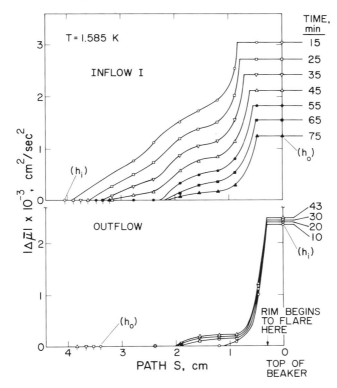

Fig. 2. Chemical potential differences along the flow path S (inner beaker wall) at various times for both inflow and outflow processes. Points at $S = 0$ represent $|\Delta\bar{\mu}_t| = g|h_o| - h_i|$ at each observation time; points with vertical tails on the $\Delta\bar{\mu} = 0$ axis represent the height of the inside or outside level (i.e., h_i or h_o) at the corresponding times.

corresponding to each $|\nabla\bar{\mu}|$ determination (and to the appropriate p or S_0) was found using the relation

$$v_s(p) = \dot{h}_i a / d(p) \times P \qquad (2)$$

where a and P are respectively the inside cross-sectional area and perimeter of the beaker, and $d(p)$ is the film thickness at the probe insertion point p or S_0 calculated from

$$d(p) = 300 \times 10^{-8} / (h_p - h_i)^{1/3} \qquad (3)$$

valid from h_i to S_0. The usual assumed value of the coefficient in (3) is, as will be seen below, reasonable for the present purposes.

Using the above procedure, plots of $|\nabla\bar{\mu}|$ vs. v_s have been prepared from both of the two experiments shown in Fig. 2 and another filling experiment, Inflow II. All three plots coincide within a few percent in v_s, with $|\nabla\bar{\mu}| = 5 \times 10^4$cm sec^{-2} and a maximum at $v_s = 55$ cm sec^{-1}, decreasing to one-hundredth of this value at $v_s = 51$ cm sec^{-1}, below which $|\nabla\bar{\mu}|$ remains small but discernable down to $v_s \sim 28$ cm sec^{-1}. Thus even though the potential profiles for inflow and outflow differ

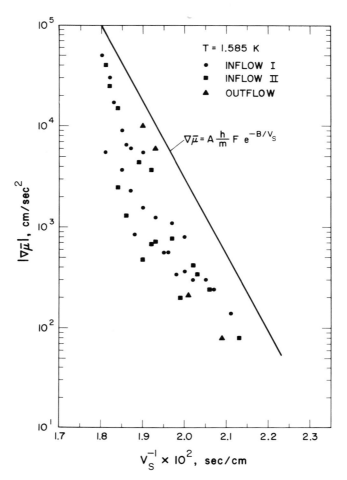

Fig. 3. Plot of $\nabla\bar\mu$ vs. v_s^{-1} for the data of three flow processes compared with the predictions (solid line) of the thermal fluctuation theory of dissipation with constants F and B determined from independent experiments.[4]

considerably, it is evident that the two processes are governed by the same dissipation function. Further, the agreement in v_s argues for the appropriateness of the same coefficient in (3) for the three experiments.

Finally, we propose to show that the $(|\nabla\bar\mu|, v_s)$ data obtained here are remarkably consistent with the dissipation function suggested by the thermal fluctuation theory,[1] a general form of which is

$$\nabla\bar\mu = A(h/m) F \exp(-B/v_s) = dv_s/dt \qquad (4)$$

Here $A = d \times P$; and F and B are strongly temperature-dependent parameters. Hence a plot of $\log|\nabla\bar\mu|$ vs. v_s^{-1} should yield a straight line if Eq. (4) represents the data. Such a plot is shown in Fig. 3. Here the solid line is drawn using the values of F and B determined from an entirely different film flow experiment at Los Alamos[4] which measured overall values of dv_s/dt in film flow, as contrasted with the point-by-point measurement of $\nabla\bar{\mu}$ as described here. No adjustment of parameters has been made to effect this comparison. However, the evaluation of F and B depends upon an assumed magnitude of d as does that of v_s in the present experiment. By only slightly exploiting the uncertainty in d [e.g., by increasing the coefficient in Eq. (3) by 10%], a very reasonable fit of Eq. (4) to the data could be effected.

The present experiments, in addition to providing insight into the nature of the potential profile for the flowing saturated He II film, lend strong confirmation to the proposal that a function of the form (4) provides a very satisfactory description of the dissipation in the film.

References

1. J.S. Langer and J.D. Reppy, in *Progress in Low Temperature Physics,* C.J. Gorter, ed., North-Holland, Amsterdam (1970), Vol. 6, pp. 1–35.
2. W.E. Keller and E.F. Hammel, *Phys. Rev. Lett.* **19**, 998 (1966).
3. W.E. Keller, in *Proc. 12th Intern. Conf. Low Temp. Phys. 1970,* Academic Press of Japan, Tokyo (1971).
4. J.K. Hoffer, J.C. Fraser, E.F. Hammel, L.J. Campbell, W.E. Keller, and R.H. Sherman, this volume.

Application of the Fluctuation Model of Dissipation to Beaker Film Flow *

L.J. Campbell

Los Alamos Scientific Laboratory, University of California
Los Alamos, New Mexico

Although the striking phenomenon of He II film flow over a gravitational barrier has received much experimental attention, there has been no theory which could successfully account for the following observations: the temperature dependence and magnitude of the flow rate; the distribution of the dissipation and the variation of the film thickness over the flow path; the formation of droplets; and the dependence of flow rate on barrier height, level difference, and substrate.

The starting point for a theory should be the recognition that film flow is a steady-state, irreversible process.† Therefore dissipation processes must be included; theories based on equilibrium thermodynamics or ideal potential flow would seem unpromising.

It is found empirically[1] that in isothermal film flow most and possibly all the dissipation can be described by a force per unit mass on the superfluid of the form

$$dv_s/dt = - A\kappa v(T)e^{-b(T)/v_s} \tag{1}$$

This is interpreted as arising from fluctuations which lead to intermediate states (small vortex rings) through which the system can decay to a state of lower free energy.[2] According to this, the factors in Eq. (1) are identified as follows: A is the cross-sectional area of the flow, κ is the unit of circulation ($= h/m = 10^{-3}$ cm^2 sec^{-1}), $v(T)$ is the effective attempt frequency, and $b(T)$ is the barrier velocity. The sign of the rhs of Eq. (1) is always opposite that of v_s. In principle, $v(T)$ and $b(T)$ are calculable but for this work the empirical values obtained from film oscillations[1] will be used.

Supposing the film thickness $\delta(y)$ to be known over the flow path, $0 \leq y \leq L$, it is simple to find from Eq. (1) the relation between the level difference H and flow rate R (cm^2 sec^{-1}). Multiplying Eq. (1) by $\rho_s v_s A$ gives the dissipation per unit length of flow path. The integral of this over the flow path must equal the gravitational power $\rho_s H v_s A$. Using the fact that $v_s(y) A(y) = \rho R p_0/\rho_s = $ const, where, by convention, p_0 is the minimum perimeter, the relation becomes

$$H = \int_0^L dy\, A(y)\, \kappa v e^{-b/|v_s(y)|} \tag{2}$$

* Work performed under the auspices of the United States Atomic Energy Commission.
† Of course, for small differences in bulk levels the dissipation is quite small and oscillation of the levels can occur.

Assuming the film thickness is given by $\delta = a/h^{1/3}$, where h is the height above bulk liquid, then Eq. (2) can be integrated analytically for beaker flow by dividing the path into three parts: inside, outside, and across the rim. The discontinuities in film thickness at the rim that this implies are, of course, unphysical, but the error should be unimportant for small level differences. Letting the subscripts 1 and 2 refer to the paths inside and outside the beaker, the result can be written as

$$gH = D_1 + D_{\text{rim}} + D_2 \tag{3}$$

where

$$D_1 = 3a\kappa v p_1 c_1^2 \left[e^{-x_1}(1 - x_1)/2x_1^2 + \tfrac{1}{2}E_1(x_1) \right] \tag{4}$$

$$x_1 = c_1 H_1^{-1/3}, \qquad c_1 = bp_s a/\rho R, \qquad c_2 = bp_s a p_2/\rho R p_1$$

The distance to the rim is H_1 and $E_1(x)$ is the exponential integral, $E_1(x) = \int_x^\infty dy\, e^{-y}/y$. In (3), D_2 is the same as D_1 apart from subscripts and D_{rim} is given by

$$D_{\text{rim}} = (2\pi a\kappa v/H_r^{1/3}c_r^2)\left[e^{-w_1}(1 + w_1) - e^{-w_2}(1 + w_2) \right] \tag{5}$$

$$c_r = 2\pi b p_s a/\rho H_r^{1/3} R p_1, \qquad w_1 = c_r r_1, \qquad p_1 = 2\pi r_1$$

where r_1 is the inner radius. The value of H_r, the effective height of the film on the rim, presumably should be something between H_1 and H_2.

For larger level differences and flow rates the film thickness profile is not known a priori and the hydrodynamic equations of motion must be used. This was first done by Kontorovich,[3] who failed, however, to take account of any dissipative force. When the hydrodynamic equation of motion containing Eq. (1) is integrated the following equation for the velocity at a point y on the flow path is obtained:

$$\frac{1}{2}\rho_s v_s^2(y) + \rho \int_{y_1}^y \mathbf{g}(x) \cdot d\mathbf{x} - \rho\, \frac{q}{\delta^3(y)} + \rho \int_{y_1}^y dx\, p(x)\, \delta(x)\, v\kappa e^{-b/\, r_s(x)} = 0 \tag{6}$$

This is the same as Kontorovich's Eq. (5) except for the last term, which is the total energy per unit volume that has been dissipated by the fluid in going from y_1 (one of the bulk liquid levels) to y. Since $v_s(y)$ and $\delta(y)$ are related by $v_s(y)\, p(y)\, \delta(y) = \rho R p_0/\rho_s \equiv M$, Eq. (6) can be rewritten in terms of $\delta(y)$ only:

$$\frac{\rho_s}{\rho}\frac{1}{2}\left(\frac{M}{p}\right)^2 \frac{1}{\delta^2} + \int_{y_1}^y \mathbf{g}\cdot d\mathbf{x} - \frac{q}{\delta^3} + \int_{y_1}^y p\delta v\kappa \exp(-b p\delta/M)\, dx = 0 \tag{7}$$

Note that the gravitational term always corresponds to the net height above the starting level, $\mathbf{g} = 980\hat{\mathbf{z}}$, where $\hat{\mathbf{z}}$ is a unit vector in the upward direction.

For sufficiently small flow rates the first (kinematic) and fourth (dissipative) terms of Eq. (7) are negligible and the usual static film thickness is obtained,

$$\delta = (q/gh)^{1/3}$$

where $q = ga^3$, $a = 3 \times 10^{-6}$ for glass substrates.

As Kontorovich points out, the derivation of Eq. (7) uses assumptions that are not valid for $y \approx y_1$. Indeed, for $y = y_1$ we have

$$\frac{1}{2}\left(\frac{M}{p}\right)^2 \frac{1}{\delta^2} - \frac{\rho}{\rho_s}\frac{q}{\delta^3} = 0 \tag{8}$$

While the physically correct thickness $\delta = \infty$ is a solution, it is not the solution obtained in the limit $y \to y_1$, which is

$$\delta = 2(\rho/\rho_s)\,(p/M)^2\,q \approx 10^{-5}\,\text{cm}$$

for typical rates and perimeters. However, this represents an error in the position of y_1 of less than 0.03 cm, which is negligible as long as y_1 is a few mm from the rim.

Equation (7) can be solved numerically by regarding it as a cubic equation for δ and progressing in small intervals dy starting from y_1. There is only one real solution at each y. The gravity force function $\mathbf{g} \cdot dy$ and the perimeter function $p(y)$ define the arbitrary size and shape of the container. The temperature fixes ρ_s and the dissipation model parameters $v(T)$ and $b(T)$. The substrate determines q, while y_1 is the arbitrary level of the bulk liquid. The dissipation integral is calculated using the previously determined $\delta(y_i)$ and a predicted value of δ at the current point y.

As y increases from y_1 the solution of Eq. (7) passes from the Kontorovich region of no dissipation to a region where the dissipation plays a major role. (This usually occurs at the top of the beaker and typically begins abruptly when integrating from outside to inside.) As y proceeds from the top of the beaker down the other side the film thickness increases, the dissipation integral D no longer grows, and a solution continues to exist until the gravitational term has become sufficiently negative to exactly cancel D. This point defines the end of the flow path y_2 since it corresponds to the bulk fluid. In Eq. (7) the direction of integration is the same as the direction of flow, from a higher to lower level. One can just as well integrate in the opposite direction, from the lower to the higher level, by merely changing the sign of the dissipation term. It is not difficult to prove that the solutions are identical. Thus Eq. (7) constitutes a theory capable of predicting the measurable features of film

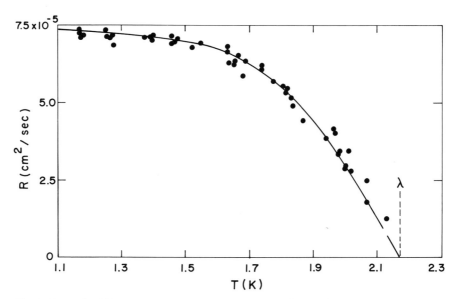

Fig. 1. Normalized flow rates vs. temperature. The data are from Ref. 4; the solid curve is the prediction of Eq. (7).

flow. It is semiphenomenological to the extent that the dissipation model's parameters $v(T)$ and $b(T)$ must be measured rather than calculated from first principles. Even so, it holds the promise of correlating a large class of film flow data.

As an example, consider the temperature dependence of flow rates observed for finite level differences. In 1950 Mendelssohn and White[4] published rate–temperature data for a variety of flow experiments. When the data from each experiment were normalized at 1.4°K the shape of the transfer curves were quite similar, having a temperature dependence of approximately $1 - (T/T_\lambda)^7$. These data are shown in Fig. 1. The solid curve in Fig. 1 is the theoretical rate–temperature relation predicted by Eq. (7) using the parameters for an arbitrary glass beaker: inside radius 1 cm, wall thickness 0.2 cm, inside level 3 cm (from rim), and outside level 5 cm. The absolute value of the rate for these parameters at $T = 1.4°K$ was $8.07 \ 10^{-5} \ cm^2 \ sec^{-1}$.

As a final example, the theory predicts for large flow rates that the film thickness can be constant and thicker than the corresponding static film over a portion of the flow path where the film is descending. This happens when the gravitational energy that is gained in falling a distance dy is exactly cancelled by the energy lost through dissipation. If the film becomes appreciably thicker than the static case, it might be energetically favorable for it to thin by forming macroscopic droplets which fall and are readsorbed below that unstable region. This is suggestive of observations of droplet formation mentioned in the review paper of Jackson and Grimes.[5]

Acknowledgment

I wish to thank Professor Michael Cohen for helpful discussions.

References

1. J.K. Hoffer, J.C. Fraser, E.F. Hammel, L.J. Campbell, W.E. Keller, and R.H. Sherman, this volume.
2. J.S. Langer and J.D. Reppy, in *Progress in Low Temperature Physics,* C.J. Gorter, ed. North-Holland, Amsterdam (1970), Vol. 6, p. 1–35.
3. V.M. Kontorovich, *Soviet Phys.—JETP* **3**, 770 (1956).
4. K. Mendelssohn and G.K. White, *Proc. Phys. Soc.* (*London*) **A63**, 1328 (1950).
5. L.C. Jackson and L.G. Grimes, *Adv. Phys.* (*Phil. Mag. Suppl.*) **7**, 435 (1958).

Film Flow Driven by van der Waals Forces *

D.G. Blair and C.C. Matheson

School of Mathematics and Physics, University of East Anglia
Norwich, England

To examine film flow driven by van der Waals forces, a film detector is required that combines a large surface area with a narrow flow path for film formation from the helium bath. Furthermore, we require that the film formation rate by superflow be far in excess of the film formation rate by condensation. Evidence that the latter criterion applies in our experiments is given below; we have also examined evaporation and condensation directly, as described in an accompanying paper.[1]

We began our experiments using a torsional oscillator film detector.[2] A glass torsion fiber supports a cone, now made from titanium foil, which oscillates in the vapor above a helium bath. The helium film flows onto the cone along the torsion fiber and is detected as a change in the resonant frequency of the oscillator. Typically the frequency is ~ 10 Hz, and changes in frequency of one part in 10^7 can be measured. The film is easily "burnt off" the conical substrate by means of miniature lamps delivering a couple of milliwatts of infrared radiation.

If the film thickness d is less than the equilibrium film thickness d_{eq} (as determined by the bath level), there will be a driving potential $\Delta\mu$ resulting in film formation and given by

$$\Delta\mu = \rho g h \left[(d_{eq}/d)^{1/\alpha} - 1 \right] \tag{1}$$

where ρ is the density of bulk He II, g is the gravitational acceleration, h is the height of the film above the bath, and α is the exponent defining the film profile $d = Ah^{-\alpha}$. The driving potential is extremely high at small film thicknesses. It was intended to relate this potential to the flow rate of the film as it formed on the oscillator. Unfortunately, the results from the torsional oscillator were difficult to analyze because film formation seemed to be masked by an unexpected effect of the vapor, discussed in the accompanying paper.[1] The *apparent* film formation rate observed using the oscillator showed only a weak dependence on potential.

To overcome these difficulties, we have observed the film on a levitated substrate. An automatic ellipsometer[3] is used to measure the helium film thickness on a plane gold substrate which is lifted by means of a superconducting film on a perspex support in a magnetic field. A thin glass fiber hanging from the substrate allows the film to form from the bath. Correct orientation of the substrate with respect to the beam of the ellipsometer is achieved by means of a small ceramic magnet situated beneath the substrate in the magnetic field determined by four superconducting solenoids; oscillations were damped by sheets of copper foil. The experimental arrangement is shown in Fig. 1.

* Supported by the Science Research Council and the University of East Anglia.

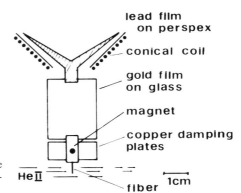

Fig. 1. Schematic diagram of the levitated substrate apparatus used for observing the helium film dissipation function.

For a helium film the time resolution of the ellipsometer is about 200 Å sec^{-1} and the thickness resolution is better than 1 Å.

Film formation has been observed over periods from 10 to 1000 sec, depending on the diameter of the flow fiber. The observed rate of film formation is directly proportional to the flow fiber diameter. Should there be a condensation contribution to the film formation, the formation rate would show a minimum value, corresponding to the rate of condensation.

Operation of the "burn-off" lamps presumably leads to a considerable loss of thermal equilibrium which is observed as an initial delay ($\sim$ 1 sec) in the formation of the film. Thereafter film formation is smooth and corresponds to flow velocities of 20 to 30 m/sec.

To analyze the ellipsometric data, we have used the thin-film approximation to the Drude equations. This involves treating the polarizer and analyzer azimuths (P and A, respectively) as independent, leading to a pair of solutions for the film thickness.[3] That these solutions lead to identical values of film thickness serves to check the validity of this approximation. In our system P is more sensitive to the helium film than A, but the linear approximation also breaks down at smaller thicknesses. For example, the exact solution deviates from the approximate solution calculated from P by more than 1% for film thicknesses greater than 80 Å. To overcome this difficulty without recourse to the exact solution of the Drude equations (which gives rise to considerable problems when applied to the low-refractive-index situation), we have derived a polynomial correction function which brings the approximate solution for P to within 1% accuracy for film thicknesses up to 400 Å.

From the observed time dependence of film thickness as the film forms on the levitated substrate, calculated as above, we can relate the superfluid velocity v_s to the driving potential $\Delta\mu$. The Langer–Fisher model for dissipation[4] predicts that the driving potential is related to the superfluid velocity as follows

$$\Delta\mu = (h/m) V_f v_0 \exp\left(-\beta\rho_s/k_B\rho T v_s\right) \tag{2}$$

where h/m is the quantum of circulation, V_f is the volume of film in which dissipation takes place, v_0 is the fluctuation attempt frequency, β is the activation energy of the fluctuations giving rise to the dissipation, k_B is the Boltzmann constant, and ρ_s and ρ are the superfluid and total fluid densities, respectively, for He II at temperature T.

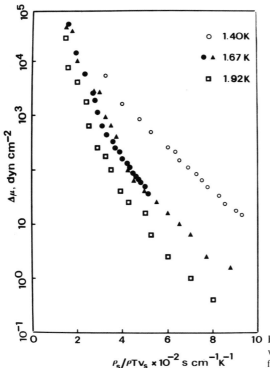

Fig. 2. The van der Waals driving potential $\Delta\mu$ versus the quantity $\rho_s/\rho T v_s$ for helium film formation on a levitated substrate.

We assume that the nonequilibrium film on the substrate has a profile which follows the law

$$d = A h_{app}^{-\alpha} \tag{3}$$

where A is determined experimentally from the equilibrium film and h_{app} is the equivalent height of the equilibrium film for the thickness measured. Then the volume rate of superflow onto the substrate is related to the rate of change of film thickness at the point of observation on the substrate by integrating the thickness over the geometry of the substrate shown in Fig. 1. The superfluid velocity is then easily determined from the known dimensions of the perimeter of the flow fiber. From our data it appears that the film thickness in the dissipation region is not given by Eq. (3) but is the equilibrium film thickness. For, if thicknesses given Eq. (3) are assumed, at high values of $\Delta\mu$, v_s becomes unreasonably large.

In Fig. 2 we have plotted the log of $\Delta\mu$ [calculated using Eq. (1)] against the quantity $\rho_s/\rho T v_s$. Clearly from Eq. (2) such a plot should be linear and independent of temperature.

Our results do not fit the Langer–Fisher theory as well as those obtained thermally by Liebenberg.[5] There is a clear temperature dependence in our data, reflected in the fact that the experimental curves at different temperatures are displaced from each other. Furthermore, it is not possible to fit a straight line to the data at a given temperature.

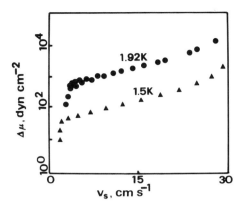

Fig. 3. The van der Waals driving potential $\Delta\mu$ versus the superfluid velocity v_s for some cases of helium film formation.

The results shown here are typical of the majority of about 100 such film formations that we have observed and analyzed. However, sometimes the observed dissipation function fails to fit the Langer–Fisher theory even to the extent described above. In these cases v_s varies over a much wider range than the Langer–Fisher theory predicts and the data are much better described by the empirical relation

$$\Delta\mu = C \exp(Dv_s) \tag{4}$$

where C and D are constants. Two typical sets of such data are shown in Fig. 3. In this case C increases from about 1.4×10^2 at $1.4°K$ to 2.2×10^3 at $1.92°K$, while D is of the order of 8×10^{-2} at both temperatures.

We are not aware of any experimental reason that would account for such apparently random differences in the behavior of the dissipation function. However, we have obtained some data when the substrate has been highly contaminated. In this situation all the data fit the form specified by Eq. (4).

It is possible that our results are affected by evaporation of the film as it forms on the substrate. Further measurements are underway in which special precautions are being taken to minimize the radiation level.

References

1. D.G. Blair and C.C. Matheson, this volume.
2. D.G. Blair and C.C. Matheson, in *Proc. 12th Intern. Conf. Low Temp. Phys. 1970*, Academic Press of Japan, Tokyo (1971), p. 111.
3. C.C. Matheson and J.L. Horn, this volume.
4. J.S. Langer and M.E. Fisher, *Phys. Rev. Lett.* **19**, 560 (1967).
5. D.H. Liebenberg, *Phys. Rev. Lett.* **26**, 744 (1971).

5
Superfluid Hydrodynamics

Decay of Saturated and Unsaturated Persistent Currents in Superfluid Helium

H. Kojima, W. Veith, E. Guyon,* and I. Rudnick

Department of Physics, University of California
Los Angeles, California

Persistent current states in superfluid helium contained in a rotating superleak have been investigated[1,2] using fourth-sound techniques.[3] These investigations showed that every phenomenon observed in rotating, locked superfluid helium has an analogous counterpart in a highly irreversible, high-field type II superconductor. In the present paper we report some experimental results concerning decay of superfluid helium persistent currents.

A narrow annular cavity (inner radius = 5.01 cm, outer radius = 5.48 cm, depth = 0.50 cm) is tightly packed with Al_2O_3 powder (grain size 170 ~ 325 Å) and forms our experimental cell. The superfluid component velocity V_s or angular velocity ω_s is determined from the Doppler-shifted splitting of an angularly dependent degenerate fourth-sound mode in the cavity.[3] If the cavity (hence the normal fluid component) rotates at angular velocity ω_n, the splitting determines the relative angular velocity, $\omega_n - \omega_s$.

Figure 1 shows the relative angular frequency for a complete cycling of ω_n when the superfluid helium is initially at rest below T_λ. This plot appears very similar to a magnetization versus applied field hysteresis curve in a highly irreversible type II superconductor.[4] If $\omega_n/2\pi$ does not exceed $\omega_{cl}/2\pi$ (analogous to H_{cl}) = 1.0 cps, $\omega_n - \omega_s$ is linear in ω_n and the curve is reversible. As ω_n is increased above ω_{cl}, $\omega_n - \omega_s$ departs from the linear behavior and becomes irreversible. As ω_n is further increased, $(\omega_n - \omega_s)/2\pi$ saturates to a maximum value, $\omega_\sigma/2\pi = 1.90$ cps, and remains constant for our largest speed of rotation of 9.0 cps. This saturated persistent current velocity is also called the "critical velocity." The hysteretic behavior of $\omega_n - \omega_s$ is clearly seen as ω_n is reduced from the saturated region.

In the initial region, where $0 < \omega_n < \omega_{cl}$, vortices are excluded. The region has been called the Landau region.[1] A Landau state is analogous to a Meissner state in superconductors. Vortices begin to enter when $\omega_n = \omega_{cl}$. Our data in Fig. 1 indicate that ω_{c2} (analogous to H_{c2}) may be inaccessibly large.

Kukich *et al.*[5] first observed the logarithmically time-dependent decay of saturated superfluid helium persistent currents and interpreted the persistent current decay in terms of Langer and Fisher's theory[6] of thermally activated processes. We have extended such measurements to include five decades (1 to 10^5 sec) in time and previously unobserved decays of unsaturated persistent currents whose initial superfluid velocities are slightly less than that of the saturated persistent current.

* Permanent address: Physique des Solides, Université de Paris, Orsay, France.

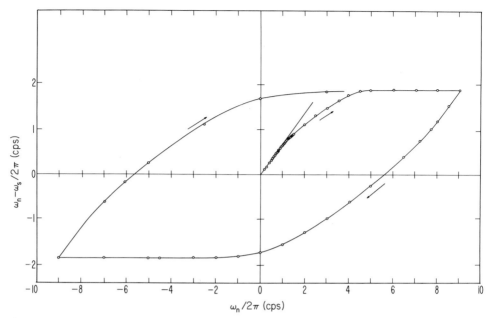

Fig. 1. A plot of $\omega_n - \omega_s$ for complete cycling of ω_n starting from rest at $T = 1.30°$K. The Landau region occurs in the initial part of the curve below 1.0 cps. The saturated relative angular frequency is 1.9 cps.

A method of measuring the Doppler-shifted splitting very rapidly by a well-known acoustical technique was used. A narrowband source excites both of the split resonances of a particular angularly dependent mode. When the source is turned off these two resonances free-decay exponentially with time. The separation of the doublet is determined from the beating during the free decay. The complete free decay took about 1/4 sec and about two measurements per second could be made. An example of free decay recording is shown in Fig. 2. The decay appears linear because sound level is plotted on a logarithmic scale. The beating in this example corresponds to 66.5 Hz.

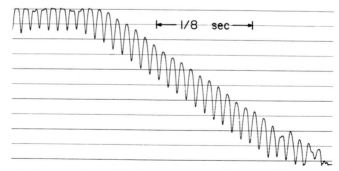

Fig. 2. A recording of the beats between the Doppler-shifted fourth-sound split modes in free decay. The beat frequency determines the superfluid component velocity. One small division corresponds to 1 dB.

A persistent current is created by cooling the rotating annular cavity around its axis at a constant initial angular frequency (depending on the persistent current velocity desired) through T_λ to a low temperature (1.30°K). Then the rotation frequency is reduced to ~ 1.0 cps. Subsequently the bath temperature is regulated (± 0.5 m°K) and when the necessary electronic adjustments are made the rotation is stopped (in less than 1 sec). The circumference of the resonator is 33 cm, and since the persistent current is near saturation even when the resonator rotates at 1 cps, the change in V_s is almost 33 cm sec^{-1} when the resonator is brought to rest. The decay in V_s while it is rotating at 1 cps is negligible compared to this. The free decays are recorded by a tape recorder continuously from the moment the rotation stops, i.e., $t = 0$, to about 10^3 sec. Thereafter decays are recorded at appropriate time intervals.

We summarize the results of decay measurements at $T = 1.30$°K in Fig. 3, in which the superfluid component velocity is plotted against log t. The different symbols refer to persistent currents with different initial velocities. The set of data A (initial velocity = 67.7 cm sec^{-1}) is the saturated current, and B (65.7 cm sec^{-1}), C (64.1 cm sec^{-1}) and D (61.7 cm sec^{-1}) are unsaturated currents. We first note that all initially saturated and unsaturated persistent currents decay at a logarithmic

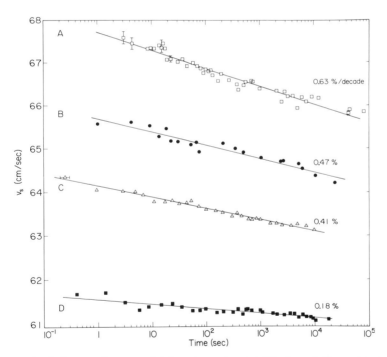

Fig. 3. Decay of saturated (A) and unsaturated (B, C, and D) persistent current velocities plotted against log t. Vertical bars on some data points in A indicate probable experimental errors in the initial parts of the data. The error is smaller for points for large t (> 20 sec) since a greater amount of data are available for averaging. The lines are the least square fit curves. The fractional decay rates per decade are indicated.

rate for $1 < t < 10^5$ sec. Persistent superconducting currents in a high-field ir-reversible type II superconductor also decay logarithmically with time.[7] Anderson has developed a theory of thermally activated "flux creep" to explain the logarithmic decay of the supercurrent.[8]

The fractional decay rate per decade $(1/V_s)\,dV_s/d(\log t)$ decreases as the initial velocity decreases and vanishes when it is about 14% below the saturated value. The highest decay rate observed is 0.63% per decade. If this decay rate continues inde-finitely, the current decays 11% during a time interval equal to the age of the earth. The persistent currents are indeed persistent!

It is clear from Fig. 3 that the magnitude of superfluid velocity alone does not determine the instantaneous decay rate dV_s/dt. One of the determinants of the in-stantaneous decay rate is the initial value of the persistent current velocity. Thus the average superfluid component velocity alone does not characterize a persistent current state.

If we are to use the Langer and Fisher theory to understand our results, then we must recognize that the simple presence of vortex lines does not ensure decay. The interaction of vortex lines with flow can only lead to decay if the vortex lines can move subject to an appropriate activation energy. The fact that unsaturated currents also decay at a logarithmic rate is perhaps more easily understood in terms of a thermally activated vortex creep (analogous to flux creep in superconductor) than in terms of the Langer and Fisher model, although that model with a distribution of critical velocities for the distribution of pore sizes could also account for the observation.

Finally, we note that a part of the decay of unsaturated persistent currents may result from a distribution of superfluid component linear velocity across the finite width (10% of radius) of the annulus. Thus the velocity along the outside rim may be saturated while that in the remainder of the annulus is unsaturated. If this saturated current decays and the unsaturated one does not, then the average velocity decays at a lower rate than that of totally unsaturated current. From this kind of effect alone we expect the decay rate to vanish at velocities 5% (14% experimentally) below the saturated current and thus other effects must also play a role. Preliminary results using an annulus 2 mm wide support this conclusion.

References

1. H. Kojima, W. Veith, S. Putterman, E. Guyon, and I. Rudnick, *Phys. Rev. Lett.* **27**, 714 (1971).
2. H. Kojima, W. Veith, E. Guyon, and I. Rudnick, *J. Low Temp. Phys.* **8**, 187 (1972).
3. I. Rudnick, H. Kojima, W. Veith, and R.S. Kagiwada, *Phys. Rev. Lett.* **23**, 1220 (1969).
4. C.P. Bean, *Rev. Mod. Phys.* **36**, 31 (1964); W.A. Fiets, M.R. Beasley, J. Silcox, and W.W. Webb, *Phys. Rev.* **136**, A335 (1964).
5. G. Kukich, R.P. Henkel, and J.R. Reppy, *Phys. Rev. Lett.* **21**, 197 (1968).
6. J.S. Langer and M.E. Fisher, *Phys. Rev. Lett.* **19**, 560 (1967); J.S. Langer and J.D. Reppy, *Progress in Low Temperature Physics*, C.J. Gorter, ed., North-Holland, Amsterdam (1970), Vol. 6.
7. Y.B. Kim, C.F. Hempstead, and A.R. Strand, *Phys. Rev.* **90**, 306 (1962).
8. P.W. Anderson, *Phys. Rev. Lett.* **9**, 309 (1962).

Rotating Couette Flow of Superfluid Helium*

H.A. Snyder

Aerospace Engineering Sciences Department
University of Colorado, Boulder, Colorado

Equations for the flow of He II when $\nabla \times v_s \neq 0$ have been proposed but have not been tested thoroughly. Here I refer to the derivations by Hall and Vinen and by Bekeravich and Khalatnikov (HVBK). There are seven terms in these equations which do not appear in the Landau–Khalatnikov equations; they may be attributed to vorticity in the superfluid. Many experiments have been described in the literature which test two or three of the vorticity terms in the HVBK equations. In these tests the experimental arrangement is chosen to make the magnitude of a particular vorticity term much larger than the others, and it is assumed that the term being tested is the only contributor to the result. Notable are the torsion-pendulum studies of Hall and the rotating second-sound attenuation measurements of Hall and Vinen. All the vorticity terms have not been tested and no experiments have been reported in which each of the seven terms contributes appreciably to the result.

The purpose of the experiment described here is to test the validity of the HVBK equations by testing the seven vorticity terms simultaneously. It is known that the most stringent test of the Navier–Stokes equations is the prediction of the onset of a secondary flow. The simplest case of a secondary flow that can be set up in the laboratory is the flow between two concentric rotating cylinders. The standard test to determine if a fluid is Newtonian is to determine the Reynolds number at which Taylor vortices (the secondary flow) appear in a double cylinder experiment. Accordingly, it appears that a measurement of the onset of Taylor vortices between concentric rotating cylinders might be an appropriate test of the HVBK equations when the fluid is He II.

As a first step the linear stability analysis for rotating Couette flow was formulated using the HVBK equations. It is found that the system of differential equations is eighth order with homogeneous boundary conditions corresponding to the vanishing of the three components of v_n at both inner and outer cylinders and the vanishing of the normal component of v_s at both cylinders. The problem for a Newtonian fluid, of course, is sixth order.

Several dimensionless parameters appear in the stability equations: the mutual friction parameters B and B'; the quantity $\lambda \rho_n / \eta$, where λ is the energy density of the vortex line, η is the viscosity, and ρ_n is the normal fluid density; ρ_s/ρ; R_1/R_2,

* The National Science Foundation is the major source of support for this work.

the ratio of the cylinder radii; Ω_1/Ω_2, the ratio of the angular velocities of the cylinders; and $\Omega_1 R_2^2/(\eta/\rho_n)$, the Reynolds number based on the normal fluid density. These dimensionless numbers are prefactors of the vorticity terms in the stability equations; some are temperature dependent and the others may be varied by changing the dimensions of the apparatus to be described below; within the range of available speed of this apparatus and within the temperature range 1.1–2.1°K each of the vorticity terms in the HVBK equations may be checked individually. If term-by-term confirmation is not desired, only one point on the stability curve is sufficient to prove or disprove the proposed equations.

The eighth-order stability equation has been reduced to eight first-order coupled equations and integrated numerically as an initial-value eigenvalue problem using the Runge-Kutta method. Our procedure in the integration follows Harris and Reid.[1] It is appropriate to check that our results approach the results for a Newtonian fluid as ρ_n approaches ρ and ρ_s approaches zero. For our apparatus $R_1/R_2 = 1/2$, and for the case $(\Omega_1/\Omega_2) = 0$ the critical Reynolds number Re for onset of Taylor vortices in a Newtonian fluid has been calculated by Roberts[2] to be 272. When $\rho_n = 0.99\,\rho$ our calculations give Re = 272 and this result is independent of B, B', and $\lambda\rho_n/\eta$. As the temperature decreases and ρ_s increases, we find that the lowest critical Re increases rapidly. At $T = 1.63°K$, where the experiments have been carried out, we find that Re $\approx$ 2000.

Before we describe the apparatus and the experimental results it is appropriate to note that the HVBK equations assume that there is a regular array of quantized vortex lines in the superfluid. Therefore the HVBK equations cannot be expected to hold for nonsteady flows during the interval of time required to create or destroy superfluid vortex lines. These equations are also invalid when the vortex array loses ordering, i.e., when the superfluid velocity field is turbulent. Therefore it must be established what interval of time is required to produce vortex lines and at what rate they decay in our test apparatus. The design of the experiment must also provide a way to distinguish a regular array from a random distributor of vortex lines.

The rotating cylinder apparatus follows the Bearden design,[3] which is reputed to have great mechanical precision and stability. Change gears permit the inner and outer cylinders to turn at fixed ratios of angular velocity; standing modes of second sound can be excited in the annular cavity between the cylinders. The second-sound transmitter and receiver are painted on the inner cylinder, which consists of anodized aluminum. Aquadag is used as a receiving material and the transmitter is silver. The radius of the inner cylinder is 0.795 cm and the outer cylinder has a radius 1.590 cm; the axial length of the resonant cavity is 12.04 cm. Some resonant Q's of the cavity exceed 5000 even when the cylinders are rotating. This indicates that the space between the cylinders is uniform to better than 0.25 μm and that this spacing is maintained during rotation.

We infer the flow between the cylinders by analyzing the resonant Q curves as a function of rotation. Since the additional attenuation of second sound by vortex lines is dependent upon the angle between the direction of propagation and the vortex line, each normal mode will react differently to a given array of vortex lines. For example, a purely axial mode will show attenuation directly proportional to the number of vortex lines with component projection on a plane perpendicular to the axis; it will not respond to axial vortex lines. A total of 20 normal modes of the cavity have been studied as a function of the Reynolds number to determine the position and orientation of the vorticity field in the superfluid. We are unable to study the flow field of the normal fluid with this method. However, the object of the experiment is to compare the onset of Taylor vortices in the superfluid with the prediction of the HVBK equations, and this can be accomplished with the method outlined here.

A major difficulty in the interpretation of the data is the assignment of a mode number to each observed peak. It was necessary to write a computer program which integrated the surface Green's function over the transmitter and receiver; the output indicated the relative height of the peaks and this helped us to distinguish one peak from another. It would be desirable to solve the complete acoustic problem for the cavity with sheared mean flow and dissipative boundary layers included, but this has not been carried out yet. I plan to do this calculation soon.

All the data reported here are for the case of $T = 1.63°K$ and only the inner cylinder rotates. The result is one point on the stability curve; this is all that is necessary to test the equations as a whole. We plan to take more data at various temperatures and various ratios of Ω_1/Ω_2 in order to test specific terms in the equations.

Figures 1 and 2 show typical data. Each mode has a different dependence of excess attenuation on Reynolds number, but all the curves have some features in common. There is a region of no additional attenuation at the lowest speeds of rotation. Apparently this region is analogous to the region free of vortices in an annulus with solid body rotation. If the energy minimization calculations of Stauffer

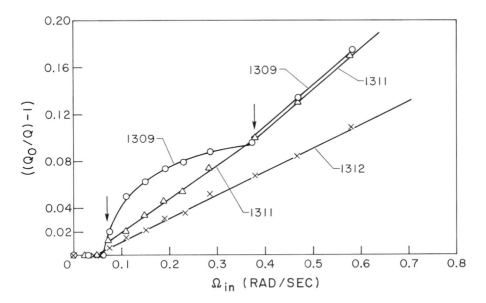

Fig. 1. The additional attenuation due to vortex lines as a function of angular velocity of the inner cylinder. Frequency of the normal modes is indicated on the figure.

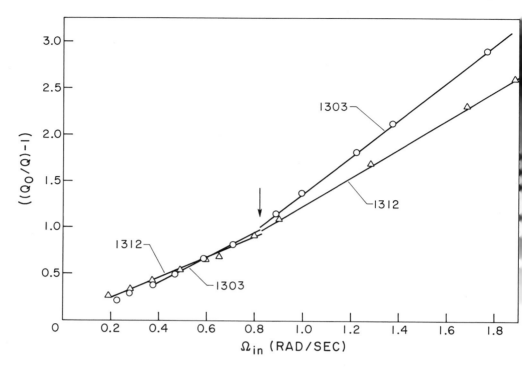

Fig. 2. The additional attenuation due to vortex lines as a function of angular velocity of the inner cylinder. Frequency of the normal modes is indicated on the figure.

and Fetter[4] are applied using the constant value of $\nabla \times v_s$ for the shear flow occurring here instead of the 2Ω for solid body rotation, then the break point at 0.06 rad sec^{-1} is an order of magnitude larger than the predicted value. However, this comparison is not valid physically because the system with shear is not in thermodynamic equilibrium.

All the modes show three break points on curves such as Fig. 1 and 2. In the region between the first and second break point the curves may be straight or curved depending upon the mode. All the curves are straight between the second and third break points; all the curves are straight above the third break point up to angular velocities of 8 rad sec^{-1}, the highest velocity tested. The first break point corresponds to a Reynolds number of 1160 and the second to 7200. By analyzing the shapes of these curves, I believe that Taylor vortices occur in the superfluid at the second break or Re = 7200. Comparing this result with the calculated value of Re $\approx$ 2000, we conclude that the HVBK equations do not describe the flow of He II adequately in our apparatus, or that the accepted values of B, B', and λ are grossly incorrect.

Further work should be undertaken to verify this conclusion with points at various temperatures and rotation ratios. By this means, it will be possible to determine which terms are incorrect. We plan further experiments along these lines if financial support is forthcoming.

Acknowledgments

The rotating cylinder apparatus was made and measured at the National Bureau of Standards, Boulder. Mr. Kevin Rudolph carried out the numerical calculations. David Baldwin, Robert Hess, and Larry Foreman did most of the experimental work and assisted with taking the data.

References

1. D.L. Harris and W.H. Reid, ONR Tech. Rep. 562 (07)/43 (1962).
2. P.H. Roberts, *Proc. Roy. Soc.* **A283**, 550 (1965).
3. J.A. Bearden, *Phys. Rev.* **56**, 1023 (1939).
4. D. Stauffer and A.L. Fetter, *Phys. Rev.* **168**, 156 (1968).

New Aspects of the λ-Point Paradox*

Robert F. Lynch and John R. Pellam

Department of Physics, University of California—Irvine
Irvine, California

The purpose of this publication is twofold: (a) to eliminate finally any equipment source for the "λ–point paradox" and (b) to report a newly observed phenomenon dependent on the past history of the liquid helium–disk system.

A previous publication[1] showed that the paradox behavior remains essentially invariant to whether the disk or the sample rotates, the "relative rotation rate" alone being significant. That work suffered the obvious flaw that relative rotation between the rotor system and the disk suspension system might introduce a hidden heat source capable of affecting the behavior. This flaw is eliminated in the present effort by taking transient measurements of disk deflections upon starting the rotor and disk systems simultaneously from rest (no relative rotation).

This circumstance is represented by the topmost data of Fig. 1 (solid curves) labeled "disk–rotor locked." Since measurements of disk deflection from equilibrium position are taken immediately upon starting from rest, these transient results correspond to the disk rotating through an otherwise undisturbed helium background. Persistence of the λ–point paradox is unmistakable, as indicated by the elliptical data points; the direction of the arrows indicate the *sense* of rotation of the system.

A further confirmation of the anomaly was made by constructing an apparatus of radically different design. It consists of a rod extending from the bottom of the dewar through the dewar cap with a bearing at each end. A 0.28-cm disk is suspended by a quartz fiber (fiber constant 1.3×10^{-5} dyn cm deg^{-1}) attached to an arm on the rod at a radial distance of 1.82 cm, 45° angle to the relative flow. The rod is rotated by a motor and thus the disk moves as before through stationary helium but in a rotor-free system. The deflection from equilibrium is observed on a scale using a strobe light source synchronized to the rotation (analogous to a previous experiment[1]).

Taking the deflection as a function of time from the start of rotation, we find a decrease of deflection to a steady lower value in approximately 1 hr. The source of this decrease is not understood; nothing similar occurs in the original apparatus. By observing at various temperatures one can obtain two quantities vs. temperature: the initial value of deflection, and deflection in the "steady state." These are given in Fig. 2, where both curves display the general shape of the anomaly below T_λ. However, the deflection is nonzero at T_λ, and the deflection above T_λ is lower than

---

* Supported in part by the National Science Foundation.

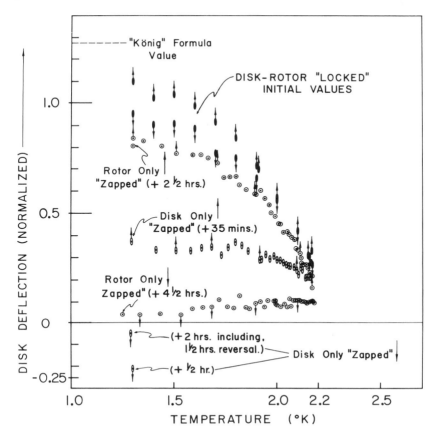

Fig. 1. Disk deflections (normalized) versus temperature. Disk deflections versus temperature (°K) obtained for various disk rotor rotational situations. The upper points (closed ellipses) represent data for initial deflection upon starting the locked disk–rotor system from rest; arrows indicate "sense" of effective disk rotation. All other data indicate temperature dependences for various "zapped" systems. Open ellipses indicate deflections for the disk only rotating (arrows indicate "sense") under conditions of "zap" duration as indicated. Open circles represent corresponding disk deflections for the rotor only rotating (see arrows) and under indicated "zap" conditions. All data are normalized to the low-temperature limit—averaged for both directions—for rotor-only rotations.

the $T \rightarrow 0°$K value of both curves. These latter observations appear more in agreement with Trela's[2] results than with the original findings.[3]

The lower points (triangles) in Fig. 2 represent a new phenomenon based upon pressure-cycling the sample. The process, which we have termed "zapping," involves the following sequence. Starting from an initial temperature equilibrium at 1.25°K, the liquid helium sample is pressurized rapidly to atmospheric pressure and held at that value for 5 min, whereupon it is again pumped down to 1.25°K. As may be observed in Fig. 2, the "zapped" values of deflection are characteristically less than the original—or "normal"—values for the otherwise identical state of the system. The effect becomes dominant enough in some cases to actually produce *negative* deflections!

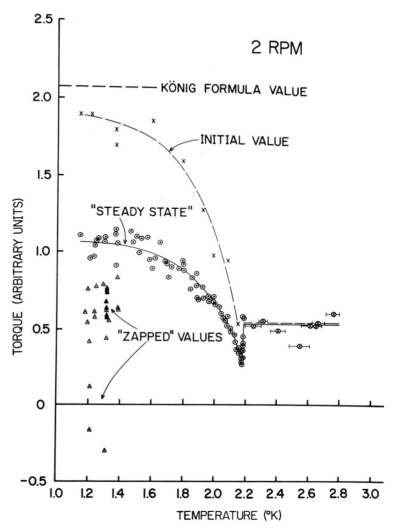

Fig. 2. Torque versus temperature for the modified apparatus. Torque (arbitrary units) as a function of temperature (°K) for a disk rotated through stationary liquid helium in modified apparatus. Crosses indicate initial values (immediately upon starting disk from rest) and circles represent steady-state torques reached after rotating approximately 1 hr. Values of torque observed for the "zapped" condition are represented by triangles, where the open triangles indicate rotation down and closed triangles indicate rotation up.

Although initially observed in the modified (rotorless) apparatus, the same effect of "zapping" manifests itself in the original system when the sample chamber itself is pressurized according to the same sequence. (Heretofore the sample within the rotor had been maintained as a sealed-off system passively following the temperature of the outer bath.) The results of "zapping" the rotor sample are summarized in the graphs of Fig. 3.

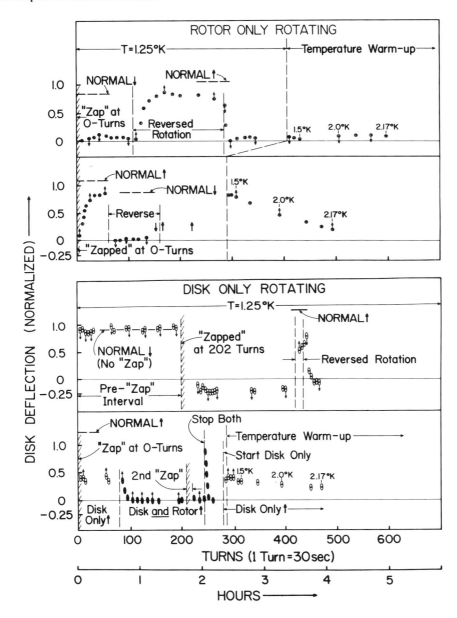

Fig. 3. Representative sequences for "zapped" sample (rotor). Normalized deflections of disk as a function of turns and elapsed time (one turn = 30 sec) for "zapped" sample in rotor. Upper two sequences illustrate the effects of "zapping" under varied conditions of pure rotor rotation; circles represent rotor rotation only and arrows indicate "sense" of rotation. Lower two sequences refer to "zapping" for varied conditions of pure disk rotation, given by ellipses with directional arrows. Time sequences and time of "zapping" are indicated in each case. The extended time duration of the "zapped" state is illustrated, including the warmup period from 1.25°K to the λ-point for both disk–rotation and rotor–rotation situations.

Various procedures are illustrated in the four sequences shown. The upper two sequences correspond to the effects of "zapping" under varied conditions of rotor rotation and the lower two sequences represent "zapping" with disk rotation only. The graphs are intended as self explanatory in regard to procedures and indicate time of "zapping" both in terms of number of turns (one turn = 30 sec), and on an elapsed hour scale. The "normal" levels represent pre-"zapped" values for the associated situations and thus provide reference levels.

Clearly the most dramatic aspect of these data involves the evident persistence of the "zapped" state. As may be noted, the situation persists following "zapping" for an apparently indeterminate period. This includes during long warmup periods as observable in Fig. 3. The "zapped" condition also survives temperature cycling—including subsequent cooldown following extended warmup periods—although this is not illustrated here.

The temperature dependence of disk deflection for the "zapped" state is shown in the remaining sets of data of Fig. 1. Here the deflection versus temperature is given for various rotational situations, including some of the warmup intervals of Fig. 2. In some cases the warmup commenced several hours following "zapping" and clearly the resultant state persists throughout the temperature rise interval.

Explanation for the "zapped" condition remains elusive. On no occasion has the zero setting (for the nonrotating system) been affected by "zapping," thus ostensibly eliminating electric charge. The effect does appear to involve a disk–helium condition in that removal of the disk from the liquid destroys the "zapped" state—or precludes it by removal prior to "zapping." The possibility accordingly arises that the phenomenon might involve some form of persistent circulation attached to the disk, although such a mechanism is not clear.

Note added in proof

A subsequent apparatus modification allowing faster delivery of gas to the sample (intended to enhance the "zapping" effect) now produces two significant results: (1) an increase, rather than a decrease, in deflection upon "zapping," and (2) visible lengthy "whiskers" of some contaminant formed on the disk following "zapping." The latter effect appears a likely source for the entire "zap" behavior and is currently under investigation as such.

References

1. J.R. Pellam, *Phys. Rev. Lett.* **27**, 88 (1971).
2. W.J. Trela, *Phys. Rev. Lett.* **29**, 38 (1972).
3. J.R. Pellam, *Phys. Rev. Lett.* **5**, 189 (1960).

Torque on a Rayleigh Disk Due to He II Flow*

W.J. Trela and M. Heller

Department of Physics, Haverford College
Haverford, Pennsylvania

The Rayleigh disk has been used in several experiments[1-7] to probe the local velocity field of He II. However, the flow pattern around the disk has not been well understood.[8] We present here a simple model for He II flow past a Rayleigh disk which assumes that superfluid potential flow and normal fluid Helmholtz flow occur. A torque which increases as the temperature decreases below T_λ is predicted. New measurements of the torque on a Rayleigh disk in nonrotating helium are then presented which are in substantial agreement with this model. Finally, we show that our model is in agreement with Pellam's Rayleigh disk measurements[1-4] in rotating helium below T_λ.

The classical flow pattern for a perfect (nonviscous) liquid around a Rayleigh disk is potential flow. The superfluid component of He II provides the only real liquid capable of perfect potential flow, and experiments[7,9] indicate that at sufficiently low velocity, potential flow does take place. A flat, rectangular disk (width w and height d) exposed to pure potential flow will experience a torque of the form[10]

$$\tau = (1/8)\,\pi w^2 d\rho v^2 \sin 2\theta \tag{1}$$

where ρ is the fluid density, v the undisturbed fluid velocity, and θ the angle between the disk and fluid velocity. There is no net drag force.

Normal viscous liquids *do not* move past a Rayleigh disk with potential flow. Instead, flow separation with a velocity discontinuity occurs at the disk edges and a stagnant region exists behind the disk.[11] By direct observation, Kitchens *et al.*[12] have seen velocity fields of this general character when the normal component of He II flows past a flat plate. This type of flow is known as Helmholtz flow and its importance and relevance to superfluid hydrodynamics has been suggested by Craig.[8] An idealized version of this type of velocity field can be analyzed quantitatively. If it is assumed that the wake is composed of dead liquid (velocity zero) and extends to infinity, then the torque on a flat disk is (Ref. 10, p. 102)

$$\tau = 3\pi w^2 d\rho v^2 \sin 2\theta\,/8(4 + \pi \sin \theta)^2 \tag{2}$$

It is important to note that for identical experimental parameters, the torque due to potential flow is a factor of $(1/3)(4 + \pi \sin \theta)^2$ larger than for Helmholtz flow. In this case there is also a net force on the disk.

From the above results we see that the torque on the disk is very sensitive to

---

* Work supported in part by the National Science Foundation and the Research Corporation.

the details of the local flow pattern. We now propose the following model for He II flow past a Rayleigh disk. At sufficiently low velocity the superfluid component exhibits pure potential flow around the Rayleigh disk. The normal fluid always displays Helmholtz-type flow. The net torque on the disk is the sum of the superfluid and normal fluid contributions:

$$\tau = \frac{\pi w^2 d \sin 2\theta}{8} \left[\rho_s v_s^2 + \frac{3\rho_n v_n^2}{(4 + \pi \sin \theta)^2} \right] \tag{3}$$

where the subscripts s and n refer to the superfluid and normal fluid components, respectively. We have assumed that no coupling occurs between the normal and superfluid components. For helium above the λ-point the torque is given by the Helmholtz expression.

Figure 1 shows torque versus temperature as predicted by this model for an attack angle $\theta = 45°$. These results have been normalized to unity at $T = 1°K$. Between 4.2°K and the λ-point the torque increases by about 15% due to the change in the helium density. At T_λ the torque is continuous. Below T_λ it increases rapidly

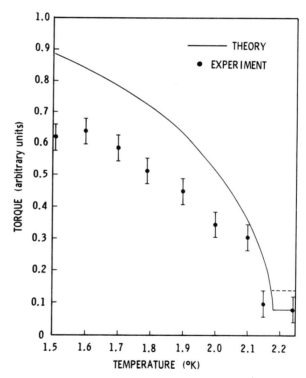

Fig. 1. Torque on a Rayleigh disk versus temperature. Solid curve represents Eq. (3) for $\theta = 45°$ and is normalized to the experimental values above T_λ. The circles indicate our experimental data at a velocity of 0.22 cm sec^{-1} and attack angle $\theta = 45°$. The dashed line represents the absolute magnitude of the torque expected for Helmholtz flow [Eq. (2)].

with decreasing temperature. This significant increase in torque below T_λ is due to the temperature dependence of the superfluid and normal fluid densities and the order of magnitude difference between the torques due to potential and Helmholtz flow.

In order to provide an experimental test of this model and to study the relevance of Helmholtz flow to He II hydrodynamics, we have measured the torque on a Rayleigh disk in He II flow. Our apparatus is shown in Fig. 2. We use a rectangular disk 0.1 mm thick, 1.8 mm wide, and 1.9 mm high, suspended vertically in a 1.0-cm-diameter channel. The suspension consists of a quartz torsion fiber (torsion constant $\sim 10^{-5}$ dyn cm deg^{-1}) attached to a rigid quartz rod on which are mounted the Rayleigh disk, a mirror, and an eddy current damping plate. The channel is connected to a vertical tube in which a displacement plunger can be moved across the helium level to cause liquid flow through the channel under a gravitational head. Torque due to liquid flow past the disk is measured by observing the deflection of a laser beam reflected by the mirror.

The results of our measurements of torque versus temperature at a velocity of 0.22 cm sec^{-1} and attack angle of 45° are also presented in Fig. 1. These results show a continuous torque at the λ-point and an increasing torque as the temperature is lowered below T_λ. We see that the experimental data are in qualitative agreement with our theory. However, we note that there is some quantitative difference between theory and experiment shown in Fig. 1. Comparing these experimental results at a velocity of 0.22 cm sec^{-1} with our previously reported work[13] at a velocity of 0.5 cm sec^{-1}, we see much closer agreement between theory and experiment. We believe

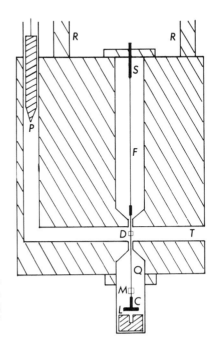

Fig. 2. Schematic diagram of the Rayleigh disk suspension system and flow channel including the following: Rayleigh disk (D), torsion fiber (F), damping plate (C), damping magnet (L), mirror (M), quartz rod (Q), plunger (P), flow channel (T), suspension point (S), and support rods (R).

that the difference which exists between our model and experiments is mainly due to superfluid velocity which is still in excess of the critical velocity for the breakdown of potential flow. We do not expect the flow pattern above critical velocity, with its vorticity and turbulence, to yield a potential flow type torque.

The absolute magnitude of the torque we measure above T_λ is about 40% less than the predicted Helmholtz value. Our previous measurements showed a similar disagreement. We believe this disagreement is not significant but is merely due to the nonideal experimental conditions as compared to the theoretical model. The experimental Rayleigh disk is not a perfectly flat plate but has a 0.010 in. fiber and a blob of epoxy on one surface. The effects of this asymmetry are observable. When liquid flow approaches the flat side of the disk the torque is 30% larger than when the fiber and epoxy side face upstream. Our measurements clearly show that above T_λ the magnitude of torque is much closer to the Helmholtz than the potential flow value.

We believe that the model presented here also explains the results of Pellam's experiments in rotating He II below T_λ. In Fig. 3 we present the results from four experiments by Pellam[1-4] in which he measured the torque on a Rayleigh disk suspended in a container of rotating He II. Pellam's results, normalized to unity at the low-temperature limit, are in excellent agreement with our model below the λ-point. Thus the previously unexplained temperature-dependent torque observed by Pellam below the λ-point can be understood by assuming potential and Helmholtz flow by the superfluid and normal fluid components, respectively.

In conclusion, we have shown that the torque on a Rayleigh disk depends strongly upon the character of the local velocity field at the disk. We propose that at sufficiently low velocity He II flow occurs with potential superflow and Helmholtz

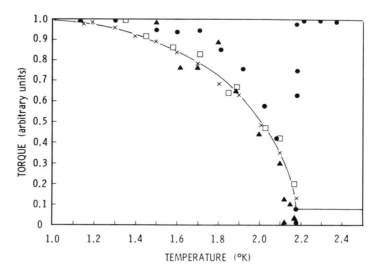

Fig. 3. Relative (normalized) torque on a Rayleigh disk versus temperature. Solid curve represents Eq. (3) for $\theta = 45°$. The circles, squares, triangles, and crosses represent data from Ref. 1–4, respectively.

normal flow. Our experimental results indicate that Helmholtz flow takes place above T_λ and that below the λ-point the torque increases as the temperature decreases in qualitative agreement with our model.

Acknowledgment

The authors would like to thank Mr. John Thorstensen for his valuable assistance in gathering the latest experimental data.

References

1. J.R. Pellam, *Phys. Rev. Lett.* **5,** 189 (1960).
2. J.R. Pellam, in *Proc. 9th Intern. Conf. Low Temp. Phys., 1964,* Plenum, New York (1965), p. 191.
3. J.R. Pellam, *Phys. Rev. Lett.* **20,** 1144 (1968).
4. J.R. Pellam, *Phys. Rev. Lett.* **27,** 88 (1971).
5. D.S. Tsakadze and L.G. Shanshiashvilli, *Zh. Eksperim, i Teor. Fiz. Pis'ma* **2,** 305 (1965) [*Soviet Phys.—JETP Lett.* **2,** 194 (1965)].
6. J.R. Pellam, in *Progress in Low Temperature Physics,* C.J. Gorter, ed., North-Holland, Amsterdam (1955), Vol. 1, p. 54.
7. T.R. Koehler and J.R. Pellam, *Phys. Rev.* **125,** 791 (1962).
8. P.P. Craig, *Phys. Rev. Lett.* **13,** 708 (1964).
9. P.P. Craig and J.R. Pellam, *Phys. Rev.* **108,** 1109 (1957).
10. H. Lamb, *Hydrodynamics,* 6th ed., Cambridge Univ. Press, Cambridge (1932), p. 86.
11. A. Sommerfeld, *Mechanics of Deformable Bodies,* Academic, New York (1950), p. 215.
12. T.A. Kitchens, W.A. Steyert, R.D. Taylor, and P.P. Craig, *Phys. Rev. Lett.* **14,** 942 (1965).
13. W.J. Trela, *Phys. Rev. Lett.* **29,** 41 (1972).

Superfluid ^{4}He Velocities in Narrow Channels between 1.8 and 0.3°K

S.J. Harrison and K. Mendelssohn

Clarendon Laboratory
Oxford, England

We have made an accurate determination between 1.8 and 0.3°K of superfluid velocities through submicron pores in irradiated and etched mica sheets of the type first used by Notarys.[1] The advantages of these narrow channels over previous types (Millipore filters, Vycor glass, etc.) are considerable. Each irradiating particle leaves a straight train of dislocations behind it in the mica, which provides an etching path for the HF used, normal to the crystal planes. Once the radiation damage has been removed the activity of the acid in the normal direction is greatly reduced, and a large number of parallel-sided, singly connected channels can be "grown" to a reproducible diameter. The channels were trapezoidal in shape, and had a 10% taper along their length. The spread in size was about 20% and each 5-μm-thick mica sample contained between 10^4 and 10^7 pores in a 1-cm^2 area in the middle of a 1-in.^{2} sheet, depending on irradiation time. During an experiment the number of open pores was less precisely known, but all the data were extracted from each sample without letting it warm above the λ-point, so that the results were consistent for each channel size and only the absolute values of the superfluid velocity were subject to error. In calculating the velocities, it was assumed that all the pores were open, so that the experimental values provided a lower limit.

Figure 1 shows a schematic diagram of the apparatus. Two 6-cm-long coaxial capacitors were used to monitor the helium level on either side of the mica sheet. Each of these capacitors formed a part of one of the tank circuits of two back-diode oscillators. The resonant frequencies changed with level difference, and the two signals were mixed with a 50-MHz standard, amplified, and finally recorded alternately on punched paper tape every second. The system could resolve a change in level of 5×10^{-6} cm. A hollow copper cylinder could be raised and lowered in the helium bath by means of a synchronous motor at the top of the apparatus. Different flow rates were achieved by changing the drive frequency to the motor, and also by changing the reduction ratio of the gearbox in the lifting mechanism. The accuracy of the measurement of the flow rate could be improved by taking readings over a period of several minutes for a given plunger speed. A valve on the part of the chamber which was isolated from the rest of the system by the mica sample allowed the whole experimental chamber to be evacuated and filled with liquid helium above the λ-point without rupturing the fragile mica sheet.

Figure 2 (a) shows a typical family of curves of superfluid velocity in 200-Å pores as a function of level difference for several temperatures. The results were all corrected

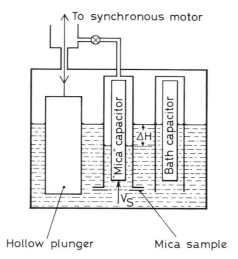

Fig. 1. Schematic diagram of the experimental chamber.

for thermal effects induced by superfluid flow since it is the chemical potential difference $\Delta\mu$ rather than the pressure difference ΔP that drives the superfluid.

If the thermal contact across the mica is defined as

$$C = \dot{Q}/\Delta T$$

where a temperature difference ΔT produces a heat flow $\dot{Q}$ then the superflow will give rise to a fountain pressure

$$\Delta P = \rho \frac{S^2 T}{C} \frac{dV}{dt}$$

where dV/dt is the volume flow rate of the superfluid. Hence there will be a low pressure region where this fountain pressure equals the hydrostatic pressure, and a corresponding linear flow regime. This effect was clearly visible in our results, and was easily corrected for.

The relation between superfluid velocity and pressure difference was systematically investigated over the temperature range 1.87–0.39°K for five different pore diameters (200, 500, 800, 1200, 2000 Å), and notable features were immediately obvious. Intrinsic velocities of up to 900 cm sec⁻¹ were observed. Above 1.0°K there was good agreement with Notarys'[1] extension of the Langer–Fisher[2] theory, which assumes thermal excitation and a 2π phase change per excitation per pore. Hence

$$\Delta\mu = (\rho/m)\hbar 2\pi f$$

where f is the Langer–Fisher fluctuation frequency

$$f = f_0 V_p \exp\left[-\frac{\beta}{k} \frac{\rho_s(T)}{V_s \rho T} \right]$$

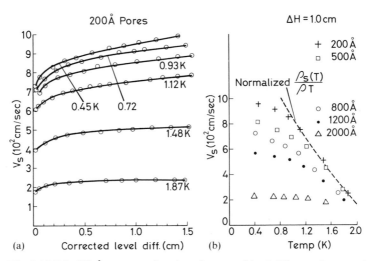

Fig. 2. (a) V_s in 200-Å pores as a function of corrected level difference for several temperatures. (b) V_s at constant level difference vs. temperature for five different pore sizes. The dashed line is the relation predicted by the LF theory.

for pores of volume V_p. This gives the predicted relation

$$\Delta\mu = \frac{h}{m}\rho V_p f_0 \exp\left[-\frac{\beta}{k}\frac{\rho_s(T)}{V_s\rho T}\right]$$

For all but the 2000-Å channels we found that this relation was obeyed above 1.0°K. Figure 2(b) shows the superfluid velocity at constant level difference as a function of temperature for all channel sizes, and the dashed line is the predicted LF variation, normalized to the results from the 200-Å pores.

Below 1.0°K we found behavior similar to that observed by Martin and Mendelssohn[3] in films. In this region the predominant dissipation mechanism would seem to be due to close vortex spacing, or superfluid "turbulence." As the channel size was decreased this effect did not become apparent until lower temperatures. For the 2000-Å channels we did not find agreement with the thermal fluctuation theory over the temperature range investigated.

Figure 3(a) shows $\log_e$ (corrected level difference) as a function of $\rho_s(T)/v_s\rho T$ for 800-Å pores. It can be seen that above 1.0°K, the LF theory holds quite well and in this case the experimental values of β and f_0 are 1.4×10^{-12} and 5×10^{26}, respectively.

There appears to be a tendency for the slope of Fig. 3(a) to change slightly with temperature, even at the higher temperatures. This effect became more pronounced for larger channel sizes—but in the case of the smallest (200-Å) channels was not at all in evidence above 1.0°K. We feel that this is a further indication that for larger channels and lower temperatures superfluid turbulence plays an increasing role. Figure 3(b) shows values for β and f_0 from the relevant data for the four smaller pore sizes. The values above 1.0°K are in substantial agreement with those of Notarys[1] and Clow and Reppy[4] and in contrast to the theoretical value of $\beta = 50 \times 10^{-12}$.

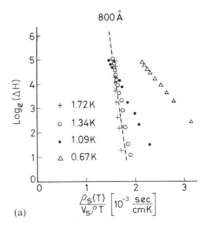

Fig. 3. (a) $\log_e$ (corrected level difference in cm × 100) vs. $\rho_s(T)/V_s\rho T$ for 800-Å pores at four temperatures. The dashed line gives β and f_0. (b) A table of experimental values for β and f_0 for four pore sizes.

	7×10^{24}	5×10^{26}	8×10^{27}	10^{26}	f_0
	1.0×10^{-12}	1.0×10^{-12}	1.9×10^{-12}	1.4×10^{-12}	β
	1200Å	800Å	500Å	200Å	

(b) $P = \dfrac{h}{m}\rho \, V_p \, f_0 \, \exp\left[-\dfrac{\beta}{k} \dfrac{\rho_s(T)}{V_s\rho T}\right]$

In conclusion, we find good qualitative agreement with the Notarys' extension of the LF thermal fluctuation theory for temperatures above 1.0°K and for channel sizes below 2000 Å. Below 1.0°K and for the largest pore size we suggest that the predominant dissipative mechanism is modified significantly by close vortex spacing.

Acknowledgment

We should like to thank H.A. Notarys for providing us with samples of irradiated mica.

References

1. H.A. Notarys, *Phys. Rev. Lett.* **22**, 1240 (1969).
2. J.S. Langer and M.E. Fisher, *Phys. Rev. Lett.* **19**, 560 (1967).
3. D.J. Martin and K. Mendelssohn, *Phys. Lett.* **30**, 107 (1969).
4. J.F. Clow and J.D. Reppy, *Phys. Rev. Lett.* **19**, 291 (1967).

Critical Velocities in Superfluid Flow through Orifices*

G.B. Hess

Department of Physics, University of Virginia
Charlottesville, Virginia

A number of recent experiments[1-3] on the flow of superfluid helium have found an "intrinsically limited" flow which behaves qualitatively as predicted by Iordanskii and by Langer and Fisher for flow limited by homogeneous nucleation of quantized vortex rings.[4-6] This theoretical model gives a relation between pressure head Δp and superfluid velocity v_s:

$$\Delta p = \rho \kappa v$$

$$= \rho \kappa V v_0(v_s, T) \exp\left[-\rho_s(T) E(v_s)/kT\right] \tag{1}$$

where v is the nucleation rate, ρ_s is the superfluid density, $\kappa = h/m$ is the quantum of circulation, V is the volume available for nucleation, and $\rho_s E$ is the energy of a critical vortex ring just large enough to remain stationary when directed against the flow:

$$E = (\kappa^3/16\pi v_s)(\eta - \tfrac{1}{2})(\eta - \tfrac{3}{2}) \tag{2}$$

with $\eta = \ln(8R/a) \approx 5$; R is the radius of the critical vortex ring and $a(T)$ is the vortex core parameter. The velocity and temperature dependences of the attempt frequency v_0 are not important compared to the exponential factor and v_0 can be regarded as a constant of order 10^{34} cm^{-3} sec^{-1}. Then the superfluid velocity is given to a good approximation by

$$v_s = (\rho_s \kappa^3/16\pi\gamma kT)(\eta - \tfrac{1}{2})(\eta - \tfrac{3}{2}) \tag{3}$$

with

$$\gamma = \ln(\rho \kappa V v_0/\Delta p) \tag{4}$$

It is important to note that γ, and hence the magnitude of the exponent in Eq. (1), is large compared to unity.

The temperature dependence of the superfluid critical velocity through several pinholes of approximately 10 μm diameter is shown in Fig. 1. The abscissa is $\rho_s/\rho T$ and the scales are logarithmic, so Eq. (3) predicts a straight line of slope one if η is constant. There is disagreement with theory in three respects: (a) The velocity is roughly ten times smaller than predicted (at 2.0°K); this was true in previous experiments also.[1,2] (b) Most of the data fit a slope of about 0.8 rather than 1. By contrast, Notarys[2] found slope one in etched fission tracks, except in the widest used, 0.2 μm. (c) The dependence on pressure head is somewhat stronger than predicted.

* Research supported in part by National Science Foundation Grant GH-32746.

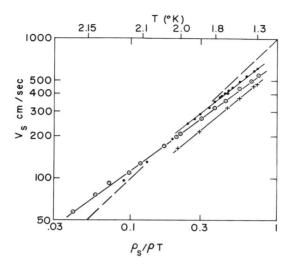

Fig. 1. Critical velocities in superfluid flow through several pinholes, plotted against superfluid fraction over temperature (in °K). Open circles: orifice C, 9.4 μm diameter, at $\Delta p = 35$ μm of He; solid circles: orifice F, 9.1 μm diameter, at 70 μm of He; crosses: orifice D, 7.2 μm diameter, at 35 μm of He. The solid lines are fits to the data and the dashed line has slope one.

Part of the present discrepancy in magnitude can be accounted for by consideration of the gross geometry of the pinhole. The hole is flared at one end, but meets the other surface at a sharp edge, with radius of curvature about 0.1 μm. The velocity at this edge should be, by rough estimate, 2.5 times the mean velocity in potential flow through the orifice. Because of the strong dependence on velocity implicit in Eq. (1), nucleation of vortex rings will occur predominantly in the high-velocity region near the edge; in fact, predominantly within a fraction of the radius of curvature, or a few hundred Å, of the edge. At the presumed velocities a critical vortex ring has a radius of about 30–300 Å. Now a vortex ring nucleating sufficiently close to the wall is likely to make contact and can then develop through a critical, stationary state consisting of a half ring terminating on the wall.* In this case the critical energy has only half the value given in Eq. (2), so that the same nucleation rate is attained at only half the velocity required for nucleation away from the wall, aside from logarithmic factors. (In the geometry described by Notarys[2] smaller arcs of rings can nucleate in the acute angle of the diamond-shaped cross section to give about a sixfold reduction in critical velocity.) It is clear that this boundary nucleation should completely dominate homogeneous nucleation. All this accounts for about a factor of five.

It is an intriguing possibility that the remaining discrepancies may be due to smaller-scale wall structure, in the range roughly 1000–30Å. For instance, either a crystal edge which is sharp down to nearly atomic scale[8] or a recess along a grain

---

* A similar mechanism for vortex ring nucleation by ions was proposed by Donnelly and Roberts.[7]

boundary might give an effect of the required magnitude. In potential flow across a sharp edge the velocity diverges as $r^{-\alpha}$ with distance r from the edge, where

$$\alpha = (\pi - \theta)/(2\pi - \theta)$$

and θ is the angle between the boundary planes on the solid side.[9] The smaller the half ring attached at the edge, the larger the local velocity it sees. This modifies the argument which led to Eq. (3), such that with nucleation at an edge[10]

$$v_s \propto (\rho_s/\rho T)^{1-\alpha} \tag{5}$$

A value $\alpha \approx 0.2\,(\theta \approx 140°)$ would account for the temperature dependence of the experimental velocities in Fig. 1. Note that for one orifice there is apparently a transition below about 250 cm sec^{-1} to a slope near one. This may indicate that a particle or boundary structure which is dominating vortex nucleation at lower temperatures has an extent of the order of 100 Å. Finally, the much reduced nucleation volume implicit in this picture largely removes the discrepancy (c) in pressure head dependence.

Banton[11] and Schofield and Sanders[12] have observed transitions between two or more critical velocities during a single "pass" consisting of relaxation of an initial chemical potential difference imposed between two reservoirs. A striking example of this behavior is shown in Fig. 2 for flow through a 25-μm-diameter pinhole at 1.37°K. Data for ten passes are superimposed. There are two reproducible modes of dissipation, and transitions between them occur more or less at random. The two modes can be followed as the temperature is changed, and the velocities at $\Delta p = 1$ dyn cm$^{-2} = 70\ \mu$m of He are shown as a function of temperature in Fig. 3. The slow mode velocity is 87 ± 4 cm sec^{-1}, independent of temperature within experimental error. The fast mode is consistent with relation (5) with $\alpha = 0.52$. Either mode could probably result from vortex nucleation at a dust particle of diameter $\lesssim 2000$ Å on the wall of the orifice,[10] but it is not understood why the slow mode dissipation mechanism operates only intermittently. One remarkable feature is that the two modes have essentially the same variation of velocity with pressure head. All of the 25-μm-orifice data are for flow in one direction (from the flared side). Velocities in the other direction are about 10% lower and the slow mode occurs predominantly. These data are not quantitatively reproducible on subsequent coolings from room temperature: From one to three modes occurred at differing velocities.

A further experiment was undertaken to test the prediction that vortex nucleation near the points of contact between solid particles in a flow channel will result in a low and only weakly temperature-dependent critical velocity.[10] A 1.8-mm-i.d. glass tube was constricted to 390 μm in an hourglass shape. No dissipation was detectable in superfluid flow through this tube up to at least 3 cm sec^{-1}. The tube was then packed with glass spheres of approximately 53 μm diameter. The predicted critical velocity for spherical particles has the temperature dependence of (5) with $\alpha = 0.586$ and magnitude 0.45 cm sec^{-1} (at 1.3°K) with uncertainty perhaps as great as a factor of two. Superfluid flow measurements were made in the temperature range 1.31–1.94°K. The superfluid velocity at $\Delta p = 5\ \mu$m of He was found to be 0.31 ± 0.01 cm sec^{-1}, *increasing* very slightly at the highest temperatures. The supercritical dependence of velocity on pressure head is stronger than the logarithmic

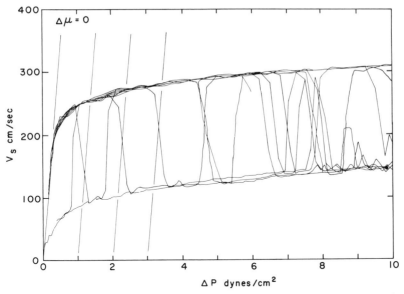

Fig. 2. Superfluid velocity against pressure head for flow through orifice *I* (25 μm diameter) at 1.37 K. The slant lines are lines of constant chemical potential difference.

dependence predicted by the thermal nucleation model; Δp varies roughly linearly with v_s^2. This perhaps can be understood as resulting from saturation of the nucleation sites due to the rather slow transport of supercritical vortices. In summary, a low critical velocity is indeed observed, but its temperature dependence is not yet understood.

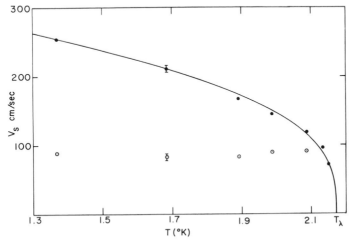

Fig. 3. Superfluid velocity against temperature for flow through orifice *I* at $\Delta p = 70$ μm of He. Open circles: slow mode; solid circles: fast mode. The solid line is proportional to $(\rho_s/\rho T)^{0.48}$.

Acknowledgment

The author is indebted to Mr. B. Broulik for assistance with the apparatus and in data reduction.

References

1. J.R. Clow and J.D. Reppy, *Phys. Rev. Lett.* **19**, 291 (1967); *Phys. Rev. A* **5**, 424 (1972); G. Kukich, R.P. Henkel, and J.D. Reppy, *Phys. Rev. Lett.* **21**, 197 (1968).
2. H.A. Notarys, *Phys. Rev. Lett.* **22**, 1240 (1969).
3. G.B. Hess, *Phys. Rev. Lett.* **27**, 977 (1971).
4. S.V. Iordanskii, *Zh. Eksperim. i Teor. Fiz.* **48**, 708 (1965) [*Soviet Phys.—JETP* **21**, 467 (1965)].
5. J.S. Langer and M.E. Fisher, *Phys. Rev. Lett.* **19**, 560 (1967); J.S. Langer and J.D. Reppy, in *Progress in Low Temperature Physics,* C.J. Gorter, ed., North-Holland, Amsterdam (1970), Vol. 6, p. 1.
6. R.J. Donnelly and P.H. Roberts, *Phil. Trans. Roy. Soc. (London)* **271** 41 (1971).
7. R.J. Donnelly and P.H. Roberts, *Phys. Rev. Lett.* **23**, 1491 (1968).
8. W.F. Vinen, in *Liquid Helium,* G. Careri, ed., Academic, New York (1963), p. 350.
9. L. Prandtl and O.G. Tietjens, *Fundamentals of Hydro and Aeromechanics,* Dover, New York (1957), Chapter 11.
10. G.B. Hess, *Phys. Rev. Lett.* **29**, 96 (1972).
11. M.E. Banton, *Bull. Am. Phys. Soc.* **16**, 640 (1971).
12. G.L. Schofield, Jr., Ph.D. Thesis, Univ. of Michigan, 1971, unpublished; T.M. Sanders, Jr., private communication.

Observations of the Superfluid Circulation around a Wire in a Rotating Vessel Containing Liquid He II*

S.F. Kral and W. Zimmermann, Jr.

School of Physics and Astronomy, University of Minnesota
Minneapolis, Minnesota

Several years ago observations in our laboratory of the superfluid circulation around a fine wire stretched along the axis of a cylindrical container filled with liquid He II gave rather clear evidence of preferential metastability at quantum levels in the range from -3 to $+3$ quantum units.[1] The quantum unit of circulation is h/m, where h is Planck's constant and m is the mass of the ^{4}He atom. Thus those observations strengthened the evidence presented by Vinen a few years earlier for the existence of quantization of circulation in superfluid helium flow around a solid obstacle.[2] Our measurements at that time were all conducted with the wire and container at rest.

It seemed of considerable interest to us to extend those measurements to the case of the wire and container in rotation. We were interested in looking for circulations larger than those seen at rest, in studying the dependence of circulation upon angular velocity of rotation, and in comparing that dependence with the equilibrium predictions of the quantum vortex model for superfluid rotation. Therefore in the present work we have made measurements of the circulation in rotation at angular velocities up to 4 rad sec^{-1} in magnitude with a cell similar to the one used in our earlier work. These measurements were made at temperatures between 1.2 and 1.4°K.

Figure 1 shows the experimental cell, consisting of a brass can containing the central wire, which is approximately 50 mm in length and from 0.066 to 0.083 mm in diameter, and a concentric quartz tube having a 5.0 mm inner diameter, which defines the outer boundary of the flow space around the wire. The can is filled with liquid helium and the entire cell can be put into rotation about its axis.

The apparent circulation κ around the wire is determined by measuring the difference $\Delta\omega_t$ between the angular frequencies of the two lowest transverse vibrational modes of the wire. With the wire and container in rotation at angular velocity ω_r, $\Delta\omega_t$ as observed in the rotating system is given to good accuracy by the expression

$$\Delta\omega_t = \left[(\Delta\omega_\kappa - \Delta\omega_r)^2 + (\Delta\omega_0)^2 \right]^{1/2}$$

where $\Delta\omega_\kappa = \kappa\rho_s/\mu$, $\Delta\omega_r = 2\omega_r$, and $\Delta\omega_0$ is the intrinsic angular frequency splitting of the wire due to asymmetries in the wire and its mounting. Here ρ_s is the superfluid density and μ is the effective mass of the wire per unit length. Hence a measurement of $\Delta\omega_t$ enables a determination of $\Delta\omega_\kappa$ and thus of κ to be made.

---

* Supported by the U. S. Atomic Energy Commission.

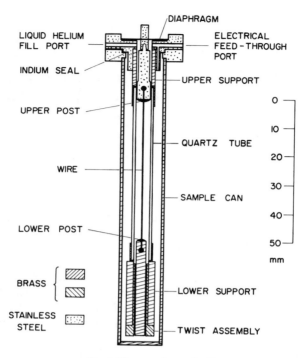

Fig. 1. The experimental cell.

In order to excite and observe the wire's vibration, the cell was situated in a transverse magnetic field which rotated with the cell. Measurements of $\Delta\omega_t$ were made recurrently at 5- or 6-sec intervals by passing an ac current pulse through the wire to excite its lowest modes of vibration. The emf induced along the wire as it then underwent damped free oscillation was detected. This emf took the form of a decaying beat pattern from which $\Delta\omega_t$ could be determined.

Circulation data were obtained using two main procedures. In the first, ω_r was varied in steplike fashion from one value to another, measurements of κ all being made at constant ω_r. Unlike our earlier experiments, very little evidence for quantization of circulation was obtained in this type of run. Both at rest and in rotation κ showed considerable metastability at a wide variety of levels, far from as well as near to quantum levels. Various attempts to drive κ to any reproducible levels that might have been preferred by the system were unsuccessful. These attempts included rapping the apparatus, giving particularly large excitation pulses to the wire, and abruptly stopping and restarting rotation.

Nevertheless, we observed that at each value of ω_r the values of κ measured appeared to lie within a restricted ω_r-dependent range the median of which increased with ω_r. This behavior is illustrated in Fig. 2, where the dots show data obtained after steplike changes in ω_r, both increasing and decreasing. We believe that these observations indicate the existence of a region in the κ vs. ω_r plane inside of which κ is metastable or stable, while outside it is unstable. Note that except for small values of ω_r the sense of κ always agrees with that of ω_r. Note also that for this wire, which had a diameter of 0.079 mm, the range of metastability at $\omega_r = 0$ runs from about

− 3 to + 3 quantum units, in good agreement with the range of circulations observed in our earlier measurements at rest with wires of similar diameter.

In the second procedure for obtaining data, measurements of κ were made while rotation was being accelerated or decelerated in a slow and steady fashion. Data taken in this way are shown by the curves in Fig. 2. Starting at zero, ω_r was accelerated in the positive sense to 1.1 rad sec^{-1}, then in the negative sense to − 1.1 rad sec^{-1}, and finally in the positive sense again to 1.0 rad sec^{-1}, the entire sequence having taken ∼ 11 hr. The data are seen to trace out a form of hysteresis loop.

We believe that such a loop gives additional support for the existence of a metastable region. We suppose that the relationship between the loop and the region of metastability in Fig. 2 is as follows. As ω_r is increased from zero and κ starts to increase, the curve followed lies somewhere near the lower limit of metastability. Then as the acceleration is reversed and ω_r decreases, κ tends to remain relatively stable as the curve cuts back through the metastable region until the upper limit of metastability is reached. Then κ tends to decrease, following near to the upper limit until ω_r reaches its most negative value, and so forth. Similar data are shown by the continuous curves in Fig. 3 for another wire of similar diameter.

A particularly noteworthy feature of the hysteresis loops shown in Figs. 2 and 3 is that at large absolute values of ω_r the circulation shows a marked preference to remain more or less constant as ω_r varies in either direction, sometimes changing from level to level in a rather steplike way. A number of these steps appear to be nearly one quantum unit in size, and some of the plateaus are very nearly at quantum levels, but there are many exceptions.

It is of considerable interest to compare the region of metastability observed for a given wire with the curve expressing the equilibrium value of κ to be expected as a function of ω_r. If we apply minimum-free-energy considerations to the usual model for the rotation of the superfluid based on quantization of circulation and the

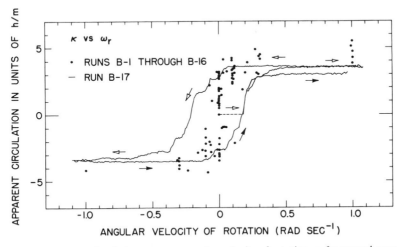

Fig. 2. Apparent circulation κ versus angular velocity of rotation ω_r for several runs with the same wire. The dots represent data taken after steplike changes in ω_r, while the curves represent data taken during nearly constant acceleration and deceleration having mean magnitude 1.3×10^{-4} rad sec^{-2} at a temperature of 1.23°K. The wire's diameter was 0.079 mm.

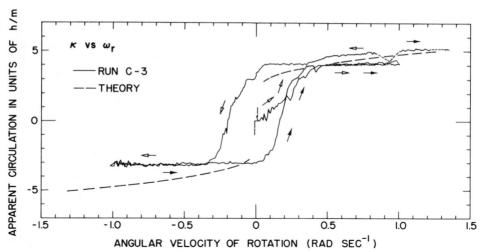

Fig. 3. Apparent circulation κ versus angular velocity of rotation ω_r. The continuous curves show data taken during nearly constant acceleration and deceleration having mean magnitude 1.1×10^{-4} rad sec^{-2} at a temperature of 1.27°K. The wire's diameter was 0.083 mm. The dashed curves show the predictions of equilibrium theory in the two limits discussed in the text.

existence of singly-quantized vortices, we expect a steplike curve for this function. As ω_r increases from zero, κ is first zero, then one quantum unit, then two units, and so forth, the transitions taking place at critical values of ω_r which depend upon the wire's diameter, the inner diameter of the quartz tube, and the vortex core parameter, and which reflect the details of the vortex distribution in the liquid.

Although a complete calculation of the equilibrium value of κ as a function of ω_r is difficult, it is easy to determine the values of ω_r at which κ should go from 0 to ± 1.[3] Further, in the limit of large ω_r one may use a simple continuum model similar to that of Stauffer and Fetter to evaluate the equilibrium behavior of κ.[4] The results of these calculations are shown by the dashed curves in Fig. 3. The steps from $\kappa = 0$ to ± 1 occur at $\omega_r = \pm 0.011$ rad sec^{-1}. Both the discrete and continuum portions of the curve seem to be well situated within the region of metastability that we would infer from the hysteresis loop.

Thus, despite the absence of clear-cut direct evidence for quantization of circulation in the current experiment and the presence of a high degree of metastability, we believe that we have a good understanding of the average dependence of κ on ω_r based on a quantum vortex model for the rotation of the superfluid component. It continues to seem likely that nonquantum values of κ are associated with quantum vortices pinned to or near to the wire, but our experiment does not give us direct evidence on this matter.

References

1. S.C. Whitmore and W. Zimmermann, Jr., *Phys. Rev.* **166**, 181 (1968).
2. W.F. Vinen, *Proc. Roy. Soc. (London)* **A260**, 218 (1961).
3. A.L. Fetter, *Phys. Rev.* **153**, 285 (1967).
4. D. Stauffer and A.L. Fetter, *Phys. Rev.* **168**, 156 (1968).

An Attempt to Photograph the Vortex Lattice in Rotating He II*

Richard E. Packard and Gay A. Williams

Department of Physics, University of California
Berkeley, California

It is by now well established that He II in rotation is threaded by quantized vortex lines.[1] It is our intent to establish the spatial distribution of vortices in a rotating cylinder. We are attempting to photograph the vortices using a technique suggested to us by T. M. Sanders.

Electrons produced in the liquid by a radioactive source form microscopic bubbles which can be trapped on the vortex lines.[2] On the application of an electric field along the axis of the line, the electron emerges from the liquid through the meniscus. The charge is then accelerated by a high voltage and finally strikes a phosphor screen thus marking the position of the line with a flash of light. In this way the spatial arrangement of the vortex lines could be recorded by photographing the image on the phosphor.

The simplicity of this method is clouded by several uncertainties which may make the experiment unsuccessful. The rotating He must be maintained at temperatures below $0.5°K$ so that the helium vapor does not defocus the electrons. At this low temperature there is very little normal fluid present so that vibrations and fluctuations in the vortex lattice will not be damped. This would tend to smear out the photograph. A second (and perhaps related) problem lies in the behavior of the trapped electron bubble at these low temperatures. It has been observed[3] that the bound lifetime for the bubble on the vortex line is much shorter than would have been expected from our present knowledge of the ion–line interaction.[4] (At $0.30°K$ we measure a lifetime of several seconds.) If the ion leaves the vortex before it is extracted from the liquid, it may not accurately mark the position of the line.

Our cryostat is shown in Fig. 1. The experimental cell is held below $0.3°K$ by an adsorption-pumped ^{3}He refrigerator. The entire system, including the dewar and the camera, is uniformly rotated on an air bearing.[5]

The experimental cell is shown in Fig. 2. Electrons are pulled from the region in front of the tritium source S and are accelerated toward the source grid box SG. At this low temperature the ion forms a vortex ring which drifts into the interior of SG. Some of the coasting ions become trapped on vortex lines in the drift space. By holding the meniscus grid GM and the repeller R at a negative potential with respect to SG the vortex lines in the drift space become charged with around 10^3 electrons per line. By switching GM positive with respect to SG, the stored, trapped

* Work supported by the National Science Foundation.

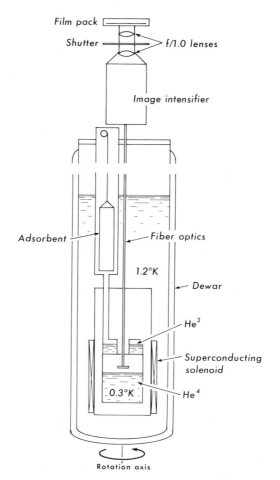

Fig. 1. A diagram of the rotating apparatus.

charge is extracted through the meniscus and is accelerated toward the phosphor. The entire cell is surrounded by a superconducting solenoid which produces a 3-kG focusing field along the rotation axis. The phosphor (ZnO:Zn) is deposited on a coherent fiber optics disk which is butted against a 1/8-in. fiber optics bundle which extends up to room temperature.

Using a photomultiplier to record the light emerging from the fiber optics, we find that the apparatus produces about ten photons per vortex line in a pulse about 50 msec long. (In principle, one could detect the presence of a single line with this signal.) The optical signal is small because the accelerating potential is only 400 V. We find that higher voltages cause electrical breakdown in the cell. Since the light signal is small, the fiber optics output is viewed by an image-intensifier camera which has an overall photon gain of around 6000. The output of the intensifier is focused on a film which is ultimately developed at an ASA speed of 8000. The overall sensitivity of this system is sufficient to record the presence of a single line and the resolu-

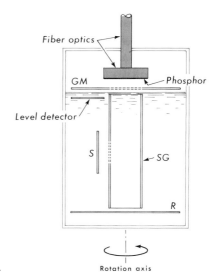

Fig. 2. A diagram of the experimental cell.

tion should be sufficient to distinguish between points separated by 0.2 mm, the expected spacing of vortex lines at a rotation speed of 1.0 sec.$^{-1}$ (We estimate the modulation transfer function of the image intensifier camera is $\sim 60\%$ at five line pairs per millimeter.)

At the present time we have recorded images due to the rotation-induced signal. However, these show no detail at all and are a complete blur (independent of the magnetic focusing field). Slower rotation speeds (i.e., fewer vortex lines) merely make the blur grow fainter but do not produce added detail. At present we do not know the ultimate mechanism which blurs the image.

We are currently trying to test the resolution of our system by directing a narrow, collimated beam of vortex rings at the meniscus, and looking at the resulting dot of light on the phosphor. These tests are not yet complete. We are hopeful that this technique will eventually allow detailed studies of vortex motion and creation.

Acknowledgment

We wish to acknowledge the great help and encouragement in this experiment given by T. M. Sanders and F. Reif.

References

1. J. Wilks, *The Properties of Liquid and Solid Helium*, Clarendon Press, Oxford (1967), Chapter 12.
2. R.J. Donnelly, *Experimental Superfluidity*, Univ. of Chicago Press (1967), Chapter 6.
3. R.L. Douglass, *Phys. Lett.* **28A**, 560 (1969).
4. R.J. Donnelly and P.H. Roberts, *Proc. Roy. Soc. (London)*, Ser. A **312**, 519 (1969).
5. R.E. Packard and T.M. Sanders, *Phys. Rev.* **A6**, (1972).

Radial Distribution of Superfluid Vortices
in a Rotating Annulus *

D. Scott Shenk and James B. Mehl

Department of Physics, University of Delaware
Newark, Delaware

An annular geometry is convenient for theoretical[1,2] and experimental[3-5] studies of the equilibrium state of superfluid helium in a rotating container. Previously we reported studies of the radial distribution of superfluid vortices in an annulus with a width of 1.79 mm.[5] Standing waves of second sound were used as probes of the vortex distribution. In this paper we report on extensions of the work to annuli with widths of 1.27 and 0.75 mm.

In a rotating annulus the vortices are expected to be straight and parallel to the axis of symmetry. The exact calculations of Fetter[1] show that the free energy of the superfluid is minimized if the flow is irrotational and vortex-free below an angular velocity Ω_0. As the angular velocity is increased slightly above Ω_0 a ring of vortices is predicted to fill quickly. The value of Ω_0 is given by

$$\Omega_0 = (\kappa/\pi D^2) \ln(2D/\pi a) \tag{1}$$

where κ is the quantum of circulation, D is the annulus width, and a is the vortex core parameter. In an approximate calculation valid at high angular velocities Stauffer and Fetter[2] have shown that the free energy is minimized if the vortices are uniformly distributed in the region near the center of the annulus and excluded from regions near the inner and outer annulus walls. The thickness of the vortex-free regions is predicted to be

$$\Delta R = s[(1/2\pi) \ln(b/a)]^{1/2} \tag{2}$$

where $s^{-2} = 2\Omega/\kappa$ is the vortex density at the center of the annulus and b is a length proportional to the average vortex spacing s. As the angular velocity decreases, ΔR increases until it reaches a maximum value of approximately $D/2$ at Ω_0. Note that with a small correction within the logarithmic term in Eq. (2) (setting $b = 2D/\pi$ at low Ω) the condition $\Delta R = D/2$ becomes equivalent to Eq. (1). This near equivalence suggests that the theoretical picture outlined above will be accurate at angular velocities extending from below Ω_0 to large Ω. Our experimental results support this.

Experiments were performed using apparatus similar to that described earlier.[5] The annuli were formed with precision Pyrex cylinders. The outer cylinder had an

* Research supported by the National Science Foundation and by the University of Delaware Research Foundation.

inner diameter of 3.81 cm and a height of 3.7 cm in all cases, while inner cylinders with different outer diameters were used. The inner walls of the annulus were coated with carbon films which were used as transducers to excite and detect standing waves of second sound. The apparatus was suspended from a small turntable at the top of the helium dewar. The temperature of the helium bath was controlled with an electronic regulator which held drifts to less than 1 μ°K/hr.

With carefully constructed annular resonators we found that fundamental and second harmonic radial resonances typically had Q's of several thousand. Because of the low losses in these resonators at rest, the additional losses introduced by vortices could be accurately determined by measuring the amplitude of standing waves.

Typical measurements of the second-sound amplitude in rotation are shown in Fig. 1, in which the inverse amplitude $1/A$ is plotted as a function of Ω. At low angular velocities $1/A$ remains constant and begins to increase at Ω_0, as first observed by Bendt and Donnelly.[3] At high angular velocities the measurements approach the straight lines drawn in Fig. 1. An analysis of the deviation of $1/A$ from linearity yields information about the vortex distribution.

According to the theory of Hall and Vinen,[6] a uniform distribution of vortices introduces an additional second-sound attenuation proportional to Ω. Since the amplitude of standing waves is inversely proportional to the total losses, the losses

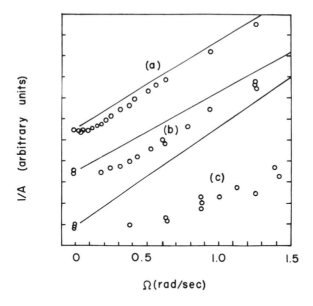

Fig. 1. Plots of $1/A$ vs. Ω, where A is the amplitude of second sound in annular resonators of widths (a) 1.79, (b) 1.27 and (c) 0.75 mm. All measurements were made at 1.80°K. One division on the vertical axis corresponds to a 5% change in $1/A$ for plots (a) and (b) and 2.5% for plot (c). At high angular velocities the points approach the straight lines.

in rotation can be described by

$$1/A - 1/A_0 = K\Omega \qquad (3)$$

where A_0 is the amplitude at $\Omega = 0$ and K is proportional to the Hall–Vinen parameter B. To take into account the spatial distribution of vortices, we modify Eq. (3) by introducing a function $y(\Omega)$,

$$1/A - 1/A_0 = K\Omega y \qquad (4)$$

where y must equal zero at low angular velocities, begin increasing at Ω_0, and approach unity as the vortex distribution becomes more uniform at high Ω_0.

The angular velocity dependence of y can be calculated if we assume the vortices are uniformly distributed within a region of width $(D - 2\,\Delta R)$ at the center of the annulus. The width should increase smoothly from zero at Ω_0 and approach the full annulus width D as ΔR approaches zero at high Ω. In order to make the width zero at Ω_0, we assume $b = 2D/\pi$ at low angular velocities. At high angular velocities it is reasonable to assume $b = 0.27s$, a result valid for a triangular vortex lattice.[7] For our analysis we have interpolated between these limits by using the expression

$$b = 0.27s + (2D/\pi - 0.27s)\exp(1 - \Omega/\Omega_0) \qquad (5)$$

The two terms in this expression are approximately equal at $\Omega = 3.5\Omega_0$, at which roughly three rows of vortices are expected in the annulus.

The energy loss introduced by vortices is proportional to $(v_n - v_s)^2$ according to the Hall–Vinen theory. This quantity is averaged over the region occupied by vortices to determine y as a function of Ω. Since a fundamental radial resonance has an antinode in $(v_n - v_s)$ at the center of the annulus, while a second harmonic mode has a node at the same position, the value of y at low Ω will differ greatly for the two modes. The value of y also depends on the core parameter a through Eq. (2) and on the form assumed for Eq. (5).

Data were analyzed by computing linear least square fits to Eq. (4) for fixed values of a, with Ωy as the independent variable and A_0 and K as free parameters. The best fits were generally obtained with values of a between about 0.1 and 2.0 nm. A value of 0.5 nm was used for a in the fits shown in Fig. 2. According to Eq. (4), the quantity plotted on the vertical axis, $\Delta(1/A)/\Omega$, is equal to Ky. Properly scaled values of y are shown as smooth curves in Fig. 2. The values of A_0 found in the numerical fits were used to calculate $\Delta(1/A) = 1/A - 1/A_0$. For the best fits the values of A_0 determined numerically agreed with the measured amplitude at low Ω to within an estimated experimental uncertainty of about 0.2%.

The overall agreement of the data in Fig. 2 with the calculated curves supports the correctness of the vortex-free-region model over the full range of angular velocities. Moreover, the agreement with theory of measurements taken with both the fundamental and second harmonic modes in the 1.79-mm annulus supports our assumption that the losses in rotation can be completely attributed to vortices.

At the highest angular velocities at which measurements were made there are approximately 4, 9, and 13 rows of vortices in the 0.75-, 1.27-, and 1.79- mm annuli. Thus it might be expected that in the case of the 1.79-mm annulus, at least at high Ω, a good fit could be obtained using $b = 0.27s$ instead of Eq. (5). We found, in fact,

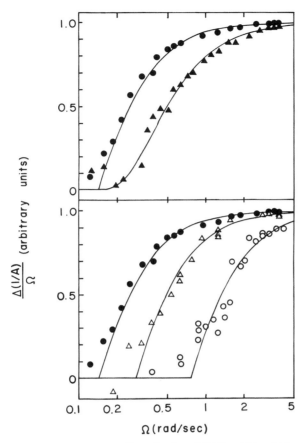

Fig. 2. Plots of $\Delta(1/A)/\Omega$ vs. Ω, where $\Delta(1/A) = 1/A - 1/A_0$. The mode and annulus width for the plots are as follows: solid circles, fundamental, 1.79 mm; solid triangles, second harmonic, 1.79 mm; open triangles, fundamental, 1.27 mm; open circles, fundamental 0.75 mm. All measurements were made at $1.80°$K. The fundamental mode data at low Ω are identical with those shown in Fig. 1. The smooth curves are proportional to Ky and are calculated for each case using $a = 0.5$ nm and the value of K determined numerically.

that the data taken with this annulus could be satisfactorily fit over the full range of Ω using $a = 0.5$ nm and $b = 0.27s$. In the narrower annuli, on the other hand, use of Eq. (5) was essential in order to get satisfactory fits with reasonable values of a.

After a change in angular velocity equilibrium was reached very quickly in the widest annulus at $1.80°$K. In the narrower annuli it was necessary to wait between 10 and 40 min to reach equilibrium at $1.80°$K, and the time appeared to increase at lower temperatures. For this reason much of the work has been performed at $1.80°$K. We have, however, observed qualitatively similar results in annuli of all three sizes at $1.65°$K, and in the widest annulus at $1.40°$K.

References

1. A.L. Fetter, *Phys. Rev.* **153**, 285 (1967).
2. D. Stauffer and A.L. Fetter, *Phys. Rev.* **168**, 156 (1968).
3. P.J. Bendt, *Phys. Rev.* **164**, 262 (1967); P.J. Bendt and R.J. Donnelly, *Phys. Rev. Lett.* **19**, 214 (1967).
4. J.A. Northby and R.J. Donnelly, *Phys. Rev. Lett.* **25**, 214 (1970).
5. D.S. Shenk and J.B. Mehl, *Phys. Rev. Lett.* **27**, 1703 (1971).
6. H.E. Hall and W.F. Vinen, *Proc. Roy. Soc.* (*London*) **A238**, 215 (1956).
7. V.K. Tkachenko, *Zh. Eksperim. i Teor. Fiz.* **49**, 1875 (1965) [*Soviet Phys.—JETP* **22**, 1282 (1966)].

Effect of a Constriction on the Vortex Density in He II Superflow

Maurice François and Daniel Lhuillier

Laboratoire de Mécanique des Fluides, Université de Paris
Orsay, France

Michel Le Ray

Laboratoire d'Hydrodynamique Superfluide, Centre Universitaire
Valenciennes, France

and

Félix Vidal

Groupe de Physique des Solides, Ecole Normale Supérieure
Paris, France

We have performed experimental studies of the influence of a constriction on the vortex density in He II superflow. The vortex density in the flow is analyzed by measuring the extra attenuation $\alpha' = \alpha - \alpha_0$ (α_0 is the static attenuation) of a second-sound resonant wave. The experimental apparatus is similar to that described in Ref. 1 except that the constriction and the superleak are now allowed to stand at different places; in all of our experiments the superflow, thermomechanically driven by a heater, goes from the second-sound resonator (SSR) to the constriction and thus the second sound always probes the vortex density in the part of the tube which is beyond the constriction. The apparatus maintains a constant hydrostatic pressure difference along the fountain tube (typically 15 cm He) so the temperature difference between the heater chamber and the bulk (or SSR) is constant and corresponds to this hydrostatic fountain pressure as long as there is no important dissipation along the superflow.

The second-sound extra attenuation due to a superflow with velocity V_s in a large tube (10×13 mm^2) is represented on curve A in Fig. 1; as usual, besides a critical velocity V_{s1}, α' grows as V_s^2. The effect of a constriction ($\phi = 2.5$ mm or $R = 26.5$, where R is the ratio of the cross sections) on the vortex density appears on curve B of Fig. 1. One can see that the vortex density is always less than its corresponding value without the constriction and that a break appears in this curve; such a break has already been observed in other experiments[2] involving a constriction. Furthermore, temperature measurements along the flow indicate the existence of a temperature difference, and thus dissipation, between the constriction's extremities (see curve B of Fig. 2).

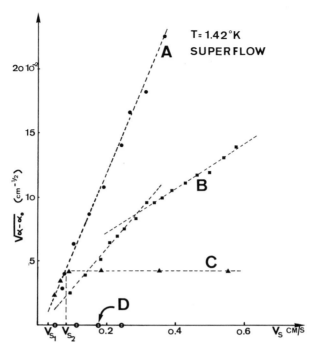

Fig. 1. Square root of the second-sound extra attenuation beyond the constriction versus V_s, the superflow velocity in the SSR. Curve A (open circles): without constriction; curve B, (squares): with $\phi = 2.5$ mm constriction ($R = 26.5$); curve C (triangles): with $\phi = 2.5$ mm constriction filled with 0.3-μm powder; curve D (open circles): with $\phi = 2.5$ mm constriction filled with 70-Å powder.

The role of this localized dissipation has been analyzed in other experiments where the constriction has been filled with powder (constricted superleak) to modify the critical velocities and dissipation laws in the constriction. In this case the main results can be summarized as follows.

First, using the above constriction ($\phi = 2.5$ mm) filled with a 0.3-μm powder (jeweler's rouge) we find (curve C, Fig. 1) that the constriction now induces a saturation of the vortex density in the SSR. The velocity of the flow beyond which this saturation occurs is defined as V_{s2} (in the SSR), to which V'_{s2} corresponds in the constricted superleak. Up to V_{s2} curve C is identical to curve A, indicating no constriction effect; above V_{s2}, α' reaches its saturation value α'_m and simultaneously a temperature difference begins to develop across the constricted superleak (curve C, Fig. 2 and Fig. 2 of Ref. 1).

Second, using the same powder compressed in the same way but with a different diameter d for the constriction, we find to within a few percent that $V'_{s2} d = $ const at a fixed T (see Fig. 2 of Ref. 1).

Moreover, using a different resonator with the same ratio of constriction R, we obtain the same value of the maximum attenuation α'_m; and for R large enough to get

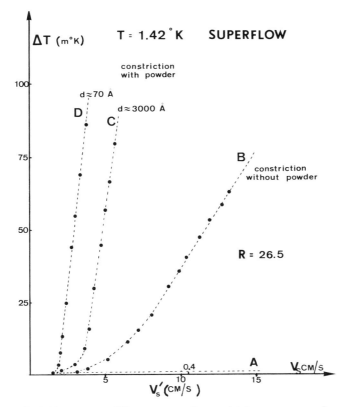

Fig. 2. Temperature difference across the constriction versus superflow velocity V_s in the SSR and V'_s in the constriction. Curves A–D correspond to the same experimental situations as in Fig. 1.

$V_{s2} < V_{s1}$ we have observed that $\alpha'_m = 0$. In fact, with $R = 75$ and jeweler's rouge powder it has been impossible to detect any attenuation of the second-sound amplitude for V_s up to 10 V_{s1} (beyond which ΔT becomes prohibitively large); and with Carbolac I powder (grain size $\simeq 70$ Å), the same effect is already found for $R \simeq 25$.

The most natural idea is that the appearance of ΔT across the constricted superleak is due to vortex development, and thus that V'_{s2} is the classical critical velocity in the powder. But in this case we ought to find that whatever d, $V'_{s2} =$ const, since we have used the same powder. Equation (1) shows quite different behavior and suggests that the flow properties in the constricted superleak are determined at T fixed not only by the mean pore diameter but also by the diameter d of the constriction,[3] or perhaps more generally by the ratio of constriction R. Actually, R seems to be better able to express the influence of the inhomogeneous velocity field (due to the constriction) on the critical velocity V'_{s2}.

In order to get more information on the specific role of the powder in the saturation effect, we have used a quite different powder (Carbolac I, grain size $\simeq 70$ Å). Figure 3 plots ΔT across the constricted superleak and $(\alpha')^{1/2}$. Note, first, the presence of an important hysteresis in the development of vortices in the powder, which has been found in other types of experiments using fourth-sound attenuation measurements.[4] Second, as in the former case, until V_{s2} the results for $(\alpha')^{1/2}$ and ΔT are absolutely identical to those obtained without constriction. But when V_s just exceeds V_{s2} the vortex density in the SSR disappears abruptly, while that in the powder reaches a very high value. The complete correlated hysteresis behavior of the dissipation in the constricted superleak and in the bulk part of the flow allows us to claim that the modifications of vortex development in a region before a constriction are controlled by the dissipation in the constriction. Third, from Figs. 1 and 2 (curves C and D) and from Ref. 4 one can notice that the value of the constant appearing in Eq. (1) depends on the powder; the same conclusion is drawn for α'_m (e.g.,[4] for rouge $R = 10$, $\alpha'_m = 1.2 \times 10^{-2}$; for Carbolac $R = 10$, $\alpha'_m = 2.6 \times 10^{-3}$). Moreover, investigations of the saturation effect at different temperatures ranging from 1.3 to 2.05°K indicate that the constant of Eq. (1) and α'_m are increasing functions of temperature: typically with $R = 10$, jeweler's rouge, $\alpha'_m = 1.21 \times 10^{-2}$ at $T = 1.4$°K and $\alpha'_m = 2.3 \times 10^{-2}$ at $T = 1.9$°K.

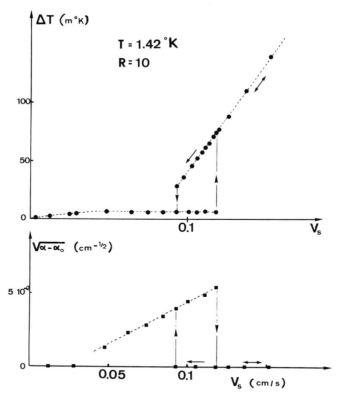

Fig. 3. Lower curve: square root of the second-sound extra attenuation beyond the constriction versus V_s, the superfluid velocity in the SSR. Upper curve: corresponding values of the temperature difference across the $\phi = 4$ mm constriction ($R = 10$) filled with 70-Å powder, versus V_s.

In conclusion, even for small value of R the presence of a constriction considerably affects the development of vorticity in a superflow and may play a dominant role in various experiments involving it, as in Refs. 2 and 5. The experimental results reported above, and especially curves $A-D$ of Figs. 1 and 2, show that the constriction effect (i.e., smaller or saturated vortex density or even complete absence of vorticity) in the region beyond the constriction is related to the very nonuniform vortex density along the flow. One can say that the greater the vortex density in the constriction, the greater the deviation from Vinen's law[6] in the tube. We must emphasize that the critical velocity V'_{s2} in the constriction is not the only characteristic parameter of the effect; we must consider also the law of vorticity development in this region, as one can see from Fig. 3. Lastly, it is possible, with R large enough or with fine powder, to completely localize the dissipation in a small region of the flow.[5]

References

1. M. François and D. Lhuillier, *Phys. Lett.* **40A**, 89 (1972).
2. H.C. Kramers, *Superfluid Helium*, J.F. Allen, ed., Academic Press, London and New York (1966), p. 206.
3. H. Kojima, W. Veith, S.J. Putterman, E. Guyon, and I. Rudnick, *Phys. Rev. Lett.* **27**, 714 (1971); A.L. Fetter, *Phys. Rev.* **153**, 285 (1967).
4. M. François, M. Le Ray, and J. Bataille, *Phys. Lett.* **31A**, 563 (1970); and in *Proc. 12th Intern. Conf. Low Temp. Phys. 1970*, Academic Press of Japan, Tokyo (1971).
5. R. de Bruyn Ouboter, K.W. Taconis, and W.M. van Alphen, in *Progress in Low Temperature Physics*, C.J. Gorter, ed., North-Holland, Amsterdam (1967), Vol. 5, p. 60.
6. W.F. Vinen, *Proc. Roy. Soc.* **A242**, 493 (1957).

AC Measurements of a Coupling Between Dissipative Heat Flux and Mutual Friction Force in He II

Félix Vidal

Groupe de Physique des Solides, Ecole Normale Supérieure
Paris, France

and

Michel Le Ray,* Maurice François, and Daniel Lhuillier

Laboratoire de Méchanique des Fluides, Université de Paris
Orsay, France

Previous measurements on the variation of second-sound velocity in the presence of superfluid vortices suggest the existence of a coupling between the dissipative heat flux and the mutual friction force.[1] This paper presents our latest results on the second-sound measurements and preliminary results on the fourth-sound velocity in presence of a supercritical superflow.

In the case of the second-sound measurements the superfluid vortices are created by uniform rotation and supercritical counterflow or superflow. Typical results in counterflow are presented in Fig. 1. The second-sound velocity u_2 is measured using an original method[1,2] with a relative accuracy of 10^{-5}. In order to avoid entrainment effects,[3] the counterflow is created by a dc heat flux normal to the second-sound wave vector $\mathbf{k}$. The dependence of all measured quantities upon second-sound bolometer and emitter powers levels was checked to ensure the absence of nonlinear effects. The decreasing of u_2 was found to be independent of second-sound amplitude provided the precautions mentioned previously were taken. In addition, the temperature of the helium bath was controlled with an electronic regulator capable of holding drifts to less than $10^{-5}\,^\circ$K. The only u_2 variation that is predicted by the two-fluid model is that due to the mutual friction force of Gorter–Mellink–Vinen:[4]

$$\delta u_2/u_2(0) = [u_2(V_s) - u_2(0)]/u_2(0) \approx - \alpha'^2/2k^2$$

where α' is the second-sound extra attenuation. From the data of Fig. 1 one can verify that the measured u_2 variation is at least 100 times greater than the above effect.

These measurements confirm our previous results[1] and to interpret them we have suggested the possible existence of a coupling between the dissipative heat flux

---

* Permanent address: Laboratoire d'Hydrodynamique Superfluide, Centre Universitaire, Valenciennes, France.

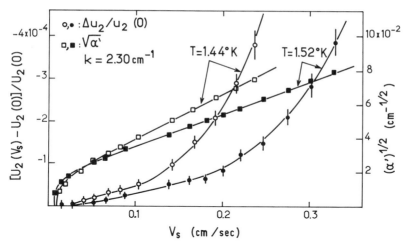

Fig. 1. Relative velocity decrease and square root of the extra attenuation of second sound versus the superfluid velocity V_s in counterflow experiments.

$\mathbf{q}$ and the mutual friction force $\mathbf{f}$:

$$\mathbf{f} = \beta T \rho_s (\mathbf{V}_s - \mathbf{V}_n) - \beta v \, \nabla T \tag{1}$$

$$\mathbf{q} = - (K - \beta v^2) \nabla T - \beta v T \rho_s (\mathbf{V}_n - \mathbf{V}_s) \tag{2}$$

where v is the coupling coefficient, V_n and V_s are the normal and superfluid velocities, ρ_s is the superfluid density, T is the temperature, K is the heat conductivity coefficient, and β is connected with Vinen's coefficients A (counterflow) and B (rotation) by

$$T \rho_s \beta = A \rho_n (\mathbf{V}_n - \mathbf{V}_s)^2 = B \rho_n \Omega / \rho$$

where ρ is the density of He II, $\rho_n = \rho - \rho_s$ the normal fluid density, and Ω the angular velocity of rotation.

The first term on the right hand side of Eq. (1) is the classical result obtained by Vinen,[4] giving rise to an extra attenuation

$$\alpha' = \rho \rho_s T \beta / 2 \rho_n u_2^0 \tag{3}$$

where u_2^0 is the second-sound velocity without mutual friction force ($\beta = 0$).

The new terms in Eqs. (1) and (2), those which contain v, lead to a first order decrease in u_2,[5]

$$\Delta u_2 / u_2^0 \equiv (u_2 - u_2^0)/u_2^0 = - \beta v / s_0 \tag{4}$$

where s_0 is the equilibrium entropy per unit mass. From Eqs. (3) and (4) we may deduce

$$v = (\rho C_v / 2 s_0) |\Delta u_2| / \alpha'$$

(where C_v is the heat capacity per unit mass), so that from simultaneous measurements of Δu_2 and α' numerical values of v can be deduced.

Typical results obtained in rotation with the second sound propagating normally to the vortices has been shown in Fig. 1 (b) of Ref. 1. One can see that v is independent of the angular velocity of rotation Ω (i.e., of the vortex density). The temperature dependences of v/Ω and of the mutual friction coefficient B_{exp} are shown in Fig. 2.

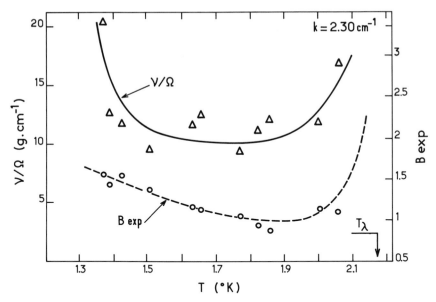

Fig. 2. The ratio of the coupling coefficient v to the angular velocity of rotation and the mutual friction coefficient B_{exp} versus the temperature, for $k = 2.30$ cm^{-1}.

There is agreement, within experimental error, between our values of B_{exp} and the data reported by previous authors.[4,6] Measurements made at various second-sound frequencies show that v is a decreasing function of the frequency while, as discovered by previous authors,[4] B_{exp} was found to be independent of the second-sound frequency. For instance, using two second-sound cavities with $k = 2.30$ and 0.62 cm^{-1} in the fundamental mode, we obtain, respectively, $v/\Omega = 12.6$ and 60 g cm^{-1} at $T = 1.65°$K.

It is possible that the experimental frequency dependence of v can be interpreted with a vortex model: we see, for instance, from Eq. (2) that as soon as dissipation occurs, that is to say, as soon as $\beta \neq 0$, some entropy is transported with V_s; but usually the dissipation is associated with vortices, and we may wonder if there exists a transport of entropy associated with the vortex motion in superfluid helium. When the second sound is tending to modify the vortex motion, it will do so more easily if its frequency approaches an eigenfrequency of the array of vortices: Our results suggest the existence of a very low eigenfrequency which, in the case of rotation, may simply be Ω, the angular velocity. Another point of view is that the increase of Δu_2 when the resonant frequency decreases can be related to an increase of the viscous penetration depth $[\delta_n = (2\eta_n/\rho_n\omega)^{1/2}$, where η_n is the normal fluid viscosity] of the shear waves induced in the normal fluid by the vortex vibrating at the frequency of the second sound.

The first-sound velocity is unaffected by the coupling between $\mathbf{f}$ and $\mathbf{q}$ resulting in a fourth-sound velocity decrease following

$$\frac{\Delta u_4}{u_4^0} \equiv \frac{u_4 - u_4^0}{u_4^0} = -\frac{\rho_n u_2^0 \beta v}{\rho u_4^0 s_0} \tag{5}$$

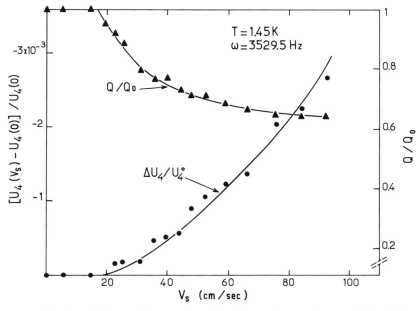

Fig. 3. Velocity variation and attenuation of fourth sound versus the superfluid velocity in superflow experiments.

where u_4^0 is the fourth-sound velocity when $\beta = 0$. Using an apparatus analogous to that described elsewhere,[7] we have performed preliminary measurements of the fourth velocity in presence of superflow. First results (Fig. 3) show a decrease of u_4 whenever fourth sound is attenuated, as is expected from Eq. (5).

References

1. F. Vidal, M. Le Ray, and M. François, *Phys. Lett.* **36A**, 401 (1971); F. Vidal, *Compt. Rend. Ac. Sc.,* **275B**, 609 (1972).
2. F. Vidal and Y. Simon, *Phys. Lett.* **36A**, 165 (1971); F. Vidal, M. Le Ray, M. François, and D. Lhuillier (to be published).
3. I.M. Khalatnikov, *Zh. Eksperim. i Teor. Fiz.* **30**, 167 (1956) [*Soviet Phys.—JETP* **3**, 649 (1956)].
4. W.F. Vinen, *Proc. Roy. Soc. (London)* **A240**, 114 (1957); H.E. Hall and W.F. Vinen, *Proc. Roy. Soc. (London)* **A238**, 204 (1956); **A238**, 215 (1956).
5. D. Lhuillier, F. Vidal, M. François, and M. Le Ray, *Phys. Lett.* **38A**, 161 (1972).
6. P. Lucas, *J. Phys. C: Solid State Phys.* **3**, 1180 (1970).
7. M. François, M. Le Ray, D. Lhuillier, and J. Bataille, in *Proc. 12th Intern. Conf. Low Temp. Phys. 1970*, Academic Press of Japan, Tokyo (1971), p. 67.

The He II–He I Transition in a Heat Current*

S.M. Bhagat, R.S. Davis, and R.A. Lasken

Department of Physics and Astronomy, University of Maryland
College Park, Maryland

Simple thermodynamic arguments[1,2] lead one to believe that in the presence of a counterflow velocity the transition from He II to He I should occur at temperatures somewhat lower than the λ temperature at saturated vapor pressure. The most convenient method of generating a counterflow velocity is to apply a heat flux Q in an otherwise undisturbed column of He II. However, the application of a heat current has certain concomitant effects which have not been included in the theoretical studies so far and which may alter the existing results both qualitatively and quantitatively. First, when a heat flux is present there is invariably a temperature gradient, and second, for moderate Q values the thermal resistance is in large part due to a tangled mass of vortex lines in the fluid. Unfortunately, there does not appear to be any simple way to modify the existing theory to take account of these effects.†

In this paper we present the results of two sets of experiments designed to measure the temperature at which He II and He I will be in equilibrium when a steady heat current is flowing through the fluid.

Experiment 1

Details of this technique have been reported earlier.[4,5] To summarize, a long column of He II was maintained in a thin-walled, stainless tube (mounted inside a high-vacuum chamber) with several thermometers attached to it. The temperature–time curve for each thermometer was obtained in the presence of a steady Q by allowing the main helium bath to warm up slowly towards the λ-point. For temporal correlation of the data two thermometers, as well as the voltage across the heater (at constant current), were monitored simultaneously. As explained in Ref. 4, the temperature at the He II–He I boundary, designated as $T_\lambda(Q)$, was identified by the point at which a sharp rise in the slope of the thermogram occurred (Fig. 1). The present data, together with our earlier results, are well represented by the equation

$$\Delta T_\lambda = T_\lambda(0) - T_\lambda(Q) = bQ^{(1.00 \pm 0.01)} \tag{1}$$

Values of b obtained by least-squares fits to the data are given in Table I, where we have also listed other relevant parameters. It is interesting to note that for a given heater-to-thermometer distance l, ΔT_λ is sensibly independent of channel cross section. However, b increases somewhat as l is increased. It should be noted that in

* Work supported by the National Science Foundation, Grant No. GH34434.
† An oversimplified attempt to include the effects of vortex lines was made in Ref. 5.

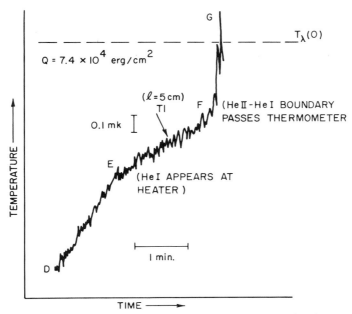

Fig. 1. Typical temperature vs. time curve for a thermometer placed 5 cm away from the heater. As explained in the text, such thermograms were obtained by allowing the main helium bath to warm up slowly toward the λ-point. (See more detailed descriptions in Refs. 4 and 5.)

our interpretation l essentially represents the length of the He I region. Data were also taken for $l = 0.5$ and 1.5 cm. Although the precision was not enough to deduce quantitative results at small l, all the data are consistent with

$$b = \alpha l^{1/n} \qquad (2)$$

where $n \sim 5$ or 6. If this relationship is true for very small l values, it implies that

Table I. Channel Specifications and Values of the Parameter b in Eq. (1).

Channel i.d. cm	l,* cm	Range of Q, mW cm^{-2}	b, K cm^2 W^{-1}
0.29	2.5	$4 < Q < 24$	0.036 ± 0.006
	5	$4 < Q < 24$	0.042 ± 0.006
	10	$4 < Q < 24$	0.045 ± 0.008
0.45 [†]	3	$2 < Q < 61$	0.051 ± 0.003
	5	$3 < Q < 61$	0.057 ± 0.005
	12	$4 < Q < 61$	0.064 ± 0.004
0.58 [†]	5	$2 < Q < 91$	0.052 ± 0.004
	10	$2 < Q < 91$	0.059 ± 0.004
1.17	3	$1 < Q < 25$	0.046 ± 0.004

* Heater to thermometer distance.
† Data from Ref. 4.

$T_\lambda(Q) \to T_\lambda(0)$ as $l \to 0$; that is, when the He I first appears in the system (in the immediate vicinity of the heater) the temperature is indeed $T_\lambda(0)$. However, when one has a two-phase regime, i.e., a length l of He I and the rest of the tube filled with He II, the temperature at the He II–He I boundary depends upon the heat flux through the boundary. Although we do not pretend to understand the l dependence, $\Delta T_\lambda \simeq 0$ for $l \simeq 0$ is plausible on the basis of an argument (developed in Ref. 5) involving the negative expansion coefficient of He I. The $l^{1/n}$ dependence should not be confused with the l dependence exhibited in Eq. (5) of Ref. 5. The $T_\lambda(Q)$ values in Eq. (1) are measured on different thermometers at *different* instants of time.

   The observed values of b are nearly an order of magnitude larger than those predicted by existing theories.[1,2]

Experiment 2

   In this experiment the values of $T_\lambda(Q)$ were inferred by extrapolating our measurements on the speed of second sound.[6] In the presence of a colinear dc heat flux it is found that u_2 reduces linearly with Q^2 for T close to T_λ. If one defines the He II–He I transition point by the relation

$$u_2(Q, T_\lambda(Q)) = 0 \qquad (3)$$

a value for $T_\lambda(Q)$ can be derived from the data. Unfortunately, our measurements extend over a rather small Q range and the extrapolation is over a very large range. It is truly valid only if any dependence on higher powers of Q^2 is negligible. The interesting point is that we find

$$\Delta T_\lambda(Q) = 0.059Q \qquad (4)$$

for a square channel of cross section 1 cm $\times$ 1 cm and $l = 1$ cm. The agreement with Eq. (1) is remarkable. However, Eq. (3) gives values for ΔT_λ which are nearly one order of magnitude larger than those predicted by theory.[2] In order to check that our data are not somehow spurious or subject to a large systematic error, we have repeated the experiment with the dc heat Q flowing at right angles to the direction of the second-sound wave. In this case entrainment effects will be absent and one expects that the variation in u_2 would be considerably reduced.[2,7,8] Preliminary data indeed confirm that for $\mathbf{Q} \perp \mathbf{k}$ the change in u_2 is more than one order of magnitude smaller than when $\mathbf{Q} \parallel \mathbf{k}$. In fact, for reasons described below, in the case of $\mathbf{Q} \perp \mathbf{k}$ we were able to make measurements at much larger values of Q and yet observed no significant ($\lesssim 0.5\%$) change in u_2.

   The experiments with $\mathbf{Q} \perp \mathbf{k}$ also shed light on another problem. As described in Ref. 6, for $\mathbf{Q} \parallel \mathbf{k}$, increasing Q led to a drastic reduction in the second-sound signal even though the quality factor of the resonance was not affected strongly. Our explanation for this effect was that since dc and ac heating were generated in the same element, the local temperature rise due to Kapitsa resistance would cause a thin layer of He I to be formed at the heater surface, thereby reducing the second-sound amplitude. Thus in the $\mathbf{Q} \perp \mathbf{k}$ case one should expect no such drastic reduction in the signal when Q is increased. Indeed, it was observed that one could get to Q values three or four times larger than in the colinear case before the reduction in amplitude became comparable.

Discussion

Experiments similar to experiment 1 but with a somewhat different geometry have been reported by other authors.[9,10] Their results are qualitatively similar to ours but they claim significantly smaller values of ΔT_λ for comparable Q values. In Ref. 5 we have presented a detailed discussion to account for these discrepancies. To summarize, we feel that neither the experiments of Ref. 9 nor those of Ref. 10 were designed in such a way as to measure the temperature at the boundary of He II and He I when there exists a finite heat flux through that boundary. In either case, the authors were measuring essentially the temperature at the heater end, i.e., corresponding to our $l \simeq 0$. It is therefore not surprising that they found ΔT_λ values much smaller than those indicated in Table I for sizable values of l. As further proof that there is no systematic error in our method we presented in Ref. 5 measurements which are identical to those of Ref. 9 (see Fig. 2 of Ref. 5) and very similar to those of Ref. 10 (see Fig. 3 of Ref. 5, curve marked T_1) with, of course, identical results.

In conclusion, we feel that our interpretation is indeed valid and that when a heat current is present the temperature at the interface of He II and He I is less than $T_\lambda(0)$. In this sense T_λ is a function of the heat flux Q in the fluid, as represented by Eq. (1). At present we are trying to improve our precision so that we can check our Eq. (2) at small l values.

References

1. Yu. G. Mamaladze, *Soviet Phys.—JETP* **25**, 479 (1967).
2. H.J. Mikeska, *Phys. Rev.* **179**, 166 (1969).
3. S.M. Bhagat and R.A. Lasken, *Phys. Rev.* **A4**, 264 (1971).
4. S.M. Bhagat and R.A. Lasken, *Phys. Rev.* **A3**, 340 (1971).
5. S.M. Bhagat and R.A. Lasken, *Phys. Rev.* **A5**, 2297 (1972).
6. S.M. Bhagat and R.S. Davis, *J. Low Temp. Phys.* **7**, 157 (1972).
7. D. Lhuillier, F. Vidal, M. François, and M. Le Ray, *Phys. Lett.* **38A**, 161 (1972).
8. F. Vidal, M. Le Ray, M. François, and D. Lhuillier, this volume.
9. P. Leiderer and F. Pobell, *J. Low Temp. Phys.* **3**, 577 (1970).
10. G. Ahlers, A. Evenson, and A. Kornblit, *Phys. Rev.* **A4**, 804 (1971).

Pumping in He II by Low-Frequency Sound*

G.E. Watson

Department of Physics, American University of Beirut
Beirut, Lebanon

Since 1965, when Richards and Anderson[1] first demonstrated successfully the analog of the Josephson effect in helium, there have been several successful improvements of this experiment.[2] In all of the successful experiments an essentially identical arrangement of the crystal sound generator and orifice has been used. In all of these experiments a pumping effect has been observed where the helium is pumped toward the crystal generator. The pumping action seems essential for the observation of the Josephson effect. In the course of a similar experiment a pumping effect has been observed that seems to have rather different characteristics than those in Refs. 1 and 2. The experimental apparatus is shown in Fig. 1.

The whole chamber is immersed in a pumped helium bath. The inner space is filled with liquid by condensation through two fine capillary tubes from room temperature so as to half fill the capacitors. One capacitor serves to measure the helium level (with a resolution of 0.5 μm). The other capacitor (identical in construction) was not used in this experiment. Most experiments were carried out at a temperature of $T = 1.16°$K. Audio sound waves are generated by a capacitive transducer and detected by a second, similarly constructed capacitive microphone. Standing wave resonances are built up in the two 3-cm cylindrical chambers in order to enhance the somewhat low efficiency of the generator. The two chambers are separated by a 25-μm-thick nickel disk containing a single 12.5-μm-diameter orifice. Sound coupling between the two chambers takes place through the orifice and by vibration of the orifice disk (which has a fundamental clamped-edge resonance of about 10.4 kHz). The chambers are 5 mm in diameter. The generator and receiver diaphragms have a fundamental free resonance of about 40 kHz at low temperatures so that their response could be expected to be relatively linear up to 20 kHz.

Pumping effects were observed as follows. With liquid in the chambers and the generator on, maxima of the received signal were sought while sweeping the frequency from 200 Hz to 20 kHz. Because the generator was operated without a dc bias voltage, all received sound signals were double the generator frequency. Each of these maxima will be referred to as a resonance. There were many resonances observed. In particular, longitudinal standing wave resonances in the two chambers were readily identified. A pumping of helium through the orifice was observed at many resonances. Table I shows some of the resonances, their identification, and whether pumping was observed for one particular run.

In the first few experiments the pumping was always away from the generator.

---

* Work supported by an Arts and Sciences Research Grant.

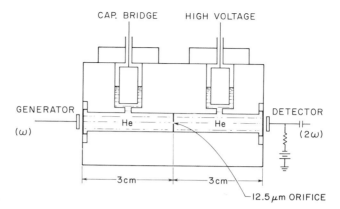

Fig. 1. Experimental chamber.

This is in the opposite direction from that observed by the authors in Ref. 2. In later experiments the direction switched and remained in that direction until the last few experiments, when it again switched back to the original direction. Typical pumping speeds through the orifice were 3–12 cm sec^{-1}, whereas when the generator was shut off the gravity-forced return flow was at a speed of about 20–30 cm sec^{-1}, which is typical for this size of orifice. The return flow showed little "trail off" as equilibrium was approached. Pumping was only observed at resonances. In Fig. 2 we show the initial pumping speed through the orifice as a function of the sound frequency, and for comparison the received sound amplitude. The double resonance is to be expected both because of the coupling between the 3-cm chambers and because their dimensions were not identical. Pumping sometimes occurred at one of a resonance pair but not at the other; for example, on April 26 good pumping was observed at 3470 Hz but no pumping at 3420 Hz.

The pumping effect always required a minimum generator voltage for initiation.

Table I. Sound Resonances (April 19)

Resonant frequency, kHz	Calculated resonance and identification	Pumping
0.614	Helmholtz resonator?	Yes
3.384	3.5 kHz—fundamental in 3-cm chamber including one side hole	Yes
5.816	5.9 kHz—third harmonic in 6-cm chamber	Yes
6.788	?	Yes
7.524	?	Yes
8.272	8.3 kHz—second harmonic in 3-cm chamber in- cluding one side hole	Yes
10.514	10.4 kHz—orifice plate fundamental resonance	No
17.156	?	Slow
18.612	?	No
20.230	?	Yes

This voltage varied from run to run, most likely because of changes in the generator efficiency. Above this threshold the pumping speed through the orifice quickly saturated and became independent of generator voltage. It was also independent of generator frequency. The "pumping strength" (the helium head difference between the two capacitors at which the pumping saturated), however, increased very rapidly above the threshold voltage. In Fig. 3 the pumping strength is shown as a function of generator voltage for two different runs. As can be seen, it increases approximately exponentially with voltage.

Finally, attention should again be brought to the direction of the pumping. As mentioned before, pumping in both directions has been observed. In almost all runs the pumping was in one direction throughout the run. In one run, however, flow in both directions was observed. The change from one direction to the other seemed quite spontaneous, though it never occurred during the pumping. In one case opposite direction flows were observed consecutively at the same frequency after letting the helium level come to equilibrium in between.

There were a few runs where no pumping was observed. In these runs, however, if flow were induced through the orifice by other means, it could be arrested by turning on the sound generator. Because of this lack of pumping in some runs an attempt was made to determine whether the helium preparation influenced the pumping effect. Helium was condensed into the chamber both above and below T_λ and it was left to settle overnight at $4°K$. There seemed to be no obvious effect on the pumping.

No pumping was observed in He I. The helium level was observed for about 30 min with a resolution of about 1 μm. This would set an upper limit of about 10^{-2} cm sec^{-1} pumping speed through the orifice.

It should be mentioned that pumping in both directions through an orifice at ultrasonic frequencies has been seen by Schofield.[3]

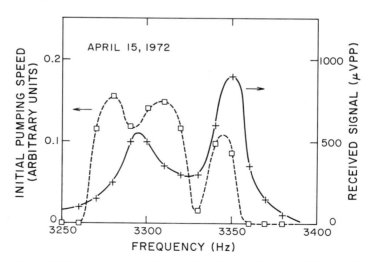

Fig. 2. Initial pumping speed through the orifice as a function of the sound frequency. For comparison the right-hand scale shows the received sound amplitude.

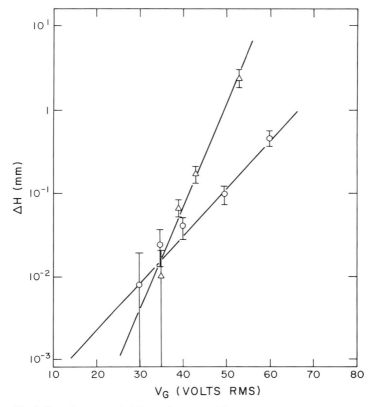

Fig. 3. Pumping strength ΔH as a function of the sound generator voltage V_G.
Triangles: March 25. Pumping away from generator; $T = 1.16°$K. Circles:
July 8. Pumping away from generator, $T = 1.16°$K.

Conclusions

In previous experiments at ultrasonic frequencies the pumping has been attributed to the Bernoulli effect. This would seem difficult in this experiment. There
should be no transverse velocities at the orifice. On the other hand, there will be
velocities transverse to the entrance holes to the capacitors. If this were the cause,
however, it should be easy to identify by adjusting the sound resonances to be first
in one chamber, then the other, and to observe the effect on the pumping. No
such effects were observed. The exponential increase of pumping strength with generator voltage would also be unexpected. Finally, it is very difficult to understand the
reversal in pumping direction with no apparent geometric changes in the chamber. I
therefore think it unlikely that the Bernoulli effect is the cause of the pumping in this
experiment.

An idea of the possible complexity of the effect can be seen in the experiments
of Ingard and Labate.[4] They investigated the propagation of sound in air through an
orifice in the low frequency range up to 700 Hz. By visual observation (with smoke

particles) they confirmed* that steady circulation of the gas was present even at the lowest sound amplitudes. At higher sound intensities transitions to more complex streaming took place, with a jet of vortex rings being present at the highest sound levels (140 dB). At some intensity levels the streaming was asymmetrical, being predominately on one side of the orifice or the other. Such streaming, if it were present in this experiment, could be responsible for the pumping action.

A classical theory of this streaming has been given by Eckart.[6] He has shown that the streaming is a second-order effect related to the production of vorticity by viscous forces. A rather unusual result is that the steady-state streaming velocity is independent of the fluid viscosity, a nonzero viscosity only being necessary to create the vorticity initially. Order of magnitude calculations for the effect give finite velocities only at high frequencies in liquids ($f > 1$ MHz), but at audio frequencies in gases (as substantiated by Ingard and Labate) because of the much lower acoustic impedance ρc where ρ is the density of the gas and c the sound velocity. Thus, in He I at the frequencies ($f \sim 20$ kHz) and sound pressure levels ($P < 100$ dB most likely) in this experiment one would expect very small streaming. The effect might be enhanced in He II and thus explain the pumping effect that has been observed. Any quantitative comparison awaits a calculation similar to Eckart's for He II.

It should be emphasized that the original purpose of the work reported on in this paper was to look for phase synchronization effects (the ac Josephson effect) in a geometry and at a frequency very different from previous workers. The pumping effect was useful in this regard and was unexpected. No structure in the pumping curves was observed that was reliably reproducible or could be positively related to the Josephson equation $\hbar \omega = mg\Delta h$ where $\hbar$ is Planck's constant, $\omega/2\pi$ the frequency, m the helium mass, and Δh the helium head difference between the chambers.

References

1. P.L. Richards and P.W. Anderson, *Phys. Rev. Lett.* **14,** 540 (1965).
2. P.L. Richards, *Phys. Rev.* **2A,** 1532 (1970). B.M. Khorana, *Phys. Rev.* **185,** 299 (1969). J.P. Hulin, C. Laroche, A. Libchaber, and B. Perrin, *Phys. Rev.* **5A,** 1830 (1972).
3. G.L. Schofield, Jr., Ph.D. Thesis, Univ. of Michigan, 1971, unpublished.
4. U. Ingard and S. Labate, *J. Acoust. Sci. America,* **22,** 211 (1950).
5. Rayleigh, *Theory of Sound,* MacMillan Co. Ltd., London (1896), Vol. II, p. 217.
6. C. Eckart, *Phys. Rev.* **73,** 68 (1948).

* Acoustic streaming at an orifice was known already by Rayleigh (see Ref. 5).

Optical Measurements on Surface Modes in Liquid Helium II

S. Cunsolo and G. Jacucci

Gruppo Nazionale di Struttura della Materia, Istituto di Fisica
Università degli Studi di Roma, Italia

The properties of the surface modes both of the bulk liquid helium II and of the film have been studied theoretically and experimentally by Atkins et al.[1,2] Their measurements, based on the optical detection of the local elevation of the surface, mostly refer to values of the wavelength of 1 mm and above.

Measurements in a much shorter wavelength region may be carried out with a light-beating technique based on the scattering of coherent light be excited surface waves.

Seeking to check at higher frequencies the hydrodynamic predictions for the dispersion relation of these modes at various values of the depth of the liquid and thickness of the film, we have set up one such apparatus. We report here preliminary results regarding the surface waves on bulk liquid helium.

The phase velocity of the surface wave of wave number k is given by[2]

$$v^2 = \frac{\rho - \rho'}{\rho + \rho'} \frac{g}{k} + \frac{\sigma k}{\rho + \rho'} \tanh(dk) \tag{1}$$

where ρ and ρ' are the liquid and vapor densities, respectively; σ is the surface tension; g is the acceleration of gravity; and d is the depth of the liquid. When the wave number is greater than 100 cm^{-1} one may neglect the gravitational term, and a measurement of the dispersion relation directly yields the value of σ as for pure capillary waves. In our experiments d is 1 mm and k ranges from 67 to 800 cm^{-1}, so that the hyperbolic tangent may be replaced by one. Neglecting the vapor density, Eq. (1) then reduces to

$$v^2 = (\sigma/\rho)\, k \tag{2}$$

The experimental apparatus is shown schematically in Fig. 1, where the scattering angles θ and ϕ are also indicated. A 1-mW He–Ne gas laser beam is shined on the liquid helium free surface in a cell included in a thermal bath inside an optical dewar. A diaphragm selects the light scattered by a surface wave of wave number $k = k_{\text{light}}[\cos\theta - \cos(\theta + \phi)]$ which mixes on the cathode of a photomultiplier tube with light scattered by dirt and window imperfections. A beat note is generated at the frequency of the surface wave and detected by a HP 310 A wave analyzer.

The excitation of the surface wave is accomplished, as in the cited experiment,[2] by periodic heat dissipation. On a glass surface held horizontally under the liquid vapor interface a thin evanohm wire 0.01 mm in diameter and 1 cm long is stretched

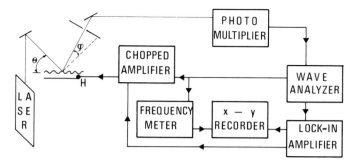

Fig. 1. Scheme of the scattering geometry and of the detection system.
The heater is indicated by H.

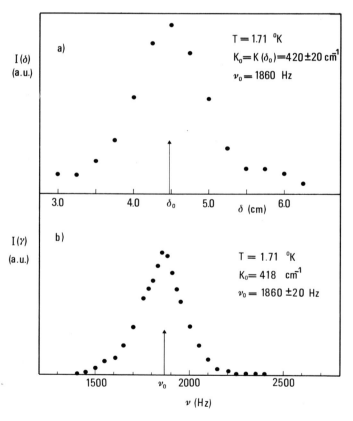

Fig. 2. Signal amplitude (a) at a fixed frequency value as a function of
the variable δ, which is proportional to k; and (b) at a fixed k value as
a function of frequency.

normally to the plane of incidence. The small diameter of the wire ensures a good efficiency of the method of excitation even at the short wavelengths reached.

The power dissipated in the wire has been typically of the order of 0.2 mW. No dependence of the experimental results on the power input has been found on doubling this value.

The measurements are obtained by recording the beat note amplitude as a function of the frequency at fixed values of the scattering angles θ and ϕ, i.e., of the wave number k. Alternatively, we have operated at a fixed value of the frequency by varying the value of k and have obtained coincident results. In both procedures the linewidth is dominated by the low-frequency mechanical vibrations of the surface, which limit the resolution in the experiment by producing random temporal fluctuations of the scattering angles. Thus measurements on the attenuation of the surface waves have not been possible. Figure 2 shows the results obtained with the two procedures for the same point of the dispersion curve. The points shown are obtained with an integration time of 1 min.

Figure 3 shows the experimental results for the dispersion relation obtained in a single run at the temperature of 1.7°K with the procedure of varying the frequency. The angle θ was 12° and ϕ ranged from 3.3 to 36 mrad. The solid line shown in the

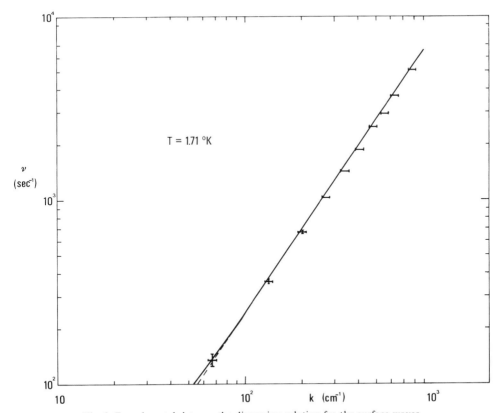

Fig. 3. Experimental data on the dispersion relation for the surface waves.

figure represents Eq. (2) taking for σ and ρ their measured[3] values at $T = 1.7°K$:

$$\sigma = 0.327 \quad \text{erg cm}^{-2}, \qquad \rho = 0.1456 \quad \text{g cm}^{-3}$$

The dashed line is the continuation of the asymptotic pure capillary behavior.

The larger error in our data is due to the uncertainty in the angle θ. This angle is held constant during the measurement and this error reflects approximately in a constant percentage error in k, so that the slope of the v versus k line in the log-log plot is unaffected by this error. A least squares fit of our data for $k > 200$ cm^{-1} yields a value of 1.51 for this slope, in excellent agreement with the three-halves power low implied in Eq. (2), thus proving that the capillary waves in helium II behave as in an ordinary liquid.

The gravitational contribution is noticeable only at very low k values in the range of our experiment. Hence this method should be feasible for accurate measurements of the surface tension and of its dependence on the temperature. Preliminary runs at a fixed k value in the range $T = 1.3–2.1°K$ show a temperature variation of σ in good agreement with the available data[4] from capillary rise experiments.

References

1. C.W.R. Everitt, K.R. Atkins, and A. Denenstein, *Phys. Rev.* **136 A**, 1494 (1964).
2. K.A. Pickar and K.R. Atkins, *Phys. Rev.* **178**, 399 (1969).
3. K.N. Zinoveva, *Zh. Eksperim. i Teor. Fiz.* **29**, 899 (1955) [*Soviet Phys.—JETP* **2**, 774 (1956)].
4. K.R. Atkins and Y. Narahara, *Phys. Rev.* **138 A**, 437 (1965).

6
Helium Bulk Properties

Measurement of the Temperature Dependence of the Density of Liquid ^{4}He from 0.3 to 0.7°K and Near the λ-Point*

Craig T. Van Degrift and John R. Pellam

Department of Physics, University of California – Irvine
Irvine, California

The total density of ^{4}He has a particularly interesting temperature dependence. It has a minimum at about 1.14°K, a near-infinite slope at T_λ, and a maximum at a temperature slightly above T_λ. Direct measurements of the density at saturated vapor pressure have been made by Kerr and Taylor.[1] Indirect measurements, inferred from the index of refraction and dielectric constant, have been reported by a number of workers, the more recent being by Chase et al.,[2] Boghosian and Meyer,[3] and Harris-Lowe and Smee.[4] These indirect measurements can be related to the density through the use of the Clausius–Mosotti relation and the atomic polarizability. The questions of whether the Clausius–Mosotti relation is valid and whether the atomic polarizability may be considered to be temperature independent have been considered by these authors and by Kerr and Sherman.[5] Very recently Roach et al.[6] have measured the density from 0.1 to 0.7°K using the velocity of sound and have verified a T^4 dependence.

We report in this paper preliminary measurements of the temperature dependence of the dielectric constant at saturated vapor pressure from 0.3 to 0.7°K and in the region within 10 m°K of the λ-point. These measurements were made using an improved tunnel diode oscillator circuit described elsewhere in these proceedings.[7] The sensitivity was such that density changes as small as 0.05 ppm could be resolved. In the present measurements, however, the temperature resolution was not sufficiently precise to take full advantage of this sensitivity.

A diagram of our apparatus is shown in Fig. 1. It has a massive copper block on which are mounted thermometers, a heat exchanger, and the inductor of the resonant circuit. The capacitance is between a central copper plate and the walls of a cavity inside the copper block. The capacitor plate is supported by 13 0.4-mm-diameter sapphire spheres partially imbedded in the copper and is thereby spaced about 0.12 mm from the cavity walls. The cavity was created by the use of a press-fitted plug which, for good measure, was sintered and then soldered. The wire connecting the coil to the capacitor passes through a sapphire–copper–indium vacuum feedthrough. When making measurements near the λ-point, 99% of the capacity influenced by the helium occupies a region only 3 mm in height; accordingly, the

* Supported in part by the National Science Foundation, Grant No. GP 10699.

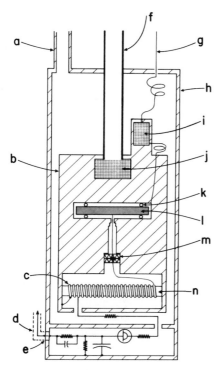

Fig. 1. Diagram of apparatus: (a) Exchange gas pumping tube, (b) Cu block with thermometers, (c) high-purity Cu wire coil, (d) oscillator coaxial cable, (e) glass feedthrough, (f) ^{3}He pumping tube, (g) 0.03-mm-i.d. sample filling tube, (h) vacuum wall surrounded by ^{4}He bath, (i) sintered Cu heat exchanger, (j) sintered Cu evaporator, (k) 0.4-mm-diameter Al_2O_3 spherical spacers, (l) capacitor plate, (m) Al_2O_3–Cu–In feedthrough, and (n) fused quartz tube coil form.

hydrostatic pressure dependence of the λ transition should blur the data over a temperature range of only about 0.3 $\mu°$K in width.

The sample chamber is filled through a 0.03-mm-i.d. capillary tube. During the run the helium liquid–vapor interface can be positioned in a small reservoir (not shown in Fig. 1) situated just above the capacitor, or it can be positioned in the sintered copper heat exchanger. In either case the 300-cm-long section of capillary passing from the sample chamber to the ^{4}He bath contains only a helium film during the measurements. The heat leak through this film does not exceed 30 ergs sec^{-1} at 0.3°K (bath at 1.14°K) and appears less than 1 erg sec^{-1} when the bath temperature is 2.171°K, providing the sample is within 10 m°K of the λ-point. Nevertheless, to ensure that the helium in the capacitor is at the same temperature as the capacitor, we installed the heat exchanger and another 150 cm of capillary tubing so that the film is heatlagged to the copper block.

The experimental procedure was to first evacuate the capacitor and heat exchanger at room temperature through a "solder valve" (since evacuation through the capillary filling line was too time-consuming). After cooling, the background temperature dependence of the frequency was measured over the temperature ranges of interest. With this particular oscillator the background provided only a minor correction near the λ-point, but was ten times greater than the effect of the helium at 0.3°K and equal to it at about 0.45°K. The capacitor was then filled, the frequency variation with temperature measured, and the background subtracted. The conversion from relative frequency change to relative density change was accomplished by first determining the fraction of total capacity affected by the helium (96.962%) from the known dielectric constant of 1.0572484 at 1.072°K (the other ~ 3% corresponds to stray capacitance) and from the very large frequency shift observed during the filling of the capacitor. Second, the changes in frequency and hence dielectric constant were related to changes in density by evaluating the Clausius–Mosotti relation using the value of[4] 0.123296 cm³/mole⁻¹ for the atomic polarizability (assumed to be temperature independent).

Our thermometry was done using dc measurements of a germanium thermometer having a resistance of 3350 Ω near the λ-point and 18 MΩ at 0.297°K. It was calibrated using ⁴He vapor pressure above 1°K and ³He vapor pressure below 1°K. These calibrations were made in haste and much remains to be done before our temperature scale errors are reduced to a magnitude compatible with the frequency errors. Very close to the λ-point the calibration of the temperature scale is not so crucial, but even during these measurements our resolution of 10 μ°K was inadequate.

In Fig. 2 we show the results near the λ-point. The data are presented as the fractional deviation of density from its λ-point value, $(\rho - \rho_\lambda)/\rho_\lambda$. The λ-point is clearly distinguishable: It separates an upward arching portion of the curve for $T < T_\lambda$ from a downward arching portion for $T > T_\lambda$. A maximum fractional increase of 59.8 ppm is evident at 6.75 m°K above T_λ. The inset shows a 25 × enlargement of the region around the λ-point and we can see that the λ-point is still easily identifiable. The scatter in the graph in both the main curve and the inset is attrib-

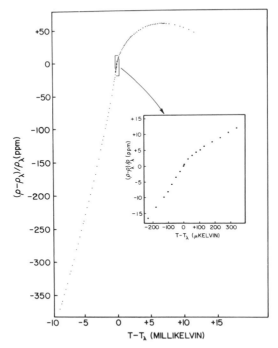

Fig. 2. Relative density change near the λ-point: The raw data are plotted as $(\rho - \rho_\lambda)/\rho_\lambda$ vs. $T - T_\lambda$.

utable either to temperature errors of a few $\mu°$K or slight plotting errors; the frequency noise *per se* is not visible on either scale. These data have not yet been fitted to the appropriate Pippard–Buckingham–Fairbank[8] relation but are known to follow the measurements of Kerr and Taylor[1] to within about 5%.

In Fig. 3 we show the temperature dependence of the density from 0.3 to 0.7°K. The raw data are plotted as solid circles versus temperature, representing the fractional density decrease $(\rho_0 - \rho)/\rho_0$ below the estimated value ρ_0 for absolute zero (right-hand scale). One finds that these follow a smooth curve which rapidly increases with temperature. To test for a T^4 dependence we have divided the data by T^4 and plotted $(\rho_0 - \rho)/\rho_0 T^4$, using the scale at the left. From 0.4 to 0.55°K these data seem to follow the relation $(\rho_0 - \rho)/\rho_0 = 253 T^4$ ppm with a scatter of about 0.05 ppm error in density and about 3 m°K error in temperature. Again we believe most of the scatter originates in temperature errors. Above 0.55°K, $(\rho_0 - \rho)/\rho_0$ rises less rapidly with temperature, falling by about 10% below the T^4 behavior upon reaching 0.7°K.

A careful analysis of deviations from a T^4 behavior should indicate whether or not the elementary excitation curve for ^{4}He first deviates from its phonon-like portion by arching upward. The interpretation of thermal expansion data in this respect, however, is not completely straightforward, as pointed out by Roach et al.,[6] and we will defer such conclusions until we have measured the dependence up to 1.5°K. Furthermore, we must eliminate the systematic deviation of up to 40% from a T^4 behavior evident below 0.4°K, which we are certain is an experimental anomaly

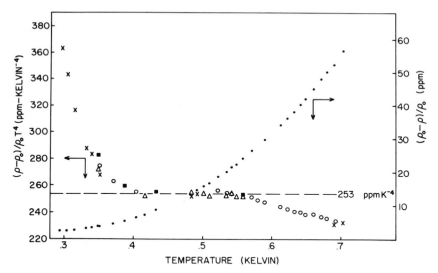

Fig. 3. Temperature dependence of density in the 0.3°K to 0.7°K range. The solid circles give the relative density data $(\rho_0 - \rho)/\rho_0$ and show the rapid increase with temperature. The other points (different symbols for data taken on different days) represent these data divided by T^4 to illustrate deviation from the expected T^4 behavior. Note that from 0.4 to 0.57°K these points lie within 2% of the value of 253 ppm °K^{-4}. At higher temperatures the deviation results from the increasing roton contribution. At low temperatures the deviation, up to 40%, results from the imperfect removal of a rapidly increasing background temperature dependence.

related to the very rapidly changing background temperature dependence. We have not yet made a careful analysis of possible systematic errors such as might arise from the slight change in helium pressure head as the liquid–vapor interface level changes, or from inadequate thermal equilibrium between the sample and the thermometers.

We conclude by pointing out that although our preliminary data have in general about 50 times less scatter than previous measurements of the density, our temperature scales, at present, do not justify precise comparison with theory. Furthermore, the potential precision of the results, as represented by the noise, will be realized only after the previously mentioned defects are corrected. When the full capabilities of this method of density measurement are thus attained, very accurate comparison may be made with the Pippard–Buckingham–Fairbank relations near the λ-point and with the Landau model and its variants below 1.5°K.

Acknowledgment

We would like to acknowledge the expert technical assistance from Mr. R.T. Horton.

References

1. E.C. Kerr and R.D. Taylor, *Ann. Phys.* **26**, 292 (1964).
2. C.E. Chase, E. Maxwell, and W.E. Millett, *Physica* **27**, 1129 (1961).
3. C. Boghosian and H. Meyer, *Phys. Rev.* **152**, 200 (1966).
4. R.F. Harris-Lowe and K.A. Smee, *Phys. Rev. A* **2**, 158 (1970).
5. E.C. Kerr and R.H. Sherman, *J. Low Temp. Phys.* **3**, 451 (1970).
6. P.R. Roach, J.B. Ketterson, B.M. Abraham, and M. Kuchnir, *Phys. Lett.* **39A**, 251 (1972).
7. C.T. Van Degrift, this volume.
8. M.J. Buckingham and W.M. Fairbank, in *Progress in Low-Temperature Physics,* C.J. Gorter, ed. North-Holland, Amsterdam (1961), Vol. III, Chapter III.

Second-Sound Velocity and Superfluid Density in ⁴He Under Pressure and Near T_λ

Dennis S. Greywall and Guenter Ahlers

Bell Laboratories
Murray Hill, New Jersey

In this paper we present values of the superfluid fraction ρ_s/ρ which were derived from precise measurements of the second-sound velocity u_2 in pure ⁴He.[1] These results extend to temperatures very near the superfluid transition temperature T_λ and cover the entire pressure range over which the superfluid exists. The data were taken in order to provide further tests of scaling and universality near this line of critical singularities.* The results for ρ_s/ρ could *not* be accurately described at all pressures in terms of a simple power law of the form

$$\rho_s/\rho = k(P)\,\varepsilon^\zeta, \qquad \varepsilon \equiv 1 - T/T_\lambda \qquad (1)$$

even for $\varepsilon \lesssim 10^{-3}$. However, the asymptotic behavior could be inferred after making reasonable general assumptions about contributions to ρ_s/ρ which are of higher order in ε than the leading term given by Eq. (1). Our results are consistent with both scaling and universality.

The second-sound velocity was determined with a precision of 0.1% by measuring the resonant frequencies of a cavity 1 cm high and 1 cm in diameter. The signal was excited and detected using superleak condenser transducers [3,4] which dissipated little energy and thus permitted the temperature of the thermally well-isolated sample to be held constant to within $\pm 10^{-7}$ °K for periods of time sufficiently long to measure a resonance frequency. The λ temperature was determined with high precision by noting the onset of thermal resistance in the liquid. Near T_λ the present results for u_2 agree well with previous measurements[5-7] which are available only at saturated vapor pressure (SVP). At temperatures far from T_λ ($\varepsilon > 10^{-2}$) our data agree at all pressures to within 2% with the velocities measured by Peshkov and Zinov'eva[8] in the temperature range $1.6°\mathrm{K} \leq T < 2.0°\mathrm{K}$.

In order to determine the temperature dependence of ρ_s/ρ, we used the result[9]

$$\rho_s/\rho = u_2^2(u_2^2 + S^2 T/C_p)^{-1}$$

of linear two-fluid hydrodynamics. Here C_p is the heat capacity at constant pressure and S is the entropy. Since we desired ρ_s/ρ along isobars, the u_2 data collected along isochores were first corrected to velocities along isobars. The results for ρ_s/ρ at three of our pressures are shown as a function of ε in Fig. 1. It can be seen that the three

---

* For a recent review of critical phenomena near the superfluid transition see Ahlers,[2] who also gives additional references.

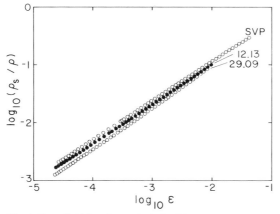

Fig. 1. Superfluid fraction ρ_s/ρ derived from measurements of u_2 versus $\varepsilon \equiv 1 - T/T_\lambda$. The numbers give the pressure in bars.

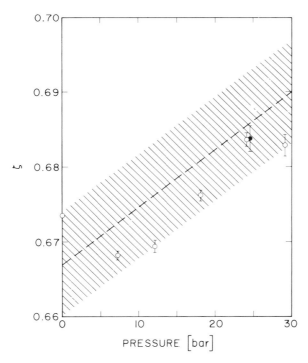

Fig. 2. Apparent critical exponent for ρ_s/ρ resulting from a fit of the data with $\varepsilon < 10^{-2.3}$ to Eq. (1). All results pertain to isobars, but the open circles are based on values of u_2 derived from measurements along isochores, whereas the solid circle is based on measurements along an isobar.

curves shown are not parallel. If Eq. (1) is adequate to describe the data, then the slopes of the lines in Fig. 1 are equal to ζ. Thus it may appear that ζ, although approximately equal to 2/3, is slightly dependent upon pressure. Such behavior would be in disagreement with the concept of universality, according to which critical exponents should be independent of inert variables, like the pressure in this case, which do not affect the symmetry of the system. The results of fitting the data at each pressure with $\varepsilon < 10^{-2.3}$ to the simple power law, Eq. (1), are shown in Fig. 2. This figure reveals a pressure dependence of the apparent exponent which is in overall agreement with the recent results of Terui and Ikushima.[10] These authors also obtained ρ_s/ρ from u_2 and fitted their results to Eq. (1). The general trend and approximate range of uncertainty of their exponent are indicated by the shaded area in the figure.

The precision of the present results permits a more detailed analysis of the data. In order to show graphically in some detail the ε dependence of ρ_s/ρ, we multiply ρ_s/ρ by $\varepsilon^{-2/3}$ and present in Fig. 3 a high-resolution plot of this variable as a function of ε on logarithmic scales. At SVP our results agree well with values measured with other methods by Tyson,[11] Clow and Reppy,[12] and Kriss and Rudnick.[13] The results of Tyson are represented by the solid line. At $P \lesssim 20$ bars and at larger ε our results are compared with those of Romer and Duffy.[14] The permitted systematic error in their results increases rapidly with decreasing ε. The range of permitted values is shown by the shaded area; it overlaps the present high-pressure results.

If ρ_s/ρ is accurately described by Eq. (1), then the data in Fig. 3 should fall on straight lines with slopes equal to the difference between ζ and 2/3. However, with the exception of the data at SVP, curvature is clearly revealed at each pressure over any reasonable range of ε. This indicates that the data must be described by a

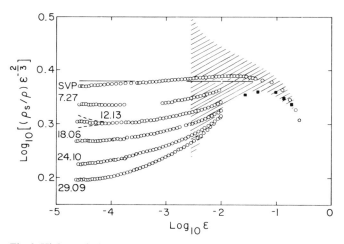

Fig. 3. High-resolution plot of the superfluid fraction ρ_s/ρ along isobars derived from measurements of u_2. The numbers give the pressure in bars. The solid squares correspond to u_2 at 25.3 bars measured by Peshkov and Zinov'eva.[8] The effect of a shift in T_λ of ± 2 $\mu°$K upon the results is demonstrated by the dashed curves in the figure. The solid line represents values of ρ_s/ρ at vapor-pressure measured by Tyson.[11] The shaded area corresponds to the range of values of $(\rho_s/\rho)\,\varepsilon^{-2/3}$ under pressure permitted by the results of Romer and Duffy.[14]

function which is more complicated than a simple power law. Unfortunately, it is not possible to determine from experimental measurements the correct functional form of ρ_s/ρ. We have thus assumed that the departures of ρ_s/ρ from pure power law behavior can themselves be described by a power law with an exponent which is larger than ζ. Explicitly, we have compared the data with the function

$$\rho_s/\rho = k(P)\,\varepsilon^\zeta [1 + a(P)\,\varepsilon^x]; \qquad x > 0 \tag{2}$$

Fitting the results with $\varepsilon \lesssim 10^{-2}$ at each pressure separately to Eq. (2), we find that within permitted systematic and random errors ζ and x are independent of pressure and $k(P)$ and $a(P)$ can be described by simple polynomials in P. A combined, least-squares fit of all of the data with $2 \times 10^{-5} \lesssim \varepsilon \lesssim 10^{-2}$ yields*

$$\zeta = 0.677 \pm 0.02; \qquad x = 0.36 \pm 0.1$$

where the uncertainties include our best estimate of permitted systematic errors. The asymptotic exponent ζ is in good agreement with the scaling law

$$\zeta = (2 - \alpha')/3$$

and the experimental result[2,15] that the C_p exponent α' is near zero. The absence of a pressure dependence for ζ is in agreement with universality.

References

 1. D.S. Greywall and G. Ahlers, *Phys. Rev. Lett.* **28**, 1251 (1972).
 2. G. Ahlers, in *Proc. 12th Intern. Conf. Low Temp. Phys., 1970*, E. Kanda, ed., Academic Press of Japan, Tokyo (1971), p. 21.
 3. R. Williams, S.E.A. Beaver, J.C. Fraser, R.S. Kagiwada, and I. Rudnick, *Phys. Lett.* **29A**, 279 (1969).
 4. R.A. Sherlock and D.O. Edwards, *Rev. Sci. Instr.* **41**, 1603 (1970).
 5. C.J. Pearce, J.A. Lipa, and M.S. Buckingham, *Phys. Rev. Lett.* **20**, 1471 (1968).
 6. D.L. Johnson and M.J. Crooks, *Phys. Rev.* **185**, 253 (1969).
 7. J.A. Tyson and D.H. Douglass, *Phys. Rev. Lett.* **21**, 1380 (1968).
 8. V.P. Peshkov and K.N. Zinov'eva, *Zh. Eksperim. i Teor. Fiz.* **18**, 438 (1948).
 9. I.M. Khalatnikov, *Introduction to the Theory of Superfluidity*, Benjamin, New York (1965), Chapter 10.
10. G. Terui and A. Ikushima, *Phys. Lett.* **39A**, 161 (1972).
11. J.A. Tyson, *Phys. Rev.* **166**, 166 (1968).
12. J.R. Clow and J.D. Reppy, *Phys. Rev. Lett.* **16**, 887 (1966); *Phys. Rev.* **5A**, 424 (1972).
13. M. Kriss and I. Rudnick, *J. Low Temp. Phys.* **3**, 339 (1970).
14. R.H. Romer and R.J. Duffy, *Phys. Rev.* **186**, 255 (1969).
15. G. Ahlers, *Phys. Rev.* **A8**, 530 (1973).
16. D.S. Greywall and G. Ahlers, *Phys. Rev.*, **A7**, 2145 (1973).

---

* See, however, Ref. 16.

Superfluid Density Near the Lambda Point
in Helium Under Pressure

Akira Ikushima and Giiuchi Terui

The Institute for Solid State Physics
The University of Tokyo, Tokyo, Japan

The second-sound properties and the superfluid density deduced therefrom should be most fundamental in the study of the λ transition because they are inherently related to the critical modes and the coherence of the λ transition, respectively.[1] Furthermore, there seems to exist a widely held precept of critical-point theory which indicates that the character of the critical behavior is independent of the material provided the symmetry properties of the Hamiltonian are the same. The precept has been supported to some extent by a number of theoretical studies,[2] while Baxter[3] has recently carried out a detailed calculation of the critical exponents in a rather special system, the two-dimensional Ising lattice, giving a conclusion against the above precept. No direct experimental study has been done of this problem, and, in this connection, it should be quite effective to look at critical exponents associated with the λ transition as a function of pressure because the pressure does not change the symmetry of the system as long as helium is in the liquid state. The present paper reports the change of the critical exponent of the superfluid density as a function of the pressure.

The superfluid density ρ_s was deduced from the second-sound velocity determined by measuring the resonant frequencies of the CW second-sound in a cavity and by using previously published data of the specific heat and the entropy.[4,5] For the specific heat the relation $C_p(P) = B'(P) - A'(P) \log_{10}[T_\lambda(P) - T)]$ was adopted with experimentally determined $A'(P)$, $B'(P)$, and $T_\lambda(P)$. The values of $A'(P)$ and $B'(P)$ for pressures higher than ~ 23 atm were not available and were extrapolated from the measured values.

Figure 1 shows plots of ρ_s/ρ thus deduced versus $\varepsilon = 1 - T/T_\lambda$, where plots are not made for $\varepsilon \gtrsim 10^{-2}$ because the data for $C_p(P)$ are not available at this temperature range. Moreover, the plots for $\varepsilon \lesssim 10^{-4}$ are thought to be less reliable since some gravitational effect should be expected there. Figure 2 shows the corresponding values of the critical exponent of the superfluid density ζ determined from ρ_s/ρ at $10^{-4} \lesssim \varepsilon \lesssim 10^{-2}$, as a function of pressure.

First, it should be noted that ζ seems to depend, even though slightly, on the pressure up to almost the solidification limit. Usual statistical analysis leads to the empirical relation

$$\zeta(P) = a + bP$$

$$a = 0.667 \pm 0.004$$

$$b = 0.0008 \pm 0.0002 \text{ kg}^{-1} \text{ cm}^2$$

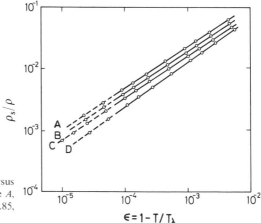

Fig. 1. Typical plots of the superfluid densities versus $\varepsilon = 1 - T/T_\lambda$ at various constant pressures. Here A, B, C, and D indicate results at $P = $ SPV, 9.06, 19.85, and 28.35 kg cm^{-2}, respectively.

where the errors quoted are the probable errors. The error in b becomes twice as large if we use a 95% confidence limit, and the increase of ζ with increasing pressure is still not meaningless.

If one appreciates the change of ζ, the present result is a very interesting deviation from the belief mentioned in the opening paragraph.

Baxter's treatment was followed by the work of Kadanoff *et al.*[6] and Suzuki,[7] which show that, given the Hamiltonian $H = H_0 + \lambda H_1$, where λ is a gradually varying parameter with an external variable, the critical exponent should vary *continuously* as a function of λ if the following conditions are satisfied: (1) The specific heat diverges logarithmically in the unperturbed system. (2) The perturbing Hamiltonian H_1 is written as a product of the energy densities of the unperturbed system.

The first condition appears to be satisfied in the λ transition, while it is not clear whether the second one is satisfied in the present case. Furthermore, it should be noted here that one cannot discard the possibility of the confluent singularity which would be appearing at higher pressure. The change of the interaction range under the pressure would also be another possibility, although this may not be the

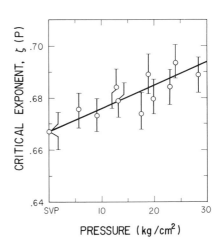

Fig. 2. Critical exponent of the superfluid density versus pressure.

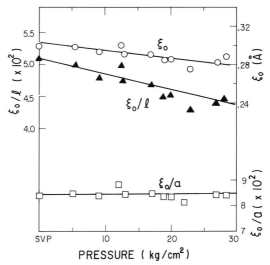

Fig. 3. Variation of ξ_0 ($\bigcirc$), ξ_0/a ($\square$), and ξ_0/l ($\blacktriangle$) as functions of pressure, where a and l are the mean interatomic distance and the so-called "thermal wavelength," respectively.

case because the interaction between helium atoms is short range, and the change of the interaction range would not essentially affect the short-range nature, and thus the coherence, of the system.

Very recently Greywall and Ahlers[8] reported a result similar to the present one, proposing from a small bending in plots of the second-sound velocity u_2 or ρ_s at especially smaller ε that u_2 or ρ_s cannot be described by a single power law in ε. However, as noted previously, one should take into account the gravity effect on the measured results* and this effect is easily shown to be from a few to several percent for $\varepsilon = 10^{-5}$, depending on the position of the temperature sensor. The effect is larger for higher pressure. Therefore the result at $\varepsilon \lesssim 10^{-5}$ should always be considered carefully. Furthermore, the physical meaning of the seven parameters in their expression used to analyze the result should be clarified.

Writing $\rho_s/\rho = A\varepsilon^\zeta$, the variation of A with pressure is also interesting. The superfluid density measures the extent to which the helium atoms in the system are in a coherent state. As a result, the quantity A is related to ζ_0 in the coherence length, $\zeta = \zeta_0 \varepsilon^{-\nu}$, as[1,9]

$$\zeta_0 = m^2 k_B T/(4\pi\hbar^2 A\rho)$$

where m is the helium-atom mass, k_B is the Boltzmann constant, and $\hbar$ is the Planck constant divided by 2π.

Figure 3 shows ζ_0 thus deduced versus pressure together with the variation of

---

* Ahlers has indicated that the lambda temperatures in Ref. 8 were the averaged values corresponding to the middle point of the cell although they put the temperature sensor at the bottom. Therefore the correction in u_2^2 due to the gravity is 0.13% at $P \simeq 28$ atm and $\varepsilon = 1 \times 10^{-5}$, and is less for lower pressures.

ζ_0 divided by the mean interatomic distance a. This result means that ζ_0/a is constant versus pressure, and this is worth noting because a should not be unique as the characteristic length of the problem. For example, consider $l = h/(2\pi m k_B T_\lambda)^{\frac{1}{2}}$, the so-called "thermal wavelength" or the de Broglie wavelength for a particle of mass m at T_λ. The quantity ζ_0/l decreases approximately 20% with increasing pressure up to the solidification limit, which exceeds the present experimental uncertainty. The present result is simple but important; the number of atoms related to ζ_0 is kept constant even if the volume changes up to about 30%.

Acknowledgment

The authors would like to express sincere thanks to Dr. M. Suzuki for several enlightening discussions.

References

1. R.A. Ferrell, N. Menyhard, H. Schmidt, F. Schwabl, and P. Szepfalusy, *Ann. Phys.* **47**, 565 (1968).
2. M.E. Fisher, *Rep. Prog. Phys. (London)* **30**, 615 (1967); M. Suzuki, *Prog. Theor. Phys.* **46**, 1054 (1971).
3. R.J. Baxter, *Phys. Rev. Lett.* **26**, 832 (1971); *Ann. Phys.* **70**, 193 (1972).
4. G. Ahlers, in *Proc. 12th Intern. Conf. Low Temp. Phys., 1970* Academic Press of Japan, Tokyo (1971), p. 21.
5. C.J.N. van den Meijdenberg, K.W. Taconis, and R. de Bruyn Ouboter, *Physica* **27**, 197 (1961).
6. L.P. Kadanoff and F.J. Wegner, *Phys. Rev.* **B4**, 3989 (1971).
7. M. Suzuki, *Phys. Rev. Lett.* **28**, 507 (1972).
8. D.S. Greywall and G. Ahlers, *Phys. Rev. Lett.* **28**, 1251 (1972).
9. B.I. Halperin and P.C. Hohenberg, *Phys. Rev.* **177**, 952 (1969); H. Schmidt and P. Szepfalusy, *Phys. Lett.* **32A**, 326 (1970).

Hypersonic Attenuation in the Vicinity of the Superfluid Transition of Liquid Helium*

D.E. Commins and I. Rudnick

Department of Physics, University of California
Los Angeles, California

The attenuation of first sound has been measured in the vicinity of the lambda transition of liquid helium in a number of experiments and at frequencies between 16.8 kHz and 1 GHz. The experiments of Williams and Rudnick[1] have shown that for frequencies between 600 kHz and 3.17 MHz the attenuation maximum occurs at a temperature T_p below T_λ such that $\omega |T_\lambda - T_p|^{-1} = $ const, where ω is the angular frequency. The temperature relative to T_λ at which the maximum occurs and the amplitude of the maximum are both proportional to the frequency in the range 16.8 kHz to 3.17 MHz. The data were interpreted by assuming that the attenuation was due to independent mechanisms in addition to the classical absorption due to thermal and viscous losses: (1) a relaxation process occurring only below the lambda point, described by Pokrovskii and Khalatnikov,[2] in which the relevant relaxation time is given by ξ/C_2 where ξ is a coherence length and C_2 is the velocity of second sound [near T_λ this time is proportional to $(T_\lambda - T)^{-1}$]; and (2) a critical attenuation which is nonsingular and symmetric about T_λ due to inherent fluctuations of the order parameter. Other experiments were performed at a frequency of 1 GHz[3] over a wide temperature range. In the vicinity of the lambda transition they showed that the attenuation undergoes a very sharp maximum and that the peak is nonsingular and occurs 3–4 m°K below the lambda point. These measurements, however, were performed in an open bath and it was not possible to reach the submillidegree accuracy which is necessary to firmly establish confidence in the location of the peak.

The purpose of the present experiment was to measure the attenuation of first sound at a frequency of 1 GHz and to measure the temperature with greater accuracy. A fixed-path interferometer, using a pair of cadmium sulfide transducers, was weakly thermally linked to the main helium bath by a partial vacuum and its temperature was allowed to drift at rates as slow as 3×10^{-6} °K sec^{-1}. The temperature of the helium sample was measured in the immediate vicinity of the acoustic field. The temperature and the amplitude of the received acoustic signal were simultaneously recorded. Such a recording is shown in Fig. 1; it demonstrates that the attenuation maximum (MAX) occurs below the lambda temperature. Several slow drifts were performed upward as well as downward and the amplitude of the input signal as well as the pulse repetition frequency were widely varied without any noticeable

* Supported in part by the Office of Naval Research.

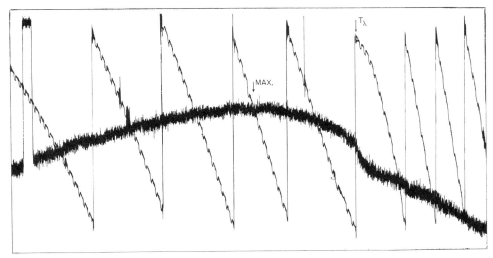

Fig. 1. Simultaneous recording of the received amplitude signal (thick line) and of the temperature (thin line). The temperature is increasing from left to right. The zero-amplitude level is at the top of the figure and the amplitude is increasing from top to bottom.

change in the results. The maximum absorption occurs 3.26 ± 0.20 m°K below T_λ, confirming with greater accuracy the earlier results.

The relative measurement could then be used to determine the absolute attenuation for each value of the temperature. The results are displayed in Fig. 2 for temperatures between 2.0 and 2.3°K. The average of the maximum value of the attenuation is 2470 cm^{-1}. Figure 3 presents a semilogarithmic plot of the total attenuation versus $|T - T_\lambda|$ for the temperature region $|T - T_\lambda| < 10^{-2}$ °K. After the contribution of classical losses due to the viscosity and thermal conductivity have been subtracted, the temperature dependence of the attenuation is investigated, on both sides of the temperature T_p of the maximum, in a region $|T_p - 0.02| < \Delta T < |T_p - 0.125|$ °K.

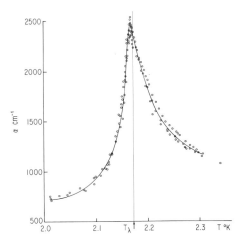

Fig. 2. Ultrasonic attenuation near T_λ. Typical data points are shown and the continuous line is the best-fit curve for ten runs.

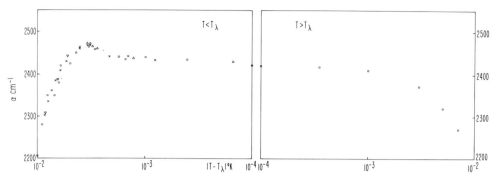

Fig. 3. Semilogarithmic plot of the total ultrasonic attenuation *vs.* $|T - T_\lambda|$.

The average values for the exponent n of

$$\alpha = a/|T - T_p|^n$$

were found to be

$$\bar{n} = 0.49 \pm 0.06 \quad \text{for} \quad T > T_\lambda$$

$$\bar{n} = 0.51 \pm 0.05 \quad \text{for} \quad T < T_\lambda$$

These compare with the respective values of 0.46 and 0.48 in the earlier work.[3] The sharp attenuation maximum, occurring about 3.26 m°K below the lambda transition, cannot be explained by the Pokrovskii–Khalatnikov relaxation mechanism since the relaxation time leads one to expect a maximum of only 180 cm^{-1}, 94 m°K below T_λ. This conclusion does not exclude the possibility that another relaxation time might be involved, such as $\tau_1 = \xi/u_1$, where u_1 is the velocity of first sound and ξ a coherence length.

It is worth noting also that the attenuation at T_λ agrees with the earlier values[3] and scales as $\omega^{1.22}$ as reported by Williams.[4]

To summarize: The attenuation of first sound at 1 GHz has been determined between 2.0 and 2.3°K; it has a very large, nonsingular peak of about 2500 cm^{-1}, with its maximum at 3.26 $\pm$ 0.20 m°K below the lambda point. The attenuation peak cannot be explained by the Pokrovskii–Khalatnikov theory.

References

1. R.D. Williams and I. Rudnick, *Phys. Rev. Lett.* **25**, 276 (1970).
2. V.L. Pokrovskii and I.M. Khalatnikov, *JETP Lett.* **9**, 149 (1969); I.M. Khalatnikov, *Soviet Phys.—JETP* **30**, 268 (1970).
3. J.S. Imai and I. Rudnick, *Phys. Rev. Lett.* **22**, 694 (1969).
4. R.D. Williams, Doctoral Dissertation, Department of Physics, University of California, Los Angeles, (1970).

Superheating in He II

R.K. Childers and J.T. Tough

Department of Physics, The Ohio State University
Columbus, Ohio

Introduction

While there is considerable evidence that liquid He II can exist in metastable states produced by the isothermal reduction of the pressure below the vapor pressure $P_v(T)$ (tensile strength measurements),[1,2] the complementary process of superheating (isobaric increase of temperature beyond the vaporization temperature) has not been observed. Several recent papers suggest that metastable superheating in fact does not occur.[3] This is difficult to reconcile with the finite tensile strength, unless one considers the extreme sensitivity of the tensile strength to mechanical vibrations. A very small mechanical shock is sufficient to nucleate vapor phase in the super-stressed liquid. Our observations of superheating are on small samples of He II which are highly vibration-isolated.

Apparatus and Methods

The liquid He II sample is contained in a 0.03-cm^3, disk-shaped copper and epoxy reservoir which is connected to a thermally regulated bath of He II by a long ($l = 10$ cm), thin (i.d. $= 10^{-2}$ cm), vertical stainless steel tube (Fig. 1). Heat can be supplied to the sample at the rate $\dot{Q}$ by an electrical heater in the sample. The temperature of the bath and of the sample reservoir is measured with carbon resistance thermometers T_1 and T_2, respectively.

Where $\dot{Q} = 0$ the sample will be at the pressure $P_v(T_1) + \rho g(h + l)$. As $\dot{Q}$ is increased (keeping the bath temperature T_1 constant) the change in the sample pressure and temperature will at first be related by the London equation $\Delta P = \rho S \, \Delta T$ (see Fig. 1). The nature of heat flow in the connecting tube is such that when $\dot{Q}$ is increased beyond a critical heat current $\dot{Q}_c$ the temperature will increase much more rapidly ($\Delta T \propto \dot{Q}^3$) than the pressure.[4] A crude estimate of the pressure increase obtained from the Allen–Reekie rule ($\Delta P \propto \dot{Q}$) indicates that at a limiting heat current $\dot{Q}^*$ the sample will become superheated. That is, if metastable superheated states of He II occur, they will be for $\Delta T > \Delta T^*$ (see Fig. 1). A more rigorous estimate of $\dot{Q}^*$ and ΔT^* can be obtained by extrapolating the data of Brewer and Edwards,[5] who measured *both* ΔP and ΔT across a tube with the same dimensions as ours. Figure 2 compares the two estimates of ΔT^* at $T_1 = 1.56°$K and shows that they only differ by about 0.025°K. Since we actually observe values of ΔT much greater than either estimate of ΔT^*, its exact value is somewhat irrelevant to the question of superheating. Consequently we have used the Allen–Reekie rule to

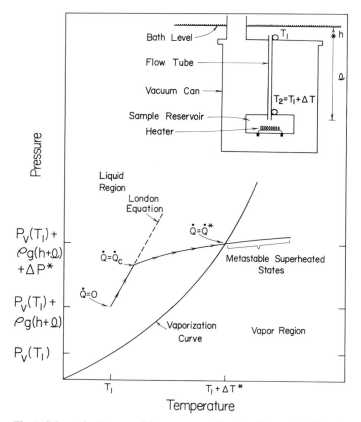

Fig. 1. Schematic diagram of the apparatus (inset) and the path followed
by the sample in the $P-T$ plane.

indicate the expected path of the sample in the $P-T$ plane (Fig. 2) for $T = 1.23$ and
$1.90°K$ and we find $\Delta T^* = 0.370$ and $0.040°K$, respectively. Our method of examining
the question of superheating was simply to measure ΔT vs. $\dot{Q}$ at various fixed values
of T_1 and determine whether $\dot{Q}$ could be increased substantially beyond $\dot{Q}^*$ without
evidence of vaporization of the sample.

Data and Observations

Data were obtained for various temperatures between 1.23 and 1.9°K. For
simplicity we display only the data taken at the lowest and highest temperatures
as a double log plot of ΔT versus $\dot{Q}$ (Fig. 3). Our observations are as follows: As
$\dot{Q}$ is increased ΔT follows the well-established mutual friction curve ($\Delta T \propto \dot{Q}^3$)
from the critical heat current $\dot{Q}_c$ well past the limiting heat current $\dot{Q}^*$ at ΔT^*, where
a "lower branch" is found to intersect. Any point on the "upper branch," which
is an extension of the $\dot{Q}^3$ curve, is metastable in that the sample temperature falls
rapidly to a point on the "lower branch." This may occur either naturally (after
waiting a time anywhere from 1 to 60 min) or after a mechanical shock is applied

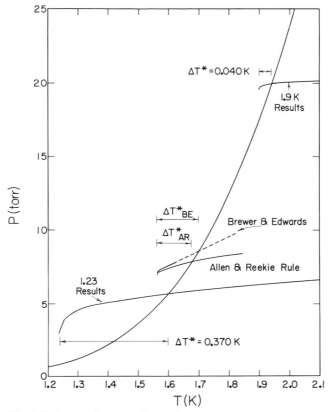

Fig. 2. Estimates of the sample path in the $P-T$ plane and the satura-tion temperature difference based on the Allen–Reekie rule ΔT_{AR}^* and the data of Brewer and Edwards ΔT_{BE}^* at 1.56°K, and the Allen–Reekie rule at 1.23 and 1.9°K.

to the apparatus. Any point on the "lower branch" is a steady state of a finite lifetime in that after a time τ the temperature of the sample reservoir begins to increase extremely rapidly. The time τ is very reproducible and is found to vary approximately as $\tau \simeq \text{const}/\dot{Q}$ for all bath temperatures T_1. In the "lower branch" $\Delta T \propto \dot{Q}$.

The $\dot{Q}^3$ curve for $T_1 = 1.23°K$ in Fig. 3 exhibits a slight curvature for high heat currents due to the large temperature gradient along the tube. In this regime the thermal resistance, which is related to the slope of the curve, changes dramatically from one end of the tube to the other. Our measured values are an average thermal resistance for the entire tube, appropriate to some intermediate temperature between T_1 and T_2. An extreme example of this behavior is found in the experiments of Craig, Hammel, and Keller on heat flow in very narrow channels.[4] We make no attempt to correct the data for this effect, since the evidence for superheating is quite apparent without the correction. The data obtained at intermediate temperatures T_1 resemble those at 1.9 and 1.23°K in every respect, with the curvature becoming less at the higher temperatures.

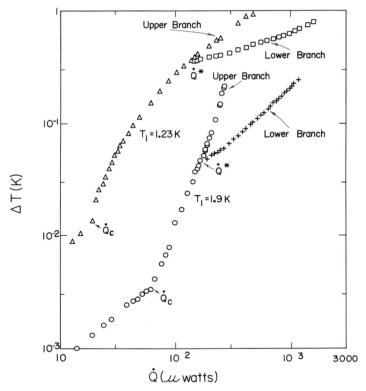

Fig. 3. Experimental data at 1.23 and 1.9°K. Along the "upper branch" the
system is in a steady state of finite lifetime τ.

Interpretation and Discussion

We interpret the data on the "upper branch" in Fig. 3 as representing metastable
superheated states of He II. Our interpretation is based primarily on the close agreement between the observed values of ΔT^* (Fig. 3) and those estimated from the Allen–
Reekie rule (Fig. 2). Furthermore, the states of the system in the "upper branch"
have all of the qualitative features of metastability noted in the tensile strength
measurements[2] (decaying "naturally" but randomly in time, or induced by external
mechanical shock).

The key to this interpretation is the identification of the features of the "lower
branch" with the growth of vapor phase in the sample reservoir. If the transition
from the "upper branch" to the "lower branch" signals the nucleation of vapor in
the reservoir, then a steady state can be reached where the heat input $\dot{Q}$ is balanced
by the flow of *vapor* through the tube. This steady state will then last for a time τ
sufficient to vaporize the entire sample, $\tau = nL_v/\dot{Q}$, where n is the number of moles
of sample and L_v is the latent heat of vaporization. As mentioned, the lifetime of
the states on the "lower branch" has precisely this dependence on $\dot{Q}$. Further, the
experimental values of $\tau\dot{Q}$ have approximately the same value and temperature

(T_2) dependence as nL_v. During the time τ the temperature difference ΔT should be steady, as observed, and should be related to $\dot{Q}$ through the Poiseuille formula for the transport of vapor through the tube under the pressure difference $\Delta P = P_v(T_2) - P_v(T_1)$. That is, the "lower branch" should be defined by

$$\Delta T/\dot{Q} = 8\pi\eta RT/(\pi a^2)^2 P_v L_v(dP_v/dT)$$

where η is the vapor viscosity and R is the gas constant, and ΔT is assumed small. Evaluating this expression at $1.9°\mathrm{K}$ gives $\Delta T/\dot{Q} \simeq 150°\mathrm{K}\ \mathrm{W}^{-1}$, which compares favorably with the experimental value of $210°\mathrm{K}\ \mathrm{W}^{-1}$. Comparison at other temperatures is made rather difficult by the large values of ΔT.

No attempt was made to determine the maximum degree of superheating (i.e., an upper limit to the "upper branch") so that we expect a comparison with tensile strength data to show a smaller but comparable degree of superheating in our experiments. Estimates of the free energy change in both cases show this to be true.

Conclusions

We conclude that our interpretation of the data is substantially correct, and that superheating of He II does occur. The production of the metastable superheated state is made possible in this experiment by the nonlinear thermomechanical properties of the He II in the thin connecting tube. Failure to observe superheating in previous experiments[3] probably is a result of excessive vibration in the apparatus. One of the fascinating unsolved problems of liquid helium physics is the rather *small* value of the tensile strength. The problem is intimately linked to the mechanism for vapor-phase nucleation in the metastable state. Our data would seem to pose serious objections to those theories [1,2] that rely upon concentrations of vorticity as the nucleating agent, since our sample is in intimate contact with the enormous level of vorticity giving rise to the mutual friction in the connecting tube.

References

1. J.R. Shadly and R.D. Finch, *Phys. Rev.* **A3**, 780 (1971), and references cited therein.
2. K.L. McCloud, Ph.D. Dissertation, The Ohio State University, 1968, unpublished.
3. R. Eaton, W.D. Lee, and F.J. Agee, *Phys. Rev.* **A5**, 1342 (1972), and references cited therein.
4. W.E. Keller, *Helium-3 and Helium-4*, Plenum Press, New York (1969).
5. D.F. Brewer and D.O. Edwards, *Phil. Mag.* **6**, 1173 (1961).

Evaporation from Superfluid Helium [*]

Milton W. Cole [†]

Department of Physics, University of Toronto
Toronto, Ontario, Canada

Interest in evaporation from liquids arises because the measured variable, the angular and spectral distribution of atoms leaving the liquid surface, may serve as a probe of the properties of a single atom in the liquid. We have found[1] for superfluid helium that the liquid–vapor interface plays a mediating role in the microscopic process of evaporation which tends to obscure information about properties of the bulk liquid. However, in so doing, it allows us to obtain information about the surface which is also of intrinsic interest.

Our model assumes that the evaporation process occurs when a single quasiparticle arrives at the surface and transfers all of its energy to an atom, which enters the vapor region. This neglect of inelastic and multi- (quasi-) particle processes is a plausible first approximation,[2] and experiment (e.g., with monochromatic phonons generated by a superconducting tunnel junction) may soon determine its validity.

Invariance of the system with respect to translation parallel to the liquid–vapor interface implies that the parallel component of momentum is conserved in the transfer process. In combination with energy conservation, this guarantees that the energy E and momentum $\mathbf{p}$ of an evaporating atom are uniquely determined by the energy ω and momentum $\mathbf{q}$ of the incident quasiparticle (we take $\hbar = 1$). The reverse is not true,[‡] however, because q is a multivalued function of ω if ω exceeds the roton minimum energy $\Delta = \omega(q_0)$. The conservation laws are thus $p_{\parallel} = q_{\parallel}$ and $E = p^2/2m = \omega(q) - L$, where L is the binding energy per atom of the liquid.

The rate of evaporation per unit solid angle Ω at θ and per unit energy, $d^2\dot{N}/dE\, d\Omega$, is calculated[1] from a gas-kinetic analysis of the He quasiparticles to be

$$(d^2\dot{N}/dE\, d\Omega)(E, \theta) = \left(\sum_q t_{\mathbf{qp}} \right) e^{-\beta(E+L)} mE\, (\cos\theta)/4\pi^3 \qquad (1)$$

where m is the atomic mass and the parameter $t_{\mathbf{qp}}$ is the probability that an incident quasiparticle $\mathbf{q}$ transfers to a state of outgoing atom $\mathbf{p}$. The sum in Eq. (1) is over all branches of the excitation spectrum which can contribute to the particular final state under consideration. Only phonons contribute if E is less than $E_0 = \Delta - L \approx 1.5°\text{K}$, but for $E > E_0$ the two roton branches can contribute as well.

Previously I conjectured[1] that "anomalous" rotons ($q < q_0$, group velocity

---

[*] Research supported by the National Research Council of Canada.
[†] Present address: Physics Dept., University of Washington, Seattle, Washington.
[‡] Such an asymmetry occurs also for tunneling to a superconductor, which has two quasiparticle branches above the energy gap.

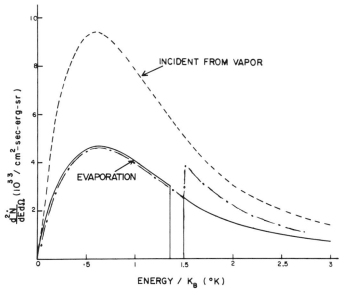

Fig. 1. Possible forms of the predicted evaporation spectrum, compared with the spectrum of atoms incident on the surface ($T = 0.6°$K, $\theta = 0$). The dash-dot curve includes all possible elastic processes. The full curve does not include a contribution from either anomalous rotons or high-energy phonons.

opposite to momentum) do not contribute to evaporation; the effective force at the surface is directed downward, as is $q_\perp$ of the incident quasiparticle, so that a final state of an outgoing atom ($p_\perp$ upward) is unlikely. A similar heuristic argument rules out a high-energy phonon contribution for $E > E_1 \approx 1.4°$K, for which the momentum **p** of an emitted atom would exceed the quasiparticle momentum **q**. If these exclusions are correct, a gap appears in the evaporation spectrum between E_1 and E_0 and the sum in (1) reduces to a single term arising from the "ordinary" roton branch ($q > q_0$).

The result (1) is remarkable in that no sign of the structure of the He excitation spectrum is evidenced; the result is independent of $\omega(q)$. This occurs because of the presence in the derivation of (1) of the quasiparticle group velocity, as well as because of an important geometric factor, consequent from parallel momentum conservation, which strongly favors low-q quasiparticles.*

Since the parameter t'_{pq} for the reverse process (incident vapor atom creating quasiparticle) equals t_{qp}, by time reversal symmetry, we find from (1) and gas-kinetic theory explicitly

$$d^2\dot{N}/dE\,d\Omega = d^2\dot{N}_{\mathrm{con}}/dE\,d\Omega \qquad (2)$$

where by the right-hand side we mean the distribution function for atoms condensing into any of the available quasiparticle states of the liquid. The sum in (1) can therefore never exceed one. Equation (2) is simply the requirement of detailed balance between liquid and vapor subsystems in equilibrium. This relationship can be exploited to allow study of evaporation indirectly by scattering atoms from the liquid surface.[3]

---

* This is analogous to the spectrum of thermionic emission from a metal, which does not exhibit the Fermi degeneracy of the metal's electrons.

Since we do not expect significant variation in the t parameters, Eq. (1) predicts a smooth spectrum except for the possible gap and steps discussed above. In Fig. 1 we depict various possibilities, taking t to be a different constant for each branch of the quasiparticle spectrum.

Direct measurement of this spectrum has been performed by King and co-workers.[4,5] The first results indicated that the evaporating atoms were on average considerably more energetic than vapor atoms incident on the surface. The analyses of Widom and co-workers[6,7] and Anderson[2] attributed this to a relatively intense roton contribution. However, the preliminary result of the second experiment[5] is a spectrum very similar to that of atoms incident from the vapor. This is qualitatively in agreement with our prediction, as seen in Fig. 1.

The evaporation problem can be explored with the transfer Hamiltonian approach [1,6,8] which has been a powerful tool for investigating electronic surface problems of solids. The ingredients we must supply to this formalism are the single-particle spectral weight functions[9] of liquid and vapor subsystems and certain matrix elements W_{qp} for the transfer through the inhomogeneous interfacial region ($\sim 10\,\text{Å}$ wide). What we obtain[1] in this way is an identification of the unknown parameter t,

$$t_{pq} = Z_q W_{qp} \tag{3}$$

where Z_q is the strength of the quasiparticle pole in the single-particle spectral weight function. Evaluation of (3) requires knowledge both of this function and of the surface density profile (to calculate W_{qp}). We note, however, that these two factors serve only to modulate the simple behavior shown in Fig. 1. It would be incorrect to conclude[10] from the hypothesis that the order parameter vanishes at the surface that the excitation picture we have described (based on bulk properties) is invalid. Of course the surface does weaken restrictions on scattering and decay processes imposed by isotropy of the bulk medium, and this aspect must be taken into account eventually. The same is true of inelastic processes ignored here which may in fact be important.[3]

References

1. M.W. Cole, *Phys. Rev. Lett.* **28**, 1622 (1972).
2. P.W. Anderson, *Phys. Lett.* **29A**, 563 (1969).
3. J. Eckardt, D.O. Edwards, F.M. Gasparini, and S.Y. Shen, this volume.
4. J.G. King and W.D. Johnston, *Phys. Rev. Lett.* **16**, 1191 (1966).
5. J.G. King, J. McWane, and R.F. Tinker, *Bull. Am. Phys. Soc.* **17**, 38 (1972).
6. A. Widom, *Phys. Lett.* **29A**, 96 (1969).
7. D.S. Hyman, M.O. Scully, and A. Widom, *Phys. Rev.* **186**, 231 (1969).
8. M.H. Cohen, L.M. Falicov, and J.C. Phillips, *Phys. Rev. Lett.* **8**, 31 (1969).
9. L.P. Kadanoff, in *Many Body Problem*, E.R. Caianiello, ed., Academic Press, New York (1964), Vol. 2.
10. A.A. Sobyanin, *Zh. Eksperim. i Teor. Fiz.* **61**, 433 (1971) [*Soviet Phys.—JETP* **34**, 229 (1972)].

Angular Distribution of Energy Flux Radiated from a Pulsed Thermal Source in He II below 0.3°K*

C.D. Pfeifer

Department of Physics, University of Wisconsin
Madison, Wisconsin

and

K. Luszczynski

Department of Physics, Washington University
St. Louis, Missouri

Introduction

The transfer of thermal energy from heated solids to He II has been investigated extensively and the basic features of the experimental data can be accounted for by various phonon coupling models.[1] Investigations of the energy flux radiated from pulsed point heaters to low-temperature He II ($T_0 < 300$ m°K) have produced data which are not entirely consistent with the predictions of these models.[2] Therefore a *helium blackbody source* has been proposed to explain the low-temperature data. According to this model, a layer of hot helium is generated at the heater surface. This helium source, at temperature T_s, absorbs the phonon radiation from the solid and reradiates excitations into the ambient bath. A calculation of the special distribution of the flux radiated by such a source,[2] assuming long lifetimes of the excitations, indicates that for T_s less than 600 m°K the source radiates isotropically a flux of phonons which propagate at the first-sound velocity. For T_s in excess of 600 m°K the flux contains additional components which propagate at 131 ± 15 m sec^{-1}. Effects due to the finite lifetime and dispersion of the excitations can, of course, modify this spectral distribution. The following experiment was designed to measure the angular distribution of excitations emitted by two heaters with different surfaces over a wide range of heater power densities to provide further information concerning the flow of energy from heated solids to low-temperature He II.

Experimental Considerations

A sample cell containing 100 ml of helium condensed from Grade A well helium gas was cooled to 230 m°K with a dilution refrigerator. The two heaters and six detectors which were immersed in the sample were carbon film strips with

* Supported by a grant from U.S. Army Research Office (Durham) No. DA-ARO-D-31-124-G922 and by the National Science Foundation No. GP-24572 and No. GP-15426, A-No.2.

367

active areas of 1 mm^2. One heater (H_1) and all the detectors were made from 5 KΩ per square IRC resistance strip material. The other heater (H_2) was made from a carbon film vacuum-deposited on a glass substrate. The thermal time constants of the heaters were of the order of microseconds, and that of heater H_1 was nearly twice that of H_2. The detectors were positioned at distances D of 1 and 2 cm from the heaters and at angles Φ ranging from 7 to 89° with respect to the heater normals. The heaters were supplied with dc pulses of 300 μsec duration which produced heater energy fluxes W_H ranging from 0.1 to 5 W cm^{-2} and heater film temperatures T_H from 6 to 30°K. A small dc current was maintained in each of the detectors and the voltage drop across the detector was monitored. As the energy flux propagated through the sample the detector films were heated slightly by the power absorbed, causing a transient decrease in the resistance of the film and a resultant decrease in the voltage drop. Using an empirical steady-state calibration relating the energy flux absorbed by each film W_A to the decrease in the voltage drop across the film, a flux of 8 $\times$ 10^{-9} W cm^{-2} could be observed with a signal-to-noise ratio of one.

Data

For heater energy flux values W_H less than a characteristic value W_H^{SE} the detected signals reproduced the shape of the input heater pulse with some rounding of the leading and trailing edges due to the thermal time constants of the system. The onset of the detected pulse traveled at velocity $v_1 = 235 \pm 10$ m sec^{-1} with no delay at the heater. Above W_H^{SE} the signal developed a second edge structure associated with a slower flux component. Some typical oscilloscope displays of this signal at two of the six detectors used, $C(D = 1$ cm, $\Phi = 7°)$ and $F(D = 1$ cm, $\Phi = 86°)$, are shown in Fig. 1 for heater H_1. The heater energy flux for these photographs was $W_H = 4.08$ W cm^{-2}. The vertical sensitivity to absorbed power and the horizontal sweep are indicated on the photographs. The second edge signal traveled at velocity $V_2 = 130 \pm 30$ m sec^{-1} and it was delayed at the heater a time Υ which increased

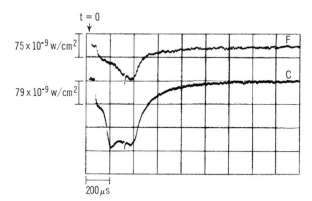

Fig. 1. Photographs of the signals observed at detectors C ($D = 1$ cm, $\Phi = 7°$) and F ($D = 1$ cm, $\Phi = 86°$) for energy flux $W_H = 4.08$ W/cm^{-2} at heater H_1. The input pulse duration was 300 μsec and the sample temperature was 0.23°K. The horizontal sweep rate and vertical detector sensitivities to absorbed power are shown.

with detector angle ϕ; for example, $\Upsilon(7°) = 60 \pm 20\,\mu\text{sec}$ and $\Upsilon(86°) = 123 \pm 48$ μ sec. The value $W_H^{S,E}$ at which the second edge was first observed was different for the two heaters. For H_1, $W_H^{SE} = 1.5\,\text{W cm}^{-2}$ and for H_2, $W_H^{SE} = 0.7\,\text{W cm}^{-2}$. Furthermore, the heater energy flux value at which the second edge was first observable was higher at larger detector angles, and the second edge structure was less pronounced at these large angles. As W_H was increased above W_H^{SE} the amplitude of the plateau between the signal onset and the second edge remained constant at each detector.

Observations of the power absorbed at the end of the signal pulse W_A indicated that for W_H less than 2 W cm^{-2} the energy flux decreased with distance D from the source as $1/D^2$. For larger W_H the energy flux decreased less rapidly with distance, approaching a $1/D$ dependence. Plots of the observed signal W_A as a function of W_H are shown in Fig. 2 for detectors $C(D = 1\,\text{cm}, \Phi = 7°)$ and $F(D = 1\,\text{cm}, \Phi = 86°)$ and heater H_1. Below $W_H = 1.5\,\text{W cm}^{-2}$ the plots of $W_A(W_H)$ for each heater–detector pair exhibited two linear regions. For $W_H < 0.4\,\text{W cm}^{-2}$ (region 1) $W_A = m_1 W_H$ and for $W_H > 0.8\,\text{W cm}^{-2}$ (region 2) $W_A = m_2 W_H + b$. The solid lines in Fig. 2 are least squares fits to the data in the two regions. For both heaters the relative magnitude of slope m_1 with respect to that of m_2 decreased with increasing detector angle. In particular, $m_1 > m_2$ for $\Phi < 60°$ and $m_1 < m_2$ for $\Phi > 60°$.

The absorbed flux W_A exhibited a strong angular dependence which varied as $(\cos \Phi)/D^2$ in region 1. This dependence is displayed in Fig. 3, which shows a plot of the *ratio* of the energy flux absorbed at each detector to that absorbed at detector

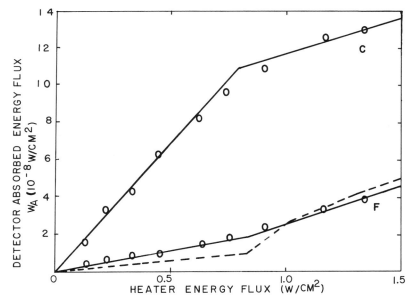

Fig. 2. Plots of the energy flux absorbed near the end of the detected pulse at detectors C and F as a function of heater flux W_H for heater H_1. The solid lines are least squares fits to the experimental data in region 1 ($W_H < 0.4\,\text{W cm}^{-2}$) and region 2 ($W_H > 0.8\,\text{W cm}^{-2}$). The dashed line is the flux absorbed at detector F predicted from the source evolution model described in the text.

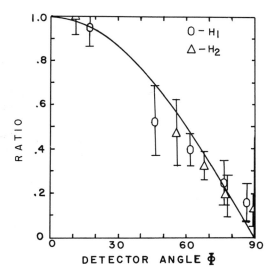

Fig. 3. Plot of the ratio of the energy flux absorbed at each detector to that absorbed at detector C in region 1 for both heaters. The ratio has been normalized as indicated in the text. The solid line is a plot of cos Φ, where Φ is the detector angle measured from the heater normal.

$C(\Phi = 7°)$, given by $W_A(\Phi) D_C^2(\cos \Theta_\Phi)/W_A(C) D_\Phi^2(\cos \Theta_C)$, where Θ_Φ is the inclination of the *detector surface* to the incident flux. The solid line is a plot of cos Φ.

Discussion

As mentioned above, the experimental observations can be divided into low- and high-energy flux regions. In region 1 the observed signals are consistent with diffuse radiation from the heater plane of a flux of phonons which interact at most through small-angle scattering processes. In region 2 the angular dependence of the flux changes, and a number of new effects become apparent. The experimental data have been interpreted in terms of the evolution of a *helium blackbody source* which grows from the heater surface. The angular distribution of the energy flux radiated from such a source has been calculated using a formalism similar to that used in the electromagnetic case.[3] In region 1 the source conforms to the heater plane, while in region 2 it evolves to a hemispherical bubble with its base on the heater surface. Using the observed signal at the $\Phi = 7°$ detector to normalize the calculation, this semiempirical model yields absorbed power plots which are in good agreement with the experimentally determined values for $W_H < 1.2 \text{W cm}^{-2}$. In particular, the Φ dependence of the slopes m_1 and m_2 discussed in relation to the $W_A(W_H)$ plots of Fig. 2 is reproduced. [The calculated values of $W_A(W_H)$ for detector F are shown by the dashed line in Fig. 2.] It is interesting to note that the heater flux value $W_H = 0.5$ W cm^{-2}, at which the angular distribution of the radiated flux begins to change appreciably, corresponds to a helium blackbody source temperature of 600 m°K.

This is the temperature at which rotons begin to contribute significantly to the energy density of the blackbody source.

The second edge structure can be understood in terms of a rapid increase in the temperature of the helium source due to a second-sound pulse *within* the source volume during the early portion of the heater pulse. According to this model, the source can radiate a significant number of excitations at 130 m sec^{-1} before it expands and cools to its steady-state configuration determined by the heater power density W_H. A detailed discussion of this model is contained in Ref. 3.

Acknowledgments

The suggestions of Prof. E. Feenberg and Prof. R.W. Guernsey, Jr. and the support of Prof. R.E. Norberg are gratefully acknowledged.

References

1. G.L. Pollack, *Rev. Mod. Phys.* **41**, 48 (1969).
2. R.W. Guernsey and K. Luszczynski, *Phys. Rev.* **3A**, 1052 (1971).
3. C.D. Pfeifer, Ph.D. Thesis, Washington University, St. Louis, 1971, unpublished; *Dissertation Abstracts* **B32**, 5399B (1972).

Electric Field Amplification of He II Luminescence Below 0.8°K*

Huey A. Roberts and Frank L. Hereford

Department of Physics, University of Virginia
Charlottesville, Virginia

The passage of alpha particles through liquid He II produces luminescence, part of which derives from metastable atomic or molecular systems with lifetimes which become increasingly longer as the temperature is reduced below T_λ.[1] This is caused by the increased diffusion coefficient of the metastables ($\sim e^{\Delta/kT}$) which enables them to escape prompt collision-induced destruction near the alpha-particle "track," where a high density of collision partners is initially available. The resulting delay of their radiative destruction causes the intensity of scintillation pulses from the fluid (radiation emitted within approximately 10^{-6} sec of alpha emission) to decrease below about 1.7°K to a constant value attained at about 0.8°K. At this temperature and below all of the metastables escape prompt destruction. The, T dependence can be accounted for quantitatively by calculating the fraction of metastables which undergo prompt radiative destruction,[2] assuming the diffusion coefficients of both metastables and their collision partners (ions or other metastables) to be porportional to $e^{\Delta/kT}$.

We report here measurements of the intensity of the *total luminescence* (time integrated) as well as the *pulse* intensity ($\tau = 1.25 \times 10^{-6}$ sec) for temperatures down to 0.3°K, and a study of the effect of an electric field applied in the region of the alpha-particle "tracks." The results show that the variation of the *total* intensity with temperature and its behavior under an applied field are strikingly different from the corresponding variations of the *pulse* intensity.

It is important to recognize that weak alpha sources ($\sim 10^3$ sec^{-1}) were employed so that luminescence involved processes associated with single alphas. Other workers[3] have studied the He II luminescence spectrum with much stronger electron beams ($\sim 10^{13}$ sec^{-1}). These spectroscopic observations disclosed the presence of a$^3 \sum_u^+$ and other metastable states,[3] but exhibited no strong temperature dependence, presumably because of the "steady-state" nature of the luminescence emanating from the high density of excitation in the source region.

The cryogenic equipment and sample chambers have been described previously.[2] In a typical experiment a Po^{210} alpha source was deposited on a 0.3-cm segment of a nichrome wire of 0.015 cm diameter mounted on the axis of a cylindrical sample chamber which was filled with liquid helium. The chamber walls were POPOP-coated to shift the vacuum-UV helium luminescence to the visible region for detection

* Work supported by the U.S. Army Research Office and the National Science Foundation (through the Center for Advanced Studies, University of Virginia).

through a sapphire window by a photomultiplier. An appropriate configuration of cylindrical, gold-plated grids surrounded the source wire for application of electric fields. Scintillation pulse intensity was measured by standard pulse-height analysis methods and the total luminescence by a voltage-to-frequency conversion technique.

The results can be summarized as follows (discussion is limited to $T < T_\lambda$).

(1) With no applied field the *pulse* intensity decreases as described above for $T \lesssim 1.7°K$, but the *total* intensity remains roughly constant down to $T \approx 1.2°K$ (Fig. 1). As T is lowered further it drops to a minimum near 0.6°K and then rises again. If the chamber radius and diameter are doubled, the pulse intensity is unaffected; however, the total intensity below 0.8°K rises significantly (data not shown), indicating increased radiative destruction in the fluid of metastables, which apparently are quenched at the walls in the case of the smaller chamber.

The solid curve for the pulse intensity in Fig. 1 represents the calculation mentioned previously of the prompt destruction fraction with $\Delta/k = 8.65°K$. The solid curve for the total intensity represents an extension of this destruction vs. diffusion model to a quasistationary diffusion process, assumed to govern delayed radiative destruction of metastables between the source and the walls. It accounts only for the

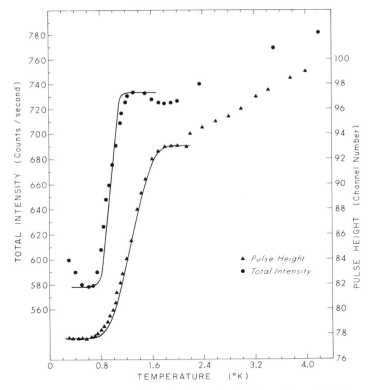

Fig. 1. Total luminescence intensity (left scale, circles) and the scintillation pulse intensity (right scale, triangles) vs. temperature. Absolute values of the two curves cannot be compared. The solid curves were calculated as described in the text.

gross features of the total luminescence, leaving unexplained the increase below 0.6°K and the weak maximum near 1.4°K. The fact that the 22% drop in total intensity is greater than the 16% drop in pulse intensity indicates a contribution to the total luminescence from an additional metastable species which, even at T_λ, survives beyond the pulse time τ, and undergoes quenching at the walls at the lower temperatures.

(2) The application of an electric field in the alpha source region reduces the pulse intensity at all temperatures. The effect of the electric field on the total intensity, however, is very different above and below approximately 0.8°K. At higher temperatures a continuous decrease with increasing field strength was observed (Fig. 2), similar to that for the pulse intensity. We assume that extraction of ions from the alpha "track," thereby preventing their fast recombination within the pulse time, is responsible.

In contrast to this behavior, below 0.8°K the field is seen to increase the total intensity. The results in Fig. 2 were obtained with a source wire diameter of 0.035 cm, for which an applied potential of 1000 V produced a field of approximately 13,300 V cm^{-1} at the center of the alpha- "track" region. A positive V_s corresponded to pulling positive ions away from the source. Changing the field direction (data not shown) produced no effect above 1.6°K but intensified the peak near 1.4°K. We believe that this reflects the fact that near 1.4°K only negative ions form stable charged vorticity in the presence of a high field,[4] thereby providing an increased

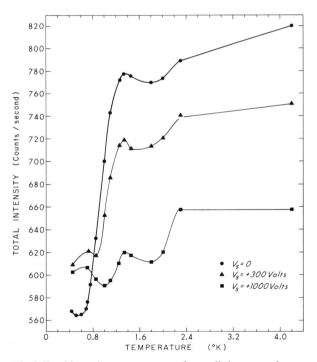

Fig. 2. Total intensity vs. temperature for applied source voltages (relative to the first grid) of 0, +300, and +1000 V.

charge density in the fluid for collision-induced destruction of metastables as discussed below.

The variation of both the pulse and total luminescence with applied field is
shown for $T = 0.68°K$ in Fig. 3. The source wire diameter was 0.015 cm in this case,
yielding approximately 16,500 V cm^{-1} at the center of the source region for $V_s = 1000$
V. The lower half of the figure shows that the total intensity rises to a maximum near
250 V; however, the pulse intensity (upper half) decreases continuously with applied
voltage.

(3) The field effects above and below 0.8°K are different in another important
respect. Rectangular pulses of equal width and spacing were applied to the source
region, and the pulse width was varied to determine how fast the luminescence
responded. The results showed that at 1.24°K the intensity dropped by one-half the
"constant field decrease" (the pulsed field on "on" one-half the time) within about
10^{-5} sec. On the other hand, at 0.68°K approximately 8×10^{-4} sec was required for
the intensity to *increase* by one-half the "constant field increase." This latter time
is the order of that required for the generation of charged vortex rings and their
traversal of centimeter distances.[5]

We interpret the field-induced amplification of the total luminescence below
0.8°K as resulting from increased radiative destruction of metastables in collisional
interaction with charged vorticity. Applied fields are known to produce both positively and negatively charged vorticity at these low temperatures, and the requirement

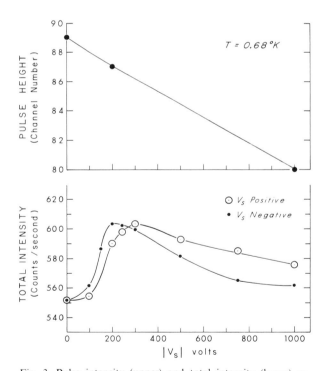

Fig. 3. Pulse intensity (upper) and total intensity (lower) vs.
applied source voltage at 0.68°K.

of about 10^{-3} sec for the amplification effect to become established is consistent with this interpretation. Since the metastables survive the pulse time τ below $0.8°K$ in any case, the decreasing pulse intensity under an applied field reflects only the extraction of ions that would otherwise undergo recombination yielding prompt luminescence.

Whether the metastable systems are captured in the tangle of charged vortex lines and rings produced by the field cannot be determined from the results. However, we suggest that the increase in luminescence *for zero field* below $0.6°K$ (Fig. 1) may be due to metastable capture by vorticity produced by the alpha particles. Charged vortex rings emanating from alpha sources have, in fact, been observed below $0.5°K$ with no applied electric field.[5] If the metastables produce bubble states, as is thought to be the case, their vortex capture below $0.6°K$ would inhibit their diffusion to the walls and enhance the probability of their radiative destruction in the fluid.

References

1. M.R. Fischbach, H.A. Roberts, and F.L. Hereford, *Phys. Rev. Lett.* **23**, 462 (1969).
2. Huey A. Roberts, William D. Lee, and Frank L. Hereford, *Phys. Rev.* **A4**, 2380 (1971).
3. M. Stockton, J.W. Keto, and W.A. Fitzsimmons, *Phys. Rev.* **A5**, 372 (1972), and references cited therein.
4. L. Bruschi, P. Mazzoldi, and M. Santini, *Phys. Rev. Lett.* **21**, 1738 (1968), and references cited therein.
5. G.W. Rayfield, *Phys. Rev.* **168**, 222 (1968).

Correlation Length and Compressibility of ^{4}He Near the Critical Point

A. Tominaga and Y. Narahara

Department of Applied Physics, Tokyo University of Education
Bunkyoku, Tokyo, Japan

Measurements of the light scattering at a right angle by ^{4}He near the critical point T_c have been made as a function of temperature. From these measurements we estimated the critical index γ', which characterizes the temperature dependence of the isothermal compressibility times the density squared along the liquid side of the coexistence curve, and also derived the correlation length of helium as a function of temperature.

Our experimental setup consists of a single-mode helium–neon laser of 30 mW (wavelength = 6328 Å), an attenuator, and the associated focusing optics, a sample cell in the cryostat, and a cooled photomultiplier (Hamamatsu 1P21) under a photon-counting mode. We reduced the laser power to about 3 mW, using the variable attenuator in front of the cryostat. During experiments we monitored and manually adjusted this power to better than 1%. As shown in Fig. 1, the sample cell was made of OFHC copper with four plane parallel windows coated with antireflection dielectric. The diameter of each slit is 2 mm. A germanium resistor as a thermometer and a heater resistor are embedded in the copper block. The temperature of the cell was regulated to better than 20 $\mu°$K by an electronic temperature controller.[1] The pressure of the sample was measured with a fused quartz bourdon gauge with sensitivity 25 μm Hg.

We determined the critical pressure P_c and temperature T_c by observing maximum scattered intensities. Thus determined, T_c and P_c are 5189.863 $\pm$ 0.03 m°K and 1706.008 $\pm$ 0.025 mm Hg in T_{58}, respectively. These values are closest to values obtained by Edwards.*

According to the Einstein and Ornstein–Zernike theories, the scattered intensities are proportional to

$$(\partial\rho/\partial\mu)_T/[1 + (K\xi)^2]^{1-\eta/2}$$

where ρ and μ are density and chemical potential of the sample, K is the scattering wave vector, ξ is the long-range correlation length, and η is a measure of the departure from Ornstein–Zernike behavior proposed by Fisher.[3]

The incoming beam was focused just below the liquid–vapor phase boundary. We measured the scattered light intensities in the temperature region $2 \times 10^{-5} < 1 - T/T_c < 4 \times 10^{-2}$. Our results are shown in Fig. 2. Since η is expected to be very

---

* See Table 1 of Ref. 2.

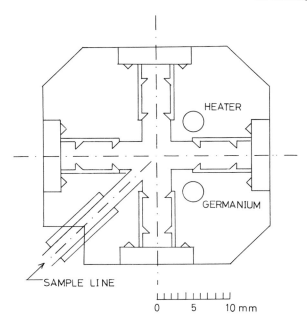

Fig. 1. The horizontal cross section of the sample cell made
of OFHC copper.

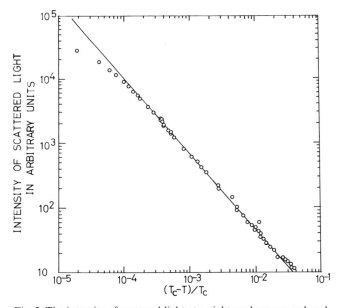

Fig. 2. The intensity of scattered light at a right angle versus reduced
temperature along the liquid side of the coexistence curve. The solid
line corresponds to the zero-angle limit and $\gamma' = 1.180$.

small according to Fisher and our present data are not sufficiently accurate to detect it directly, we assumed η is equal to zero. Assuming $(\partial \rho / \partial \mu)_T = \Gamma' (1 - T/T_c)^{-\gamma'}$ and that the scattered intensities are directly proportional to $(\partial \rho / \partial \mu)_T$ in most of our temperature region where $K\xi \ll 1$, $[(1 - T/T_c) > 4 \times 10^{-4}]$, we obtain $\gamma' = 1.180 \pm 0.01$. We calculated $(\partial \rho / \partial \mu)_T = \rho^2 K_T$, where $K_T = (1/\rho)(\partial \rho / \partial p)_T$, using results due to Roach[4] and plotted it against $(1 - T/T_c)$ on a log-log graph. By drawing a straight line of $(1 - T/T_c)^{-1.180}$ in the most reasonable fashion to his results, we estimated Γ' as $(6.5 \pm 2.5) \times 10^{-9} \, \text{g}^2 \, \text{dyn}^{-1} \, \text{cm}^{-4}$. Comparing this with the value of Γ' obtained by Wallace and Meyer[5] on ³He, $8.8 \times 10^{-9} \, \text{g}^2 \, \text{dyn}^{-1} \, \text{cm}^{-4}$, we may say that it is a reasonable value.

Since our estimation of the attenuation[6] of the incoming beam in the sample of helium is negligibly small (about 1% at our temperature closest to T_c) and also since the beam bending due to the gravitational effect is small, we assume that deviations from the straight line in Fig. 2 for $(1 - T/T_c) < 4 \times 10^{-4}$ are entirely due to the temperature dependence of ξ (Fig. 3). Assuming $\xi = \xi_0 (1 - T/T_c)^{-\nu'}$, we obtain $\nu' = 0.59^{-0.04}_{+0.14}$ and $\xi_0 = 1.59 \pm^{0.95}_{1.3} \, \text{Å}$. Within the experimental uncertainty, we find $\nu' = \gamma'/2$. Our value of $\gamma' = 1.180$ is approximately consistent with 1.14 ± 0.10 of Roach[4] for ⁴He and 1.17 ± 0.03 of Wallace and Meyer[5] for ³He. Our value is far from 1.4, the preliminary results of Garfunkel et al.[7] for ⁴He if $\gamma = \gamma'$. This discrepancy may be caused by a difference of the background scattering. We do not agree with their argument that all the scattering at $T/T_c - 1 \simeq 10^{-2}$ is parasitic and can be regarded as background (see the scattered intensity at $1 - T/T_c = 10^{-2}$ in Fig. 2). If we apply this argument to our data, the slope of our data, namely γ', becomes larger up to about 1.4, which is the value obtained by Garfunkel et al. We took as background scattering the scattering from the empty cell near the critical temperature and the helium gas at about 2 atm at nitrogen temperature. The values in these two cases are equal (about 10.4 counts sec⁻¹). Webb and Fairbank[8] obtained $\gamma = 1.26 \pm 0.06$ for ³He. If we accept these data, it seems that the critical index γ' of helium is a little smaller than the 1.20,[9] 1.21,[10] 1.23,[6] and 1.219,[11] of Ar, Xe, SF_6, and CO_2, respectively (we assume $\gamma = \gamma'$ for some cases).

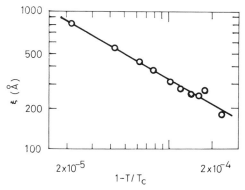

Fig. 3. Temperature dependence of the correlation length along the liquid side of the coexistence curve. The data points were obtained assuming the Ornstein–Zernike theory and $\gamma' = 1.180$. The solid line shows $\xi = 1.59 (1 - T/T_c)^{-0.59}$ in angstroms.

Among critical indices there are the inequalities

$$\alpha' + 2\beta + \gamma' \geqq 2 \qquad \text{and} \qquad \gamma' \geqq \beta(\delta - 1)$$

Using the 0.127 of Moldover[12] or the 0.128 ± 0.008 of Barmatz[13] for α', 0.352 ± 0.003 of Roach and Douglass[14] for β, and our value for γ', we can satisfy the first relationship. From the second inequality we obtain $\delta \leqq 4.35$, which agrees with $\delta = 4.218$ of Kiang's calculation[15] on the liquid droplet model.

For $T > T_c$ our preliminary experiment along the critical isochore shows $\gamma = 1.21 \pm 0.05$ for $3 \times 10^{-4} < T/T_c - 1 < 10^{-2}$.

Further measurements are in progress.

References

1. A. Tominaga, *Cryogenics* **12**, 389 (1972).
2. H.A. Kierstead, *Phys. Rev.* **A3**, 329 (1971).
3. M.E. Fisher, *J. Math. Phys.* **5**, 944 (1964).
4. P.R. Roach, *Phys. Rev.* **170**, 213 (1968).
5. B. Wallace, Jr., and H. Meyer, *Phys. Rev.* **A2**, 1563 (1970).
6. V.G. Pugrielli and N.C. Ford, Jr., *Phys. Rev. Lett.* **25**, 143 (1970).
7. M.P. Garfunkel, W.I. Goldburg, and C.M. Huang, in *Proc. 12th Intern. Conf. Low Temp. Phys., 1970,* Academic Press of Japan, Tokyo (1971), p. 59.
8. J.P. Webb, Ph.D. Thesis, Stanford Univ., 1968.
9. H.D. Bale, W.C. Dobbs, J.S. Lin, and P.W. Schmidt, *Phys. Rev. Lett.* **25**, 1556 (1970).
10. I.W. Smith, M. Giglio, and G.B. Benedek, *Phys. Rev. Lett.* **27**, 1556 (1971).
11. J.H. Lunacek and D.S. Cannel, *Phys. Rev. Lett.* **27**, 841 (1971).
12. M.R. Moldover, *Phys. Rev.* **182**, 342 (1969).
13. M. Barmatz, *Phys. Rev. Lett.* **24**, 651 (1970).
14. P.R. Roach and D.H. Douglass, *Phys. Rev. Lett.* **17**, 1083 (1966).
15. C.S. Kiang, *Phys. Rev. Lett.* **24**, 47 (1970).

Coexistence Curve and Parametric Equation of State for ^{4}He Near Its Critical Point*

Henry A. Kierstead

Argonne National Laboratory
Argonne, Illinois

Recent advances in the theory of critical phenomena have emphasized the need for high-quality experimental data and, through the scaling law, have provided a means for analyzing the data in a simple fashion. Available data on the thermodynamic properties of ^{4}He fall short of satisfying this need. Therefore we have measured the pressure coefficient $(\partial P/\partial T)_\rho$ along 29 isochores in the density range $-0.2 < \Delta\rho < 0.2$ and the temperature range $-0.02 < t < 0.032$, where $\Delta\rho = (\rho - \rho_c)/\rho_c$ and $t = (T - T_c)/T_c$.

The apparatus and experimental techniques were the same as were used in the author's study of the critical isochore.[1] They were designed to achieve the highest possible resolution in pressure and temperature, minimize gravity effects, and maintain a reasonably short thermal time constant. Temperatures were measured with germanium resistance thermometers calibrated on the NBS Provisional Scale $2-20°$K (1965)[2] (the acoustical thermometer scale). The temperature resolution was $0.3~\mu°$K but the calibration accuracy of the absolute temperatures was ± 2 m°K. The accuracy of the density measurements was about $\pm 0.02\%$. The accuracy of the pressure coefficient was limited by the resolution of the differential pressure gauge to about $\pm 0.1\%$. Numerical calculations indicate that gravity effects are negligible in these measurements. The data are presented in Figs. 1 and 2.

The abrupt changes in slope of the isochores (except for the critical isochore) as they passed from the two-phase region to the one-phase region were used to locate 28 points on the coexistence curve. They are tabulated in Table I and plotted in Fig. 3. They were fitted by the method of least squares to the equation

$$|\Delta\rho| = B(-t)^\beta \qquad (1)$$

The best values for the parameters were found to be

$$B = 1.395 \pm 0.020 \qquad (2)$$

$$\beta = 0.3555 \pm 0.0028 \qquad (3)$$

$$\rho_c = 69.580 \pm 0.020 \qquad (4)$$

---

* Work performed under the auspices of the U.S. Atomic Energy Commission.

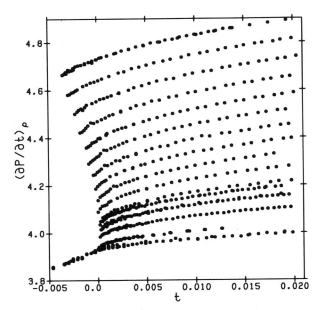

Fig. 1. Pressure coefficient $(\partial P/\partial T)_\rho$ of ⁴He in units of P_c/T_c
=328.209 Torr/°K. The values of $|\Delta\rho$ for the isochores shown are,
from top to bottom: 0.1993, 0.1844, 0.1696, 0.1546, 0.1400, 0.1265,
0.1127, 0.0994, 0.0858, 0.0720, 0.0571, 0.0532, 0.0437, 0.0305,
0.0114, and 0.0015. The isochore for $\Delta\rho = 0.0114$ is one reported
in Ref. 1.

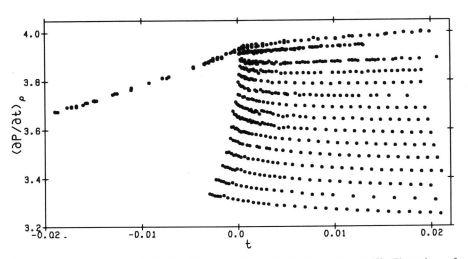

Fig. 2. Pressure coefficient $(\partial P/\partial T)_\rho$ of ⁴He in units of $P_c/T_c = 328.209$ Torr/°K. The values of
$\Delta\rho$ for the isochores shown are, from top to bottom: 0.0015, − 0.0094, − 0.0279, − 0.0418, − 0.0555,
− 0.0703, − 0.0839, − 0.0977, − 0.1112, − 0.1249, − 0.1386, − 0.1525, − 0.1670, − 0.1811, and
− 0.1953. The isochore for $\Delta\rho = -0.0094$ is one reported in Ref. 1.

Table 1. Coexistence Curve of ^{4}He*

$\Delta\rho$	$10^3 t$
− 0.195319	− 3.9440
− 0.181091	− 3.2223
− 0.167007	− 2.5613
− 0.152491	− 1.9810
− 0.138551	− 1.4823
− 0.124897	− 1.1287
− 0.111244	− 0.8178
− 0.097734	− 0.5549
− 0.083937	− 0.4131
− 0.070284	− 0.1796
− 0.055481	− 0.1353
− 0.041828	− 0.0276
− 0.027887	− 0.0686
0.030463	− 0.1349
0.043541	− 0.0414
0.043828	− 0.0828
0.053170	− 0.0968
0.057051	− 0.1223
0.071997	− 0.2390
0.085794	− 0.4065
0.099448	− 0.5888
0.112670	− 0.8393
0.126467	− 1.1674
0.139976	− 1.5448
0.154636	− 2.0518
0.169582	− 2.6481
0.184385	− 3.3809
0.199332	− 4.1968

* Densities and temperatures have been reduced using $\rho_c = 69.580$ mg cm^{-3} and $T_c = 5.19828°$K.

This value of ρ_c is more accurate and slightly smaller than our earlier value[1] (69.64 ± 0.07) and agrees exactly with Moldover's value[3] (69.58 ± 0.07). It is consistent with that of el Hadi et al.[4] (69.76 ± 0.20) but not with those of Edwards[5] (69.451 ± 0.069)* and Roach[6] (69.0). The values of B and β given above are consistent with those of Edwards[5] (1.4166 ± 0.0032 and 0.3598 ± 0.0007) and Roach[6] (1.42 and 0.354 ± 0.010).† Edwards's $\Delta\rho$'s are 0.001–0.002 smaller in magnitude than ours, whereas Roach's are 0.002–0.005 larger.

Agreement of our measurements with Eq. (1) is shown in Fig. 3. The greater errors for small values of $\Delta\rho$ result from the fact that the discontinuity in $(\partial P/\partial T)_\rho$ approaches zero as $\Delta\rho$ approaches zero.

The 1551 pressure coefficient measurements were fitted to the parametric equations of the "linear model" proposed by Schofield et al.[7] This model is defined

---

* We have increased Edwards' value of ρ_c by 0.128 mg cm^{-3} because of a change in the value for the molar polarizability used by Edwards. We are indebted to him for supplying this information.

† We are indebted to Dr. Roach for extensive discussions of his experiment.

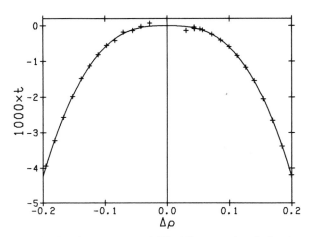

Fig. 3. Coexistence curve of ⁴He. The curve is calculated
from Eq. (1).

by the equations

$$t = r(1 - b^2\theta^2) \tag{5}$$

$$\Delta\rho = m\theta r^\beta \tag{6}$$

$$\Delta\mu \equiv \mu - \mu_c = ar^{\gamma+\beta}\theta(1 - \theta^2) \tag{7}$$

in which μ_c is the chemical potential on the critical isochore, β and γ are critical exponents,[8] and r and θ are the parametric variables. The constants b and m are defined by

$$b^2 = (\gamma - 2\beta)/[\gamma(1 - 2\beta)], \qquad m = B(b^2 - 1)^\beta \tag{8}$$

The pressure can be shown to be given by

$$P = -A_0(t) + a\theta(1 - \theta^2)r^{\gamma+\beta} + (am/2b^4)(L_1 + L_2\theta^2 + L_3\theta^4)r^{\gamma+2\beta} \tag{9}$$

where

$$L_1 = (\gamma - 2\beta - b^2\alpha\gamma)/[\alpha(1 - \alpha)(2 - \alpha)] \tag{10}$$

$$L_2 = \{[b^4\alpha(1 - 2\beta) - b^2(\gamma - 2\beta)]/[\alpha(1 - \alpha)]\} + 2b^4 \tag{11}$$

$$L_3 = b^4(1 - 2\beta - 2\alpha)/\alpha \tag{12}$$

$$\alpha = 2 - \gamma - 2\beta \tag{13}$$

and $A_0(t)$ is the nonsingular part of the Helmholtz free energy. It was represented by a cubic function of t:

$$-A_0(t) = 1 + C_1 t + C_2 t^2 + C_3 t^3 \tag{14}$$

The best-fit values of the parameters are given in Table II. The standard deviation of the fit was 4.9×10^{-3}, about 1.5 times the experimental error. The fit was generally

Table II. Parameters of the Model Equations

Parameter	Value	Parameter	Value
Coexistence curve [Eq. (1)]			
ρ_c	69.580 ± 0.20 mg cm^{-3}	B	1.395 ± 0.020
β	0.3554 ± 0.0028		$-$
Linear model [Eq. (5)–(14)]			
γ	1.1743 ± 0.0005	a	6.053 ± 0.016
b	1.1683 ± 0.0042	C_1	3.93125 ± 0.00015
α	0.1148 ± 0.0056	C_2	-1.724 ± 0.31
m	0.975 ± 0.018	C_3	-5.6 ± 0.6
Best parametric model [Eq. (15)–(20)]			
γ	1.2223 ± 0.0017	G_1	6.921 ± 0.317
b	1.2030 ± 0.0051	G_2	-17.232 ± 0.863
α	0.0668 ± 0.0058	G_3	11.235 ± 0.719
m	1.048 ± 0.020	G_4	2.052 ± 0.571
F_1	9.053 ± 0.119	G_5	-0.676 ± 0.288
F_2	0.514 ± 0.029	C_1	3.93023 ± 0.00020
H_1	-4.291 ± 0.188	C_2	-7.02 ± 0.37
H_2	-1.248 ± 0.053	C_3	-1.7 ± 0.9

good, but there were systematic deviations, and the deviations were not symmetric in $\Delta\rho$. Therefore, to obtain a better fit, it was necessary to abandon Eqs. (7) and (9) and construct a trial equation similar to Eq. (9) with more general functions of θ:

$$P = -A_0(t) + \theta(1 - \theta^2) F(\theta) r^{\gamma+\beta} + [G(\theta^2) + \theta(1 - \theta^2) H(\theta^2)] r^{\gamma+2\beta} \quad (15)$$

Since at constant temperature

$$dP = (1 + \Delta\rho) d\mu \quad (16)$$

it is easily shown that the singular term of lowest order in r must be the same in $\Delta\mu$ as in P. Hence

$$\Delta\mu = \theta(1 - \theta^2) F(\theta) r^{\gamma+\beta} + O(r^{\gamma+2\beta}) \quad (17)$$

After trying various functions for F, G, and H we obtained a good fit with the polynomials

$$F(\theta) = F_1 + F_2\theta \quad (18)$$

$$G(\theta^2) = G_1 + G_2\theta^2 + G_3\theta^4 + G_4\theta^6 + G_5\theta^8 \quad (19)$$

$$H(\theta^2) = H_1 + H_2\theta^2 \quad (20)$$

$A_0(t)$ was fitted by a cubic as before. Best values of the 14 parameters are given in Table II. The standard deviation was 3.0×10^{-3}, about equal to the experimental error, so no significantly better fit can be obtained by adding more terms. In fact, nearly as good a fit can be obtained without the two highest terms in G and the highest term in H. The improvement in fit over that for the linear model is due principally to the term in $F_2\theta$ in Eq. (18). Because of this term, the chemical potential and the pressure are not antisymmetric in $\Delta\rho$ even very close to the critical point. In this respect the model does not obey the scaling hypothesis.

The value of the exponent α calculated from this fit (0.067 ± 0.006) agrees very well with that obtained by Barmatz and Hohenberg[9] (0.07 ± 0.005) directly from the heat capacity data. The value of α derived from our linear model fit (0.115 ± 0.006) is nearly twice as large, and this difference illustrates the fact that the values of critical exponents derived from experimental data are model dependent.*

The three isochores reported earlier by us agree very well with Eq. (15), using the same values of the parameters. The pressure measurements of Roach deviate from Eq. (15) by no more than 0.1 %, provided Roach's values of the critical constants ($\rho_c = 69.0$ mg cm^{-3}, $T_c = 5.189^\circ$K, $P_c = 1705.0$ Torr) are used in reducing his data.

References

1. H.A. Kierstead, *Phys. Rev.* **A3**, 329 (1971).
2. H. Plumb and G. Cataland, *Metrologia* **2**, 127 (1966).
3. M.R. Moldover, *Phys. Rev.* **182**, 342 (1969).
4. Z.E.H.A. el Hadi, M. Durieux, and H. van Dijk, *Physica* **41**, 305 (1969).
5. M.H. Edwards, in *Proc. 11th Intern. Conf. Low Temp. Phys., 1968*, Univ. of St. Andrews, Scotland (1969), p. 231.
6. P.R. Roach, *Phys. Rev.* **170**, 213 (1968).
7. P. Schofield, J.D. Litster, and J.T. Ho, *Phys. Rev. Lett.* **19**, 1098 (1969).
8. M.E. Fisher, *Rep. Progr. Phys.* **30**, 615 (1967).
9. M. Barmatz and P.C. Hohenberg, *Phys. Rev. Lett.* **24**, 1225 (1970).

---

* The true values of the exponents are not model dependent since they are defined in terms of the limiting behavior at the critical point. See Ref. 8.

Effect of Viscosity on the Kapitza Conductance

W.M. Saslow*

Department of Physics, Texas A & M University
College Station, Texas

The problem of the Kapitza conductance h_K can be stated as follows. For a given temperature difference between the surface of a solid and liquid helium (^{3}He, He I, or He II), much more heat flows from the hotter to the colder material than can be explained theoretically.[1] In addition, surface preparation of the solid may significantly affect the Kapitza conductance, and this is another problem which is not well understood.[2]

Certain idealizations are always made in theoretical calculations of h_K. One set of idealizations pertains to the solid: It is usually taken to be perfectly smooth and regular (see, however, Ref. 2), so that it is described by a continuum model for the lattice vibrations. In addition, the solid is taken to be isotropic. We will also make these assumptions about the solid, so there will be no attempt to explain the effects of surface preparation. Another idealization is that the liquid has no transverse response. We have included such a response in order to determine its effects on the energy transfer between solids and liquid helium.

We will summarize our results. First, for sound waves incident from the liquid or solid, viscosity does not significantly affect the transmission to the other side. This appears to result from the enormous acoustic mismatch between the liquid and solid. Second, sound waves incident from the solid can generate viscous waves in the liquid. At thermal frequencies this effect can be comparable to the generation of longitudinal waves in the liquid. However, viscous waves damp out near the interface, so that it is not clear that such energy reaches the bulk of the liquid. One must determine where this dissipated energy ultimately resides. Nevertheless, even if all of the dissipated energy of viscous waves goes to the liquid, not enough energy would be transferred from solid to liquid to explain the measured values of h_K. Third, thermal frequency longitudinal waves incident from the liquid can transfer, on reflection, an appreciable amount (e.g., 20%) of their energy to viscous waves. Again the question arises as to the ultimate distribution of the dissipated energy. If only a small amount of this energy gets across to the solid, it may explain the large experimental values of h_K.

In calculating the energy transfer from liquid to solid and vice versa, a model is needed for the response of liquid helium at thermal frequencies. At 1°K, $\omega_T (= K_B T/h)$ is about 1.4×10^{11} sec^{-1}, which is rather high. For ^{3}He at tempera-

* On leave of absence at the Department of Physics, University of Nottingham, Nottingham, England, supported by the Science Research Council.

tures above the Fermi liquid regime and for He I one can most likely describe the shear response by a complex viscosity.[3] Such a response corresponds to viscous behavior at low frequencies and propagating shear waves at ultra high frequencies. For He II the picture is not so clear, but we adopt a similar viscosity here also. Our reasoning is as follows. First, the two-fluid model is a hydrodynamic model, and as such it is valid only at small frequencies compared to the relevant quasiparticle collision frequency. One finds that collision frequencies are typically less than thermal frequencies,[4] so the two-fluid model can be expected to break down. One might hope to use the results of recent work by Ma, which indicates that the weakly inter-acting Bose gas shows viscoelastic behavior, with the two-fluid density ρ_n, for fre-quencies above the hydrodynamic regime.[5] However, his work is not expected to apply at frequencies as high as ω_T. If one does assume such viscoelastic behavior, one concludes that the transverse modes of He II are propagating (i.e., long-lived) at thermal frequencies. In this case they would be expected to contribute to the specific heat. However, for $T \lesssim 0.6°K$, where the effect of rotons is negligible, the specific heat can be explained with longitudinal modes alone.* Therefore, at least for $T \lesssim 0.6°K$, it seems safe to assume that the transverse modes in He II are not propagating. We will assume that is true at all relevant temperatures. Since at ultra-high frequencies He II should behave like an amorphous (albeit quantum) solid,† it is tempting to assume that for somewhat lower frequencies it behaves like an amorphous liquid with its full density ρ. We make this assumption, taking $\eta = \eta_0/(1 - i\omega\tau)$ as the viscoelastic response of He II, and using an atomic model to estimate η_0 and τ. Thus, since the atoms in ^{3}He and ^{4}He are so similar, estimates of η_0 and τ for He II should be applicable to ^{3}He (and He I).

We note that a viscoelastic atomic model for the high-frequency shear response of He II is consistent with the lack of a roton onset in h_K at 0.6°K.[1] (If the viscous dissipation mechanism is to be an important contributor to h_K, it should not show a roton onset at 0.6°K.) In addition, the similar h_K values for ^{3}He and He II indicate that an important mechanism applicable to one should be applicable to the other, and our description of the viscous dissipation satisfies this.

To be specific, we take $\eta_0 \approx (1/3)\,\rho V_T l$ and $\tau = \eta_0/\rho c_s^2$. Here ρ is the liquid density, $V_T \approx (K_B T/m)^{1/2}$ is the thermal velocity of an atom of mass m, l is an atomic mean free path, and c_s is the velocity of propagation of shear waves at ultrahigh frequencies.

We take $c_s = 0.5c$, since this is approximately true for most other liquids,[3] and we take l to be of the order of an atomic spacing. For He II at 1°K we find $\eta_0 \approx 6 \times 10^{-6}$ P and $\omega_T\tau \approx 0.04$. Due to the uncertainty in the value of η_0, it was increased and decreased by a factor of ten in the calculations. Since even for $\omega_T\tau \approx 0.4$ the viscous behavior dominates, the calculations should not be very dependent upon c_s, the "elastic" parameter of our model. This was borne out by specific calculations in which c_s was varied from $0.25c$ to c. For this reason we will present results only for $c_s = 0.5c$.

It should be noted that we have neglected effects due to the finite wave vector.‡

---

* See Wilks[6] for a general presentation of the properties of He I, He II, and liquid ^{3}He.
† See Ref. 7 for a discussion of ordinary fluids at high frequencies.
‡ See Ref. 8 for the effect of wave vector on the shear response of ordinary fluids.

These might be considerable, for we find that $\text{Im}(k^{-1})$ for our calculations is typically of the order of 10^{-8} cm.

To compute the stresses $(\sigma_{xz}, \sigma_{zz})$ and velocities (v_x, v_z) in the solid and liquid, which must be matched at the interface, we used the standard formulas of elasticity[9] and fluid mechanics.[10] Waves of thermal frequency at $T = 1^\circ\text{K}$ were considered, incident from both solid and liquid, and with a uniform distribution in phase space. The normalization was chosen such that the integrated weight of the incident energy flux,

$$\int_0^{\pi/2} \left[\frac{d^2E}{d\theta\, dt} \right]_{\text{incident}} d\theta$$

was unity.

Here I present only the results for a longitudinal wave incident from the solid. (The complete equations, derivations, and calculations will be presented in another publication.[11]) They are of primary interest because they show the large effect of viscous dissipation. For $\eta_0 = 6 \times 10^{-6}$ P the $(d^2E/d\theta\, dt)_{t,l,d}$ are plotted in Fig. 1 for Cu–He II. The subscripts t, l, and d refer to transverse waves in the solid, longitudinal

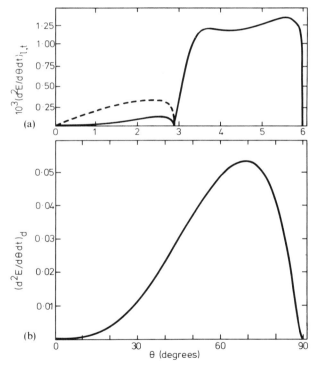

Fig. 1. In (a) the solid curve is $(d^2E/d\theta\, dt)_t$ and the dashed curve is $(d^2E/d\theta\, dt)_l$. In (b) we have plotted $(d^2E/d\theta\, dt)_d$. These curves apply to Cu–He II, with the phonons in the He II corresponding to thermal phonons at 1°K. The parameters characterizing Cu and He II correspond to the first row in Table I.

waves in the solid, and dissipation in the liquid, respectively. The dissipation is essentially the same for Cu and Pb (see Table I) because they are both so heavy and incompressible compared to He II. Further, the dissipation and transmission are nearly uncoupled, in part because each dominates a different range of angles. We note that $(d^2E/d\theta\,dt)_t \propto \sin\theta$ and $(d^2E/d\theta\,dt)_{t,d} \propto \sin^3\theta$ for small θ. The transmission goes to zero at $\theta_l \equiv \sin^{-1}(c/c_l)$ because the effective impedance of the solid is infinite at that angle. (Here c and c_l are the longitudinal sound velocities in the liquid and solid, respectively.)

For both Cu and Pb, $(dE/dt)_t$ dominates $(dE/dt)_l$. More striking, however, is the importance of $(dE/dt)_d$. The dissipation for $\eta_0 = 6 \times 10^{-6}$ P is larger than the total direct transmission by a factor of about 5000 for Cu and by a factor of about 200 for Pb. If only a small amount of this energy gets across to the solid, it may explain the large experimental values of h_K.

In Table I we show how changes in ω and in η_0 affect $(dE/dt)_{t,l,d}$ for Cu. Both $(dE/dt)_{t,l}$ are unaffected, but $(dE/dt)_d$ shows a dependence close to that predicted for $\omega\tau \ll 1$ and angles of incidence not too close to $90°$[10]:

$$(dE/dt)_d \propto (\omega\eta_0/\rho c^2)^{1/2}$$

We note that for zero frequency our equations reduce to those of Khalatnikov[12] and for $\omega\tau \gg 1$ they are equivalent to those of Kolsky[13]. In addition, energy conservation was checked explicitly, to seven-place accuracy.

This investigation leaves a number of unresolved questions:

1. What is the high-frequency shear response of He II? This problem can be studied both theoretically and experimentally. Measurement of the reflectivity of longitudinal sound in the liquid should provide a valuable probe of the high-frequency shear response of He II. For this purpose Fig. 2 presents the reflectivity of Cu as a function of incident angle for very low ($\omega\tau \ll 1$), $1°$K thermal ($\omega\tau = 0.04$), high ($\omega\tau = 0.4$), and ultrahigh ($\omega\tau \gg 1$) frequencies. Although absolute reflectivities may be difficult to measure, the angular variation should be much easier to obtain.

2. What is the form that energy takes when it is dissipated over a distance much shorter than the relevant mean free path? In particular, how does one answer this question for He II? We believe that for phenomena involving such very short dis-

Table I. Power Dissipated in the Liquid $(dE/dt)_d$ and Transmitted to the Solid in Longitudinal and Transverse Waves $(dE/dt)_{l,t}$

System[a]	$(dE/dt)_d$	$(dE/dt)_l$	$(dE/dt)_t$	$(dE/dt)_l + (dE/dt)_t$
Cu–He II ($\eta_0 = 6 \times 10^{-6}$ P)	0.242	0.628×10^{-5}	0.382×10^{-4}	0.445×10^{-4}
Cu–He II ($\eta_0 = 6 \times 10^{-5}$ P)	0.494	0.633×10^{-5}	0.381×10^{-4}	0.444×10^{-4}
Cu–He II ($\eta_0 = 6 \times 10^{-7}$ P)	0.098	0.627×10^{-5}	0.383×10^{-4}	0.446×10^{-4}
Cu–He II ($\omega = 0.35 \times 10^{11}$ sec^{-1})	0.143	0.627×10^{-5}	0.383×10^{-4}	0.445×10^{-4}
Pb–He II	0.242	0.506×10^{-4}	0.114×10^{-2}	0.119×10^{-2}

[a] For Cu we take $D = 8.9$ g cm^{-3}, $c_l = 4.7 \times 10^5$ cm sec^{-1}, and $c_t = 2.26 \times 10^5$ cm sec^{-1}.[15] For Pb we take $D = 11.4$ g cm^{-3}, $c_l = 2.16 \times 10^5$ cm sec^{-1}, and $c_t = 0.7 \times 10^5$ cm sec^{-1}.[15] For He II we take $\rho = 0.145$ g cm^{-3}, $c = 2.38 \times 10^4$ cm sec^{-1}, $c_s = 0.5c$, and $\eta_0 = 6.0 \times 10^{-6}$ P (see text), unless otherwise noted.[6] For ordinary ^{3}He the values in this table would not be very different. Also, $\omega = 1.4 \times 10^{11}$ sec^{-1}, unless otherwise noted.

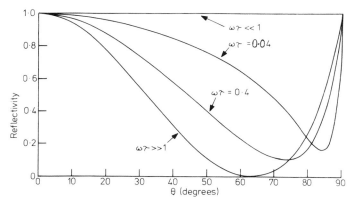

Fig. 2. The reflectivity of sound from He II off Cu. The frequency is varied as shown in the figure. All other parameters correspond to the first row of Table I.

tances an atomic model should be used to describe even He II, in the absence of a proper theory. Even for an atomic picture, the form that the energy takes is very complicated and may depend strongly on the details of the atomic collisions as incorporated into solutions of the transport equation.[14] We note that an atomic picture for the dissipated energy is consistent with the mechanism being the same for ^{3}He, He I, and He II.

3. How does one explicitly satisfy the principle of detailed balance? This is a question of theoretical importance quite independent of the effect of this mechanism on the Kapitza conductance. If unattenuated sound waves from a liquid are incident on a solid at an angle greater than the critical angle, the solid will heat up from the viscously dissipated energy generated when the sound wave is reflected at the interface. By detailed balance, some (presently unknown) inverse mechanism must exist. We note that if heat diffuses in the solid, phonon collisions must be taking place, so that a treatment of the phonon transport equation in the solid may be necessary for an understanding of this problem. Furthermore, the fact that the dissipated energy in the liquid goes into higher solutions of the transport equation* means that one must determine the coupling between these modes and the modes of the solid. It does not appear that the answer to this question will be easy to find. We remark that surface preparation may affect this coupling.

To summarize, we have pointed out a source of energy which is available for transfer from liquid to solid, and with the aid of model calculations we have shown that it can be sufficiently large to explain the measured values of h_K. From the point of view of detailed balance, we have shown that our description of the liquid (and perhaps the solid) must be made more complex.

Acknowledgment

I would like to thank L. J. Challis for his support, encouragement, and suggestions.

* For a monatomic gas with N atoms the Boltzmann equation has as its eigenmodes two propagating longitudinal modes corresponding to ordinary sound, two transverse modes corresponding to viscous waves, a heat diffusion mode, and on the order of $3N$ higher modes which cannot be described by the equations of hydrodynamics. For large wave vector these higher modes cannot be neglected. See Ref. 14.

References

1. G.L. Pollack, *Rev. Mod. Phys.* **41**, 48 (1969).
2. R.E. Peterson and A.C. Anderson, *Sol. St. Comm.* **10**, 891 (1972); H. Haug and K. Weiss, *Phys. Lett.* **40A**, 19 (1972).
3. K.F. Herzfeld and T.A. Litovitz, *Absorption and Dispersion of Ultrasonic Waves,* Academic Press, New York (1959).
4. I.M. Khalatnikov and D.M. Chernikova, *Zh. Eksperim. i Teor. Fiz.* **49**, 1957 (1965) [*Soviet Phys.— JETP* **22**, 1336 (1966)].
5. S. Ma, *Phys. Rev. A***5**, 2632 (1972).
6. J. Wilks, *The Properties of Liquid and Solid Helium,* Oxford Univ. Press, London (1967).
7. J. Hubbard and J.L. Beeby, *J. Phys. C* **2**, 556 (1969).
8. A. Akcasu and E. Daniels, *Phys. Rev. A***2**, 962 (1970).
9. L. Landau and I.M. Lifshitz, *Elasticity,* Pergamon Press, London (1959).
10. L. Landau and I.M. Lifshitz, *Fluid Mechanics,* Pergamon Press, London (1959).
11. W.M. Saslow, in preparation.
12. I.M. Khalatnikov, *Introduction to the Theory of Superfluidity,* Benjamin, New York (1966).
13. H. Kolsky, *Stress Waves in Solids,* Oxford Univ. Press, London (1953).
14. J.D. Foch Jr. and G.W. Ford, *Studies in Statistical Mechanics,* J. De Boer and G.E. Uhlenbeck, eds., North-Holland, Amsterdam (1970), Vol. V.
15. C. Kittel, *Introduction to Solid State Physics,* 2nd ed., Wiley, New York (1956).

Liquid Disorder Effects on the Solid
He II Kapitza Resistance*

C. Linnet,† T.H.K. Frederking, and R.C. Amar

School of Engineering and Applied Science, University of California
Los Angeles, California

At temperatures T below the λ-point $T < T_\lambda$ several Kapitza resistance results $R_K \propto T^{-3}$ indicate phonon transmission difficulties and provide some support for acoustic mismatch models (proposed in the past and reviewed in Ref. 1) and for more recent modifications[2-4] of the early models. Some measurements, however, show departures from the T^{-3} law, particularly at increased heat fluxes q. The departures appear to be consistent with some aspects of early thermohydrodynamic theories of the solid–He II Kapitza resistance[5,6]; however, the He II contributions appear to be very small and outside of the error limits of early data. More recent experiments indicate that the liquid He II temperature difference ΔT may cause noticeable departures from phonon transmission laws at large heat fluxes. Then the liquid contribution to the entire Kapitza temperature "drop" may reach several percent. Therefore we have conducted experiments at heat fluxes of the order of magnitude 0.5–5 W cm^{-2} with the purpose of resolving further details of the liquid He II thermohydrodynamic aspects of the solid–He II Kapitza resistance.

Heated horizontal cylinders were used at zero net mass flow as thermohydrodynamic probes of the He II layer adjacent to the solid in order to find the function $q(\Delta T)$; (Ga cylinder: diameter 0.120 cm; Pt–10% Rh wire: diameter 17.8 μm). The liquid temperature difference ΔT was obtained from phase transition thermometry, which determines the onset of vaporization in "hot" liquid near the solid, on the basis of the saturation hypothesis for the cessation of solid–He II contact associated with liquid–vapor phase conversion. According to the hypothesis, incipient vapor formation eliminates liquid–solid contact exactly at the phase boundary for He II– vapor equilibrium (without metastable superheating). Support for the saturation hypothesis has been reported recently, primarily for simple configurations.[7-9] According to the evidence, this thermometric method appears to be useful primarily for certain special geometries (e.g., horizontal heater of relatively small thickness at standard gravity), and particular conditions (e.g., He II vapor film onset flux significantly larger than the corresponding He I flux).

The heat flux is shown in Fig. 1 as a function of the liquid temperature difference for the Pt-alloy wire in the temperature range $2.10 < T < 2.16°$K. The flux is a nonlinear function of ΔT. At small ΔT, q rises initially; however, at increased power q

* Supported by the National Science Foundation.
† Present address: Aeroject Electrosystems Co., Azusa, Calif.

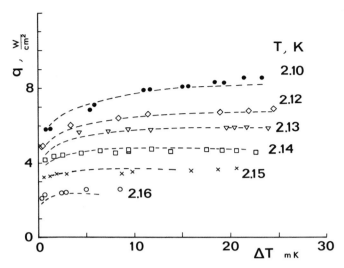

Fig. 1. Heat flux versus temperature difference of the liquid at Pt-
alloy wire [dashed curves calculated from Eq. (2)].

tends to become a weak function of ΔT. We note that these nonlinear functions represent flow resistance which is nonclassical. The results are reminiscent of the channel flow response to an applied pressure difference in "isothermal" channel flow beyond the critical velocity.

Thermohydrodynamic data for zero net mass flow have been discussed frequently in terms of power law approximations. Figure 2 plots the data of Fig. 1 in a more general form in order to assess the usefulness of this approach. The relative velocity $w_r = q/\rho_s S T$ is shown as a function of the applied potential difference $S \Delta T$ (S is the entropy per unit mass, ρ_s is the superfluid density; data reduction is based on physical properties at the arithmetic mean temperature). Adoption of a constant power law exponent $m = d(\log w_r)/d[\log(S \Delta T)]$ indeed provides for a fair account of the data. This description with $m = $ const, however, is restricted to a rather small range of the temperature and ΔT. We compare the present results with two phenomenological accounts for the nonlinear thermohydrodynamics of He II, the Gorter–Mellink model and its extension, respectively,[10] and the similarity approach.[11]

The Gorter–Mellink theory[12] is concerned with zero net mass flow in tubes (length L). The transport rate of the extended theory[10] may be expressed as

$$q = \rho_s S T w_r \tag{1}$$

with

$$w_r = \zeta_m (S \Delta T)^{1/m} \tag{2}$$

The factor ζ_m contains the Gorter–Mellink transport coefficient A_{GM}

$$\zeta_m = (L A_{GM} \rho_n)^{-1/m} \tag{3}$$

(ρ_n is the normal fluid density). The original Gorter–Mellink theory deals with the special case $m = 3$, i.e., absence of a preponderant length which characterizes the thermohydrodynamic flow pattern. The modified Gorter–Mellink approach[10] has

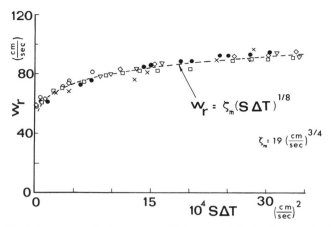

Fig. 2. Transport velocity w_r versus driving potential $S \Delta T$ (Pt-alloy data).

been considered primarily for data in the vicinity of T_λ, where $3 < m < 4$. In contrast, the present results are described by $m = 8$. Accordingly, m is considerably larger than channel flow exponents for zero net mass flow. Equation (1), in conjunction with $\zeta_m = 19$ cgs units in Eq. (2), has been used to approximate the data of Fig. 1. The data indicate that m decreases as T is lowered.

Qualitatively similar flow resistances have been obtained with the Ga cylinder. The power law exponent m, however, is about four at $2.14°K$ and about three at and below $2.10°K$. At the same time the increase in the diameter by about two orders of magnitude causes a change from $\zeta_8 = 19$ cgs units to ζ_4 of about 1 cgs unit. This size effect is not accounted for by the Gorter–Mellink theory.

Similarity considerations permit inclusion of the size effect on the thermohydrodynamics. According to the similarity model,[10] we have

$$w_r = \zeta'_m (S \nabla T)^{1/m} \qquad (4)$$

with

$$\zeta'_m = C_m (\eta_n/\rho)^{(m-2)/m} L_c^{-(m-3)/m} \qquad (5)$$

(η_n is the normal fluid viscosity). The characteristic length L_c in Eq. (5) is equal to the cylinder diameter for the present investigations. The limited experimental evidence available so far indicates that Eq. (5) permits a reasonable description, provided the experimentally found exponents m are inserted.

The size effect is reminiscent of the diameter influence in classical convection of Newtonian liquids. Accordingly, the observed behavior may be traced in part to the viscous fluid retardation of the normal fluid of He II. Therefore we interpret the data on the basis of the two-fluid model for zero net mass flow. Of particular interest is the special case $m = 4$ (thin, normal fluid layer of thickness l), for which flow development phenomena (acceleration and time-dependent effects) are relatively insignificant, aside from the vicinity of separation points.

Starting from the continuum equation[13] for the normal fluid of He II, we obtain the normal fluid flow rate of a thin layer. When the normal fluid is locked at the solid wall because of its viscous forces and moves unimpeded at its outer edge the flow rate becomes

$$\rho_n \bar{v}_n = (1/3)\, l^2 (\rho_n/\eta_n)\, \rho_s S\, \nabla T \qquad (6)$$

where $l = l(x)$ is the local layer thickness, with x the distance from the origin. The related mass flow increment is $j_n = (d/dx)(\rho \bar{v}_n l)$

The superfluid–normal fluid conversion rate enters the early thermohydrodynamic theory of Gorter *et al.*[6] as an unknown parameter. In the present approach we make use of zero net mass flow ($j_n = -j_s$) and superfluid constraints. The superfluid transport has to match the normal fluid flow rate. Near T_λ, however, j_s cannot be arbitrarily large. Because of quantization conditions ($\oint \mathbf{v} \cdot d\mathbf{s} = N\Gamma_0$; $\Gamma_0 = h/m$ = quantum of circulation), we expect a superfluid flow rate j_s limited by $j_s = C_j \Gamma_0 \rho_s / l$, where $C_j \sim 10^{-1}$. From the zero net mass flow constraint and an account for the development of $l(x)$ we obtain a mean layer thickness

$$\bar{l} = C_l (\Gamma_0 L_c \eta_n/\rho_n)^{1/4}/(S\,\nabla T)^{1/4} \qquad (7)$$

($C_l \sim 1$). Then the transport rate may be expressed in terms of $w_r/(S\,\nabla T)^{1/4}$ as

$$\zeta'_m = \zeta'_4 = (C_j/C_l)(\Gamma_0^3/L_c)^{1/4}(\rho_n/\eta_n)^{1/4} \qquad (8)$$

(at the wall $v_n \to 0$; $|w_r| \to |v_s|$). For the Ga cylinder at 2.14°K, ζ'_4 is of the order of magnitude of 10^{-2} cgs unit. It is only a weak function of temperature in the range $2.0 < T < T_\lambda$. The factor ζ'_4 [Eq. (2)] is in approximate order of magnitude agreement with the value found for Ga when $\nabla T \sim \Delta T/\xi$; ($\xi \sim 10^{-6}$ cm, $T = 2.14°K$). We note that the similarity model[11] yields $\zeta'_4 \sim (\eta_n/\rho)^{1/2}/L_c^{1/4}$. Its layer thickness $\bar{l}$ is of the order of

$$l = (\eta_n/\rho)/w_r \qquad (9)$$

It agrees, in order of magnitude, with $\bar{l}$ of Eq. (7).

The layer thickness has been evaluated from the experiments on the basis of Eq. (9). It is shown in Fig. 3 along with the flux q of the Ga experiments. For the Pt-alloy wire Eq. (9) yields l values of about 100 Å. The thickness of each sample is within

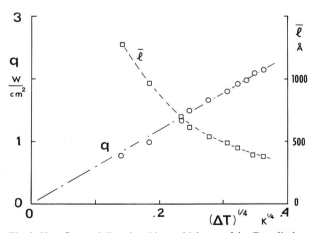

Fig. 3. Heat flux and disordered layer thickness of the Ga cylinder.

the upper bound given previously.[11] Further, it is pointed out that an extrapolation to low ΔT on the basis of $m = $ const is not meaningful.

According to these investigations, the He II thermohydrodynamics near thin cylinders is affected strongly by the size. The present results appear to support the possibility that Kapitza resistance departures from phonon transmission laws might be traced in part to liquid impedances. Examples are the pitted surface and porous coatings. These effects are geometry dependent.

Acknowledgments

We acknowledge constructive criticism and discussions with I. Rudnick, W.A. Little, G. Ahlers, R.A. Satten, and S. Putterman.

References

1. G. Pollack, *Rev. Mod. Phys.* **41**, 48 (1969); N.S. Snyder, *Cryogenics* **10**, 89 (1970).
2. M. Vuorio, *J. Phys. C* **5**, 1216 (1972).
3. H. Haug and K. Weiss, *Phys. Lett.* **40A**, 19 (1972).
4. R.E. Peterson and A.C. Anderson, *Sol. St. Comm.* **10**, 891 (1972).
5. R. Kronig, in *Proc. 2nd Intern. Conf. Low Temp. Phys., 1951*, Oxford, England, p. 99; R. Kronig and A. Thellung, *Physica* **18**, 749 (1952).
6. C.J. Gorter, K.W. Taconis, and J.J.M. Beenakker, *Physica* **17**, 841 (1951).
7. R. Eaton, W.D. Lee, and F.J. Agee, *Phys. Rev. A* **5**, 1342 (1972).
8. J.E. Broadwell and H. Liepmann, *Phys. Fluids* **12**, 1533 (1972).
9. C. Linnet, Ph.D. Thesis, Univ. of California at Los Angeles, 1971.
10. G. Ahlers, *Phys. Rev. Lett.* **22**, 54 (1969).
11. R.C. Chapman, *J. Low Temp. Phys.* **4**, 485 (1971); R.L. Haben, R.A. Madsen, A.C. Leonard, and T.H.K. Frederking, *Adv. Cryog. Eng.* **17**, 323 (1972).
12. C.J. Gorter and J.H. Mellink, *Physica* **15**, 285 (1949).
13. K.R. Atkins, *Liquid Helium*, Cambridge Univ. Press, Cambridge (1959), p. 102.

The Kapitza Resistance between Cu(Cr) and ^{4}He(^{3}He) Solutions and Applications to Heat Exchangers*

J.D. Siegwarth and R. Radebaugh

Cryogenics Division, NBS Institute for Basic Standards
Boulder, Colorado

Anderson and Johnson[1] have measured the Kapitza resistance between liquid ^{3}He and copper for the same copper surface in both the work-hardened and annealed states. Below 0.2°K the Kapitza resistance of work-hardened copper was found to be ten times smaller than that of annealed copper. This result probably explains the high Kapitza resistivities measured in dilution refrigerator heat exchangers compared to previous resistivity measurements on unannealed bulk copper.[2] High-surface-area copper heat exchangers are generally sintered together at temperatures high enough to completely anneal copper.

Anderson *et al.*[1] qualitatively explain the results in terms of an interaction between phonons and damage in the surface layers of the solid. Interactions between the phonons in a solid and surface damage layers have been observed experimentally.[3] Haug and Weiss[4] have treated interactions between ^{4}He phonons and surface dislocations in copper theoretically.

In general, heat exchangers cannot be work-hardened after manufacture in order to reduce the Kapitza resistance, so another means of hardening the heat transfer surface must be found. Also, work-hardened copper anneals to some extent at room temperature.[1] Irradiation of the assembled exchanger is one possibility, but radiation-induced defects anneal out even below room temperature.

Oxygen annealing of copper-containing aluminum, beryllium, or silicon in small quantities hardens the metal but the thermal processes required to harden such alloys are not very compatible with exchanger construction techniques, particularly with sintering in hydrogen atmospheres. Though the incompatibilities may be resolvable, a technique called precipitation-hardening of copper is more compatible with exchanger manufacturing.

Copper is precipitation-hardened by adding alloying elements to molten copper that have low solubility in the solid. The amounts added are completely soluble in the solid only at temperatures just below the melting point. The alloy is held at this temperature for "solution treating," that is, to uniformly disperse the impurity. Hardening is accomplished by "aging" the alloy at 450–550°C. Here the alloying elements are much less soluble and precipitate out as a dispersion of particles of the impurity metal or an alloy that elastically distorts the lattice.

Though numerous copper alloys can be precipitation-hardened,[5] Cu(Cr) and Cu(Zr) are the only ones considered here since good thermal conductivity is desired. The alloys used in this work contained 0.6 wt.% chromium in OFHC copper and 0.18 wt.% zirconium in 99% pure copper. Measured values of the resistance ratio are 7–10 for Cu(Cr) (the lower the aging temperature, the higher the conductivity) and about ten for the Cu(Zr).[6] The conductivities of these alloys are approximately one-tenth that of OFHC copper and should still be sufficiently high so that most of the temperature drop in an exchanger will occur at the copper–helium interface.

The Kapitza boundary resistance between various copper cells and the dilute stream of a dilution refrigerator were measured by a continuous flow technique.[7] The results are shown in Fig. 1, where $R\sigma T^3$ is shown as a function of the temperature T. Here R is the Kapitza resistance and σ is the surface area. The surface exposed to the liquid was the inside of a 3-mm-diameter hole about 2 cm long that was polished with crocus cloth. The Kapitza resistances of work-hardened copper (curve A_3) and annealed copper (curve B_3) to pure ^{3}He are shown for comparison purposes.[1] Curve B_D shows our measurement of the Kapitza resistance between annealed OFHC copper and the dilute ^{3}He stream of our dilution refrigerator. A Cu(Cr) cell (curve A_D) was solution-treated in hydrogen and briefly in a small amount of air at 950°C,

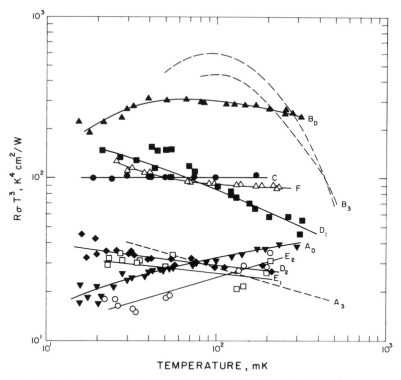

Fig. 1. Kaptiza resistance of copper and copper alloy cells to the dilute ^{3}He stream in a dilution refrigerator. Curves A_3 and B_3 are the Kapitza resistances between copper and pure ^{3}He liquid. The details are described in the text.

held at 750°C in a hydrogen atmosphere to simulate sintering, and then aged at 450°C for $4\frac{1}{2}$ hr in a forepump vacuum. The Kapitza resistance is a factor of ten lower than that of annealed OFHC copper. A second cell (curve D_1) was aged at 435°C for 17 hr in hydrogen but never heated above this after the initial melting. The Kapitza resistance approaches that of annealed OFHC copper at low temperature. This cell was then heated to 750°C for $\sim 1/2$ hr in hydrogen and aged for 5 hr at 435°C in a slightly oxidizing atmosphere. The boundary resistance (curve D_2) then agreed in magnitude but not in temperature dependence with curve A_D.

The above results suggest that perhaps aging in the presence of oxygen is the important factor leading to low Kapitza resistance. Some oxygen annealing of the chromium could result from aging in oxygen. Two more cells were made to test this. They were solution-treated at 975°C for 1 hr in hydrogen and "sintered" at 700°C in hydrogen for about 1 hr. They were aged for about 5 hr at 460°C, one in hydrogen (curve E_1) and the other in the presence of some oxygen (curve E_2). The cell aged in hydrogen shows somewhat lower values for $R\sigma T^3$ than the cell aged in oxygen. The principal difference, however, is the temperature dependence of $R\sigma T^3$. This hydrogen-aged sample has the same temperature dependence as the earlier sample aged in a slight oxidizing atmosphere (curve A_D); thus, the variation of temperature dependence cannot be attributed to the atmosphere during the aging process.

The Kapitza resistance for these aged alloys is significantly lower than our measured value of the resistance between the dilute solution and Cu–30% Ni (curve C).

The Kapitza resistance between Cu(Zr) and the dilute solution was measured and is shown in Fig. 1 also (curve F). It is considerably larger than for Cu(Cr), which may be because Cu(Zr) does not harden much by simply aging.

These measurements are not sufficient to determine a treatment that would give the lowest Kapitza resistance. There are many contributing factors to the final result; the heat-treating temperatures and times, quenching rates, the order in which the various treatments are done, furnace atmospheres, and subsidiary impurities all may have strong effects on the end result.[5] The whole process of precipitation-hardening is complex and not understood in detail. This series of measurements on Cu (Cr) indicates that solution-treating in hydrogen at about 1000°C, heating in hydrogen at 700–750°C for about 1 hr, and then aging at 450°C for several hours should reliably reduce the Kapitza resistance of Cu(Cr) significantly below that of annealed OFHC copper.

References

1. A.C. Anderson and W.L. Johnson, *J. Low Temp. Phys.* **7**, 1 (1972).
2. J.C. Wheatley, R.E. Rapp, and R.T. Johnson, *J. Low Temp. Phys.* **4**, 1 (1971).
3. P.D. Thatcher, *Phys. Rev.* **156**, 975 (1967); J.K. Wigmore, *Phys. Lett.* **37A**, 293 (1971).,
4. H. Haug and K. Weiss, presented at the 4th Intern. Cryogenic Engineering Conf. Eindhoven Univ., The Netherlands (to be published).
5. W.D. Robertson and R.S. Bray, *Precipitation from Solid Solution,* American Society of Metals, Cleveland, Ohio (1959), p. 328.
6. A.F. Clark, G.E. Childs, and G.H. Wallace, *Cryogenics* **10**, 295 (1970).
7. R. Radebaugh and J.D. Siegwarth, *Phys. Rev. Lett.* **27**, 796 (1971).

Heat Transfer between Fine Copper Powders and Dilute ^{3}He in Superfluid ^{4}He*

R. Radebaugh and J.D. Siegwarth

Cryogenics Division, NBS Institute for Basic Standards
Boulder, Colorado

The use of high-surface-area materials in dilution refrigerators has prompted several studies of the thermal resistance between such materials and dilute or concentrated ^{3}He in ^{4}He. Roubeau *et al.*[1] first reported anomalously high values for the Kapitza resistance of 20-μm-thick copper foil as compared with the Kapitza resistance of bulk copper. They proposed that this was a result of a thermal resistance between phonons and electrons. Wheatley *et al.*[2] also saw this higher resistance when using 32-μm copper foil. We looked for this effect in sintered copper powders of 27, 20, and 2 μm diameter and also found high resistances.[3] Our results were in semiquantitative agreement with Seiden's calculations[4] on the phonon–electron resistance.

We discuss here a reanalysis of some of our earlier work, along with new data which cast doubt upon the interpretation that the phonon–electron resistance was responsible for the observed high resistance.

First, krypton adsorption surface area measurement† gave a considerably smaller surface area than expected for the 2-μm-diameter powder (determined from approximate nitrogen adsorption measurements†). The surface area instead corresponds to 16-μm-diameter particles. Also, the observed T^{-4} behavior below 50 m°K for that sample was probably due to thermometry errors and unknown heat leaks. Other measurements on the same sample show no sign of a T^{-4} dependence in the thermal resistance. Those results, based on the new surface area measurements, are shown by the solid squares in Fig. 1. In this figure R is the measured thermal resistance and σ is the surface area exposed to the liquid. Adsorption studies† on our 27-μm-diameter (photomicrographs) powder[3] gave a surface area corresponding to 7-μm-diameter spheres. With this increased area the previously reported[3] $R\sigma T^3$ of 83 cm^2 K^4 W^{-1} is changed to 322 cm^2 K^4 W^{-1}, in good agreement with that for the 16-μm-diameter powder.

We then measured the Kapitza resistance to the dilute stream of bulk OFHC copper which was polished with crocus cloth (Fe_2O_3 abrasive of about 3 μm diameter), cleaned, and then annealed in hydrogen at 800°C for 1 hr. The results are shown as the solid circles in Fig. 1 and agree well with the previously measured powder samples. The value of $R\sigma T^3$ for this annealed bulk copper sample is about five

* Contribution of the National Bureau of Standards, not subject to copyright.
† Performed by P. D. Coulter of Union Carbide Corp., Cleveland, Ohio.

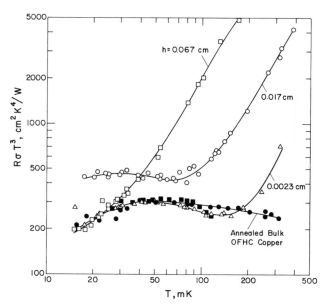

Fig. 1. The measured thermal boundary resistance to dilute ^{3}He in ^{4}He of bulk copper (solid circles) and several copper powder samples. The solid squares are for 16-μm-diameter powder and the rest of the data are for 1.8-μm powder of different depths.

times that reported by Wheatley et al.[2] for an unannealed sample. Such behavior is in agreement with the results of measurements by Anderson and Johnson[5] on the Kapitza resistance to pure ^{3}He of annealed and work-hardened copper. Our results then support the contention of Anderson and Peterson[6] that the annealing process is responsible for the high thermal resistances seen thus far in high-surface-area materials and is not due to a phonon–electron resistance. The phonon–electron resistance has recently been measured directly[6] and the result shows that much higher surface-to-volume ratios are needed before that resistance tends to dominate the Kapitza resistance.

In this paper we report on our new measurements of the thermal resistance between much finer copper powder[7] and the dilute ^{3}He stream of a dilution refrigerator. From adsorption studies[8] on the powder the characteristic spherical diameter is 1.8 μm. With this fine powder the Kapitza resistance should be small enough that both the phonon–electron resistance[6] and the phonon–^{3}He quasiparticle resistance[9] can easily be observed. These resistances are proportional to the volume of metal in the sponge and the volume of liquid in the sponge, respectively.

Three samples of this powder were measured using the configuration described in Ref. 3. The powder depths for the first through third samples were, respectively, 0.0023, 0.0170, and 0.0673 cm. Packing fractions were 40%. The sintering was done in hydrogen at 650°C for 1/2 hr. The measured thermal resistance between each sample and the dilute stream is shown in Fig. 1. The high-temperature behavior is dominated by the thermal resistance of the stationary liquid in the sponge. Because of its higher resistance, the 0.0170-cm-deep sample was remeasured, with identical results. The reason for its higher resistance is unknown but we are pursuing this point. In any case, the results on all three samples are qualitatively the same and give no hint of any resistances other than the Kapitza resistance at low temperatures.

Figure 2 shows an approximate schematic diagram of all the thermal resistances which may play a role in the transfer of heat from the electron system in the bulk metal to the ³He quasiparticle system in the bulk liquid. The bulk liquid and metal are separated by a sponge of sintered copper powder. The resistances shown between the two inner vertical dashed lines represent heat paths from the metal sponge to the liquid inside the voids (Kapitza resistance). The resistances shown dashed have seldom been considered. The other resistances occur in the process of transferring heat through the sponge, either in the liquid or in the metal. The terms R_e, R_p, $R_{p'}$, and R_q refer to thermal resistances to heat flow by electrons and phonons through the metal sponge, and by phonons and ³He quasiparticles through the liquid in the sponge channels, respectively. The resistances with subscripts of the type $x-y$ refer to the interaction between x and y, and the subscript b refers to the bulk metal or liquid.

The values for all of these thermal resistances were evaluated for the sample with 0.0673 cm powder depth and are shown in Fig. 3. This figure shows the dependence of RV_lT^3 on temperature, where V_l is the volume of liquid in the sponge voids. The bulk liquid volume in this cell was about the same as that in the voids so that $(R_{p'-q})_b = R_{p'-q}$. The resistances R_p and $R_{p'}$ were calculated by assuming a 1 μm phonon mean free path. Surface modes and wavelength effects could reduce R_p about a factor of two at 10 m°K. Both $R_{p'}$ and R_q are for the case of zero mass flow of the liquid in the fine channels. The Kapitza resistance of the bulk OFHC sample is used for $R_{p-p'}$. The curve of R_{total} shows the total thermal resistance calculated from the various individual resistances. The data begin to deviate from R_{total} considerably below 50 m°K. The data for this sample, as well as that for the other two powder depths, show no evidence for a contribution from the phonon–electron resistance R_{p-e} or the phonon–³He quasiparticle resistance $R_{p'-q}$.

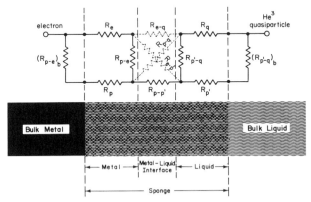

Fig. 2. Schematic diagram of the various thermal resistances involved in the transfer of heat from the electrons in bulk metal through a sintered sponge to ³He quasiparticles in the bulk liquid. The symbols are explained in the text.

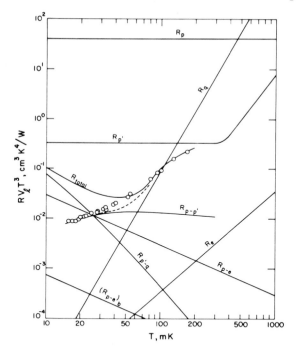

Fig. 3. Temperature dependence of the various resistances shown in Fig. 2. The data for the 0.067-cm-depth
powder is shown for comparison.

A possible explanation for the discrepancy is that both $R_{p\text{-}e}$ and $R_{p'\text{-}q}$ are at least an order of magnitude smaller than shown in Fig. 3, or that just one is lower and one of the diagonal resistances in Fig. 2 is active. The resistance $R_{p'\text{-}q}$ could be lower, since it is based solely on theoretical calculations,[9] but this seems unlikely for $R_{p\text{-}e}$ since its value is based on measurements.[6] Another explanation is that the Kapitza resistance of bulk copper is mainly due to $R_{e\text{-}q}$ instead of $R_{p\text{-}p'}$. Thus in Fig. 3 $R_{p\text{-}p'}$ would be relabeled $R_{e\text{-}q}$, and $R_{p\text{-}p'}$ would be much higher. Either explanation would give a total resistance shown by the dashed curve in Fig. 3. Phonon wavelength effects ($\lambda \approx 0.06\ T^{-1}$ °K μm in copper and 0.004 T^{-1} °K μm in ^{3}He– ^{4}He) are not observed, which would be an argument for either $R_{e\text{-}q}$ or $R_{e\text{-}p'}$ trans- porting the heat across the interface. However, heat transport by $R_{e\text{-}q}$ appears dubious since in the dilute ^{3}He phase a film of pure ^{4}He covers the metal surface and would weaken the interaction between electrons and quasiparticles. Instead, we find that the Kapitza resistance of annealed copper in the dilute solution is about half that found[5] for annealed copper in pure ^{3}He, at least for $T \approx 0.1$°K. The resistance $R_{e\text{-}q}$ may become effective only below about 30 m°K, where $R\sigma T^3$ is decreasing.

In conclusion, the thermal boundary resistance in sintered copper powder agrees well with the Kapitza resistance of bulk copper. For the 1.8-μm-diameter powder such behavior is anomalous below 50 m°K and allows much greater heat transfer per unit liquid volume than was previously thought possible.

References

1. P. Roubeau, D. LeFur, and E.J.A. Varoquaux, in *Proc. 3rd Intern. Cryogenic Engineering Conf. West Berlin, 1970,* p. 315.
2. J.C. Wheatley, R.E. Rapp, and R.T. Johnson, *J. Low Temp. Phys.* **4**, 1 (1971).
3. R. Radebaugh and J.D. Siegwarth, *Phys. Rev. Lett.* **27**, 796 (1971).
4. J. Sieden, *Compt. Rend.* **B271**, 5, 1152 (1970); **B272**, 5, 181, 355, 505, 1100 (1971); **B274**, 831, 1111 (1972).
5. A.C. Anderson and W.L. Johnson, *J. Low Temp. Phys.* **7**, 1 (1972).
6. A.C. Anderson and R.E. Peterson, *Phys. Lett.* **38A**, 519 (1972).
7. D.L. Goodstein, W.D. McCormick, and J.G. Dash, *Cryogenics* **6**, 167 (1966).
8. J.G. Daunt and E. Lerner, *J. Low Temp. Phys.* **8**, 79 (1972).
9. J.C. Wheatley, O.E. Vilches, and W.R. Abel, *Physics* **4**, 1 (1968).

Thermal Boundary Resistance between Pt and Liquid ^{3}He at Very Low Temperatures*

J.H. Bishop, A.C. Mota,† and J.C. Wheatley

Physics Department, University of California—San Diego
La Jolla, California

For many materials immersed in either liquid ^{3}He or liquid ^{4}He the low-temperature thermal boundary resistance is found to vary approximately as $1/T^3$. This behavior is often discussed in terms of the acoustic mismatch theory[1] in which phonons are assumed responsible for transferring thermal energy between the solid and the liquid helium. At least one system has been found for which the phonon resistance is not the limiting low-temperature resistance. That system is cerium magnesium nitrate (CMN) immersed in liquid ^{3}He, for which the very low-temperature thermal boundary resistance is found experimentally to vary linearly with the temperature[2] and to decrease in the presence of a small magnetic field.[3] This resistance has been discussed[4,5] in terms of thermal energy exchange via a surface magnetic coupling between the cerium spins of the CMN and the ^{3}He nuclear spins. In this paper we present data which suggest that the thermal boundary resistance between Pt and liquid ^{3}He is qualitatively different from the phonon and magnetic boundary resistances mentioned above.

The Pt sample principally studied consisted of a tight bundle of noninsulated 0.001-in-diameter 99.99% Pt wire. The following type of experiment was performed. A Pt sample was immersed in a thermal reservoir of liquid ^{3}He whose temperature was known. The Pt was heated by eddy currents to a constant and somewhat higher temperature than the ambient liquid ^{3}He. The ac field was then removed and the subsequent exponential decay of the Pt temperature to the temperature of the ambient liquid ^{3}He was observed. From this measurement two quantities were obtained: a time constant τ and the temperature difference ΔT between the Pt and the liquid ^{3}He which resulted from the eddy current heat input. This latter quantity is proportional to the thermal boundary resistance. Because of the unknown sample geometry, the magnitude of the heat input cannot be calculated and hence the magnitude of the thermal resistance is not determined in this experiment. However, by maintaining the same heat input from one data point to another, useful information about the thermal boundary resistance may be obtained. The success of this technique depends upon being able to measure the temperature of the Pt. Figure 1 shows the temperature-dependent 17-Hz susceptibility for two samples studied in a dc magnetic field of less

* Work supported by the U.S. Atomic Energy Commission under Contract No. AT(04-3)-34, P.A. 143.
† Supported by the Air Force Office of Scientific Research, Air Force Systems Command, USAF, under AFOSR Contract No. AFOSR/F-44620-72-C-0017.

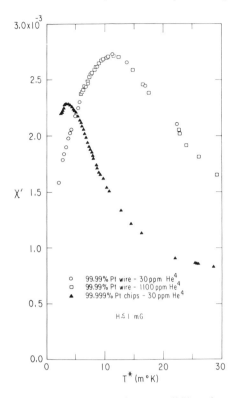

Fig. 1. The 17-Hz magnetic susceptibility of two Pt samples plotted against the magnetic temperature of a right-circular cylinder of powdered CMN which was spatially separated from, but in good thermal contact with, the Pt samples. The measurements were made in a dc magnetic field of less than 1 mG.

than 1 mG plotted against the magnetic temperature of a CMN thermometer which was spatially separated from, but in good thermal contact with, the sample measuring volume.[3] Susceptibility maxima were observed near 11 m°K for the 99.99% Pt wire and near 3.6 m°K for the 99.999% Pt chips. Measurements of τ and ΔT were made only at temperatures below the susceptibility maximum for the 99.99% Pt wire, whereas measurements of τ and ΔT for the 99.999% Pt chips were made both above and below the susceptibility maximum. This temperature-dependent susceptibility served as the Pt thermometer.

The observed temperature differences for the 99.99% Pt wire samples are shown in Fig. 2 for various concentrations of ⁴He in the liquid ³He. The temperature differences are plotted against the average Pt temperature during the time in which the measurement was made. All temperatures and temperature differences reported here are in terms of the Johnson noise temperature scale. Conversion from the CMN magnetic temperature scale to the Johnson noise temperature scale was made using the data of Webb et al.[6] The data presented in Fig. 2 may be separated into three groups according to the ⁴He concentration. Throughout this paper one must remem-

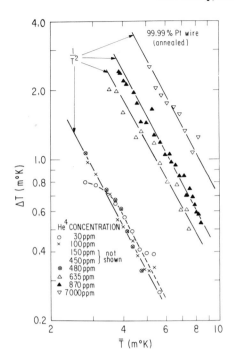

Fig. 2. The temperature difference established between the 99.99% annealed Pt wire and a thermal reservoir of liquid ^{3}He when the Pt was heated by eddy currents, plotted against the average Pt temperature during the time of the measurement. This temperature difference is proportional to the thermal boundary resistance.

ber that when a solid is placed in a ^{3}He–^{4}He mixture the ^{4}He is expected to go to the solid surface. For ^{3}He containing 100, 150, 450, and 480 ppm ^{4}He, the Pt–liquid ^{3}He temperature difference and hence thermal resistance is proportional to $1/T^2$ and independent of ^{4}He concentration. For higher ^{4}He concentrations, 635, 870, and 7000 ppm ^{4}He in ^{3}He, the Pt–liquid ^{3}He temperature difference is still found to be proportional to $1/T^2$ but is now ^{4}He concentration dependent, the temperature difference increasing as the ^{4}He concentration is increased. At very low concentrations of ^{4}He in ^{3}He the Pt–liquid ^{3}He temperature difference appears to be no longer proportional to $1/T^2$ but rather increases less rapidly than $1/T^2$ below 4 m°K. The same ^{4}He concentration behavior was observed in the time constants.

The quantitative effect of ^{4}He in the liquid ^{3}He on the time constants and temperature differences for several Pt samples studied is summarized in Fig. 3, where the ratio τ/τ_{NORM} and $\Delta T/\Delta T_{NORM}$ are plotted against the ^{4}He concentration. All data reported in this figure were obtained at 5 m°K. The data for the annealed 99.99% Pt wire are normalized to the time constant and temperature difference observed when the sample was placed in ^{3}He containing 100 ppm ^{4}He, whereas the data for all the other samples studied are normalized to their respective 30 ppm ^{4}He time constant

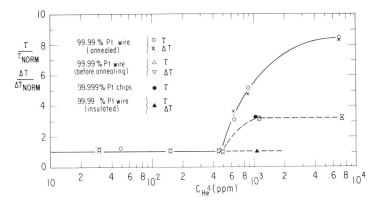

Fig. 3. Summary of the effect of ⁴He impurity in the liquid ³He on the time constants and temperature differences for several Pt samples. (See text for further explanation.)

and temperature difference. For example, the open circles represent for the annealed 99.99 % Pt wire sample the ratio of the time constant at 5 m°K for a given ⁴He concentration divided by the time constant at 5 m°K for 100 ppm ⁴He, plotted against the given ⁴He concentration. In Fig. 3 any deviation from a horizontal line through $\tau/\tau_{\text{NORM}}= 1$ or $\Delta T/\Delta T_{\text{NORM}}= 1$ indicates a dependence on ⁴He concentration. It is seen that for the annealed (1300°C for 3 hr) 99.99 % Pt wire τ and ΔT are ⁴He concentration independent below about 500 ppm ⁴He and strongly ⁴He concentration dependent above about 500 ppm ⁴He, τ and ΔT for 7000 ppm ⁴He being 8.5 times larger than τ and ΔT for 100 ppm ⁴He. The 99.99 % Pt wire was also studied before being annealed and the results are shown in Fig. 3, where it is seen that the effect of ⁴He is considerably greater on the annealed Pt wire than on the nonannealed Pt wire. Another sample consisting of 99.999 % Pt chips which were not annealed was found to have about the same ⁴He effect on the time constant as the nonannealed 99.99 % Pt wire. Furthermore, the 99.999 % Pt sample showed the same τ and ΔT temperature dependence for 30 ppm ⁴He as the 99.99 % Pt sample. Finally, a sample of heavy formex insulated 99.99 % Pt wire was studied and, as shown in Fig. 3, no ⁴He effect was observed.

The ⁴He effect for the noninsulated Pt samples and the lack of a ⁴He effect for the insulated Pt sample suggests that there exists some mechanism for coupling thermal energy out of the Pt which is greatly hindered when the Pt surface is not in direct contact specifically with the liquid ³He. A similar sensitivity to ⁴He impurity has previously been observed in the CMN–liquid ³He thermal time constants[5] and is thought to be due to the ⁴He preventing the formation of a compressed layer of ³He adjacent to the CMN surface, this compressed ³He having an enhanced magnetic coupling to the CMN. It is quite tempting to suggest that a similar coupling exists between the Pt, or something in the Pt, and the liquid ³He. However, this is only a speculation and further experimental and theoretical work is necessary to obtain an understanding of the thermal energy exchange between Pt and liquid ³He.

References

1. R.E. Peterson and A.C. Anderson, *Phys. Lett.* **40A**, 317 (1972).
2. W.R. Abel, A.C. Anderson, W.C. Black, and J.C. Wheatley, *Phys. Rev. Lett.* **16**, 273 (1966).
3. J.H. Bishop, D.W. Cutter, A.C. Mota, and J.C. Wheatley, *J. Low Temp. Phys.* **10**, 379 (1973).
4. A.J. Leggett and M. Vuorio, *J. Low Temp. Phys.* **3**, 359 (1970).
5. W.C. Black, A.C. Mota, J.C. Wheatley, J.H. Bishop, and P.M. Brewster, *J. Low Temp. Phys.* **4**, 391 (1971).
6. R.A. Webb, R.P. Giffard, and J.C. Wheatley, *Phys. Lett.* **41A**, 1 (1972).

The Leggett–Rice Effect in Liquid ^{3}He Systems

L.R. Corruccini, D.D. Osheroff, D.M. Lee, and R.C. Richardson

Department of Physics, Cornell University
Ithaca, New York

In a recent series of papers Leggett and Rice[1] and Leggett[2,3] showed that it should be possible to measure spin-wave effects in a Fermi liquid by using spin-echo magnetic resonance techniques. Spin-wave effects become important in a Fermi liquid in the "collisionless regime," when $\omega_0 \tau \gtrsim 1$, where ω_0 is the Larmor frequency and τ the quasiparticle collision time. Leggett derived coupled equations for the spin current $\mathbf{J}$ and macroscopic magnetization $\mathbf{M}$ of the liquid under these conditions, from which the behavior of the spin echoes may be derived:

$$\frac{\partial \mathbf{M}}{\partial t} + \sum_{i=1}^{3} \frac{\partial \mathbf{J}_i}{\partial x_i} - \gamma \mathbf{M} \times \mathbf{H} = 0 \tag{1}$$

$$\frac{\partial \mathbf{J}_i}{\partial t} + \frac{v_F^2}{3}(1 + F_0^a)\left(1 + \frac{F_1^a}{3}\right)\nabla_i \mathbf{M} - \gamma \mathbf{J} \times \mathbf{H} + \left(\hbar \frac{dn}{d\varepsilon}\right)^{-1}\left(F_0^a - \frac{F_1^a}{3}\right)\mathbf{J}_i \times \mathbf{M}$$

$$= \left(\frac{\partial \mathbf{J}_i}{\partial t}\right)_{\text{coll}} \tag{2}$$

Here v_F is the Fermi velocity, the F_i^a are Fermi liquid interaction parameters, and γ is the gyromagnetic ratio. The echo amplitudes are then described by the following equation:

$$\ln h_n - \delta(1 - h_n^2) = -\frac{n}{12}\frac{D_0(\gamma G)^2 t_0^3}{1 + \alpha^2 \cos^2 \phi} \tag{3}$$

where

$$\alpha = \lambda \omega_0 \tau_D$$

$$\lambda = (1 + F_0^a)^{-1} - (1 + F_1^a/3)^{-1}$$

$$\delta = \tfrac{1}{2}\alpha^2 (\sin^2 \phi)/1 + \alpha^2 \cos^2 \phi)$$

ϕ is the initial NMR pulse angle, G is the magnetic field gradient, h_n is the height of the nth echo relative to its value at $G = 0$ or $t = 0$, and D_0 is Hone's result[4] for the spin diffusion coefficient when $\omega_0 \tau_D \ll 1$.

As can be seen, the echo amplitude envelope is not exponential, as is the standard classical result due to Torrey:

$$\ln h_n = -(n/12)D(\gamma G)^2 t_0^3, \qquad \omega_0 \tau \ll 1 \tag{4}$$

In the $\phi = 90°$ limit (strong excitation) large undamped spin currents $\mathbf{J}$ are generated, maximizing this nonexponential behavior. It may be traced directly to the term $\mathbf{J} \times \mathbf{M}$ in Eq. (2), which is absent in the classical formulation. A striking example of this behavior is shown in Fig. 1. In the $\phi \to 0$ limit (weak excitation) spin currents are small and the echo attenuation is exponential. However, Eq. (3) agrees with Eq. (4) only if an effective diffusion coefficient $D_e = D_0/(1 + \alpha^2 \cos^2 \phi)$ is used in Eq. (4). It may be shown that this change in effective D is due to the presence of spin-wave excitations.

Spin-echo data were obtained for liquid ^{3}He at pressures of zero and 27 atm and for a 6.4% ^{3}He–^{4}He solution at $P = 0$. Data were taken as a function of initial pulse angle ϕ, temperature T, and field gradient G. The highest Larmor frequency attained was 36.7 MHz, and the lowest temperature about 6 m°K, well within the collisionless regime for all three samples. Cooling was provided by a Pomeranchuk effect ^{3}He compression cell.

Within experimental accuracy the observed echo attenuation is described accurately by Leggett's Eq. (3). The effective diffusion coefficient D_e is also observed to have the predicted dependence on T and ϕ, exhibiting a maximum at a temperature given by $\lambda \omega_0 \tau_D \cos \phi = 1$. This is quite different from the behavior of D_0, the spin diffusion coefficient measured in low fields, which is proportional to $1/T^2$.

From the dispersion relation for spin waves in a Fermi liquid

$$\omega = \omega_0 + iD^*k^2, \qquad \text{where} \quad D^* = D_0/(1 + i\lambda\omega_0\tau_D)$$

it will be seen that the D_e measured in this experiment is essentially the real part of D^*. This, in turn, is directly proportional to the spin-wave attenuation Im ω. The additional $\cos \phi$ dependence in D_e comes from the finite amplitude of the rf exciting field. The similarity between the present results for $D_e(\phi = 18°)$ and the closely related phenomenon of zero sound is apparent in Fig. 2.

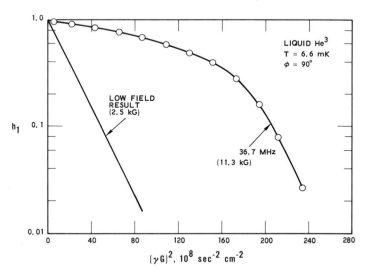

Fig. 1. Spin-echo amplitudes h_1 in liquid ^{3}He at 6.6 m°K for two different magnetic fields. Data have been plotted on semilogarithmic scale to emphasize nonexponential behavior of high field data.

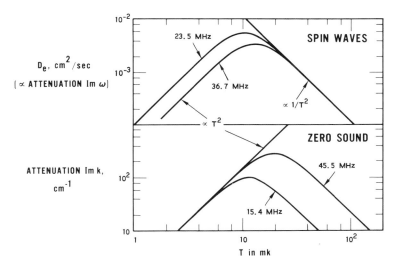

Fig. 2. Zero-sound attenuation (from Abel *et al.*[5]) compared with spin-wave attenuation coefficient $D_e(\phi = 18°) \approx$ Re D^*, liquid ³He at $P = 0$.

Finally, values of the Fermi liquid parameter λ are extracted, from which the previously unknown F_1^a may be found. Measured values of λ are: for ³He$(P = 0)$, 1.95; for ³He$(P = 27$ atm$)$, 2.9; and for 6.4% solution, 0.025. A more complete discussion of these results is contained in Ref. 6.

References

1. A.J. Leggett and M.J. Rice, *Phys. Rev. Lett.* **20**, 586; **21**, 506 (1968).
2. A.J. Leggett, in *Proc. 11th Intern. Conf. Low Temp. Phys., 1968*, St. Andrews Univ. Press, Scotland (1969), Vol. 1, p. 400.
3. A.J. Leggett, *J. Phys. C* **3**, 448 (1970).
4. D. Hone, *Phys. Rev.* **121**, 669 (1961).
5. W.R. Abel, A.C. Anderson, and J.C. Wheatley, *Phys. Rev. Lett.* **17**, 74 (1966).
6. L.R. Corruccini, D.D. Osheroff, D.M. Lee, and R.C. Richardson, *J. Low Temp. Phys.* **8** (3/4), (1972).

Helium Flow Through an Orifice in the Presence of an AC Sound Field

D. Musinski and D.H. Douglass

Department of Physics and Astronomy
University of Rochester, Rochester, New York

We report here an attempt to observe the ac Josephson effect in liquid helium. Previous investigators[1-5] have reported positive observation of the effect by monitoring the level difference between two baths separated by a small orifice. Our experiment was designed so that a temperature difference could be introduced and measured as well. For two superfluid helium baths that are weakly coupled by a small orifice the chemical potential difference is given by the expression

$$\Delta\mu = mg\,\Delta Z - S\,\Delta T + \cdots \tag{1}$$

where ΔZ is the level difference, ΔT is the temperature difference, and the other terms have their usual meaning. In the theory of the ac Josephson effect[6] the chemical potential difference is synchronized to an impressed ac chemical potential of frequency v, and anomalies in the flow through the orifice should occur whenever

$$l\,\Delta\mu = nh v \tag{2}$$

where l and n are integers. In the search for the ac Josephson effect the gravitational term $mg\,\Delta Z$ has received the most attention,[1-5] while the entropy term $S\,\Delta T$, has just begun to become of interest.[7] In this work both the gravitational and entropy terms are considered in detail.

The experimental apparatus (Fig. 1) is constructed to allow for the independent variation of several of the parameters that apply to the Josephson effect. A coaxial capacitor is used to monitor the bath level confined within its annular region. A plunger allows the outside bath level to be changed by a known amount or to set the initial $\Delta Z = 0$ level at different values of Z, the level at which the inner bath falls within the capacitor. The two resistance thermometers R_{in} and R_{out} are monitored separately to follow the temperature of the two baths and give a resultant value of ΔT, the temperature difference between the two baths; and a heater inside the capacitor can be used to supply a dc temperature bias to the inside bath. By convention ΔZ is defined to be positive when the inside bath is higher than the outside bath. Likewise ΔT is defined to be positive when the inside bath is hotter than the outside bath. Two different transducers were used to provide the ac sound field, one a 100-kHz frequency standard, the other a PZT ceramic transducer. During an experimental run ΔZ and ΔT were monitored simultaneously and chart recordings were made of ΔZ vs. time (Fig. 2a) and ΔZ vs. ΔT (Fig. 2b).

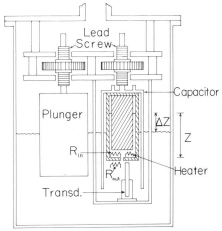

Fig. 1. Experimental apparatus. The level of the helium within the capacitor is indicated by Z and is measured from the bottom of the capacitor. The level difference ΔZ is positive if the inner level is higher than the outer level. The transducer is either a quartz crystal or a PZT ceramic transducer.

We found that it is important to measure ΔT at all times because the transducer, in addition to producing a level difference, also produces a temperature difference between the two baths, and this temperature difference changes when the bath level changes. Figure 2b illustrates this variation of ΔT with ΔZ. The section of Fig. 2b from $\Delta Z = \Delta T = 0$ to the point labeled I is the response of the system when the transducer is turned on. As the transducer pumps helium into the capacitor the temperature difference between the two baths varies with ΔZ. This "structure," or variation of ΔT, is reproducible and depends on Z rather than ΔZ. The various features of the ΔT vs. ΔZ curve occur when the liquid surface is at a particular point within the capacitor, independent of the point from which ΔZ is measured. Thus for a particular set of experimental parameters (i.e., temperature of the bath, voltage V, frequency ν, and orientation of the transducer) the system exhibits a unique Z–ΔT structure. In addition, portions of this Z–ΔT "structure" can be traced out with a dc temperature bias applied to the inner bath in the presence of the transducer's sound field.

By monitoring ΔZ vs. time we have observed many instances of "stable steps," values of ΔZ which are stable in time. Of those steps seen, only a few could be interpreted as Josephson "steps" based on the values of ΔZ at which they occur. However, by simultaneously monitoring the ΔZ–ΔT plane an additional criterion, that Eqs. (1) and (2) be satisfied, may be applied to those "steps" seen. Once a "step" has occurred and stabilized, a dc temperature bias may be applied to the inner bath, and the resultant response of the system monitored. In Fig. 2a three "steps" labeled I, II, and III are observed and their response to an impressed ΔT is indicated in Fig. 2b. For small ΔT, ΔZ changes very little, which agrees with the previous observations.[7] However, to satisfy Eqs. (1) and (2) the path in the ΔZ–ΔT plane should have been a straight line parallel to the fountain effect curve, which is observed and measured when the transducer is off. As more ΔT bias is applied a point is reached where ΔZ breaks from its stable "step" and proceeds to some other state. The amount of ΔT bias necessary to cause ΔZ to break varies, but it has been seen to be as high as 0.5 m°K.

To be candidates for the ac Josephson effect, any observed "steps" must depend

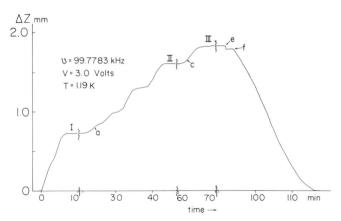

Fig. 2a. Pumping curve showing ΔZ vs. time. Three stable "steps" I, II, and III are seen. The points a, c, e, and f are the same points as indicated in Fig. 2b.

on ΔZ and not on Z. Consider Fig. 3a, which shows two emptying curves that are taken under the same experimental conditions except that the initial $\Delta Z = 0$ level has been shifted by 0.398 mm by the plunger for the second curve. Considered separately as ΔZ vs. time plots, they appear quite dissimilar and no correspondence between "steps" (labeled $A, B, \ldots$ in the left plot and I, II, $\ldots$ in the right plot) can be made. However, considered as Z vs. time curves, there is a strong correspondence between stable steps. This correspondence is made unambiguous when the concurrent ΔZ vs. ΔT plots measured simultaneously are considered. Figure 3b shows the ΔZ vs. ΔT curves corresponding to the ΔZ vs. time curves of Fig. 3a and the points corresponding to the "steps" are indicated by $A, B, \ldots$

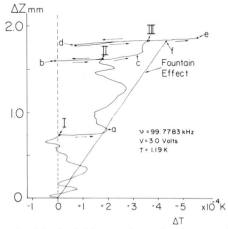

Fig. 2b. Pumping curve showing ΔZ vs. ΔT. When the signal to the crystal was first turned on the system moved along the curve from $\Delta Z = \Delta T = 0$ to I, the first stable "step." After this "step" stabilized, the dc temperature bias was turned on and increased, with the system moving along the path I $\rightarrow$ a. At a the dc temperature bias was held constant. The system now moved to II, the second stable "step." After "step" II stabilized, the dc temperature bias was reduced to zero, the system proceeding along path II $\rightarrow$ b. When the dc bias was increased again the system retraced path b $\rightarrow$ II and moves to c at which point the bias was again held constant. The system then moved to III, the third stable "step." Again the dc bias was reduced to zero, with the system going from III $\rightarrow$ d. When the bias was increased again the system moved from $d \rightarrow e$, the maximum bias attained at that time. At e the oscillator was turned off and the system moved immediately to point f. As the dc bias was then reduced to zero the system moved along the fountain effect curve $f \rightarrow \Delta Z = \Delta T = 0$.

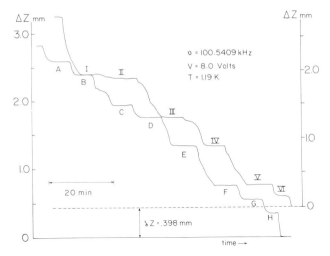

Fig. 3a. Two emptying curves taken under the same experimental conditions except that the $\Delta Z = 0$ point has been adjusted slightly to a higher value of Z by means of the plunger for the second curve. Several stable "steps" are seen and labeled for each curve. Notice that when the variation in Z, δZ, is taken into account there appears to be a correspondence between "steps" I and B, III and D, IV and E, V and F, and VI and G.

and I, II, ... Individual "steps" occur at particular points on the Z–ΔT structure and this establishes the value of Z at which the "steps" occur. A comparison of the two Z–ΔT planes with the individual "steps" superimposed on the Z–ΔT structure shows that the "steps" match on the Z–ΔT plane but not on the ΔZ–ΔT plane. Thus, these "steps" occur when the liquid level is at a particular level within the capacitor independent of the point from which ΔZ is measured. These steps appear not to be the Josephson effect.*

In addition, we have also varied the frequency of the crystal and have found that the "step" size varies opposite to that expected by Eq. (2).†

In conclusion we have found stable states of zero flow through an orifice in the presence of an ac sound field which are clearly not the Josephson effect. These states depend on the total height of the fluid in the inner bath and not on the level difference. In addition they are insensitive to a temperature difference for small differences. Careful inspection of the various curves yielded no evidence for the ac Josephson effect.

---

* One could mistake some of these "steps" as the Josephson effect if the helium level were placed at the "right" value. For example, state E in Fig. 3a would appear to be the $n = 1$ Josephson step if the initial $\Delta Z = 0$ level were raised by 0.38 mm.

† A more complete account of this will appear in a later publication.

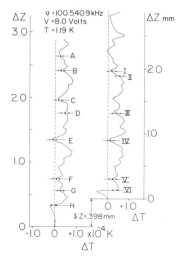

Fig. 3b. ΔZ vs. ΔT for the two emptying curves of Fig. 3a. This curve illustrates the fact that the structure in the ΔZ vs. ΔT curve produced by the transducer is a $Z-\Delta T$ structure rather than a $\Delta Z-\Delta T$ structure. A particular variation in ΔT occurs when the liquid level is at a particular value of Z independent of where ΔZ is measured from. The stable "steps" seen in Fig. 3a are superimposed on the $Z-\Delta T$ structure of this figure and labeled accordingly. The correspondence of "steps" I and B, III and D, IV and E, V and F, and VI and G shows that these stable "steps" occur at particular values of Z regardless of where ΔZ is measured from. Thus these "steps" are Z rather than ΔZ dependent.

References

1. P.L. Richards and P.W. Anderson, *Phys. Rev. Lett.* **14**, 540 (1965).
2. B.M. Khorana and B.S. Chandrasekhar, *Phys. Rev. Lett.* **18**, 230 (1967).
3. B.M. Khorana, *Phys. Rev.* **185**, 229 (1969).
4. P.L. Richards, *Phys. Rev.* **A2**, 1534 (1970).
5. J.P. Hulin, C. Laroche, A. Libeheber, and B. Perrin, *Phys. Rev.* **A5**, 1830 (1972).
6. P.W. Anderson, *Rev. Mod. Phys.* **38**, 298 (1966).
7. B.M. Khorana, D.H. Douglass, in *Proc. 11th Intern. Conf. Low Temp. Phys., 1968,* St. Andrews Univ. Press, Scotland (1969), Vol. 1, Section A3.2.

7
Ions and Electrons

Vortex Fluctuation Contribution to the Negative Ion Trapping Lifetime*

J. McCauley, Jr.

Physics Department, Yale University
New Haven, Connecticut

Introduction

In the early work of Donnelly and Roberts[1] on the trapping of negative ions by vortex lines in He II (bubble radius R, vortex circulation $\kappa = h/m$, and core parameter a) it was noted in passing that the trapping lifetime did not exhibit quite the right temperature dependence required by experiment. We have shown[2] that Donnelly and Roberts's theoretical result is valid only in the limit $\mathscr{E} \to \infty$ ($\mathscr{E}$ in V cm^{-1}), whereas the experimental data fall within the range $\mathscr{E} \lesssim 100$ V cm^{-1} and are properly described by the weak-field approximation[3]

$$\tau \approx v_B(0)/2\pi D b_0 \tag{1}$$

where

$$b_0 = 3/2[0.577 - \log \beta_\varepsilon q]^{-1}, \qquad \beta_\varepsilon q = 1.3 \times 10^{-2} \mathscr{E} (\rho_s/T^3)^{1/2}$$

where $q \propto R^{3/2}$ and we have set $R = 16$ Å; $D = \omega k T$, where ω is the ion's mobility ($\omega \sim e^{\Delta/T}, \Delta = 8.1°$, for the temperature range of interest: $1 \lesssim T \lesssim 2°$K); and $v_B(0)$, the bound-state partition function, is given by[2]

$$v_B(0) = \int\limits_{r \gtrsim q} e^{-u_v(r)} d^2 r \tag{2}$$

where $q \sim 100$ Å is the capture radius[2] and $u_v(r)$ is the ion-vortex potential energy divided by kT (these results follow from Brownian motion theory and are valid whenever $\mathscr{E} \lesssim 70$ V cm^{-1}). To an excellent approximation we have

$$\log \tau \approx -u_v(0) - (\Delta/T) + \text{const} \tag{3}$$

where the "constant" varies only logarithmically with T. The temperature variation for both the weak- and strong-field limits is thus essentially the same and for a hollow core of radius a is given by

$$\log \tau \approx \frac{\kappa^2 \rho_s \beta R}{2\pi} \log \frac{R}{a} - \frac{\Delta}{T} + \text{constant} \tag{4}$$

$$\frac{\partial}{\partial \beta} \log \tau \approx \frac{\kappa^2 \rho_s}{2\pi} R \log \frac{R}{a} \left[1 + \frac{\Delta \rho_n}{T \rho_s} \right] - k\Delta \tag{5}$$

* Work supported by NIH Grant No. 13190.

$(\beta = 1/kT)$ where $\rho_n = \rho - \rho_s \sim e^{-\Delta/T}$ (rotons dominate). The results of assuming a solid-body rotating core of radius a differ only negligibly from Eq. (4) and Eq. (5). We will eventually see that it is this *classical* treatment of the "core" which is the cause of the error committed. The discrepancy (plotting $\log \tau$ vs. β) may be stated as follows: For no choice of the arbitrary parameters R and a can both Eq. (4) and Eq. (5) be made to fit the data.[4] The following observations are an essential step in our analysis: bubble radii of 12–17 Å are indicated by reliable considerations[5] and we assume that $a \gtrsim 1.75$ Å since the ^{4}He interparticle spacing is roughly 3.5 Å. With these two restrictions Eq. (4) intersects the data at one point, say T_i, and Eq. (5) is everywhere too small.

Thermodynamic Analysis of the Discrepancy

We have shown elsewhere that[6]

$$u_v(r) = \beta(F(r) - F(\infty)) = \beta \, \Delta F(r) \tag{6}$$

where $F(r)$ is the free energy of the liquid with line–bubble separation r and $\Delta F(r)$ is just the drop in free energy of the liquid due to the extra boundary condition represented by the bubble. Denoting internal energy and entropy by $U(r)$ and $S(r)$, respectively, and using $F = U - TS$ and $U = (\partial/\partial\beta)(\beta F)$, we then obtain the very general results

$$\log \tau \approx -\beta \, \Delta F(0) - (\Delta/T) + \text{constant} \tag{7}$$

and

$$(\partial/\partial\beta)\log \tau \approx -\Delta U(0) - k\Delta \tag{8}$$

Consider the point T_i where Eq. (4) intersects the data. We see immediately that the slope error is due entirely to a neglected entropy, implying that we have neglected important degrees of freedom of the liquid. In this respect we have thus far neglected degrees of freedom of the vortex itself and so we naturally turn to the consideration of a fluctuating vortex line. We also assert the following: The "missing entropy" resides in thermal vortex fluctuations with wavelength $\lambda \lesssim R$ since these are not transmitted by the trapped bubble.[7] In order to give evidence for our assertion, we will employ the quantum hydrodynamic model of a vortex line first suggested by Fetter.[8] An interesting by-product of this work is that by a slight extension of Fetter's model we are led to suggest an entirely new and nonclassical definition for the "core parameter" of a quantized vortex. First, let us give an estimate for the size of the discrepancy to be accounted for: With $R \lesssim 17$ Å and $a \gtrsim 2$ Å, $T_i \gtrsim 1.8°$ and, denoting the missing entropy by $\delta[-\Delta S(0)]$,

$$\delta[-\Delta S(0)] = \frac{1}{RT_i}\left\{\left[\frac{\partial}{\partial\beta}\log\tau\right]_{\text{exp}} - \left[\frac{\partial}{\partial\beta}\log\tau\right]_{\text{theor}}\right\}_{T_i} \gtrsim 25k$$

where $[(\partial/\partial\beta)\log\tau]_{\text{theor}}$ is given by (5). We will now briefly outline Fetter's model and our extension of it.

Quantized Vortex Fluctuations

Consider a classical line with distributed vorticity at zero temperature:

$$\nabla \times \mathbf{v}_s = \boldsymbol{\omega} \tag{9}$$

where $\boldsymbol{\omega} = \omega\mathbf{k}$ is the vorticity. Fetter's model is obtained by assuming

$$\omega = \begin{cases} \kappa/\pi a^2, & r \leqq a \\ 0, & r > a \end{cases} \tag{10}$$

and we point out that this is the simplest thing one could get away with had one set out to represent ω by a probability distribution. For an incompressible fluid $\mathbf{v}_s = \nabla \times \mathbf{A}$, so that the energy of the liquid may be written

$$\tfrac{1}{2} \rho_s \int v_s^2 dV = \tfrac{1}{2} \rho_s \oint \mathbf{v}_s \times \mathbf{A} \cdot d\mathbf{S} + \tfrac{1}{2} \rho_s \int \mathbf{A} \cdot \boldsymbol{\omega} dV$$

where the second integral on the right-hand side is the "core" energy. With no internal boundaries present in the liquid

$$\mathbf{A}(\mathbf{r}) = \frac{1}{4\pi} \int \frac{\omega(\mathbf{r}')}{|\mathbf{r} - \mathbf{r}'|} dV' \tag{11}$$

and the core energy is given by

$$E = \frac{1}{(\pi a^2)^2} \frac{1}{2} \int_{r \leqq a} d^2 r_1 \int_{r \leqq a} \mathscr{E}(\mathbf{r}_1, \mathbf{r}_2) d^2 r_2 \tag{12}$$

where

$$\mathscr{E}(\mathbf{r}_1, \mathbf{r}_2) = \frac{\rho_s \kappa^2}{4\pi} \int \frac{d\mathbf{z}_1 \cdot d\mathbf{z}_2}{|\mathbf{R}_{12}|} \tag{13}$$

is the interaction energy of two classical vortex filaments, $d\mathbf{z} = dz\mathbf{k}$, $dV = d^2 r dz$, $\mathbf{R}_{12} = \mathbf{r}_1 - \mathbf{r}_2 + (z_1 - z_2)\mathbf{k}$; and we are using cylindrical coordinates. Fetter began by writing down (13) and defining (12) to be the self-energy of the line, having introduced the parameter a in direct analogy with the self-inductance calculation of classical electromagnetic theory.[8] The "core" is thus seen as the small area πa^2 over which the filament interacts with itself.

In order to study a fluctuating line, we make the replacements[8] $\mathbf{r}_{12} \to \mathbf{r}_{12} + \varepsilon[u_1(z_1) - u_2(z_2)]\hat{R}$ and $d\mathbf{z}_i \to dz_i[1 + \varepsilon(du_i/dz_i)]\hat{k}$, where $\mathbf{r}_{12} = \mathbf{r}_1 - \mathbf{r}_2$, and expand (12) to second order in the fluctuations:

$$E(\varepsilon) \approx E(0) + \varepsilon^2 \cdot \frac{1}{2}\left(\frac{\partial^2 E}{\partial \varepsilon^2}\right)_{\varepsilon = 0} = E(0) + \Delta E \tag{14}$$

The second-order term ΔE is the fluctuation energy quantized by Fetter, the cannonically conjugate variables being $(\rho_s \kappa)^{1/2} u_{x,i}$ and $(\rho_s \kappa)^{1/2} u_{y,i}$. Fetter's result, after Fourier-analyzing the $\mathbf{u}_i(z_i)$, is

$$\Delta E = \sum_k \left(n_k + \frac{1}{2} \right) \hbar\omega(k) \tag{15}$$

We now come to our new interpretation for the "core." Classically, $\Delta E = 0$ at $T = 0$ but quantization has introduced a zero-point energy which conceptually replaces the purely classical term $E(0)$. We thus discard the irrelevant term $E(0)$, noting that its introduction has been merely a trick which has permitted us to obtain the fluctuations to be quantized, and we accordingly identify a^2 as the mean square fluctuation in line position due to zero-point motion. The vortex singularity is thus "smeared" in accordance with the Heisenberg principle and the "core" is defined by the everpresent, *purely quantum* fluctuations. We have been exceedingly fortunate in that the classical "extension" initially introduced in order to make the self-energy finite readily admits of a quantum interpretation.

Denoting Bessel functions of imaginary argument by I_μ and K_μ, Fetter has also given the results

$$\omega(k) = (\kappa/4\pi)\, k^2 \langle K_0 \rangle \qquad (16)$$

and

$$\langle K_0 \rangle = [1/(\pi a^2)^2] \int_{r_1 \leq a} d^2 r_1 \int_{r_2 \leq a} d^2 r_2\, K_0(k|\mathbf{r}_{12}|) \qquad (17)$$

which he has evaluated only for the case $ka \ll 1$.[8] We have obtained the exact result[6]

$$\omega(k) = (\kappa/\pi a^2)\, [\tfrac{1}{2} - I_1(ka)\, K_1(ka)] \qquad (18)$$

Denoting the wavenumber cutoff by k_{max}, the zero-point mean square fluctuation by $\langle \mathbf{u}^2 \rangle_0$, and asserting that $a^2 = \langle \mathbf{u}^2 \rangle_0$, we have

$$a^2 = \hbar k_{max}/\pi \rho_s \kappa \qquad (19)$$

The cutoff k_{max} is undetermined by present theory ($\lambda_{min} = 2\pi/k_{max}$) but we assume that $\lambda_{min} \sim (m/\rho_s)^{1/3}$ in accordance with the assumption that, at least to a first approximation, the normal fluid does not "circulate." At $T = 1.7°$ this gives $\lambda_{min} = 4.0$ Å and $a = 2.3$ Å. We are now ready to estimate the missing entropy.

Fluctuation Contribution to the Lifetime

The correct treatment of this problem requires first the solution of a difficult hydrodynamic boundary value problem, treating the bubble as a boundary condition in the liquid in the presence of a deformed vortex line. Fortunately, we find it possible to estimate the desired quantity without the help of the solution to the above-mentioned problem. If the missing entropy is to be found in vortex fluctuations with $\lambda_{min} < \lambda < R$, then there are two main effects to consider: (1) the set of standing waves is reduced, owing to the additional boundary conditions to be satisfied; and (2) for modes with $\lambda < R$ the line is effectively shorter by $\Delta L \sim 2R$. The first is presumably a smaller effect, but its proper treatment should provide a density of states which negates the present necessity for introducing the arbitrary upper cutoff $\lambda_{max} = R$.[7] We can estimate the main effect 2 as follows: For a line of length L the free energy is

$$F = \frac{L}{\pi\beta} \int_0^{k_{max}} \left[\beta\frac{\hbar\omega}{2} + \log(1 - e^{-\beta\hbar\omega}) \right] dk \qquad (20)$$

and the "missing entropy" is roughly

$$\delta[-\Delta S(0)] \sim -\frac{2R}{\pi a} \int_{2\pi a/R}^{k_{max}} \left[\log(1 - e^{-\beta\hbar\omega}) - \frac{\beta\hbar\omega}{e^{\beta\hbar\omega} - 1} \right] d(ka) \qquad (21)$$

With $T_i = 1.8°$ and $R = 16$ Å, $a \approx 2.3$ Å, $\lambda_{min} \approx 4.0$ Å, and

$$\delta[-\Delta S(0)] \sim 18k$$

which is clearly in the right ballpark.* Stronger statements, however, must await the solution of the previously noted boundary value problem.

Acknowledgments

The author is indebted to Professor Onsager, who first suggested that "it may be of interest to consider a 'bent' vortex" (this work is to be submitted for publication jointly with Professor Onsager) and also to Drs. Alexander Fetter and Timothy Padmore for helpful correspondence and criticism.

References

1. R.J. Donnelly and P.H. Roberts, *Proc. Roy. Soc. London* **A312**, 519 (1969).
2. J. McCauley and L. Onsager, Electrons and Vortex Lines in He II, I: Brownian Motion Theory of Capture and Release, Yale Univ. Preprint (1972).
3. L. Onsager and J. McCauley, Electrons and Vortex Lines in He II, II: Theoretical Analysis of the Experiments, Yale Univ. Preprint (1972).
4. W.P. Pratt, Thesis, Univ. of Minnesota, 1967.
5. J. Jortner, N. Kestner, S. Rice, and M. Cohen, *J. Chem. Phys.* **43**, 2614 (1965).
6. J. McCauley and L. Onsager, Vortex Fluctuations and the Negative Ion Trapping Lifetime: Quantum Theory of the Vortex "Core," Yale University Preprint (1972).
7. A.L. Fetter and I. Iguchi, *Phys. Rev.* **2A**, 2067 (1970).
8. A.L. Fetter, *Phys. Rev.* **162**, 143 (1967).

* The corresponding change in free energy is relatively small.

The Question of Ion Current Flow in Helium Films

S.G. Kennedy and P.W.F. Gribbon

Department of Physics, University of St. Andrews
North Haugh, St. Andrews, Scotland

Maraviglia[1] has reported briefly some observations which suggest that an ion current could exist in the saturated film of He II. The film existed on the insulated dielectric surfaces above the gas–liquid interface in an ion cell. He found the same dc characteristics in the film and the bulk liquid and concluded that both positive and negative ions could move freely along the film. The possibility of a current in the gas rather than in the film was ruled out, but our paper questions the validity of his conclusion. Later Bianconi and Maraviglia[2] reported that periodic oscillations in the moving superfluid in a film modulated the positive ion current in the gas at the same frequency ω and the film current at 2ω. Our results with positive ions cannot be reconciled with their conclusions.

We have measured the ion currents as a function of temperature from 0.9 to 4.2°K for various applied electric fields from 10 V cm^{-1} to 200 V cm^{-1} in both perspex and glass-walled diode and triode cells. A triode cell is shown in the Fig. 1 insert: the current I_g in the gas was measured by a plate C_g, the current I_f "in the film" by a ring C_f; care was taken that $I_f = 0$ when the cell was full of liquid. The $\ln I$ vs. $1/T$ characteristics were obtained for various fields[3-5]; a typical characteristic for our negative ion current flow in a perspex-walled triode cell is shown in Fig. 1. It shows that while most of the source emission current I_s was collected as a grid current I_g, some current was transmitted through the grid and crossed the gas–liquid interface to reach C_g in the gas or "passed up the film" to reach C_f.

The important question was whether the collector C_f collected I_f by a preferential current path through the film or by a path through the main body of the gas and which only entered the film where it covered C_f. We are inclined to the second view, but to discuss our observations we will outline first the nature of the gas–liquid interface. It is well known[6,7] that the interface presents a potential energy barrier $\phi_b = 44°$K to a negative ion approaching the interface. This is shown in Fig. 2, where $2(AeE)^{1/2}$ comes from the applied field E across the interface and Δ/k from the ion mobility. The "apparent" barrier θ^* can be measured from the linear portion of the $\ln I^-$ vs. $1/T$ characteristic, but the "true" barrier θ is only obtained when the ion space charge density was constant, and the effective field was the same as the applied field in the potential well under the interface. No distinction in the nature of the barrier can be seen by ions in the bulk liquid or in the liquid comprising the film. Ions entered the gas by a thermal activation process in both cases. The current flow through either the bulk or film interface therefore should show a similar dependence on field and temperature.

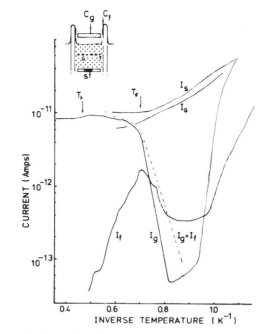

Fig. 1. A schematic current vs. reciprocal temperature characteristic for a perspex-walled triode cell (shown in the insert) for an applied field of 100 V cm^{-1}, with an interface–grid distance 0.4 cm and a grid–collector distance 1.5 cm.

This was confirmed by our observations. Both I_g^- and I_f^- appeared to respond in the same way, and indicated that the same process had contributed to the current flow to both collectors. From the λ-point T_λ to a temperature T_σ the current was maintained while space charge built up to an equilibrium density and reduced the field within the liquid. Irregular decreases in I_g^- were partially, but never completely, balanced by increases in I_f^-. An increase in I_f^- was thought to be due to a change in the field distribution above the interface throughout the cell caused by a change in the space charge distribution under the interface. Below T_σ the thermal activation process controlled I_g^-, which decreased exponentially with temperature. At the same temperature the "film" current I_f^- began to decrease: A preferential film path would have meant an increase, not a decrease, in I_f^-. We conclude that in these two temperature regions the consistent behavior of both I_g^- and I_f^- suggested a common origin in a thermally activated process through the interface.

At low temperatures $T \lesssim 1.1°$K both I_g^- and I_f^- showed a similar sharp current increase to a maximum I_g^- or $I_f^- > I_s^-$ at $T \simeq 0.9°$K. This was attributed to pre-breakdown ion multiplication in the low-pressure gas. Both I_g^- and I_f^- could show heavily damped oscillations with the same frequency as the mechanical film oscillations in and out of a closed cell. This occurred only when there was rapid cooling to low temperatures: I_g^+ and I_f^+ oscillations were not detected.[2] These observations again suggested a common origin for I_g^- and I_f^-.

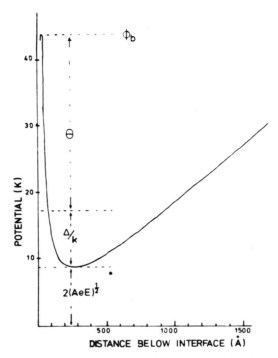

Fig. 2. The potential barrier of the gas–liquid interface,
after Schoepe and Rayfield.[7] The current through the
barrier depends exponentially on θ and Δ/k.

Further support comes from a comparison of the barrier values θ^* obtained in
our results and those obtained by other workers[3–5,8] in different types of cells. For
example, the slope of the I_g^- vs. $1/T$ characteristics gave high values $\theta^* \simeq 31.5°$K
associated with the current I_g^- in dielectric-walled cells where I_f^- had been ignored;
our value was $\theta^* = 31 \pm 2.5°$K averaged over the higher fields. Lower values $\theta^* =
27°$K were obtained from the total current measured[7] in cells whose walls were
guarded by a repulsive field which diverted all the current through the interface to a
single collector C_g: Our average value from the $(I_g^- + I_f^-)$ characteristics was
$\theta^* = 26.6 \pm 0.8°$K. Our values are quoted for a given interface position for the higher
fields: The interface position affected the current distribution and therefore the slope
of the I_g^- vs. $1/T$ characteristic, while the higher field values were known to be more
reliable.[4] Our observations were consistent with a current redistribution in the gas
rather than with a preferential current path in the film.

Our results with positive ions were inconsistent with those of Maraviglia[1]
and Bianconi and Maraviglia.[2] I_g^+ and I_f^+ increased slowly with temperature from
their minimum values above T_λ. There was no evidence for either a preferential film
current,[1] a current peak, or oscillations.[2] We attributed I_g^+ and I_f^+ to photoelectrons
from C_g emitted by ultraviolet radiation coming from atom deexcitation and ion
recombination close to the source. Our experiments gave no conclusive evidence
for a current flow in a saturated film of He II.

References

1. B. Maraviglia, *Phys. Lett.* **25A**, 99 (1967).
2. A. Bianconi and B. Maraviglia, *J. Low Temp. Phys.* **1**, 201 (1969).
3. L. Bruschi, B. Maraviglia, and F.E. Moss, *Phys. Rev. Lett.* **17**, 682 (1966).
4, W. Schoepe and C. Probst, *Phys. Lett.* **31A**, 490 (1970).
5. R. Williams and R.S. Crandall, *Phys. Rev.* **A4**, 2024 (1971).
6. G.W. Rayfield and W. Schoepe, *Phys. Lett.* **34A**, 133 (1971).
7. W. Schoepe and G.W. Rayfield, *Z. Natur.* **26A**, 1392 (1971).
8. R. Williams, R.S. Crandall, and A.H. Willis, *Phys. Rev. Lett.* **26**, 9 (1971).
9. F. Hereford and F.E. Moss, *Phys. Rev.* **141**, 204 (1966).

Impurity Ion Mobility in He II [*]

Warren W. Johnson and William I. Glaberson[†]

Department of Physics, Rutgers University
New Brunswick, New Jersey

We have previously[1] reported an experimental determination of the mobility of various impurity ions injected into liquid helium. The major conclusion of that investigation was that the ion mobility depends on the atomic number of the impurity atom in a way that cannot be explained simply in terms of the mass of the ion. It was clear that the ion–roton scattering interaction depends on the electronic structure of the core ion. We now report a more detailed examination of mobilities of ^{40}Ca and ^{48}Ca ions so as to isolate the effects of ionic mass on mobility, and discuss the data in terms of recent theories of roton-limited ion mobility.

Our experimental procedure involves producing a continuous electrical discharge in helium vapor above the liquid level. A fine tungsten filament is coated, prior to cooldown, with a solution or suspension of some compound containing the positive ion of interest. The filament is heated, introducing the compound into the discharge, and the appropriate ions are drawn through the helium surface by an electric field. The ions then enter a mobility cell and simultaneous measurements of the helium and impurity ion mobilities are made. A typical output of our time-of-flight mobility apparatus is shown in Fig. 1 illustrating the precision of the technique.

Some of our measured relative mobilities are shown in Fig. 2. Measurements could not be made at temperatures significantly below 1.2°K because of the power dissipated in the discharge and filament, and above 1.4°K it was difficult to keep the discharge stable—presumably because of the increased vapor pressure.

The scatter in the experimental data within a particular run was less than 0.1%, whereas run-to-run variations were sometimes as large as 0.3%. Since the difference in the mobilities of ^{48}Ca and ^{40}Ca ions was found to be only $\sim 1\%$ of the mobility, a modified technique was used to minimize errors in this difference. Two filaments were employed, each of which was coated with a different isotope of Ca. The data were then taken for ^{40}Ca and ^{48}Ca ions without having to warm up to room temperature or disassemble the cell to replace filaments between experiments. The experimental data for the two Ca isotopes are shown in Fig. 3. For proper comparison with theoretical ion mobilities the quantity $\mu^{-1}e^{\Delta/kT}$ is plotted. Here Δ is the temperature-dependent roton energy gap taken from a fit to the neutron scattering data.[3,4] The phonon contribution to the drag on helium ions in the temperature range of our experiments is calculated to be small[2] and we neglect it for the Ca ions.

[*] Supported in part by a grant from the National Science Foundation.
[†] Alfred P. Sloan Foundation Fellow.

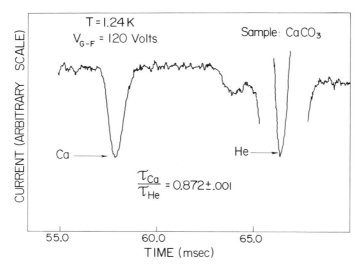

Fig. 1. Typical output of time-of-flight mobility apparatus.

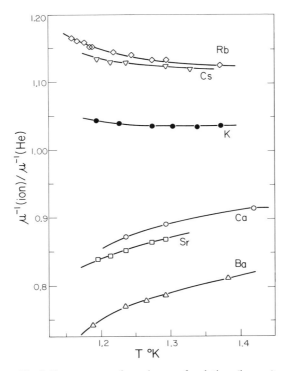

Fig. 2. Temperature dependence of relative (inverse) impurity ion mobilities.

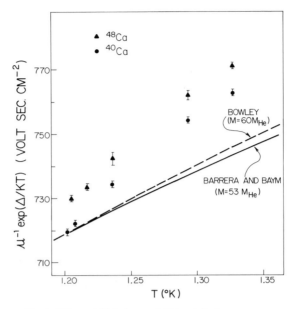

Fig. 3. ^{40}Ca and ^{48}Ca ion mobilities. Δ is the temperature-
dependent roton energy gap. The lines are an attempt to
fit ^{40}Ca ion mobilities to both the temperature and mass
dependences of the mobility with a calculation of Barrera
and Baym assuming constant total ion–roton cross section
and a calculation of Bowley assuming constant differential
cross section. The error bars do not include possible systematic errors associated with conversion of our relative
mobilities to absolute mobilities. Absolute helium ion
mobilities were taken from smoothed data of Schwartz.[9]

Bowley[5] and Barrera and Baym,[4,6] although arguing along different lines, have
found an identical expression for the mobility in terms of the transition matrix for
ion-excitation collisions which includes the effect of ion recoil (finite ion mass). By
making several different assumptions about the ion–roton transition matrix elements, these authors obtain several different explicit expressions for the roton-
limited mobility which we have attempted to fit to our Ca ion data. Both considered
the situation in which the ion–roton transition matrix element was independent of
roton momentum and obtained a mobility expression of the form

$$\mu^{-1}e^{\Delta/kT} \propto Mf_1(MT) \qquad (1)$$

where M is the ion mass. In the temperature range of our experiment f_1 decreases
with increasing argument. It is possible, by adjusting M, to fit our observed mass
dependence (for $M \sim 80M_{\text{He}}$), but the slope of the temperature dependence has
the wrong sign.

Bowley also calculated the mobility assuming that the differential scattering
cross sections were constant within the kinematically allowed range of parameters

and obtained an expression of the form

$$\mu^{-1}e^{\Delta/kT} \propto f_2(MT) \tag{2}$$

Fitting this expression to our observed mass dependence, we obtained the dashed line in Fig. 3 for ^{40}Ca ions with $M \sim 60M_{\mathrm{He}}$.

Barrera and Baym considered the assumption that the total scattering cross section (as opposed to the differential cross section) was constant. They obtained an expression of the form

$$\mu^{-1}e^{\Delta/kT} = \alpha_1 f_1(MT) + \alpha_2 f_3(MT) \tag{3}$$

where f_3 increases with increasing argument. The best fit of our data to this expression occurs for $\alpha_1/\alpha_2 = 0$ and $M \sim 53M_{\mathrm{He}}$ and is shown as the full lines in Fig. 3 for ^{40}Ca ions.

It is interesting to note that in both Eqs. (2) and (3) the mass dependence and temperature dependence of the mobility are coupled in a simple way so that Rb, for example, with a temperature dependence opposite in sign to that of Ca, should also have an isotopic mass dependence opposite in sign—a surprising result. We were somewhat surprised that, although the temperature dependence of the mobility could not be made consistent with the mass dependence under the various assumptions described above, fitting the mass dependence gave reasonable values for the Ca ionic mass under any of the assumptions.

Clearly, none of the assumptions of constant matrix element, constant differential cross section, or constant total cross section does very well in explaining our data. A more detailed picture of the ion roton interaction, perhaps along the lines suggested by Iguchi[7] or Ihas,[8] is obviously required.

Acknowledgment

We should like to acknowledge useful discussions with Professor T. Tsuneto.

References

1. W.W. Johnson and W.I. Glaberson, *Phys. Rev. Lett.* **29**, 214 (1972).
2. K.W. Schwarz (to be published.).
3. D.G. Henshaw and A.D.B. Woods, *Phys. Rev.* **121**, 1266 (1961).
4. R.G. Barrera and G. Baym, *Phys. Rev.* (to be published).
5. R.M. Bowley, *J. Phys. C: Sol. St. Phys.* **4**, 1645 (1971).
6. R.G. Barrera, Thesis, Univ. of Illinois, 1972 (University Microfilms, Ann Arbor, Michigan).
7. I. Iguchi, *J. Low Temp. Phys.* **4**, 637 (1971).
8. G. Ihas, Thesis, Univ. of Michigan, 1971 (University Microfilms, Ann Arbor, Michigan).
9. K.W. Schwarz (to be published).

Measurement of Ionic Mobilities in Liquid ³He by a Space Charge Method*

P. V. E. McClintock

Department of Physics, University of Lancaster
Lancaster, England

Introduction

The behavior to be expected of a charged impurity moving in a Fermi fluid was first discussed in detail by Abe and Aizu[1] and independently by Clark.[2] They concluded that, although mobility should be inversely proportional to viscosity at high temperatures (classical limit), a T^{-2} law should be obeyed in the low-temperature (degenerate) limit. Davis and Dagonnier have discussed the situation at intermediate temperatures in terms of quantum mechanical Brownian motion.[3] These predictions were not borne out by experiment.[4,5] In a more recent theoretical investigation Josephson and Lekner[6] show that, although the T^{-2} law may still be expected at sufficiently low temperatures, some form of weaker dependence should be observed at higher temperatures. Since there seems to be little consensus as to the form of this weaker dependence or as to the characteristic temperature at which one regime might give way to the other, it is important that accurate experimental mobility data should be obtained. For negative ions such data are now available down to 17 m°K.[5] However, for positive ions data below 1°K,[4] measured by a time-of-flight method, suffer from severe inconsistencies and hysteresis effects, apparently experimentally based. In this paper we describe mobility measurements down to 0.25°K by a completely different technique: space-charge-limited emission of ions from a sharp metal point. This method appears to be more accurate than earlier space charge techniques[7] and avoids the complication of thermal gradients due to heating at a radioactive source.

Field Emission and Field Ionization in Liquid ³He

The characteristics of field emission and field ionization in liquid ³He have been studied in detail.[8,9] In both cases there is a space-charge-limited regime for which the emission is well described by[10]

$$V = V_0 + 98(R/\alpha K\mu)^{1/2} i^{1/2} \tag{1}$$

where V is the potential in V applied to the tip, V_0 is a constant, R is the anode radius

* Work supported by the Science Research Council under contract BSR9251.

in cm, $\alpha\pi$ is the emission cone angle in steradians, K is the dielectric constant, μ is the mobility in cm^2 V^{-1} sec^{-1}, and i is the current in nA. In using Eq. (1) to determine mobilities, the largest uncertainty lies in α. For emission in vacuum $\alpha \approx 0.6$, but in liquid helium we have found $2.2 < \alpha < 3.2$, the exact value depending on the individual tip. Since mobility *changes* smaller than 1% can be resolved, it is most useful to determine the temperature dependence of the mobility by our space charge method and then scale the data to agree with the absolute value of μ at some particular temperature, obtained by another technique.

We have found that emission characteristics can be strongly influenced by the build up of thermal gradients in the liquid.[9] To avoid this effect, all data reported here were taken by applying potential to the tip only just long enough for a chart recorder to record the current, i.e., about 1.5 sec. A simple calculation shows that during this period the *average* temperature of the liquid rises by typically 30m°K. However, the maximum temperature rise probably occurs near the tip where the electric field is largest, and the change in temperature near the anode in the space charge region which controls the emission is presumably much smaller.

Details of the apparatus and experimental technique are discussed elsewhere.[9] The tungsten tips were prepared by electrolytic etching in NaOH solution and were smoothed *in situ* by field evaporation. Their radii of curvature were determined from the vacuum field-emission characteristics. Temperature measurement was by a 470-Ω Speer carbon resistor immersed in the sample within the anode assembly. It was calibrated against ^{3}He vapor pressure down to 0.45°K and the calibration was then extrapolated to 0.25°K. Care was taken to keep the sample pressure constant since we have found that α is weakly pressure dependent at low pressures.[9]

Experimental Results

Current–voltage characteristics for both positive and negative emission are presented as $i^{1/2}$–V plots in Fig. 1 for three different temperatures. Good straight lines are obtained in agreement with Eq. (1), and the intercepts are $V_0 = 410 \pm 20$ V and 490 ± 30 V for negative and positive ions, respectively. Although we can, in

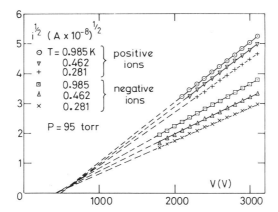

Fig. 1. Positive and negative ion emission (currents)$^{1/2}$
as a function of tip potential.

principle, determine $\mu(T)$ by measuring the gradient at a number of different temperatures, this is not a convenient method in practice because of the large number of $i(V)$ readings necessary to determine μ at a single temperature.

If we rewrite Eq. (1) in the form

$$\mu(T) = [98^2 R/\alpha K(V - V_0)^2]\, i(T) \tag{2}$$

we see that at constant tip potential, changes in mobility are directly proportional to changes in emission current. It is convenient to normalize the current by dividing it by $(V - V_0)^2$, so that data taken at different tip potentials ought to fall on the same line. In Figs. 2 and 3 we plot the normalized negative and positive emission currents, respectively, against temperature for three different tip potentials in each case. Almost all the points fall within less than $\pm 1\%$ of a "best fit" curve drawn through the data. Scaling our values of $i/(V - V_0)^2$ to agree with earlier mobility measurements[4] at 1.0°K, we obtain the mobilities indicated by the right-hand scales of Figs. 2 and 3. The error in μ relative to its value at 1.0°K is therefore indicated by the scatter of the data points and the systematic error in its absolute magnitude will be the $\pm 10\%$ quoted by Anderson et al.[4]

Discussion

From the linearity of the results shown in Fig. 1 we deduce that for both positive and negative ions the mobility is independent of electric field within our experimental range, i.e., up to several kV cm^{-1}, which is, as far as we are aware, the highest field for which the ionic mobility in liquid ^{3}He has been measured. Careful scrutiny of the data suggests that the change in mobility resulting from a 50% change in average electric field is certainly less than 1%. Therefore the fact that the field in which $\mu(T)$

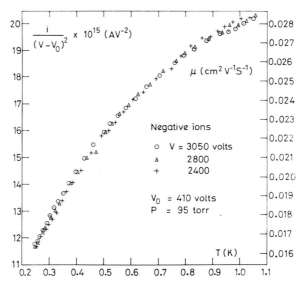

Fig. 2. Temperature dependence of the (normalized) negative ion emission current.

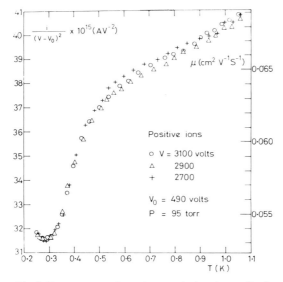

Fig. 3. Temperature dependence of the (normalized) positive ion emission current.

is measured is large and nonuniform has little effect on the reliability of our measurements.

The variation of μ_- with temperature (Fig. 2) agrees well with the earlier results of Anderson et al.[4] in the region of overlap, but the scatter of our data points appears to be smaller. The temperature dependence of μ_+ (Fig. 3) is certainly not inconsistent with the earlier results,[4] but there is such a large scatter in the latter (about $\pm 10\%$) that a meaningful comparison is difficult.

None of the existing theories predicts correctly the temperature dependence of μ_+ or μ_- in this temperature range. Only Walden's rule gives the correct sign for the temperature dependence. Between 1.0 and 2.0°K we find μ_- inversely proportional to viscosity within experimental error, using the smoothed viscosity data of Betts et al.[11] Below 1.3°K for μ_+ and 1.0°K for μ_- the observed temperature dependence is much weaker than that implied by Walden's rule.

The minimum in μ_+ near 0.3°K was unexpected and is very interesting.* A qualitative explanation can be given in terms of the "heliumberg" model of the positive ion.[13] On this model the minimum in the melting curve near 0.3°K[14] would imply a maximum in the positive ion radius at the same temperature and therefore, neglecting other effects, a minimum in mobility. However, it is hard on this basis to understand the shape of the $\mu_+(T)$ curve. In particular, the maximum value of $d\mu_+/dT$ occurs near 0.4°K, at which temperature the melting pressure is only 0.5% above its minimum value and is changing very slowly.

Our present mobility-measuring technique avoids complications arising from the heat continuously generated by a conventional radioactive source. In contrast, field emission/ionization sources generate no heat whatever except when actually

---

* Kuchnir[12] reports recent observation of this minimum with an improved time-of-flight technique.

emitting current, and they would therefore seem particularly suitable for use in future experiments at ultralow temperatures. To extend our present investigation to lower temperatures, it is clear that much shorter current pulses will be required. Preliminary experiments in He I are encouraging: we find that pulse durations of 1 msec should be quite sufficient to carry out the measurements.

Conclusion

Between 0.25 and 2.0°K the mobilities of both positive and negative ions in liquid ^{3}He at low pressure are independent of electric field up to several kV cm^{-1}. We have measured the temperature dependence of the mobilities with greater resolution than that of previously published work. For negative ions our results are in excellent agreement with the earlier work. For positive ions they are not in disagreement but, with our superior resolution, we have been able to observe a definite minimum in μ_+ at around 0.3°K. No existing theory gives a satisfactory account of the mobility in this temperature range. In view of the great simplicity and good resolution of our space charge technique it seems highly desirable to extend the measurements down into the fully degenerate temperature regime, and we propose to do so at the earliest opportunity.

References

1. R. Abe and K. Aizu, *Phys. Rev.* **123**, 10 (1961).
2. R. C. Clark, *Proc. Phys. Soc.* **82**, 785 (1963).
3. H. T. Davis and R. Dagonnier, *J. Chem. Phys.* **44**, 4030 (1966).
4. A. C. Anderson, M. Kuchnir, and J. C. Wheatley, *Phys. Rev.* **168**, 261 (1968).
5. M. Kuchnir, P. R. Roach, and J. B. Ketterson, *Phys. Rev.* **A2**, 262 (1970).
6. B. D. Josephson and J. Lekner, *Phys. Rev. Lett.* **23**, 111 (1969).
7. P. de Magistris, I. Modena, and F. Scaramuzzi, in *Proc. 9th Intern. Conf. on Low Temp. Phys. 1964.* Plenum Press, New York (1965), p. 349.
8. P. V. E. McClintock, *Phys. Lett.* **35A**, 211 (1971).
9. P. V. E. McClintock, *J. Low Temp. Phys.* **11**, 15 (1973).
10. B. Halpern and R. Gomer, *J. Chem. Phys.* **51**, 1031 (1969).
11. D. S. Betts, B. E. Keen, and J. Wilks, *Proc. Roy. Soc. A* **289**, 34 (1966).
12. M. Kuchnir, Private communication; M. Kuchnir, J. B. Ketterson, and P. R. Roach, this volume.
13. K. R. Atkins, *Phys. Rev.* **116**, 1339 (1959).
14. D. O. Edwards, J. L. Baum, D. F. Brewer, J. G. Daunt, and A. S. McWilliams, in *Proc. 7th Intern. Conf. on Low Temp. Phys. 1960,* North-Holland, Amsterdam, (1961), p. 610.

Pressure Dependence of Charge Carrier Mobilities
in Superfluid Helium

R.M. Ostermeier and K.W. Schwarz

Department of Physics and The James Franck Institute
University of Chicago, Chicago, Illinois

Accurate and extensive measurements were made of the mobility of positive and negative charge carriers in ^{4}He at pressures up to the melting pressure and temperatures in the range $0.27 < T < 1°$K. The data are analyzed in terms of the scattering of phonons and rotons by the carriers. The phonon-dominated mobility of the negatives agrees well with existing theories, and analysis of our results provides a new determination of the electron bubble radius as a function of pressure. The phonon term for the positives is calculated from the electrostriction model and is in good agreement with the experimental results. Subtracting the calculated phonon contribution from our data results in the roton-limited mobility which may be expressed in the form $(e/\mu_\pm)_{\text{roton}} = f_\pm (P, T) \exp[-\Delta(P, T)/kT]$, where $\Delta(P, T)$ is the pressure- and temperature-dependent roton energy gap derived from neutron scattering experiments. The prefactors $f_\pm(P, T)$ contain all information regarding the roton–charge carrier interaction.

For sufficiently small electric fields the drift velocity v_D is proportional to the field E, i.e., $v_D = \mu E$, and the total momentum loss per unit length e/μ is simply the sum of contributions due to the various elementary excitations. In the case of phonons Baym *et al.*[1] obtained the following expression

$$(e/\mu)_{\text{phonon}} = -(\hbar/6\pi^2) \int_0^\infty dk\, k^4 (\partial n_k/\partial k)\, \sigma_T(k) \qquad (1)$$

where the momentum transfer cross section is

$$\sigma_T(k) = \int d\Omega (1 - \cos\theta)\, \sigma(k, \theta) \qquad (2)$$

The differential cross section $\sigma(k, \theta)$ describes the interaction between phonon and carrier and can be calculated quite accurately in terms of the scattering of classical sound waves by the accepted continuum structures for the charge carriers.

The negative carrier structure consists of an electron localized in a bubble of radius a_- from which the ^{4}He atoms have been excluded. The radius is determined by minimizing the total free energy[2]

$$E = E_{el}(r, V_0) + 4\pi r^2 \sigma + \frac{4}{3}\pi r^3 P - \frac{1}{2}\left(\frac{\kappa - 1}{\kappa}\right)\frac{e^2}{r} \qquad (3)$$

where, in the idealized model, $E_{el}(r, V_0)$ is the ground-state energy of an electron trapped in a potential well of depth V_0 and radius r, σ is the surface energy density of the liquid, P is the hydrostatic pressure, and κ is the dielectric constant of liquid helium. Using this model, Celli *et al.*[3] calculated the scattering of sound waves in terms of a_- and V_0. We incorporated their results into Eqs. (1) and (2) using various values of a_- and V_0 to obtain the best fit for each of the pressures at which data were taken. The calculated inverse mobility was quite insensitive to variations in V_0 and it was therefore impossible to determine the pressure variation of the well depth. However, from the best fits we were able to deduce the bubble radius as a function of pressure, as shown by the vertical bars in Fig. 1. Along with our estimates we have also plotted the calculated bubble radius using $V_0(P)$ as determined from the Wigner–Seitz model.[2] In the lower curve σ has been scaled according to[4]

$$\sigma(P) = \sigma_{exp}(0)\left[\rho(P)c(P)/\rho(0)c(0)\right] \tag{4}$$

while the upper curve was calculated using the experimentally measured bulk surface tension throughout. Springett[5] determined $a_-(P)$ by measuring the capture cross section of the negative carrier by vortex lines. Since his analysis required a knowledge of $a_-(0)$, we have renormalized his results, indicated by the open circles, to coincide with theory at $P = 0$. From the results of their photoejection experiment Zipfel and Sanders[6] were also able to estimate the pressure dependence of a_-. Their data are shown as the solid circles. It is gratifying that such diverse experiments yield results not only in good agreement with each other but also with the theory.

Using the electrostriction model,[7] Schwarz[8] has calculated the scattering of sound waves by the positive carrier, taking into account the surrounding density and pressure gradients and the unusual boundary conditions associated with this model. Though our data do not provide a very detailed test of the pressure dependence of this theory, calculations using a liquid–solid surface tension $\sigma_{ls} = 0.135$ erg cm^{-2} independent of applied hydrostatic pressure are well within the experimental error at all pressures. This value is in good agreement with previous determinations[8,9]

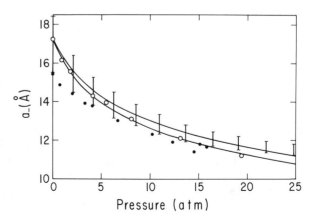

Fig. 1. Various experimental and theoretical determinations of the radius of the electron bubble as a function of pressure. These are explained in detail in the text.

and, in particular, indicates the absence of any rapid growth of the positive carrier near the melting pressure.

The roton–charge carrier interaction is not nearly as well understood as that for phonons. From the experimental standpoint this had been due to a lack of mobility data sufficiently accurate to allow one to extract reliable values of the prefactors from the dominant exponential temperature dependence. We have evaluated the prefactors $f_\pm = (e/\mu_\pm)_{\text{roton}} \exp(\Delta/k_\text{B}T)$ by subtracting the calculated phonon contribution from our data to obtain $(e/\mu_\pm)_{\text{roton}}$ and then using the recent experimental results for the roton energy gap $\Delta(P, T)$.[10]

In considering the pressure dependence of $f_\pm(P, T)$, it is interesting to note a certain correlation with the pressure variation of the carrier structures. For example, in Fig. 2 we have plotted $f_-(P)/f_-(0)$ for $T = 0.985°$K along with $a_-(P)/a_-(0)$ and the effective mass $M^*_-(P)/M^*_-(0)$ as functions of pressure. It is clear that the negative prefactor decreases monotonically and most rapidly at lower pressures, as do the bubble radius and the effective mass. This appears to be true for all temperatures below about $1.0°$K. On the other hand, the results for $f_+(P, T)$ are not as definitive. In Fig. 3 are shown $f_+(P)/f_+(0)$ for $T = 0.985$ and $0.586°$K and also $a_+(P)/a_+(0)$ and $M^*_+(P)/M^*_+(0)$ calculated for $\sigma_{ls} = 0.135$ erg cm^{-2}. Around $1.0°$K, $f_+(P, T)$ increases monotonically, but at lower temperatures its pressure dependence becomes somewhat more complex.

At present there exists no theory which can account for this behavior. The correct theory would have to include the effects of recoil of the charge carrier and an accurate dynamic description of the roton–charge carrier interaction. Recoil, important because of the comparable momenta of roton and carrier, has been considered by Bowley[11] and, more recently, Barrera and Baym.[12] On the other hand, understanding the dynamic processes of the interaction is a much more formidable problem. The simplified assumptions and conjectures which have been made concerning the interaction unfortunately provide very poor agreement with experiment.

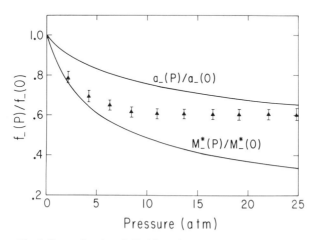

Fig. 2. Curve showing $f_-(P)/f_-(0)$ for $T = 0.985°$K as a function of pressure. Also shown are the theoretically calculated $a_-(P)/a_-(0)$ and $M^*_-(P)/M^*_-(0)$.

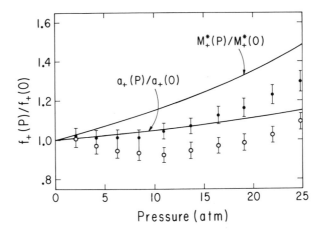

Fig. 3. Curves showing $f_+(P)/f_+(0)$ for $T = 0.586°K$ (open circles) and $T = 0.985°K$ (solid circles) as a function of pressure. Also shown are the calculated $a_+(P)/a_+(0)$ and $M_+^*(P)/M_+^*(0)$ with $\sigma_{1s} = 0.135$ erg cm^{-2}.

References

1. G. Baym, R. G. Barrera, and C. J. Pethick, *Phys. Rev. Lett.* **22**, 20 (1969).
2. B. E. Springett, M. H. Cohen, and J. Jortner, *Phys. Rev.* **159**, 183 (1967).
3. V. Celli, M. H. Cohen, and M. J. Zuckerman, *Phys. Rev.* **173**, 253 (1968).
4. D. Amit and E. P. Gross, *Phys. Rev.* **145**, 130 (1966).
5. B. E. Springett, *Phys. Rev.* **155**, 139 (1966).
6. C. Zipfel and T. M. Sanders, Jr., in *Proc. 11th Intern. Conf. Low Temp. Phys. 1968*, St. Andrews University Press, Scotland (1969), p. 296.
7. K. R. Atkins, *Phys. Rev.* **116**, 1339 (1959).
8. K. W. Schwarz, *Phys. Rev.* **A6**, 1958 (1972).
9. K. O. Keshishev, Yu. Z. Kovdrya, T. P. Mezhov-Deglin, and A. I. Shal'nikov, *Soviet Phys.—JETP* **29**, 53 (1969); R. M. Ostermeier and K. W. Schwarz, *Phys. Rev.* **A5**, 2510 (1972).
10. O. W. Dietrich, E. H. Graf, C. H. Huang, and L. Passel, *Phys. Rev.* **A5**, 1377 (1972).
11. R. M. Bowley, *J. Phys. C: Sol. St. Phys.* **4**, 1645 (1971).
12. R. Barrera and G. Baym, *Phys. Rev.* **A6**, 1558 (1972).

Two-Dimensional Electron States Outside Liquid Helium

T.R. Brown and C.C. Grimes

Bell Laboratories
Murray Hill, New Jersey

We have observed cyclotron resonance (CR) from electrons in two-dimensional (2D) surface states outside liquid helium. Their 2D character is shown clearly by the fact that the cyclotron resonant frequency depends only on the component of the applied magnetic field perpendicular to the surface. These observations have been made outside both bulk liquid helium and films of liquid helium covering a dielectric substrate. For the case of bulk helium we have also been able to determine the surface mobility of the electrons from the CR linewidths and find reasonable agreement with a simple calculation of the mobility assuming the electrons can scatter only parallel to the surface.

These surface states were predicted by Cole and Cohen[1] for surfaces of dielectrics generally and were analyzed in more detail by Cole.[2] The states are caused by the potential well formed by the attractive image potential outside a dielectric and the repulsive barrier to electron penetration at the surface. For liquid helium they are calculated to have a binding energy of 0.68 meV and a spatial extent outside the liquid surface of roughly 100 Å. Observations of these states have been reported[3,4] for bulk liquid helium, although later work[5] has questioned their existence.

Our experimental apparatus consists of a microwave spectrometer with its cavity mounted in a can partially filled with liquid helium and which is immersed in a pumped helium bath. The cavity is a right circular cylinder 1.2 cm in diameter and 0.75 cm in height split at the midplane where a Mylar sheet is inserted. This sheet is used as the dielectric backing in the film studies. For studies of the bulk liquid the liquid level is adjusted so that the free surface is at least several mm above this sheet. The cavity resonates at 23.5 GHz in the TE_{111} mode with the microwave electric field parallel to the helium surface. By applying a voltage between the halves of the cavity, electrons can either be attracted to or repelled from the surface under study. A sharp point can be inserted into either the side of the upper half of the cavity or the waveguide just above the cavity and is used to initiate a glow discharge which serves as an electron source above the liquid surface. For the bulk studies a magnetic field up to 17 kOe was produced by a superconducting Helmholtz pair which could be tilted up to 28° to the vertical. In order to observe larger angles to the surface in the film studies, the cavity was remounted so that the film was vertical and placed in a horizontal magnetic field which could be rotated more than 90°.

Figure 1 shows the power absorbed in the cavity after electrons have been trapped on the surface of a film. The different curves are for various angles of the magnetic field with the surface. As can be seen, the peak in the resonance line shifts toward

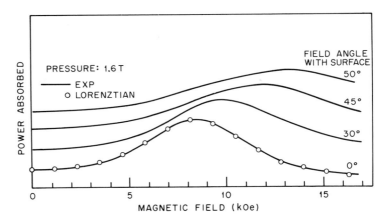

Fig. 1. Cyclotron resonance absorption lines for various angles between the applied magnetic field and the normal to the helium surface.

higher fields as the field is rotated away from the perpendicular direction. Figure 2 presents the fields at which the absorption peaks versus the secant of the angle relative to the surface. The straight line is the expected behavior for an electron in a purely 2D state. We have obtained similar curves for electrons outside of bulk liquid helium, although only up to angles of 28.° We feel these shifts to be compelling evidence for the 2D character of the electrons trapped in the surface states outside liquid helium.

In Fig. 3 we present our measurements of the surface mobility* of electrons outside bulk liquid helium versus the density of the helium gas in equilibrium with the liquid. Also presented in Fig. 3 are some of the previous measurements of the mobility†

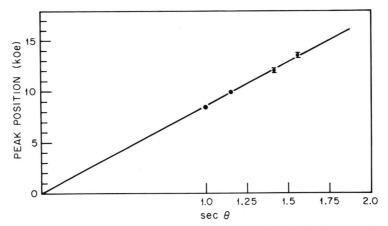

Fig. 2. Peak positions from Fig. 1 versus sec θ. The straight line is the expected behavior for a purely 2D electron.

---

* We converted our linewidths to mobilities by using $\mu = e\tau/m$, which is valid in two dimensions since τ is independent of electron velocity.

† These are taken from Ref. 3. Sommer (private communication) estimates, in addition to a 25% random error, that there may be a systematic error of comparable magnitude.

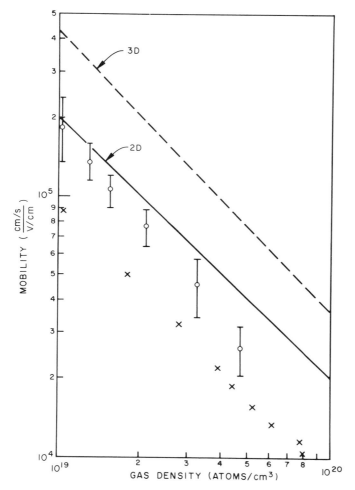

Fig. 3. Surface mobility of electrons outside bulk liquid helium versus helium gas density. ($\bigcirc$) this work; ($\times$) Sommer and Tanner[3]; ———2D theory; (- -) 3D theory.

and two theoretical estimates for the mobility, one for electrons in the gas, the other for electrons undergoing only 2D scattering on the liquid surface. We feel the agreement between our measurements and this simple theory to be reasonable and another confirmation that these electrons are in 2D states.

We are unable to report mobilities of electrons outside helium films at this time because our CR linewidths have varied from run to run in an as yet uncontrolled way. We believe this is due to variations in the film thickness and intend in the future to measure this parameter directly.

Another parameter which can be obtained from CR lines is the effective mass. For bulk helium we find this to be equal to the free mass within our uncertainties of 5%. For helium films we have observed values ranging from the free mass to roughly half that. There seems to be a correlation with the above-mentioned variations in the linewidths such that the wider lines have lower effective masses. This could also be explained by assuming a variable film thickness since for thinner films the interaction with the surface is much stronger, thus leading to smaller relaxation times and larger mass shifts. Hopefully, this situation will be cleared up when we repeat our previous measurements while also measuring the film thickness.

In conclusion, we have demonstrated the existence of 2D electron states outside liquid helium. We believe these states will prove to be a useful probe of the helium surface (and possibly other dielectrics also) and will also be interesting in their own right, since here one has a 2D electron system with a density variable over many orders of magnitude and few complicating interactions.

Acknowledgments

We would like to thank J. A. Appelbaum, E. I. Blount, and P. A. Wolff for a number of helpful discussions, and G. Adams for technical assistance.

References

1. M. W. Cole and H. M. Cohen, *Phys. Rev. Lett.* **23**, 1238 (1969).
2. M. W. Cole, *Phys. Rev.* **B2**, 4239 (1970).
3. W. T. Sommer and D. J. Tanner, *Phys. Rev. Lett.* **27**, 1345 (1971).
4. R. Williams, R. S. Crandall, and A. H. Willis, *Phys. Rev. Lett.* **26**, 7 (1971).
5. R. M. Ostermeier and K. W. Schwartz, *Phys. Rev. Lett.* **29**, 25 (1972).

An Experimental Test of Vinen's Dimensional Theory of Turbulent He II*

D.M. Sitton

Department of Aerospace Engineering and Engineering Physics, University of Virginia
Charlottesville, Virginia

and

F.E. Moss

Department of Physics, University of Missouri—St. Louis
St. Louis, Missouri

In a previous work,[1] using submerged thermionic filaments, we measured the escape probability P of negative ions trapped on vorticity in turbulent He II and demonstrated that the vorticity is identical to that in rotating helium except for spatial configuration of the lines. Above 1.75°K the results are in agreement with observations on rotating superfluid and with Donnelly's escape theory.[2] Below this temperature anomalously large values of P were measured. In order to further investigate this effect and to observe the pressure dependence of P, we have studied the space-charge-limited diode current from an 8-Ci tritium ion source.[3]

The diode is shown in Fig. 1. A heater of constantan wire wound on a copper bobbin is located in the top of a cylindrical plexiglass enclosure. The source, plated on a 1-in.-diameter by 0.005-in.-thick OFHC copper disk. is located 1.3 cm below the heater and faces the upper, large mesh grid 1/2 cm below. The entire structure is immersed in He, which enters the space between the heater and source through a leak around an electrical lead attached to the rear of the source.

Space-charge-limited currents I in the range 10^{-11}–10^{-10} A were maintained for all experimental conditions for the source voltage V shown, as established by observing that $I \propto V^2$. The diode was mounted in a pressure cell into which a sample of helium was condensed after passing through a liquid-nitrogen-cooled charcoal trap. The pressure was regulated to a precision of a few psi and the bath temperature to better than 10^{-5}°K.

During a typical run current measurements were made as T was drifted up from a low value toward T_λ while the pressure, heat flux, and diode voltage were held constant. The data are shown in Fig. 2, where the behavior of the zero-field, free ion mobility at the vapor pressure is shown by the solid lines. Notable features are that the current follows the temperature dependence of the free ion mobility near T_λ, departs from this behavior at temperatures which depend on the pressure,

* Supported in part by the Office of Naval Research.

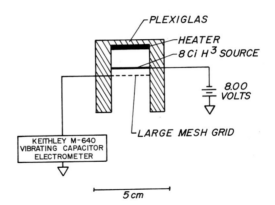

Fig. 1. Schematic diagram of the diode.

again assumes a logarithmic dependence for $T \ll T_\lambda$, and shows a saturation effect with heat flux.

These results can be explained in terms of the well-known characteristics of space-charge-limited currents in the presence of stationary charge traps:

$$I = C\mu V^2 Q_f/Q \tag{1}$$

where C is a geometric constant, μ the free ion mobility, and Q_f/Q is the fraction of free charge. In this experiment Eq. (1) is used with known values of μ to obtain information on Q_f/Q. This fraction can be calculated by equating the equilibrium rates of charge capture and escape,

$$Q_f v \sigma L_0 = PQ_c + (Q_c/L_0) \dot{L}_a \tag{2}$$

where L_0 is the equilibrium vortex line density, $\dot{L}_a$ is the vortex annihilation rate, v is the free ion thermal velocity, σ is the capture width, and Q_c is the captured charge density. The first term on the right is the escape rate due to thermal activation. The second term accounts for ion escape due to the annihilation of charged vorticity. Thus the equilibrium $\dot{L}_a$ is for the first time experimentally accessible.

In Vinen's theory[4]

$$L_0^{1/2} = F_1 \langle v_n - v_s \rangle \tag{3}$$

and

$$\dot{L}_a = F_2 L_0^2 \tag{4}$$

where F_1 and F_2 are prefactors with dimensions [time length^{-2}] and [length^2 time^{-1}], respectively. If we require that L_0 should diverge like ρ/ρ_s as $T \to T_\lambda$, then on dimensional grounds $F_1 = \beta\rho/\kappa\rho_s$, where β is a dimensionless strength constant and $\kappa = h/m_{He}$ is the circulation quantum. Using $\dot{q} = \rho ST\langle v_n \rangle$ for the heat flux with $\langle \rho_n v_n + \rho_s v_s \rangle = 0$ and $\rho = \rho_s + \rho_n$, Eq. (3) can be further reduced to

$$L_0^{1/2} = (\dot{q}\beta/\kappa)(\rho/\rho_s^2 ST) \tag{5}$$

The value of β and the nature of F_2 are to be determined by experiment.

Using Eqs. (4) and (2) with $Q = Q_f + Q_c$ results in

$$\frac{Q_f}{Q} = \frac{P + F_2 L_0}{P + F_2 L_0 + v \sigma L_0} \tag{6}$$

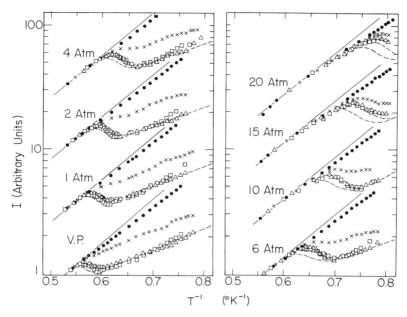

Fig. 2. Current versus inverse temperature for source mounted as shown in Fig. 1. Closed circles, 0; crosses, 19.4; triangles, 56; and squares, 98.4 mW cm^{-2}. Broken lines are calculated from Eqs. (5) and (6). Solid lines are the temperature dependence of the zero-field mobility of negative ions at the vapor pressure.

which, with Eq. (1), explains the major features of the data. In particular, for $T \to T_{\lambda}$, P becomes much larger than the other terms, so that $Q_f/Q \to 1$ and I just traces the temperature dependence of μ. At low temperatures P becomes vanishingly small, and

$$\lim_{T \to 0} (Q_f/Q) = F_2/(F_2 + v\sigma) \tag{7}$$

is independent of L_0, thus explaining the saturation effect with $\dot{q}$.

We have matched Eq. (7) to the low-temperature data with the result that

$$F_2 = (2.12 \pm 0.1) \, e^{-\Delta/kT} \quad cm^2 \, sec^{-1} \tag{8}$$

where Δ is the pressure-dependent roton gap. The error in Eq. (8) was estimated from fits to the vapor pressure, 1-, 2-, 4-, and 6-atm data only.

In order to match the data at intermediate temperatures, Eq. (6) was used with known values for P obtained from experiments on rotating helium and with L_0 specified by Eq. (5). For $\beta = (6.0 \pm 0.3) \times 10^{-3}$ the fits shown by the broken lines in Fig. 2 were obtained for $\dot{q} = 56$ mW cm^{-2} (upper curve) and 98.4 mW cm^{-2} (lower curve), and where the error was estimated from the repeatability of the data.

A striking feature of these results is the observation that F_2 is proportional to the roton density. There presently exists no model to account for such a process. We suggest that in equilibrium vortex annihilation is driven by the Magnus force. We rewrite Eq. (4) to read

$$\dot{L}_a = [\text{length}] [\text{velocity}] L_0^2 \tag{9}$$

and this formula recalls Langevin recombination. We can interpret the [velocity] as the transverse closing velocity of two vortex lines of opposite circulation lying perpendicular to the normal fluid flow. The force balance is $\rho_s(\mathbf{v}_L - \mathbf{v}_s) \times \boldsymbol{\kappa} = D(\mathbf{v}_n - \mathbf{v}_L)$, which yields the result that [velocity] $\simeq (2D/\kappa\rho_s)\langle v_n - v_s \rangle$, where D is the well-known friction coefficient measured on vortex rings.[5] The [length] can be identified as the mean spacing between lines $L_0^{-1/2}$. Allowing for the fact that only two-thirds of the total line length is actually perpendicular to the heat flux, we calculate

$$F_{2\mathrm{calc}} = (2/3)^2 \, (2D/\beta\rho) = 3.71 \, e^{-\Delta/kT} \quad \mathrm{cm}^2 \, \mathrm{sec}^{-1} \tag{10}$$

in reasonable agreement with the experimental result.

We conclude that Vinen's dimensional theory is correct in its essential feaures, and that vortex annihilation is driven by the normal fluid flow through the Magnus effect in well-developed turbulence. The annihilation model allows the turbulence theory to be cast in terms of a single dimensionless constant β, and in addition provides the first direct relation between the friction coefficient measured on vortex rings and experimentally accessible processes in turbulent He II. Finally, we conclude that the pressure dependence of P in turbulent He II is consistent with observations on rotating He.

Acknowledgments

We are grateful to Professors W. F. Vinen and R. J. Donnelly for enlightening discussions.

References

1. D. M. Sitton and F. Moss, *Phys. Rev. Lett.* **23**, 1090 (1969); and in *Proc. 12th Intern. Conf. Low Temp. Phys., 1970,* Academic Press of Japan, Tokyo (1971), p. 109.
2. P. M. Roberts and R. J. Donnelly, *Proc. Roy. Soc.* **A312**, 519 (1969), and references therein.
3. D. M. Sitton and F. Moss, *Phys. Rev. Lett.* **29**, 542 (1972).
4. W. F. Vinen, *Proc. Roy. Soc. (London)* **243**, 400 (1958), and references therein; H. E. Hall, *Advan. Phys.* **9**, 89 (1960).
5. G. W. Rayfield and F. Reif, *Phys. Rev.* **A136**, 1194 (1964).

Measurements on Ionic Mobilities in Liquid ^{4}He

G.M. Daalmans, M. Naeije, J.M. Goldschvartz, and B.S. Blaisse

Department of Applied Physics, Delft University of Technology
Delft, The Netherlands

Introduction

We have measured ionic mobilities in liquid helium in the temperature range 1.25–4.24°K. The method used[1] is based on the measurement of the pressure generated by ions moving in an electric field. We melted two electrodes, one flat and one pin-shaped, in a glass tube which is open on both ends. Creation of ions takes place in the neighborhood of and at the very sharp tungsten electrode under the influence of a potential difference of 6–30 kV. We measured the pressure generated inside the glass tube as a level difference Δh between the liquid helium levels inside and outside the tube. Application of a force balance gives a relation for the ionic current I, the level difference Δh, and the mobility μ:

$$I h_0/\mu = \rho_m g (\Delta h) S \tag{1}$$

where h_0 is the distance between the electrodes; ρ_m is the density of the liquid; S is the cross section of the glass tube; and g is the acceleration of gravity.

Experimental Results

Relation (1) only holds when the ionic mobility is field independent. In the case of field-dependent mobilities it is better to replace h_0/μ by the integral $\int_0^{h_0} dx/\mu$, thus obtaining an average mobility.

We measured at fixed temperatures the level differences Δh as functions of the current I. If μ is field independent, the $\Delta h = \Delta h(I)$ characteristic [Eq. (1)] has to be a straight line through the origin. We calculated μ from the slopes of these lines. We found different types of $\Delta h = \Delta h(I)$ characteristics, namely straight and curved lines through the origin and shifted with respect to the origin. Some typical examples are shown in Fig. 1.

The tube containing the electrodes was always mounted vertically. We used two types: type A, with the flat electrode above the pin-shaped one, and type B, in which the arrangement was reversed.

Above the lambda point all the characteristics measured with type A are straight lines, most of which cut the positive Δh axis on extrapolation. We explain this shift in terms of rising vapor bubbles produced in the vicinity of the pin-shaped emitter.

In order to compensate for the bubble effect, we also performed measurements with the type B cell. The results of both series of measurements are nearly the same,

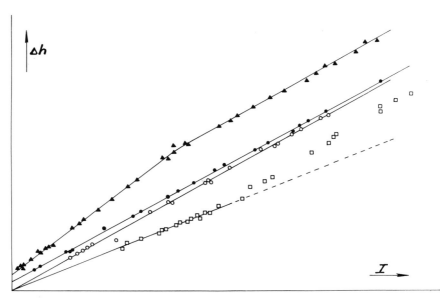

Fig. 1. Some examples of Δh as a function of I. In order to get the curves in one figure, the scales are different and are not indicated. Negative ions: (▲) $T = 1.49°$K, (●) $T = 1.79°$K. Positive ions: (○) $T = 2.155°$K, (□) $T = 1.47°$K.

as can be seen from Fig. 3. The greatest deviation from the average curve for positive and negative ions is 5%.

Below the lambda point the results with type A and type B cells were as expected. Below the lambda point we have to distinguish between the characteristics of positive and negative ions. In the case of positive ions the lines are straight as long as the temperature is higher than about 1.5°K. Below this temperature we found curved lines through the origin. The curvature of the Δh–I characteristic is positive, indicating that the average mobility decreases with I.

For negative ions we found straight lines at temperatures higher than 1.6°K. In the region between 1.3 and 1.6°K the extrapolated curves do not go through the origin, whereas they are curved in the opposite sense as for the positive ions (Fig. 2). From 1.25 to 1.3°K they pass through the origin. This negative curvature indicates an increase of the mobility at higher currents.

The average mobilities as functions of the temperature measured with the Δh method are given in Fig. 3. The values at those temperatures where the characteristics are curved are obtained from the initial slopes.

As mentioned above, in the temperature region between 1.3 and 1.6°K the extrapolated Δh–I curves for negative ions do not go through the origin. Therefore the values of the slopes at $I = 0$ are very uncertain. For this reason no experimental points of μ of the negative ions have been indicated in Fig. 3 in this temperature region.

We also performed measurements of the current I of tungsten tips as a function of V and T in a cell where the tip is in a metal box. Measurements of this kind have also been done by others.[2]

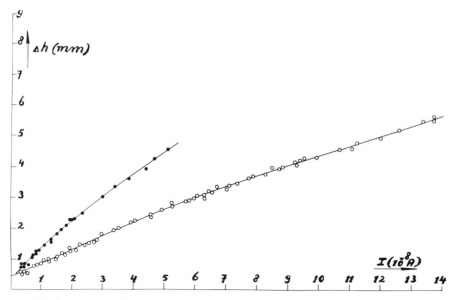

Fig. 2. Δh versus I for negative ions for $T < 1.6°$K. ($\bigcirc$) $T = 1.55°$K, ($\bullet$) $T = 1.43°$K.

Discussion

As indicated by several authors,[3] the values of μ above the λ-point should be in agreement with Stokes's law applied on a spherical particle of radius r moving through a liquid with viscosity η:

$$\mu = e/A\eta r \qquad (2)$$

For the solid positive ion $A = 6\pi$, whereas for the electron bubble $A = 4\pi$, as pointed out by Hadamard.[4] Using the bubble model, the radius r_- of the negative ion is in good approximation proportional to $\sigma^{-1/4}$, where σ is the surface tension. From the λ-point to the boiling point the radius r_- increases from 15.0 to 17.3 Å. Our data for μ_- agree well with those calculated from Stokes's law with the values of η of Zinov'eva[5] and of σ of Allen and Misener.[6]

The radius of the positive ion can be calculated using Atkins's model of a sphere of solid helium.[7] The radius changes from 4.8 to 4.1 Å from the λ-point to the boiling point. From the course of r_+ and of η as functions of T one expects a minimum in μ_+. This, however, was not clearly found.

Below the λ-point we find a sharp rise of the mobility. At temperatures of 1.60 and 1.29°K the mobility curves show maxima for negative and positive ions, respectively. This behavior is analogous to the temperature dependence of I at constant potential difference V^2. If indeed the current is space-charged-limited, one should expect I to be proportional to the mobility. According to measurements of this kind which we performed, this is true as long as μ is field independent. Therefore at lower temperatures deviations are expected to occur. So we found at those temperatures that the square root of I was not a linear function of V; the maximum in the

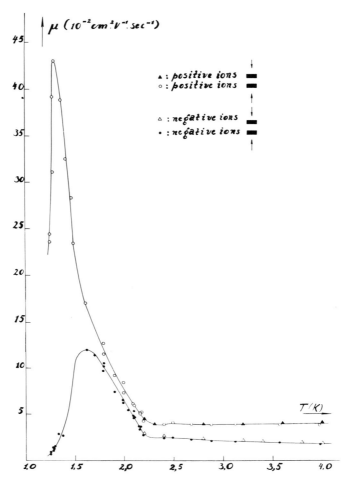

Fig. 3. Mobility μ as a function of T for positive and negative ions.
For $T > T_\lambda$ the points measured with emitter point above or below
the flat electrode, respectively, are indicated.

$I–T$ characteristic of the negative ions occurs at 1.73°K instead of 1.60°K as found
with the Δh method (Fig. 3).

At very low values of the electric field Scaramuzzi[8] found a continuously
increasing μ with decreasing T.

The large drop in the mobility at decreasing temperatures below 1.60 and 1.29°K
for negative and positive ions, respectively, can qualitatively be explained as follows.
We take as a model a small spherical electrode surrounded by a metal sphere. For
the case of space-charge-limited currents the field strength E is zero at a radius r_0.
At increasing radius r it increases, passes through a maximum, and decreases. One
finds with $I = 10^{-7}$ A, $\mu = 0.1$ cm^2 V^{-1} sec^{-1}, and $r_0 = 1$ μm that $E_{max} \approx 10^5$ V
cm^{-1}. This is much higher than the fields E_g of the giant discontinuities which occur
at sufficiently low temperatures.[9] Because the ions have to move about 10^{-8} m in

order to extract the creation energy of a vortex ring from the field, ionic ring vortices will be created near the emitter.

According to measurements of Bruschi et al.,[9] the ions stay attached to the vortices when the field decreases. The higher the energy of those ionic vortices, the lower the mobility. In this temperature region the average mobility increases with I due to the diminishing lifetime of ionic vortices.[10]

As mentioned above, the $\Delta h - I$ characteristics of positive ions between 1.2 and 1.5°K have a positive curvature, indicating a decrease of the average mobility with I. This is to be expected since, e.g., Bruschi et al.[9] found μ to decrease strongly with E in this region.

The negative curvature of the $\Delta h - I$ characteristics of negative ions for temperatures below 1.6°K indicated a higher value of the average mobility at higher currents, as mentioned above. This can be explained if the main contribution to our measured values of μ stems from ions in fields larger than the field at which the drift velocity has a minimum.[10] At those high fields μ increases with E. The order of magnitude of our values agrees with the measurements of Bruschi et al.[10]

The injection of charges into the liquid results in fluid motion (macroscopic vortex), which causes a difference between the drift velocity and the total velocity of the ions. Moreover, the glass tube exerts a viscous force on the liquid. The two effects compensate each other for the greater part.

References

1. B. S. Blaisse, J. M. Goldschvartz, and P. C. Slagter, *Cyrogenics* **10**, 163 (1970); B. S. Blaisse, J. M. Goldschvartz, and M. Naeije, *Proc. 12th Intern. Conf. Low Temp. Phys., 1970,* Academic Press of Japan, Tokyo (1971), p. 99.
2. B. Halpern and R. Gomer, *J. Chem. Phys.* **51**, 1031 (1969); A. Hickson and P. V. E. McClintock, in *Proc. 12th Intern. Conf. Low Temp. Phys., 1970,* Academic Press of Japan, Tokyo (1971), p. 95; D. M. Sitton and F. Moss, *Phys. Lett.* **34A**, 159 (1971).
3. D. T. Grimsrud and F. Scaramuzzi, in *Proc. 10th Intern. Conf. Low Temp. Phys., 1966,* VINITI, Moscow (1967), Vol. I, p. 197.
4. J. Hadamard, *Compt. Rend.* **CLII**, 1735 (1911).
5. K. Zinov'eva, *Soviet Phys.—JETP* **7**, 421 (1958).
6. J. F. Allen and A. D. Misener, *Proc. Cambr. Phil. Soc.* **34**, 299 (1938).
7. K. R. Atkins, *Phys. Rev.* **116**, 1339 (1959).
8. F. Scaramuzzi, Report LNF-63/79 (1963).
9. L. Bruschi, P. Mazzoldi, and M. Santini, *Phys. Rev. Lett.* **21**, 1738 (1968).
10. L. Bruschi, B. Maraviglia, and P. Mazzoldi, *Phys. Rev.* **143**, 84 (1966).

Influence of a Grid on Ion Currents in He II

C.S.M. Doake and P.W.F. Gribbon

Department of Physics, University of St. Andrews
North Haugh, St. Andrews, Scotland

There is little qualitative information on the influence of the structure of a grid electrode on the behavior of a current beam in an ion cell. This paper presents some work on the interaction of bare ions and charged vortex rings with grids of different mesh size.

Measurements were made on the dc characteristic of grids mounted in triode and tetrode cells in He II at temperatures near $1°K$ and at pressures up to 7 atm. Four Buckbee-Mears grids were available with apertures in the range $\sim 40-400\ \mu m$. These sizes were greater than the maximum aperture of $\sim 10\ \mu m$ used by Gamota and Sanders[1] in their work on the transmission of vortex rings through a grid.

The most interesting results were obtained from the dc characteristics of the transmitted current I_T as a function of the incident field E_I on the grid shown in Fig. 1. The negative current I_T^- had a sharp peak I_{max}^- at a certain field: This field E_{max} was greater than the vortex ring nucleation field E_c with $E_{max} \simeq 0.65E_c$, from the Huang–Olinto[2] ring stability criterion. The positive current I_T^+ had a smaller and

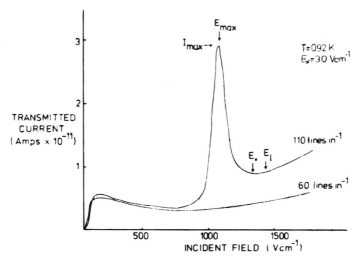

Fig. 1. The transmitted negative ion current through two grids as a function of the incident field on the grids for $T = 0.92°K$ and a constant extracting field $E_x = 30\ V\ cm^{-1}$. I_{max} and E_{max} define the current peak. E_0 and E_1 are defined in Fig. 2.

wider peak at the same field. The results could be expressed also in terms of the transmission coefficient of the grid or the ratio of transmitted to incident current $T = I_T/I_I$.

Our observations at low fields $E_I < E_c$ showed that at a constant extracting field E_x both T^- and T^+ for negative and positive ions had a similar field dependence at different temperatures. This meant that the observations did not reflect the properties of the bare ions of either species or the way in which the ions interacted with the normal fluid, but instead they reflected the transmission properties of the grid itself. This was confirmed by the fact that our T values could be fitted to an empirical relationship[3] which was known to express the properties of a grid in classical liquids.[4]

Our observations at high fields $E_I > E_c$ were supplemented by a detailed study of the properties of the current peak as a function of the grid mesh size, temperature, and pressure. The peak was observed only with the three finer grids: A critical grid size appeared to be necessary for the existence of a peak. However, the properties of the peak were dependent on the roton number density in the normal fluid. This was shown by a study of E_{max}^- as a function of temperature and pressure. The variations of E_{max}^- with temperature T was given by

$$E_{max}^- \propto e^{-\Delta/kT}$$

where Δ/k is the pressure-dependent roton energy of the dispersion relation (equal to $7.58 \pm 0.20°K$), while the variation of E_{max}^- with pressure was described by

$$E_{max}^-(P) = E(0)\, e^{\beta P}$$

where $E(0)$ is the value of E_{max}^- at SVP and $\beta = 0.063 \pm 0.005$ atm^{-1} is the pressure-dependent coefficient in the roton number density.

It was found that the peak I_{max}^- varied with E_{max}^- at constant pressure in a way similar to the temperature dependence of the ion–vortex line capture cross section curves at different pressures.[5] This suggested that a current peak was related to the

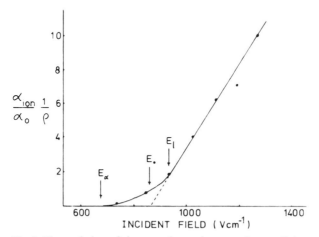

Fig. 2. The variation of the second-sound attenuation coefficient in arbitrary units as a function of the incident grid field in a triode: after Fig. 5 of Bruschi *et al.*[7] Negative ions at $T = 0.885°K$ and $E_X = 400$ V cm^{-1}. E_α, E_0, and E_1 are denoted by the arrows.

capture of ions on some form of vorticity. Further support to this idea was given by the time scales of certain hysteresis effects associated with the growth and decay of current, which were similar to the time scales associated with the changes of superfluid vorticity density due to a steady heat current and detected by Vinen's second-sound attenuation experiments.[6]

We suggest that the formation of superfluid vorticity at the grid may be responsible for our observations at $E > E_c$. The charged vortex rings approached and interacted with the grid and with each other. The rings broke down into a tangled array of short vortex lengths close to and linked with the grid. At E_{max}^- the rings had reached a size comparable with both their separation and the grid apertures, and a maximum vorticity density had developed at the grid. At $E \geq E_{max}^-$ the vorticity began to break off into the volume of liquid on the extracting side of the grid.

The ions became trapped on the vortex line cores, the negative ions being more strongly bound than positive ions at the same temperature. The ions constituted a trapped space charge which exerted a repulsive Coulomb field on the other, approaching charged rings, and so gave a higher current transmission through the grid, with maximum transmission occurring at E_{max}^-.

Further support for the suggested vorticity comes from a comparison of our fields with the fields associated with the detection of vorticity by second-sound attenuation in an ion beam measured by Bruschi et al.[7] We found reasonable agreement if we redefined their fields as shown in Fig. 2. Their onset field E_α was identified with E_{max}^-, their linear extrapolation E_0, with the field corresponding to our current minimum, and their linear increase E_1 with our linear increase beyond the current minimum. A common origin for the vorticity appeared to be present in both experiments, and we suggest that our explanation could be applicable to their results.

References

1. G. Gamota and T. M. Sanders, Jr., *Phys. Rev.* **A4**, 1092 (1971).
2. K. Huang and A. C. Olinto, *Phys. Rev.* **A139**, 1441 (1965).
3. A. Januszajtis, *Acta Phys. Polon.* **24**, 809 (1963).
4. P. E. Secker and T. J. Lewis, *Brit. J. App. Phys.* **16**, 1649 (1965).
5. R. Donnelly, *Experimental Superfluidity*, Univ. of Chicago Press, Chicago, (1967).
6. W. F. Vinen, *Proc. Roy. Soc.* **A243**, 400 (1957).
7. L. Bruschi, B. Maraviglia, and P. Mazzoldi, *Phys. Rev.* **143**, 84 (1966).

Collective Modes in Vortex Ring Beams

G. Gamota

Bell Laboratories
Murray Hill, New Jersey

Introduction

The existence of quantized vortex rings was conjectured by Feynman[1] in his discussion of superfluid helium, but it was not until several years later that Rayfield and Reif[2] first identified a charged vortex ring in He II. Since then many experiments dealing with the individual properties of vortex rings were performed and a report on these can be found in a recent review article.[3] In this work we are interested in investigating phenomena arising from the mutual interactions among charged vortices. We seek plasmalike effects that can be found in a gas made up of individual charged vortex rings. The conjecture of such a phenomenon was first made by Careri.* The initial work on the gaslike properties of charged vortices, referred to as dc experiments, was reported earlier.[5] There it was shown that a single pulse of charged vortices does not smear out as would a pulse of, say, electrons in a vacuum, but in fact compresses longitudinally until it reaches a steady-state distribution. The final width of this "bell"-shaped distribution was found to be rather insensitive to the initial width of the pulse as well as the energy. Hasegawa and Varma[6] explained this behavior by a simple one-dimensional theory which assumed that the vortices acted as particles in a gas. The theory includes the Coulomb interaction among the vortices as well as pressure due to fluctuations. They assume that the vortices are, in fact, not monoenergetic but that there exists a distribution whose average value is equal to the accelerating energy given to the beam.† The subsequent work, referred to as the ac experiments and reported here, involves the study of collective modes of oscillation of these vortex beams which also arise from the above interactions. Again Hasegawa and Varma[6] take a one-dimensional picture but use the Vlasov equation, instead of the fluid equation used earlier, to write a phase-space distribution function $f(x, p, t)$. This is solved and a wave dispersion relation is found which shows that an instability in the wave would occur if k was below a critical wave number k_c. Due to this instability the wave is expected to grow as a function of the propagation distance.

Experimental results are presented‡ which demonstrate the existence of this growing wave and a study is made to determine its frequency dependence. We will show that, in agreement with theory, the wave does grow with distance and does

---

* While not necessarily connected with the topic of vortices, Careri[4] suggested the possibility of observing plasmalike effects in ion beams in superfluid helium.
† To account for the observed effects, it is only necessary to assume that $(\Delta\mathscr{E}/\mathscr{E}) \times 100 \simeq$ a few percent.
‡ Some early results have been presented by Gamota.[7]

have the predicted amplification vs. frequency characteristic which is dependent upon the density as well as the velocity of the beam.

Theory

In this section we want to summarize the basic theoretical results pertinent to the collective mode problem. We follow Hasegawa and Varma in deriving the equations of motion.

Charged vortices are taken to be quasiparticles interacting through the Coulomb force. The logarithmic terms in the basic vortex ring equation* are neglected so that we have the following relations for velocity v, energy $\mathscr{E}$, and dynamic impulse p of a ring:

$$p = \alpha/v^2 \tag{1}$$

$$v = \beta/\mathscr{E} \tag{2}$$

where α and β are taken to be constant. Next a stream of charged vortices is assumed moving in one dimension which has a phase-space distribution characterized by $f(x, p, t)$. The one-dimensional Vlasov equation for f is then

$$\frac{\partial f}{\partial t} + v\frac{\partial f}{\partial x} + qE\frac{\partial f}{\partial p} = 0 \tag{3}$$

where the electric field E relates to particle density n through the Poisson equation,

$$\partial E/\partial x = 4\pi q n(x, t) \tag{4}$$

and by definition

$$n(x, t) = n_0 \int f(x, p, t)\, dp \tag{5}$$

where n_0 is the average density.

The Vlasov equation is linearized by writing

$$f = f^0(p) + f^{(1)}(x, p, t) + \cdots \tag{6}$$

and we obtain

$$\frac{\partial f^{(1)}}{\partial t} + v\frac{\partial f^{(1)}}{\partial x} + qE^{(1)}\frac{\partial f^0}{\partial p} = 0 \tag{7}$$

where $E^{(1)}$ and $n^{(1)}$ are related to $n^{(1)}$ and $f^{(1)}$ respectively as in Eqs. (4) and (5). We look for solutions of Eq. (7) of the form

$$f^{(1)}(x, p, t) = g(p)\, e^{i(kx - \omega t)} \tag{8}$$

and find from Eq. (7) the following expression:

$$1 = \frac{4\pi q^2 n_0}{k^2} \int_{-\infty}^{\infty} \frac{\partial f^0(p)/\partial p}{v - \omega/k}\, dp \tag{9}$$

Equation (9) has not been solved exactly at the present time but by assuming a

---

* The energy, velocity, and dynamic impulse p of a vortex ring of radius R having a hollow core of radius a and circulation κ is given by $\mathscr{E} = \frac{1}{2}\rho\kappa^2 R[\ln(8R/a) - 2]$, $v = (k/4\pi R)[\ln(8R/a) - \frac{1}{2}]$, and $p = \pi\rho\kappa R^2$; see, for example, Basset.[8] For a singly quantized vortex ring $\kappa = h/m = 10^{-3}$ cm^2 sec^{-1}.

Gaussian distribution for f^0 and treating the deviation from average velocity as small, $\Delta v/\bar{v} \ll 1$, one can approximately solve Eq. (9) and for real ω obtain the following dispersion relation:

$$(\omega - k\bar{v})^2 \approx -\omega_0^2 + 3k^2 v_T^2 \qquad (10)$$

where $\bar{v}$ is the average velocity of the beam, ω_0 is the plasma frequency, defined as $\omega_0^2 \equiv +4\pi n_0 \bar{v}^3 q^2/2\alpha$, and $v_T^2 \equiv \langle \Delta v^2 \rangle$. To obtain stable waves, we need $k > k_c$, where

$$k_c^2 = \omega_0^2/3v_T^2$$

For $k < k_c$ the waves grow exponentially. This instability, we should note, arises primarily due to the fact that p is inversely proportional to v [Eq. (1)], and thus the effective mass m^* defined as

$$\Delta p = m^* \Delta v \qquad (11)$$

is negative.

Experiment

To create the collective modes, we produced a small density and velocity perturbation by modulating the energy of the beam. This was done in the experimental cell shown in Fig. 1. The whole cell is immersed in superfluid helium which is kept at $T \sim 0.3°K$. Four grids, G_1–G_4, are spaced between a radioactive source and a guarded collector. G_1 is 5.7 mm from the source; G_2 and G_3 are separated by 0.13 mm and are situated between G_1 and G_4, the latter being 1 mm away from the collector C and acting as an electrical guard. Runs were taken with the distance L between G_3 and G_4 equal to 27.5 and 8.8 mm. Runs with the 27.5 mm length were taken with two different source strengths to study the density dependence. The electrical connections are shown in Fig. 1. G_3 and G_4 are grounded while a small sinusoidal voltage V_{ac} is applied to G_1 and G_2, which are connected together, and a dc voltage V_{dc} is applied between G_1 and the source. Thus, charged vortices which are created near the source

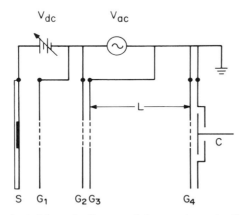

Fig. 1. Schematic diagram of the experimental cell. The cell is cylindrical in shape and is ~ 4 in. long and 3 in. in diameter.

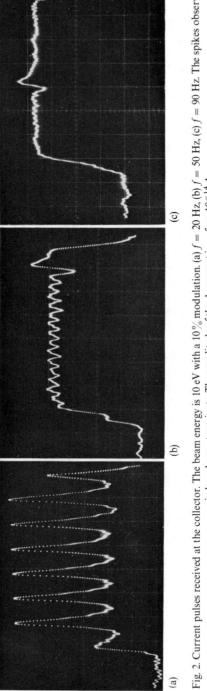

Fig. 2. Current pulses received at the collector. The beam energy is 10 eV with a 10% modulation. (a) $f = 20$ Hz, (b) $f = 50$ Hz, (c) $f = 90$ Hz. The spikes observed in b and c are transients. The amplitude of the dc current is $\sim 5 \times 10^{-14}$ A.

acquire energy eV_{dc} as they pass through G_1. They continue to move through the field-free region and pass G_2. The distance between G_2 and G_3 is very small. This is necessary so that it will be short compared to the distance a ring will travel in one period of the highest experimentally applied frequency. Under such conditions the electric field can be considered as constant and the vortices become density as well as velocity modulated as a result of their small energy change. Between G_3 and G_4 is again a field-free region where the velocity-modulated beam propagates and then, after passing G_4, arrives finally at the collector. With the aid of a fast electrometer and a signal averager one can easily observe the modulation in the signal current. Typically, the dc current I_{dc} was $\geq 2 \times 10^{-14}$ A for the weak source and 2×10^{-13} A for the strong source, while the modulation I_{ac} (pp) was less than 30% of I_{dc} to avoid operation in a nonlinear regime due to saturation.

We will first show some early results when a pulse technique was used rather than CW. This was accomplished by putting the source in place of G_2 and then gating the beam on and off with a square voltage pulse having a small sinusoidal voltage superimposed on it while the beam was on. Typical results are shown in Fig. 2. From the top trace we see that at low frequencies an initial 10% modulation of beam energy has grown to nearly 100% modulation of collected current. As the frequency is increased the modulation drops, and finally, within our sensitivity, becomes unobservable around 90 Hz. In the CW technique, as described earlier, sinusoidal modulation is applied to a continuous beam of vortices. We show examples of such data in Fig. 3. The wiggly line in the center represents the received current signal with beam off, indicating good electrical shielding between G_2 and collector.

To compare our results with theory, we introduce an amplification factor A defined as

$$A = (I_{ac}/I_{dc})/(V_{ac}/V_{dc}) \tag{12}$$

Data reduced in such a manner are shown in Fig. 4. The dashed lines have no theoretical significance and are just used to join the experimental points. This run was taken with a weak source, producing $I_{dc} = 4.6 \times 10^{-14}$ A, and as a result the error bars are quite large: $\sim 20\%$. The curves display the following characteristics. All rise almost linearly from the origin* up to a maximum and then slowly decrease. We also note that the initial slope appears to increase with energy and that the frequency where the maximum A occurs is inversely proportional to the energy.

To fit the experimental data to the theory, we need to linearize $J = nqv$ and assume a wavelike form for the current density J, the number density n, and the velocity v:

$$\begin{pmatrix} J(x, t) \\ n(x, t) \\ v(x, t) \end{pmatrix} = \begin{pmatrix} J \\ n \\ v \end{pmatrix} e^{i(kx - \omega t)} \tag{13}$$

Now defining small changes in quantities by the subscript 1, we have

$$J_1 = n_1 q \bar{v} + \bar{n} q v_1 \tag{14}$$

Using Eq. (13) and Eq. (14) as well as $\bar{J} = \bar{n} q \bar{v}$ ($\bar{J}, \bar{n}, \bar{v}$ are steady-state dc components),

---

* We assumed that the curves started at the origin since we can justify this on theoretical grounds.

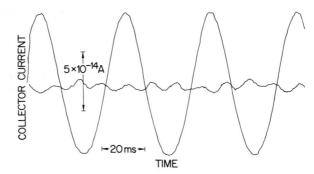

Fig. 3. Typical current signal recived at the collector. The trace in the middle is obtained under similar conditions except that the vortex beam has been shut off with dc potentials. The sine wave obtained is for a 20-eV beam, $V_{ac} = 0.2$ V and $f = 25.5$ Hz.

we find a relationship between a velocity and current density modulation:

$$\frac{J_1}{J} = \frac{v_1}{\bar{v}} \frac{1}{2} \left| \frac{e^{ik_1 x}}{1 - k_1 \bar{v}/\omega} + \frac{e^{ik_2 x}}{1 - k_2 \bar{v}/\omega} \right| \tag{15}$$

where k_1 and k_2 are roots from Eq. (9).

Since Eq. (9) has not been solved exactly, we must use the roots from Eq. (10),

$$k \simeq \frac{\omega}{\bar{v}} \left[1 \pm i \left(\frac{\omega_0^2}{\omega^2} - \frac{3v_T^2}{\bar{v}^2} \right)^{1/2} \right] \tag{16}$$

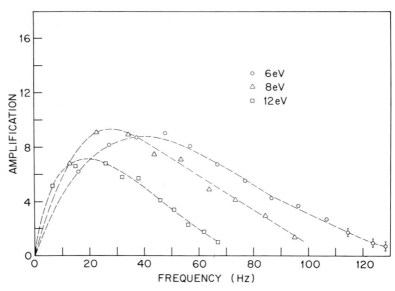

Fig. 4. Amplification vs. frequency, data from set 1. The dashed lines have no theoretical significance and are used only to connect the points.

This is valid only when $\omega < \omega_c$, where ω_c is defined as the critical frequency occurring when $k_1 = k_2$. Equation (15) now becomes simplified and we obtain

$$\frac{J_1}{\bar{J}} = \frac{v_1}{\bar{v}} \frac{1}{\Delta} \sinh \frac{\Delta \omega x}{\bar{v}} \tag{17}$$

where

$$\Delta \equiv (\omega_0/\omega_c)(\omega_c^2/\omega^2 - 1)^{1/2}$$

and

$$\omega_c \equiv \omega_0 \bar{v}/\sqrt{3}\, v_T$$

It is now easy to see that by virtue of Eqs. (2) and (12) we can rewrite Eq. (17) to read

$$A = (1/\Delta) \sinh (\Delta \omega x/\bar{v}) \tag{18}$$

and compare it directly to our data.

Discussion

Three different sets of data are to be discussed and compared to theory: (1) low source strength with long drift space; (2) high source strength with long drift space; and (3) high source strength with short drift space.

We will first compare some results from sets 1 and 2. These are shown in Fig. 5, where we have plotted A vs. f curves for two 10-eV beams, a high- and a low-density beam. The solid lines are attempts to fit Eq. (18) to our data.

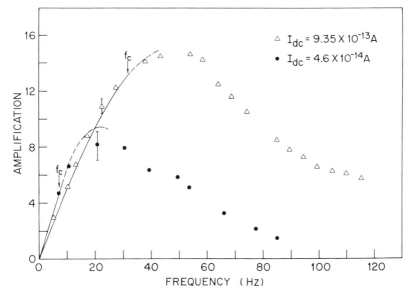

Fig. 5. Amplification vs. frequency, data from sets 1 and 2. Both curves are taken with 10-eV beams but with different source strengths, which is reflected in the dc current values. The solid lines are fits to Eq. (18) with the following parameters: ($\bullet$) $x = 23$ mm, $\omega_0 = 5$ sec^{-1}, and $\omega_c = 45$ sec^{-1}, ($\triangle$) $x = 17$ mm, $\omega_0 = 18$ sec^{-1}, and $\omega_c = 201$ sec^{-1}. The dashed lines are for $\omega > \omega_c$.

By letting x, ω_0, and ω_c be adjustable parameters we can easily approximate the linear portion of each curve with constants that are in accord with theoretical estimates. For instance, the measured L is 27.5 mm while we get $x = 17$ and 23 mm for the high and low density, respectively. We will return later to discuss why x turns out to be less than L. Also, we expect ω_0/ω_c to be a constant independent of density. Again we find agreement $\omega_0/\omega_c \approx 0.1$ for both curves. Finally, we find, as expected, that ω_0 and ω_c are proportional to the square root of the current, which we assume to be proportional to the density. The ratio of the plasma frequencies is 3.5 and that of the critical frequencies is 4.4, as compared to the expected value from the dc currents of 4.5.

Data from sets 2 and 3 are presented in Fig. 6. Here we find that the error bars are considerably smaller than in set 1 because of the larger current signals. Again we fit the data and find reasonable agreement. For instance, we find as before $\omega_0/\omega_c \approx 0.1$, and this is independent of energy and drift space, which agrees with the early dc experiment. The absolute value of the plasma frequency (e.g., taking the 20-eV beam we find $\omega_0 = 9.4 \text{ sec}^{-1}$) agrees with Hasegawa and Varma's estimate of $\omega_0 = 10 \text{ sec}^{-1}$. We also find $\omega_0 \propto \bar{v}$, although it should be pointed out that these fits are very sensitive only to the ratio of ω_0 to ω_c and the value of x and are much less sensitive to ω_0. The values of the drift space x obtained from the fit are somewhat smaller than measured, especially for data from set 2. The reason for the

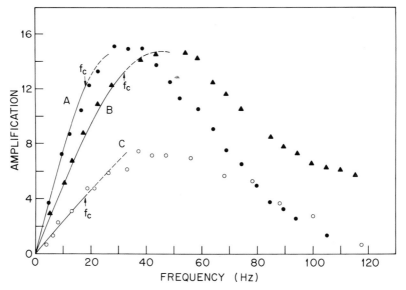

Fig. 6. Amplification vs. frequency, data from sets 2 and 3. The points are experimental and the lines are fits to Eq. (18). Curves A and B are obtained with 20- and 10-eV vortex beams, respectively, and $L = 27.5$ mm, while curve C is for a 20-eV beam and $L = 8.8$ mm. The theoretical curves are fitted with the following parameters: (A) $x = 17$ mm, $\omega_0 = 9.4 \text{ sec}^{-1}$, $\omega_c = 113 \text{ sec}^{-1}$; (B) $x = 17$ mm, $\omega_0 = 17.6 \text{ sec}^{-1}$, $\omega_c = 201 \text{ sec}^{-1}$; (C) $x = 6$ mm, $\omega_0 = 9.4 \text{ sec}^{-1}$, and $\omega_c = 113 \text{ sec}^{-1}$. The solid and dashed lines are for frequencies less than and greater than ω_c, respectively.

difference between the measured and calculated values of the drift space is most likely due to beam spreading, which will limit A to a smaller than predicted value. This is borne out by the fact that with the short drift space or weaker source the fractional discrepancy in x is smaller than for the long space and strong source, where beam spreading should be most significant.

For frequencies higher than ω_c we cannot use Eq. (18) because the k values extracted from Eq. (10) are no longer valid. We must solve Eq. (9) exactly and only then will we be able to compare the theory with data. In this regime Landau damping is to be expected, whereas Eq. (18) predicts a periodic variation of amplitude with frequency.

Finally, we show in Fig. 7 a plot of I_{ac} vs. V_{ac}/V_{dc} for several frequencies to clearly demonstrate that this phenomenon does not originate from a klystronlike effect. Even though for convenience most of the data were taken with $V_{ac}/V_{dc} = 0.02$, we have used ratios of 0.005–0.05 with no difference in the A vs. f curves. This is observed to be true for all frequencies. We see in Fig. 7 that I_{ac} is linearly proportional to V_{ac}/V_{dc} up to approximately 0.1. If V_{ac}/V_{dc} is too large, we expect and start to see saturating effects, and finally, only when $V_{ac}/V_{dc} \approx 0.25$ or larger do we see periodic behavior—but this obviously is outside the scope of this theory, which considers only small perturbations.

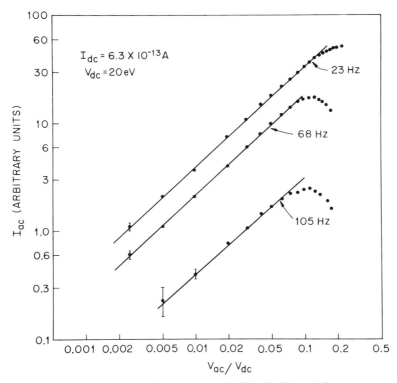

Fig. 7. Peak-to-peak ac current vs. V_{ac}/V_{dc}; data from set 2.

Conclusions

We have shown that collective modes exist in a beam of charged vortices in superfluid helium. The characteristics of these waves agree with theoretical predictions for low frequencies. Changing the density of the beam, beam energy, and drift space allowed us to determine the various parameters pertinent to the theory and good agreement was found.

At the present time the amplification appears to be limited not by the length of drift space or the source strength but by beam spreading which changes the density of the beam as it propagates through the field-free region. It would be helpful if one could investigate the problem with a collimated beam. Also of interest would be means of producing a density modulation without velocity modulation.

Acknowledgments

I wish to thank Akira Hasegawa and C. M. Varma for many helpful discussions and H. W. Dail for collecting the data.

References

1. R.P. Feynman, in *Progress in Low Temperature Physics,* C.J. Gorter, ed., North-Holland, Amsterdam (1955), vol. I, Chapter II.
2. G.W. Rayfield and F. Reif, *Phys. Rev. Lett.* **11**, 305 (1963); *Phys. Rev.* **136**, A1194 (1964).
3. G. Gamota, *J. Phys. (Paris),* Colloq. **C3**, 39 (1970).
4. G. Careri, in *Liquid Helium, Proc. Intern. School of Physics "Enrico Fermi,"* Academic Press, New York (1963), Vol. XXI.
5. G. Gamota, A. Hasegawa, and C.M. Varma, *Phys. Rev. Lett.* **26**, 960 (1971).
6. A. Hasegawa and C.M. Varma, *Phys. Rev. Lett.* **28**, 1689 (1972).
7. G. Gamota, *Phys. Rev. Lett.* **28**, 1691 (1972).
8. A.B. Basset, *A Treatise on the Motion of Vortex Rings,* MacMillan, London (1883).

Tunneling from Electronic Bubble States in Liquid Helium Through the Liquid–Vapor Interface

G. W. Rayfield and W. Schoepe

Department of Physics, University of Oregon
Eugene, Oregon

Ions have been extensively used in liquid helium to study both the superfluid and normal states. If the experimental cell is only partially filled with liquid, then the ions must pass through the phase boundary between liquid and vapor in order to reach a collector located in the vapor. The transport of charges across a phase boundary is of general interest in a variety of biological, chemical, and physical problems, helium being presumably one of the simplest cases. It has been observed that positive ions are unable to penetrate the liquid–vapor interface. However, negative ions (electronic bubbles) pass into the vapor quite easily, providing the temperature is not too low ($T > 2°$K). As the temperature is lowered, negative ions experience a rapidly increasing difficulty in penetrating the free surface, resulting in a vanishing current at moderate electric fields below $1°$K.

As the negative ion approaches the free surface it encounters an image–induced potential well.[1] This potential well results from the applied electric field and the interaction of the ion with its image above the free surface. The potential well $V(x)$ is of the form $A/x + e\mathscr{E}x$ and diverges at $x \to 0$. This form for the potential must break down near the free surface if any charge at all is to pass from the liquid into the vapor. The equation is probably valid to within a few angstroms of the free surface. The steep potential barrier and low temperature explain why positive charge is unable to penetrate the interface.

The negative ions in the potential well are sufficiently few in number so that they may be assumed to be noninteracting. If $N(t)$ is the number of ions in the well at some time t and no ion current flows from the source, then the number which escape per unit time dN/dt is proportional to the number of ions in the well: $dN/dt = -PN(t)$. Integrating, we find $N(t) = N_0 e^{-Pt}$, where N_0 is the number in the well at $t = 0$. Thus, if the flux of ions from the source is shut off at $t = 0$, then the current from the surface $j_s = -dN/dt = PN(t) = PN_0 e^{-Pt}$ decays exponentially. The experimental cell used to measure this exponential decay of the current is shown in Fig. 1. The potential well at the free surface is filled with ions by adjusting the source potential so negative charge flows from the source. The source current is then gated off, and the decay time (trapping time) of the current from the free surface is measured. In general, the trapping time $1/P$ is a function of the electric field $\mathscr{E}$ applied across the interface and the temperature T of the liquid.

The structure of the negative ion plays a key role in allowing negative charge (electrons) to pass through the free surface. The interaction potential between a

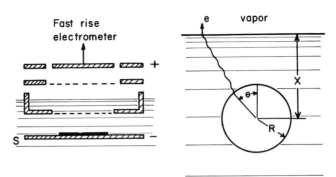

Fig. 1. The experimental cell and the electronic bubble near the free
surface of liquid helium.

helium atom and an electron is repulsive. This, in combination with the zero-point
energy, leads to the electronic bubble model for the negative ion in liquid helium
(see Fig. 1). The spherical well which traps the electron is about 1 eV deep. Near the
free surface the electrons can escape from these tiny spherical bubbles by tunneling.[2]
The analysis is similar to that used in describing α decay from radioactive nuclei.
The probability the electron tunnels a distance d through the liquid is $e^{-2\alpha d}$, where
$d = (x/\cos\theta) - R$ and $\alpha = \hbar^{-1}[2m(E_1 - E_0)]^{1/2}$, with E_1 the energy of the un-
trapped electron and E_0 the ground-state energy inside the bubble[2] (see Fig. 1).
The tunneling transition rate from the solid angle $d\Omega$ is $ve^{-2\alpha d}\,d\Omega/4\pi$, where v is
the frequency with which the electron hits the bubble wall. The total transition
rate is

$$\tfrac{1}{2}v\int_0^{\pi/2} e^{-2\alpha d}\sin\theta\,d\theta = \tfrac{1}{2}ve^{2\alpha R}w(x)$$

where $w(x) \approx e^{-2\alpha x}/2\alpha x$. In order to calculate the tunneling current, one now needs
to know the density of electronic bubbles as a function of distance from the free surface.
This is determined by the shape of the image–induced potential well. Assuming
a Boltzmann distribution, the number of electronic bubbles between x and $x + dx$
is $n(x)\,dx = N_0\exp[-V(x)/T]/I_1\,dx$, where

$$I_1 = \int_0^\infty e^{-V(x)/T}\,dx$$

The total current from the surface is

$$j_s = \tfrac{1}{2}ve^{2\alpha R}\int_0^\infty n(x)\,w(x)\,dx = NP(\mathscr{E}, T)$$

The integral can be evaluated and yields

$$P(\mathscr{E}, T) = \tfrac{1}{2}v[\exp(2\alpha R)]\,G(\mathscr{E}, T)[\exp-(8A\alpha/T)^{1/2}]\exp[2(Ae\mathscr{E})^{1/2}/T]$$

where

$$G(\mathscr{E}, T) = (e\mathscr{E})^{3/4} \Bigg/ \Bigg(\alpha^{5/4} T^{1/4} A^{1/2} 2^{5/4} \left\{ \exp \left[\frac{e\mathscr{E}}{\alpha T} \left(\frac{\alpha A}{2T} \right)^{1/2} \right] \right\}$$

$$\times \left[\exp \left(\frac{2T}{\alpha A} \right)^{1/3} \right] \left[1 + \frac{3T}{16(Ae\mathscr{E})^{1/2}} \right] \Bigg)$$

A, α, and R are chosen for a best fit to the data. The trapping time $\tau = 1/P(\mathscr{E}, T)$ is measured experimentally. Figure 2 shows the calculated and measured trapping times as a function of temperature for various electric fields across the free surface of liquid ^{4}He. The data can be plotted on a universal curve by writing

$$P/\gamma = \tfrac{1}{2} v \left[\exp(2\alpha R) \right] \exp(8 A\alpha/T)^{1/2}$$

where

$$\gamma(\mathscr{E}, T) = G(\mathscr{E}, T) \exp \left[2(Ae\mathscr{E})^{1/2}/T \right]$$

A plot of $\ln(P/\gamma)$ vs. $T^{-1/2}$ is shown in Fig. 3 for pure ^{4}He, pure ^{3}He, and two different ^{3}He–^{4}He mixtures. The results of the present experiments on pure ^{3}He and pure ^{4}He are summarized in Tables I–III.

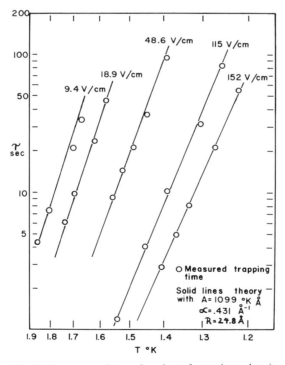

Fig. 2. The measured trapping times for various electric fields and temperatures. The solid lines are the theoretical predictions for τ.

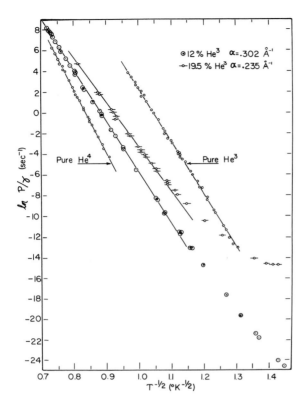

Fig. 3. Plot of ln P/γ vs. $T^{-1/2}$ for pure ^{4}He, pure ^{3}He, and two different ^{3}He–^{4}He mixtures. Solid lines are calculated values; data points are also indicated.

Table I. Evaluation of A^* from P vs. $\mathscr{E}$ at Fixed T

	Calculated	Experiment
^{4}He	1099 ($\varepsilon = 1.0572$)	1070
^{3}He	831 ($\varepsilon = 1.042$)	750

* $A = e^2(\varepsilon - 1)/4\varepsilon(\varepsilon + 1)$, in °K Å.

Table II. Evaluation of E_1^* from ln P/γ vs. $T^{-1/2}$

	Calculated (optical potential)	Experiment
^{4}He	0.87	0.82†
^{3}He	0.64	0.65‡

* $E_1 = (h^2\alpha^2/2m) + E_0$ (eV).
† $\alpha = 0.431$ Å^{-1}, $E_0 = 0.12$ eV.
‡ $\alpha = 0.384$ Å^{-1}, $E_0 = 0.09$ eV.

Table III. Evaluation of $R(\text{Å})$ from the Absolute Value of P/γ

	Calculated	Experiment
^{4}He	$15 \sim 20$*	24.8
^{3}He	—	30.1
R_3/R_4	$(\sigma_4/\sigma_3)^{1/4} = 1.21$†	1.21

* From other experiments.
† From the bubble model.

The very low value of α found for the mixture is peculiar. The P/γ plot (Fig. 3) is linear for the 12% mixture over about eight orders of magnitude with $\alpha = 0.302\,\text{Å}^{-1}$; this value of α is considerably less than that for ^{3}He ($\alpha = 0.384\,\text{Å}^{-1}$). Since

$$P/\gamma = \tfrac{1}{2}v[\exp(2\alpha R)]\exp -(8A\alpha/T)^{1/2}$$

a temperature–dependent bubble radius in the mixtures could account for the strange results in the mixtures. A detailed account of this work will be presented elsewhere.

References

1. W. Schoepe and G. W. Rayfield, *Z. Naturforsch.* **26a**, 1392 (1971).
2. G. W. Rayfield and W. Schoepe, *Phys. Lett.* **37A**, 417 (1971).

Positive Ion Mobility in Liquid ³He*

M. Kuchnir,† J. B. Ketterson, and P. R. Roach

Argonne National Laboratory
Argonne, Illinois

Previous measurements[1] of the positive ion mobility in liquid ³He below 1°K indicated the existence of a minimum with temperature; however, scatter and a hysteresis phenomena prevented a clear conclusion. A better measurement of this mobility is relevant as it should provide information on the structure of the positive ion in liquid He, a structure that is now under closer examination. Recent measurements by Johnson and Glaberson[2] revealed the effect of various impurity ions on the mobility of these structures. According to the classical snowball model of Atkins,[3] the minimum with temperature mentioned above is expected near 0.3°K as a consequence of the well-known minimum in the melting curve of ³He. In this work we suggest that the shape of the mobility minimum is related to the layered model proposed by Johnson and Glaberson.[2]

A much improved time-of-flight technique and a dilution refrigerator, rather than an adiabatic demagnetization cryostat, were employed for the present set of measurements. The apparatus was the same one used to measure electron bubble mobilities in pure ⁴He reported in a previous paper.[4] The source was kept negatively biased with respect to the first grid except during the extraction pulse, when for 10 msec an extraction field of 110 V cm⁻¹ produced a pulse of positive ions. The transit time to the collector was of the order of 1/2 sec for the fields used in this experiment. The reproducibility of the time of flight was usually within 1 msec. The error in the absolute value is estimated to be smaller than 4%.

Once the temperature of the cell was stabilized the velocity vs. field curves for the electric fields used (40–90 V cm⁻¹) were observed to be good straight lines through the origin, indicating the absence of space charge effects which were apparent in the original experiment. Above 0.5°K the agreement in absolute value of the present mobility with some of the runs of the previous experiment is unreasonably good, especially since the scatter in the first experiment was in part due to slightly different pressures used in the various runs. A pressure of 60 Torr was used in the present measurements.

Simultaneous and independent measurements by McClintock[5] confirm the results obtained here, which are presented in Fig. 1, where we have also plotted some of the previous data.[1]

The relatively high-intensity tritium source used kept the mixing chamber of the dilution refrigerator above 0.06°K and established a nonnegligible thermal

* Based on work performed under the auspices of the U.S. Atomic Energy Commission.
† Present address: Engineering Experiment Station, University of Wisconsin, Madison, Wisconsin.

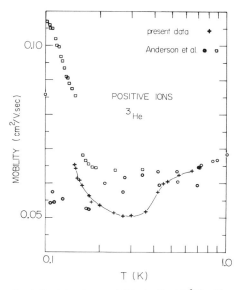

Fig. 1. Positive ion mobility in liquid ^{3}He. The line connecting the present data is drawn for the purpose of clarity.

gradient in the liquid below 0.15 K. Furthermore, the 800 cm^2 of sintered copper surface area in the cell was insufficient to overcome the Kapitza resistance below this temperature. For these reasons we have reported only data taken above 0.15 K.

Atkins's model assumes the positive ion structure to have a solid core created by the electrostriction in the polarizable liquid around it. The radius of this "snowball" is therefore a function of the charge of the ion, the polarizability, and the equation of state of the liquid.

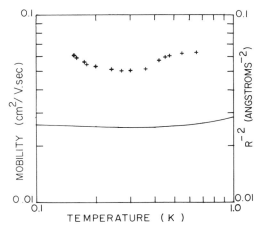

Fig. 2. The temperature dependence of the mobility compared with the temperature dependence of R^{-2} as calculated with Atkins model.

Using molar volume[6] and melting pressure[7] data, the radius R of the classical snowball was computed as a function of temperature. For hard-sphere scattering one would expect the mobility μ to be inversely proportional to the square of the radius. When we compare the temperature dependence of R^{-2} and μ in Fig. 2 we see that the increase in μ at 0.4°K is too strong, indicative perhaps of the peeling of a layer of discrete polarized atoms caused by the increase in the temperature of the system. Since theoretical calculations[8] of the positive ion mobility in terms of momentum-dependent polarization scattering do not predict this 20% change, we suggest that a structural change is relevant here.

Note Added in Proof

Subsequent measurements with a ten times weaker source indicates further need for temperature correction in these data particularly at low temperatures. In addition, we observed that the mobility rise at the lower temperatures is part of a sharp local maximum.

References

1. A. C. Anderson, M. Kuchnir, and J. C. Wheatley, *Phys. Rev.* **168**, 261 (1968).
2. W. W. Johnson and W. I. Glaberson, *Phys. Rev. Lett.* **29**, 214 (1972).
3. K. R. Atkins, *Phys. Rev.* **116**, 1339 (1959).
4. M. Kuchnir, J. B. Ketterson, and P. R. Roach, this volume.
5. P. V. E. McClintock, this volume.
6. E. R. Grilly and R. L. Mills, *Ann. Phys. (N.Y.)* **8**, 1 (1959); C. Boghosian, H. Meyer, and J. E. Rives, *Phys. Rev.* **146**, 1110 (1966).
7. A. C. Anderson, W. Reese, and J. C. Wheatley, *Phys. Rev.* **130**, 1644 (1963).
8. R. M. Bowley, *J. Phys. C: Sol. State Phys.* **4**, L207 (1971).

A Large Family of Negative Charge Carriers in Liquid Helium*

G. G. Ihas† and T. M. Sanders, Jr.

*Department of Physics, University of Michigan
Ann Arbor, Michigan*

History

In 1969 Doake and Gribbon,[1] using the Cunsolo technique[2] for measuring charged particle velocities, found a negative charge carrier with a much higher mobility than either the normal positive or negative carrier studied for the past fifteen years. They found that this new species, rather than producing vortex rings, reached a velocity apparently limited by roton creation. They were able to produce it below $1.05°$K with large electric fields and a radioactive α source. We have studied a new species having similar characteristics and have found that a large family of negative charge carriers can be produced in liquid helium.[3]

Time-of-Flight Technique

We measure ion velocities using a direct time-of-flight apparatus described elsewhere.[3, 4] Figure 1 shows the construction of the experimental cell, with ion source at the top and charge collector at the bottom. A gate pulse applied between grids $G1$ and $G2$ injects charge carriers into the drift region defined by grids $G2$ and $G3$. Equal voltages are applied across each of the five sections of the drift region created by the four field homogenizers ($H1–H4$). The brass rings shield against stray electric fields. After traversing the 65-mm drift region the ion current striking the collector is detected by a fast electrometer. Figure 2(a) shows the gate pulse and signal-averaged collected current as a function of time, with both the fast (1) and normal (N) negative carriers present. The distance between the gate and charge carrier pulses yields the transit time and thus the carrier velocity and mobility.

Production Techniques

We have used both a tritium β source and a glow discharge in the vapor above the liquid to produce the fast charge carriers (see Ref. 3). Best fast ion signals are obtained with the needle in Fig. 1 grounded, the perforated disk S held at 200–300 V below ground potential, and the helium liquid level at S. A pink to purple glow can

* Work supported in part by the U.S. Atomic Energy Commission and the National Science Foundation.
† Present address: Department of Physics, The Ohio State University, Columbus, Ohio.

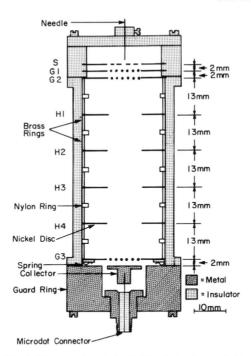

Fig. 1. The experimental cell. The electrodes are held
in a nylon tube. A spring compensates for differential
contraction. All metal parts are gold-plated and the
insulating material is nylon.

be observed around the needle under these conditions. The normal positive and
negative charge carriers can also be copiously produced under a wide range of
voltage and liquid level configurations with this discharge source.

The fast carrier signal amplitude is dependent on the temperature, the liquid
level with respect to the source, and the voltage between the source disk and needle.
As the temperature (and thus the helium vapor pressure) is raised signal strength
diminishes, with detection becoming impossible above 1.2°K. Also, all voltages in
the cell can be varied, but as long as the source disk-to-needle voltage remains constant
(along with liquid level and temperature) the signal amplitudes do not change. This
implies that the glow discharge (which depends on this voltage and the gas density in
the region of the source) is directly associated with the creation of the new charge
complex.[4]

When the liquid level is placed between the perforated disk S and grid $G1$ the
normal ion signal becomes very strong, the fast carrier signal becomes weak, and a
series of charge carriers with intermediate transit times appears. A sample data trace
taken under these conditions is shown in Fig. 2(b). The transit time of each intermediate carrier through the cell varies with drift electric field and temperature in
a reasonable fashion. An extremely short packet of charge must be used to distinguish
the various intermediate signals. We detect 13 distinct transit times, and suspect that
more might be resolved.

a)

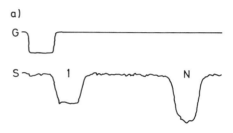

b)

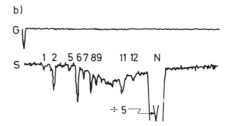

Fig. 2. Recorder traces of gate-pulse time reference (G) and signal-averaged collected current (S) for (a) the liquid helium level at the source disk and (b) the liquid helium level below the source disk. The abscissa represents time and the ordinate ion current intensity (or gate-pulse voltage). The numbers identify the various ion species plotted in Fig. 3. The fast carrier (1) and the normal negative ion (N) are seen in both traces, while the intermediate ions (2–12) can only be seen in (b).

Weak-Field Mobilities

Using the discharge source, weak-field mobilities (where drift velocity is proportional to applied electric field) were obtained. These are plotted against reciprocal temperature in Fig. 3 for all negative charge carriers studied thus far. The mobilities range from one to six times the normal negative ion mobility. The various slopes correspond to activation energies from 8.5 to 11.1°K, although scarcity of data makes some of these values quite uncertain. The intermediate ions were not followed to high fields.

Other Measurements

In an attempt to elucidate the structures of these new charge carriers, a series of experiments was run, with the following results.

We have observed the new charge carriers with transit times greater than 10 msec, indicating that these are long-lived complexes.

Magnetic fields up to 1.2 kG and applied parallel to the carrier velocity show no effect on signal intensities or carrier mobilities.

The characteristics of the new carriers were unchanged when observed in ^{4}He purified using a Vycor superleak and in helium doped with up to 100 ppm ^{3}He. The ions were also unaffected when the liquid helium in the cell was polluted with nitrogen, hydrogen, and air.

Charge Carrier Structures

We believe that these and other experimental and theoretical considerations eliminate the following models: (1) free (conduction band) electron,[3] (2) excited and multiple electron bubble states,[6] and (3) ^{3}He-dependent states.

A He$^-$ ion placed in liquid helium might be expected to form a small, non-

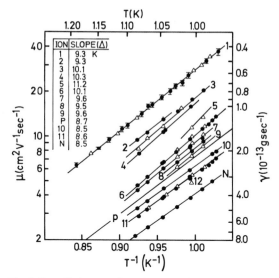

Fig. 3. Fast (1), intermediate (2–11), and normal (N) negative charge carrier mobilities μ (and friction coefficients γ)
plotted against inverse temperature T^{-1} (and temperature
T). The insert gives Δ (in °K) for each charge species.

spherical bubble the size and shape (and hence, the mobility) of which would depend
on the ion's electronic state.[7] However, vacuum properties of He$^-$ may eliminate it
as a candidate.[8]

Recent mobility measurements[9] on positive impurity ions in He II suggest
that some of our new carriers might be negative impurity ions. However, Johnson
and Glaberson did not detect any negative impurities in their experiments. Also,
under our experimental conditions when the polarities in the cell are reversed we see
only the normal positive ion. Apparently neither positive nor negative impurity
ions are easily or accidentally generated in liquid helium.

It is interesting to note that those negative charge carriers with mobilities higher
(lower) than the normal positive helium ion mobility have an activation energy Δ
larger (smaller) than the normal positive ion activation energy $[\Delta(\text{He}_n^+) = 8.7°\text{K}]$.
This trend is also reflected in the data of Ref. 9. Barrera and Baym[10] have recently
calculated the roton-limited mobility of an ion with finite mass in liquid helium.
They find that in the temperature range 1.0–1.2°K, $\Delta \gtrsim 8.7°\text{K}$ implies an ion effective
mass $m \lesssim 15m_4$ (m_4 is the helium atomic mass). This indicates that our charge carriers
have effective masses from approximately $1m_4$ to $100m_4$.

Mobility measurements in liquid helium under pressure, effective mass determinations, and photoejection spectra of these new charge carriers would help
to reveal their structures. These experiments, difficult in themselves, would be even
more challenging due to the production techniques required for the new carriers.

Acknowledgment

We wish to acknowledge the assistance of John Magerlein in acquisition and reduction of data.

References

1. C. S. M. Doake and P. W. F. Gribbon, *Phys. Lett.* **30A**, 251 (1969).
2. S. Cunsolo, *Nuovo Cimento* **21**, 76 (1961).
3. G. G. Ihas and T. M. Sanders, Jr., *Phys. Rev. Lett.* **27**, 383 (1971).
4. G. G. Ihas, Thesis, Univ. of Michigan, Ann Arbor, Michigan, 1971.
5. Roland Meyerott, *Phys. Rev.* **70**, 671 (1946).
6. D. L. Dexter and W. Beall Fowler, *Phys. Rev.* **183**, 307 (1969).
7. W. Beall Fowler and D. L. Dexter, *Phys. Rev.* **176**, 337 (1968).
8. L. M. Blau, R. Novick, and D. Weinflash, *Phys. Rev. Lett.* **24**, 1268 (1970).
9. W. W. Johnson and W. I. Glaberson, *Phys. Rev. Lett.* **29**, 214 (1972).
10. R. Barrera and G. Baym, *Phys. Rev.* **A6**, 1558 (1972).

Temperature Dependence of the Electron Bubble Mobility Below 0.3°K*

M. Kuchnir, J. B. Ketterson, and P. R. Roach

Argonne National Laboratory
Argonne, Illinois

The temperature dependence of the mobility of the electron bubble in superfluid helium has been calculated[1] using a hydrodynamic approach for the phonon–bubble scattering cross section. This calculation explains the observed T^{-3} dependence of previous experimental data[2] in the range 0.33–0.5°K as arising from a prominent s-wave resonance in the bubble cross section for thermal phonons. Furthermore, this calculation predicts that the mobility at lower temperatures should rise more sharply than T^{-3}. At very low temperature ($T < 80$ m°K) the mobility should vary as T^{-8}. This behavior is characteristic of Rayleigh scattering. We have extended the measured temperature range down to 0.15°K and have observed that the mobility indeed increases faster than T^{-3}.

The method used to measure the time of flight employs a fast electrometer. A drawing of the ion cell and the electronics is shown in Fig. 1. After a steady-state current of bare ions is established through the cell a short pulse applied to the source interrupts the ion current and simultaneously triggers the sweep of a signal averager

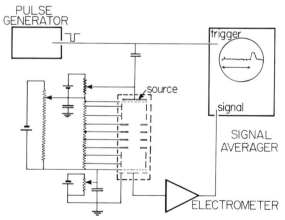

Fig. 1. Diagram of ion cell and time-of-flight measuring equipment.

* Work performed under the auspices of the U.S. Atomic Energy Commission.

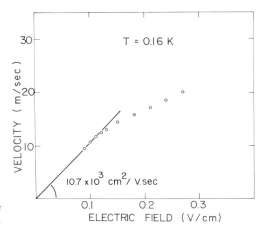

Fig. 2. The velocity vs. electric field for negative ions in liquid ⁴He at 0.16°K and at vapor pressure.

whose input receives the current detected by the fast electrometer. The time of arrival of the depletion pulse is taken as the time of flight of the ions. This time is measured as a function of the field applied to the drift space. (Care is taken not to affect the space charge condition near the source.) Figure 2 shows a typical measurement of the field–velocity relationship. For higher fields we obtain a departure from a linear field–velocity relationship, while for still higher fields only a ring pulse is observed. The slope of the straight line extrapolating the lower-field velocities

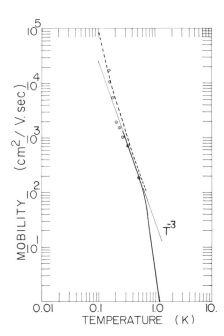

Fig. 3. The temperature dependence of the zero-field mobility of negative ions in liquid ⁴He at vapor pressure. The solid curve represents data of Ref. 2 and the dashed line the calculation of Ref. 1.

through the origin yielded the zero-field mobility. Figure 3 presents our data points along with a curve through the Schwartz and Stark data.

Only a relatively small number of points have been taken to date due to difficulty in obtaining the bare-ion currents. The time required to stabilize the space charge condition near the source was of the order of one day. The mixing chamber temperature of our dilution refrigerator (to which the cell was attached) was kept constant for extended periods of time by an electronic feedback circuit.

Due to the large mobility values electric fields as small as $50~\text{mV cm}^{-1}$ were applied. It is important that stray fields produced by a distribution of charge in the cell be much smaller than the applied field. Evidence against a slow buildup of stray charge is provided by the fact that the time of flight was observed to be independent of how long the external field had been applied. In addition, the dependence of the ion velocity on applied field is the dependence we expect from the actual field acting on the ion (i.e., linear at small fields). Based on these two observations, we believe that no stray fields masked the applied field in the drift space region. As an extra precaution, all electrode surfaces were gold-plated.

References

1. G. Baym, R.G. Barrera, and C.J. Pethick, *Phys. Rev. Lett.* **22**, 20 (1969); R.G. Barrera Perez, Ph.D. Thesis, Univ. of Illinois, Urbana, Illinois, 1972.
2. K.W. Schwarz and R.W. Stark, *Phys. Rev. Lett.* **21**, 967 (1968); K.W. Schwarz, *Phys. Rev.* **6A**, 837 (1972).

Transport Properties of Electron Bubbles in Liquid He II

M. Date, H. Hori, K. Toyokawa, M. Wake, and O. Ichikawa

Department of Physics, Osaka University
Toyonaka, Osaka, Japan

Introduction

As was found by Spangler and Hereford,[1] thermal electrons can be injected into the He II bath by heating tungsten filament immersed in the liquid. This method is useful for studying electron bubbles because it can supply a large quantity of electrons. However, many phenomena associated with this method have not been sufficiently analyzed. It is believed that the sudden onset of turbulence or a bubble near the filament surface appears above a critical heat power W_c which depends on the depth and temperature of the liquid.[2,3] The process in the vicinity of the wire resembles film boiling[4] when the He depth is shallow ($\lesssim 3$ cm) or the temperature is high ($\gtrsim 1.8°$K), while it results in a quiescent vapor bubble or a gas sheath in the deep bath ($\gtrsim 10$ cm) at low temperature ($\lesssim 1.5°$K).

Recently we tried to estimate the effective radius of the gas sheath by using microwave techniques as discussed in the next section. Although it is not yet clear that the gas sheath exists in a definite form, it seems worthwhile to try to explain the mechanical, thermal, and electrical properties of various phenomena associated with the hot filament in He II based on the gas sheath model. The electrical properties of injecting electrons were investigated by making a diode and a triode as discussed in the last section.

Mechanical and Thermal Properties of the Gas Sheath

To estimate the size of the gas sheath, i.e., the empty space volume around the wire, a tungsten filament with a diameter of 9.1×10^{-4} cm was set in a 35-GHz microwave cavity as shown in Fig. 1. The resonant frequency of the cavity shifted by the order of 1 MHz in the supercritical heat flow region because the sheath removes the same volume of liquid out of the cavity. The experiment was done in accord with standard microwave techniques,[5] and small corrections coming from the skin-depth change of the filament and the liquid density change due to a slight temperature increase were taken into account. The estimated sheath radius d_1 is shown in Fig. 1 as a function of filament temperature, which was measured by measuring the electrical resistivity of the filament and was monitored by an optical pilometer. The cavity wall was coated by superconducting metal to get a high Q value.

The mechanical and thermal stabilities of the gas sheath were considered in the following way: Helium gas atoms in the sheath transport the energy W from the wire surface to the liquid surface with accommodation coefficients α and β,

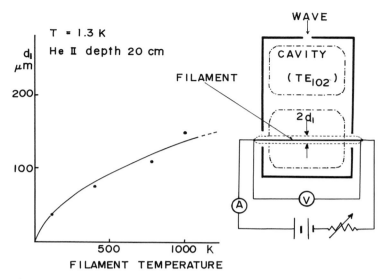

Fig. 1. Temperature dependence of gas sheath radius d_1 and a schematic view of the apparatus.

respectively. The gas pressure P was calculated by summing up the momentum change of the gas molecules at the liquid wall and balancing it against the sum of the residual gas pressure and the liquid pressure at the filament. It is noted that the usual heat transport theory developed by Knudsen[6] cannot be applied because the sheath radius is temperature dependent. The detailed calculation will be discussed elsewhere, but the result is written as follows:

$$W = CPd_1\sqrt{T} \tag{1}$$

where T is the filament temperature and C is a complex function of α and β.[7] A comparison of the theory and experiment is given in Fig. 2, where the agreement is fairly good except for the high-temperature region. It was assumed that there is no gradient in gas density and temperature within the sheath, but this assumption may not be appropriate when the sheath becomes large. This may be one reason for the deviation at high temperatures. The coefficients α and β were determined to be

$$\alpha = 0.27, \qquad \beta = 0.1 \tag{2}$$

The gas sheath model can easily explain the origin of the audible hiss.[3] The main part of the hiss frequency lies near 1 kHz and its origin was attributed[8] to the translational vibration of the sheath around the wire without changing its size and form.

The IV Characteristics of a Diode and a Triode

The electron bubble current I per unit length of a diode which consists of a coaxially arranged tungsten filament and a brass anode tube with voltage difference

V_0 is given by[7]

$$I = \mu V_0^2 / 2d_2^2 \qquad (3)$$

where d_2 is the anode radius and μ is the bubble mobility. Equation (3) was obtained by solving Poisson's equation under an assumption that the bubble moves according to the mobility model[9] in a strong electric field $E(r)$,

$$E(r) = (V_0/d_2)\left[(r^2 - d_1^{*2})^{1/2}/r\right], \qquad d_1^* \ll d_2 \qquad (4)$$

where r is the radial distance from the filament center. In our model the cathode radius d_1^* is the radial distance where the electric field is zero. It may be near the sheath surface d_1 but it is not necessary to determine it as strictly as in a usual vacuum-tube diode in which the electron emission is less than that expected from the Richardson formula. An example of the experimental results is shown in Fig. 3. The IV curve should be given by a straight line if the mobility is field independent. As is seen in Fig. 3, however, the result is not so simple. By analyzing the result in detail, the following conclusions were obtained: In region A the bubble moves with the mobility μ_R of the bubble–roton interaction,[9] but the bubble–vortex inter-action[10] should be taken into account in region B. In region C the escape probability from the vortex becomes large, so that the current I again approaches the μ_R line. Spangler et al.[1] performed measurements in the B and C regions and their results show good agreement with ours. We calculated the IV curve in the B and C regions using the escape probability in an electric field obtained by Parks and Donnelly[11]

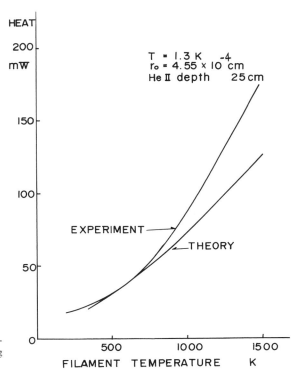

Fig. 2. Heat flow as a function of tempera-ture. Theoretical curve is drawn according to Eq. (1).

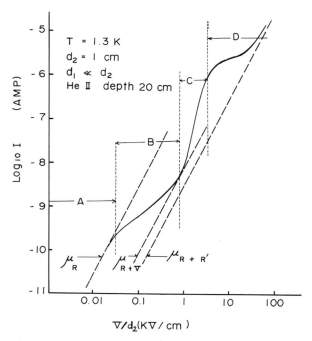

Fig. 3. Typical IV characteristic of a diode. The three broken lines μ_R, μ_{R+V}, and $\mu_{R+R'}$ correspond to the theoretical lines when the mobilities are determined by roton scattering, vortex trapping plus roton scattering, and roton scattering plus the roton-creation process, respectively.

and the result was quite satisfactory. In region D a new interaction occurs and the mobility decreases down to $\mu_{R+R'}$ line. We believe that the roton creation process begins in this region. This does not mean that the drift velocity is larger than the roton creation velocity, but rather that a bubble gets enough energy for creating a roton in the mean free path flight. Electrical breakdown was found above the D region. The grid control was also tested by making a triode. Detailed results will be published later.

References

1. G.E. Spangler and F.L. Hereford, *Phys. Rev. Lett.* **20**, 1229 (1967).
2. F.E. Moss, F.L. Hereford, F.J. Agee, and J.S. Vinson, *Phys. Rev. Lett.* **14**, 813 (1965).
3. J.S. Vinson, F.J. Agee, R.J. Manning, and F.L. Hereford, *Phys. Rev.* **168**, 180 (1968).
4. D.M. Sitton and F.E. Moss, *Phys. Rev. Lett.* **23**, 1090 (1969).
5. T. Okamura, T. Fujimura, and M. Date, *Sci. Rep. RITU* **A4**, 191 (1952).
6. M. Knudsen, *Ann. Physik* **31**, 205 (1910); **33**, 182 (1965).
7. M. Date, H. Hori, M. Wake, and O. Ichikawa, *J. Phys. Soc. Japan* (to be published).
8. M. Date, *J. Phys. Soc. Japan,* (to be published).
9. F. Reif and L. Meyer, *Phys. Rev.* **119**, 1164 (1960).
10. D.M. Sitton and F.E. Moss, *Phys. Rev. Lett.* **23**, 1090 (1969).
11. P.E. Parks and Donnelly, *Phys. Rev. Lett.* **16**, 45 (1966).

Do Fluctuations Determine the Ion Mobility in He II near T_λ?*

D. M. Sitton

Department of Aerospace Engineering and Engineering Physics
University of Virginia, Charlottesville, Virginia

and

F. E. Moss

Department of Physics, University of Missouri—St. Louis
St. Louis, Missouri

Measurements of the zero-field ionic mobility μ in the roton region and near T_λ of superfluid helium have been obtained from the characteristics of space-charge-limited discharges. The results for negative ions were obtained from the dc current $I = C\mu V^2$ of a planar diode using an 8-Ci tritium ion source. The source was plated on a 1-in.-diameter by 0.005-in.-thick OFHC copper disk which faced a collector at a distance of 0.5 cm. The source was maintained at a fixed voltage, $V = -8.00$ V, which ensured space charge limitation for all currents measured, a condition which we checked at various temperatures by observing that $I \propto V^2$. The geometric constant C was determined by matching the data to a known value of μ at a single temperature. The results for positive ions were obtained from tungsten field emission points and have been reported previously.[1] These data have been compared to the recent data of Schwarz[2] and Ahlers and Gamota.[3] Over the range from 1.4°K to T_λ tabular comparison with Schwarz's data shows a point-by-point variation of no more than $\pm 1\%$.

Of particular interest here is the anomalous decrease in the mobility of both species in the range $2.0 < T < T_\lambda$. This effect has long been known, and was first studied in detail by Scaramuzzi.[4] The decrease has traditionally been interpreted in terms of the Stokes mobility $\mu^{-1} \propto \eta R$, where η is the viscosity and R is the effective ion radius.[4] More recently Ahlers and Gamota[3] have obtained the temperature dependence of R near T_λ using the Stokes law interpretation. They observed a singularity in R for the negative ion, which they suggest may be related to the divergence of the coherence length for fluctuations in the order parameter at T_λ.

We believe that fluctuations are indeed responsible for the behavior of μ near T_λ, but that the data must not be interpreted by Stokes law. We write the inverse, roton-limited mobility as

$$\mu^{-1} = \mu_\lambda^{-1}[1 - f(T_\lambda - T)^\varsigma] \tag{1}$$

* Supported in part by the Office of Naval Research and the National Science Foundation.

Table I. The Constants in Eq. (1) with $\zeta \equiv 2/3$

	Positive ions		Negative ions	
Author	μ_λ^{-1}	f	μ_λ^{-1}	f
Schwarz	20.73	1.45	29.36	1.66
Ahlers and Gamota	20.77	1.52	30.49	1.77
Sitton and Moss	20.32	1.36	28.71	1.64
Schwarz $(T > T_\lambda)$	21.23	-0.22	29.30	-0.63

Assuming that $\zeta \equiv 2/3$, this equation is matched to the data with results shown in Table I. The graphical representations are shown in Figs. 1 and 2. It is important to note that for the positive ions $f \simeq 1.43$. Therefore, *to within experimental error*, the bracket in Eq. (1) can be replaced by $[1 - 1.43(T_\lambda - T)^{2/3}]$, and this is precisely the normal fluid fraction ρ_n/ρ found by Tyson and Douglass[5] in their version of Andronikashvili's experiment. This result argues that the mobility near T_λ, at least for the positive ions, is dominated by fluctuations of the normal fluid density.

The negative ions behave anomalously, with $f \simeq 1.7$. It is quite likely that this is yet an additional effect of the difference in structure of the positive and negative ion. It may well be that the radius of the negative ion does something strange near T_λ, perhaps as a result of critical fluctuations of the surface tension. If the behavior of the surface tension were known, $R(T)$ would be easy to calculate by minimizing the electron-bubble total energy.

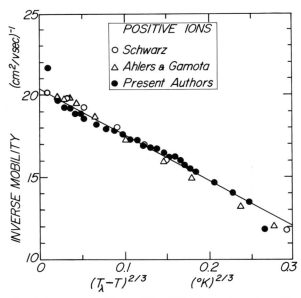

Fig. 1. Inverse mobility vs. $(T_\lambda - T)^{2/3}$ for positive ions. The solid curve is Eq. (1) matched to the data of Sitton and Moss.

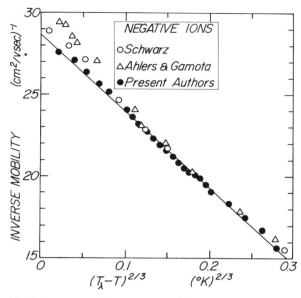

Fig. 2. Inverse mobility vs. $(T_\lambda - T)^{2/3}$ for negative ions. The solid curve is Eq. (1) matched to the data of Sitton and Moss.

Recently, in an investigation of departures from scaling and universality in He II, Greywall and Ahlers[6] made high-resolution measurements of the second-sound velocity for a range of pressures P. They obtain ρ_s/ρ from the measurements using a result of linearized, two-fluid hydrodynamics and show that their data are represented by $\rho_s/\rho = k(P)\,\varepsilon^\zeta[1 + a(P)\,\varepsilon^x]$, where $\varepsilon = 1 - T/T_\lambda$ and $k(P)$ and $a(P)$ are pressure-dependent constants. At the standard vapor pressure they find $a(P) = 0$ and $\zeta = 0.674 \pm 0.001$. This has prompted us to analyze each set of data by a least squares fit to obtain the critical exponent ζ, as shown in Table II. While the errors on the present results are smaller (due to the larger number of points), we believe that departures from $\zeta = 2/3$ as seen through the mobility results remain an open question. It is also of interest to look for remnant superfluidity above T_λ. With this in view, we have matched Schwarz's data for $T > T_\lambda$ to the same equation, with results shown by the last row in Tables I and II. Thus for $T > T_\lambda$ it seems as though $\zeta \simeq 2/3$ also, and the ion mobilities exhibit the simple scaling behavior to be expected from concentration fluctuations near critical points—at least at the

Table II. The Exponent ζ, with μ_λ As Shown in Table I

Author	Positive ions	Negative ions
Schwarz	0.691 ± 0.009	0.687 ± 0.050
Ahlers and Gamota	0.682 ± 0.014	0.672 ± 0.029
Sitton and Moss	0.666 ± 0.004	0.659 ± 0.005
Schwarz $(T > T_\lambda)$	0.613 ± 0.064	0.695 ± 0.047

vapor pressure. Caution is in order, however, since only six data points were used in the matches for $T > T_\lambda$.

Some comments regarding Stokes law are in order. It applies in the hydrodynamic regime where both the roton mean free path and the coherence length ξ are much less than the ion radius. In order to reconcile Stokes law with the data, we must have $\eta = \eta_\lambda[1 - f(T_\lambda - T)^{2/3}]$ if we discount any anomalous behavior of the ionic radii. A recent and careful search due to Webeler[7] has, however, failed to reveal a power law divergence, or indeed any critical behavior, of η in the hydrodynamic regime. In any case, over much of the temperature range of interest $\xi > R$, even assuming the small value $\xi_0 \simeq 1$ Å. It is not clear how the ion mobility is influenced by fluctuations when $\xi \gg R$; thus the results presented here pose an interesting theoretical problem.

Acknowledgments

We are grateful to J. Ruvalds and V. Celli for suggesting the search for the 2/3 power law and for many enlightening discussions. We thank G. Ahlers for providing us with his mobility data. Special thanks are due K. Kawasaki for a valuable discussion.

References

1. D.M. Sitton and F. Moss, *Phys. Lett.* **34A**, 159 (1971).
2. K.W. Schwarz, *Phys. Rev.* (to be published).
3. G. Ahlers and G. Gamota, *Phys. Lett.* **38A**, 65 (1972).
4. D.T. Grimsrud and F. Scaramuzzi, in *Proc. 10th Intern. Conf. Low Temp. Phys., 1966*, VINITI, Moscow (1967), Vol. I., p. 197.
5. J.A. Tyson and D.H. Douglass, *Phys. Rev. Lett.* **17**, 472 (1966).
6. D.S. Greywall and G. Ahlers, *Phys. Rev. Lett.* **28**, 1251 (1972).
7. R.W.H. Webeler and G. Allen, *Phys. Rev.* *A***5**, 1820 (1972).

8
Sound Propagation
and
Scattering Phenomena

Absence of a Quadratic Term in the
^{4}He Excitation Spectrum*

P. R. Roach, B. M. Abraham, J. B. Ketterson, and M. Kuchnir

Argonne National Laboratory
Argonne, Illinois

For many years it has been assumed that the long-wavelength portion of the ^{4}He elementary excitation spectrum could be described by

$$\varepsilon(k) = c\hbar k(1 + \alpha_2 k^2 + \cdots) \tag{1}$$

where ε is the energy of the excitation, k is its wave number, and c is the sound velocity; α_2 is traditionally taken to be negative. Recently, however, Barucchi et al.[1] predicted that the excitation spectrum could be described by

$$\varepsilon(k) = c\hbar k(1 + \alpha_1 k + \alpha_2 k^2 + \cdots) \tag{2}$$

and Molinari and Regge[2] suggested that an improved fit to the inelastic neutron scattering data[3] for the excitation spectrum could be achieved with this expression; α_1 was positive in the resultant fit.†

Since the frequency-dependent sound (phase) velocity according to Eq. (2) is just $c(1 + \alpha_1 k + \alpha_2 k^2 + \cdots)$, then a measurement of the frequency dispersion of the velocity is a direct indication of the higher-order terms in the excitation spectrum. Anderson and Sabisky[4] recently reported just such sound velocity measurements obtained on helium films at frequencies between 20 and 60 GHz. At their experimental temperature of 1.38°K they found a dispersion whose leading term is linear (i.e., a quadratic term in the excitation spectrum) and is of the same sign and magnitude as suggested by Molinari and Regge.[2] However, they pointed out[5] that this result is only the simplest interpretation of their data and that other expressions involving higher-order terms in the momentum, but without the linear term, also can describe their data.

In an attempt to measure the dispersion in a way that is independent of temperature effects and of possible peculiarities of He films, we have measured the dispersion in bulk ^{4}He below 100 m°K to an accuracy sufficient to allow the observation of a linear term if it is as large as predicted. Our method involves the direct measurement of the difference in velocity of simultaneous 30- and 90-MHz sound waves in the liquid. To obtain the required accuracy, the 30- and 90-MHz oscillators are phase-locked to each other so that the relative phase delay of the two signals after they have passed through the helium can be directly measured. Since it is only feasible to de-

---

* Based on work performed under the auspices of the U.S. Atomic Energy Commission.
† We have adopted the sign convention of Eq. (2) of Ref. 2.

termine changes in this relative delay, we measure the delay as a function of sound path length which is varied by moving a reflector in the helium.

A block diagram of the electronics for the experiment is shown in Fig. 1. The two oscillators were phase-locked to each other by a Hewlett-Packard model 8708A synchronizer. The normal use of the synchronizer is to lock the frequency of an oscillator to a 20-MHz internal reference frequency in the synchronizer; however, the synchronizer lends itself very well to the present application where the internal reference is replaced by the output of the 30-MHz oscillator. The switches at the oscillator outputs gated rf pulses into the sound cell containing a 30-MHz quartz piezoelectric transducer which could also be driven at 90 MHz. The variable delay (AD-YU model 20B1) in the 30-MHz branch compensated for (and measured) the difference in transit time of the two sound signals. It consisted of a variable-length coaxial line plus fixed lengths of coaxial line, all of which were calibrated to 0.01 nsec. The switch at the amplifier inputs kept the original rf pulses from disturbing the amplifiers and only passed the signals returning from the sound cell after transit through the helium. The 30-MHz signal was then frequency-tripled by a non-linear element and its phase and amplitude initially adjusted to exactly cancel the 90-MHz signal at the input to the final amplifier–detector. Then as the sound path in the helium was changed any difference in the velocities of the two signals would show up as a phase delay of one signal with respect to the other which would be exactly proportional to the change in sound path length.

Figure 2 shows the sound cell used in the experiment. The main feature of the cell is the fused quartz V-block and slider which were lapped to optical accuracy in order to allow the reflecting surface to be moved up and down above the transducer while maintaining exact orientation. Precise alignment is required to ensure that

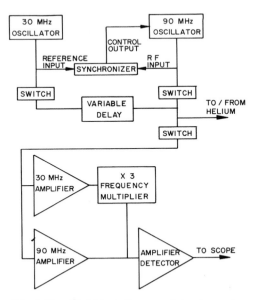

Fig. 1. Simplified block diagram of the electronics
of the experiment.

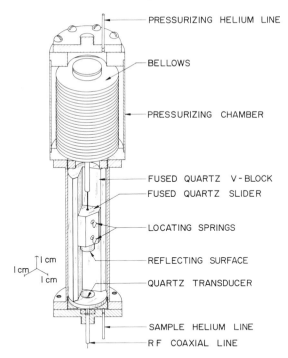

Fig. 2. Moving reflector sound cell.

the reflected wavefront is accurately parallel to the transducer. The slider is moved up and down by a shaft which is attached to the top of the metal bellows. A differential pressure of liquid helium between the inside and the outside of the bellows causes it to collapse or expand by several centimeters, thus moving the slider by this amount. The slider is lightly pressed against the V-block by springs which contact the inside of the lower chamber. The cell was attached to the copper mixing chamber of a ³He–⁴He dilution refrigerator. This allowed us to perform the experiment at temperatures below 60 m°K where there is essentially no temperature dependence to either the sound velocity or attenuation.

According to Molinari and Regge,[2] the value of α_1 in Eq. (2) is 0.275 Å. This implies that the fractional difference in velocity between 30- and 90-MHz sound waves is 4.35×10^{-5} and that for every cm of travel the 90-MHz wave will gain approximately 1.8 nsec over the 30-MHz wave. Since we can change our round-trip path length by nearly 8 cm and since our resolution is about 0.1 nsec, we should have no difficulty observing the predicted effect if it exists. Figure 3 shows the results of our measurements of the differential delay vs. total path length. The solid line is the expected result according to the prediction of Molinari and Regge[2] and the dashed line is the expected result from the measurements of Anderson and Sabisky.[4] Clearly, we saw nothing approaching the expected magnitude for α_1. At the frequencies of our experiment any reasonable value for α_2 would lead to an effect that is completely unobservable within our resolution. The advantage of this experiment is that the only variable is the path length; as long as the amplitude of the reflected sound signal

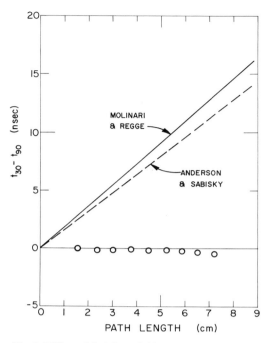

Fig. 3. Differential delay of 30- and 90- MHz sound waves vs. round-trip path length (circles). Solid line is the result expected for the prediction of Molinari and Regge.[2] Dashed line is the result according to the measurements of Anderson and Sabisky.[4]

remains constant as the reflector is moved, then such undesirable effects as attenuation in the liquid, beam spreading, or tilting of the reflector must be absent. The very small amount of delay that we do occasionally observe does not tend to be reproducible or proportional to path length and is usually associated with some such amplitude effect.

The large amount of dispersion found by Anderson and Sabisky[4] is perhaps due to the relatively high temperature at which they performed their experiment. At their temperature of $1.38°K$ the velocity can differ from its zero-temperature value by about 1%. It would not be surprising, therefore, if the difference in the *temperature* dependence of the velocity for the different frequencies used by Anderson and Sabisky could account for the effect they observed.

The present experiment thus establishes that the magnitude of the coefficient of a quadratic term in the energy spectrum of ^{4}He must be less than 0.01 Å.

References

1. G. Barucchi, G. Ponzano, and T. Regge (to be published).
2. A. Molinari and T. Regge, *Phys. Rev. Lett.* **26**, 1531 (1971).
3. R.A. Cowley and A.D.B. Woods, *Can. J. Phys.* **49**, 177 (1971).
4. C.H. Anderson and E.S. Sabisky, *Phys. Rev. Lett.* **28**, 80 (1972).
5. E.S. Sabisky, private communication.

Theoretical Studies of the Propagation of Sound in Channels Filled with Helium II

H. Wiechert and G. U. Schubert

Institut für Physik der Johannes Gutenberg Universität
Mainz, Germany

The two-fluid nature of liquid helium II leads to the possibility of several types of wave motion, e.g., first sound, second sound, fourth sound, and the fifth wave mode. First and second sound appear in the bulk liquid. The characters of these wave modes alter materially if helium is bounded by solid walls, e.g., between two plane-parallel plates forming a channel. This paper reports upon the calculation of the properties of wave modes which can exist in such channels.

The theory is based on the complete linearized set of the thermohydrodynamic equations, the so-called Khalatnikov equations.[1] In order to solve these equations in the case of wave motion in channels, the following boundary conditions at the walls have been taken into account: vanishing of the tangential component of the velocity of the normal fluid of liquid helium II, vanishing of the normal components of the total mass current, and the heat flux. There are two ways to determine the solutions of this problem: (1) to solve the boundary value problem in a mathematically exact manner [2-5] and (2) to solve the equations averaged over the width of a channel.[6,7]

Since the first method is extremely complicated it has only been possible to apply it in special cases, usually neglecting the coefficients of second viscosity, the thermal conductivity, and the coefficient of thermal expansion. In an interesting paper Adamenko and Kaganov[3] have shown that this method leads to a transcendental equation from which the dispersion relations of the wave modes can be calculated by virtue of an approximation method. It turned out that the first- and second-sound waves are modified in channels. Wehrum[5] has recently solved the transcendental equation by numerical methods and has computed the frequency dependence of the phase velocities and the absorption coefficients for the wave modes which propagate in channels of arbitrary thickness. These numerical calculations have largely confirmed the results of Adamenko and Kaganov. Moreover, it has been shown that in channels the viscous wave mode branches off in an infinite number of wave modes which are very strongly damped.[5] These wave modes correspond to different types of motion of the velocity field of the normal fluid. In channels there can exist not only wave modes with a simple parabolic profile of the velocity field of the normal fluid, but also higher modes in which parts of the normal fluid move against each other.

When solving for the wave motion in channels by averaging the Khalatnikov equations over the thickness of the channel these viscous wave modes do not appear. The calculations with this method have been carried out in the limit of very narrow channels, where the thickness of a channel is small compared to the penetration depth of a viscous wave, defined as $\lambda_v = (2\eta/\omega\rho_n)^{1/2}$, but greater than the mean free path of the elementary excitations. It can be shown that in such channels first sound converts into fourth sound[8] and second sound into the fifth wave mode[2,7] which has also been called "no sound"[9] or "thermal wave."[3,10]

Taking into account all processes which cause a dissipation of energy, as well as the coefficient of thermal expansion, one obtains the following results for the phase velocities and absorption coefficients of the fourth sound and the fifth wave mode:

$$u_4^2 = \frac{\rho_s}{\rho} u_1^2 + \frac{\rho_n}{\rho} u_2^2 \left(1 - \frac{2u_1^2 \alpha_p}{\sigma\gamma} \right) \tag{1}$$

$$\alpha_4 = \frac{\omega^2}{2u_4^3} \left\{ \frac{d^2}{3\eta u_4^2} \frac{\rho_s \rho_n^2}{\rho^2} \left[u_1^2 - u_2^2 \left(1 - \frac{u_1^2 \alpha_p}{\rho_s \sigma\gamma} (\rho_s - \rho_n) \right) \right]^2 \right.$$

$$\left. + \rho_s \zeta_3 + \frac{\kappa \rho_n u_2^2}{\rho^2 c_v u_4^2} \left(1 - \frac{u_1^2 \alpha_p}{\sigma\gamma} \right)^2 \right\} \tag{2}$$

$$u_5^2 = \frac{2\omega\rho_n u_1^2 u_2^2}{u_4^2 \gamma} \left(\frac{d^2}{3\eta} + \frac{\kappa}{\rho^2 \sigma^2 T} \right) \tag{3}$$

$$\alpha_5 = \frac{u_4}{u_1 u_2} \left(\frac{\omega\gamma}{2\rho_n} \right)^{1/2} \left(\frac{d^2}{3\eta} + \frac{\kappa}{\rho^2 \sigma^2 T} \right)^{-1/2} \tag{4}$$

Here u_4 and u_5 are the phase velocities of fourth sound and the fifth wave mode and α_4 and α_5 are their attenuation coefficients; ρ_s and ρ_n are the densities of the superfluid and normal fluid components; ρ is the total density; u_1 and u_2 are the velocities of the first- and second-sound modes; α_p is the coefficient of thermal expansion; σ is the specific entropy; γ is the ratio of the specific heats c_p and c_v at constant pressure and constant volume; ω is the angular frequency; $2d$ is the width of a channel; η is the coefficient of first viscosity; ζ_3 is a coefficient of second viscosity; κ is the thermal conductivity; and T is the absolute temperature.

From Eq. (1) it can be seen that the phase velocity of fourth sound is independent of frequency and channel width. An evaluation according to Eq. (2) yields the result that for low frequencies fourth sound is only a weakly damped wave (except near the lambda point).

In contrast to fourth sound the phase velocity of the fifth wave mode [see Eq. (3)] shows dispersion. It depends on the frequency and the channel width. The fifth wave mode turns out to be a very strongly damped wave analogous to a diffusion or thermal wave in a classical fluid. Computing the decay of the amplitude of the fifth wave mode over the distance of one wavelength, one finds that it has the constant value $e^{-2\pi} \approx 1/540$.

Except for including the important influence of the thermal conductivity κ and the coefficient of thermal expansion α_p, Eqs. (1)–(4) have also been derived by other authors[2,6,8,9] and can be confirmed by our calculation.

In order to investigate the properties of the two wave modes by experiment it is very important to consider the problem of exciting the wave modes by various types of transmitters. As Lifshitz[11] did in the case of first and second sound we have solved this problem in the case of fourth sound and the fifth wave mode for two situations: the excitation of the wave modes by periodically heating the surface of a solid body (e.g., a resistance layer) and by vibrating a plane solid surface (e.g., the diaphragm of a condenser microphone). From an extended calculation in which for simplification the influence of the thermal conductivity κ has been neglected one obtains the result[12] that with both methods fourth sound and the fifth wave mode are produced with amplitudes which should be measurable.

Using a heater as a sound transmitter, the boundary conditions require the vanishing of the normal component of the mass current at the surface and the continuity of the heat flux through the surface. Under these conditions, one finds approximately the following relations. When a periodic heat flow density $\bar{q}'$ passes through the surface of a heater the generated pressure amplitudes $\bar{p}'_4$ and $\bar{p}'_5$ of fourth sound and the fifth wave mode can be described at a distance x from the heater by the relations

$$\frac{\bar{p}'_4}{\bar{q}'} = -\frac{\rho_s \sigma u_1^2}{\rho c u_4^3}\left(1 - \frac{u_1^2 \alpha_p}{\sigma}\right) e^{-\alpha_4 x} \tag{5}$$

$$\frac{\bar{p}'_5}{\bar{q}'} = \frac{\rho_s \sigma u_1^2}{\rho c u_5 u_4^2}(1 - i)\, e^{-\alpha_5 x} \tag{6}$$

The quantities marked by a bar are averaged over the thickness of the channel. It is noteworthy that for $x = 0$ the ratio $\bar{p}'_4/\bar{q}'$ is independent of the width of the channel and the frequency, whereas $\bar{p}'_5/\bar{q}'$ is a function of these two variables (due to u_5). Equation (5) depends substantially on the term $u_1^2\alpha_p/\sigma$, which is connected with the coefficient of thermal expansion α_p. This term is of the order of magnitude unity for most temperatures. It has the value of plus one at a temperature of nearly $1.01°\text{K}$, where it cancels the contribution of the first term and $\bar{p}'_4/\bar{q}'$ becomes equal to zero. For lower temperatures the ratio $\bar{p}'_4/\bar{q}'$ is positive; for higher temperatures it is negative. Therefore $\bar{p}'_4/\bar{q}'$ suffers a phase reversal at this temperature. An evaluation of the heat fluxes associated with fourth sound or the fifth wave mode leads to the result that the thermal excitation is more favorable to the production of the fifth wave mode than of fourth sound.

In a second case we have considered the excitation of the wave modes by a plane solid surface which is vibrating as a whole in a direction perpendicular to itself. The boundary conditions in this case require that the normal components of the averaged velocities $\bar{v}_s$ and $\bar{v}_n$ of the superfluid and the normal fluid adjacent to the surface equal that of the surface. Under these conditions the ratios of the radiated pressure amplitudes $\bar{p}'_4$ and $\bar{p}'_5$ to the generating pressure amplitude $\bar{p}'$ are approximately given by

$$\frac{\bar{p}'_4}{\bar{p}'} = \frac{1}{1 + (\rho_n u_4 u_2^2 / \rho_s u_5 u_1^2)(1 - i)}\, e^{-\alpha_4 x} \tag{7}$$

$$\frac{\bar{p}'_5}{\bar{p}'} = \frac{1}{1 + (\rho_s u_5 u_1^2 / 2\rho_n u_4 u_2^2)(1 + i)}\, e^{-\alpha_5 x} \tag{8}$$

Due to u_5, both ratios (7) and (8) depend on frequency and channel width. Comparing the relations (5) and (7), one notices that in this approximation the latter is not a function of the coefficient of thermal expansion. Therefore, in contrast to $\bar{p}'_4/\bar{q}'$, the ratio $|\bar{p}'_4/\bar{p}'|$ does not vanish at a definite temperature.

From these calculations it can be concluded that it should be possible to excite fourth sound and the fifth wave mode mechanically as well as thermally. One of these results, the thermal excitation of fourth sound, has recently[13] been confirmed by experiment.

References

1. I.M. Khalatnikov, *Introduction to the Theory of Superfluidity*, Benjamin, New York (1965), Chapter 9.
2. L. Meinhold-Heerlein, *Phys. Lett.* **26A**, 495 (1968); *Z. Physik* **213**, 152 (1968).
3. I.N. Adamenko and M.I. Kaganov, *Soviet Phys.—JETP* **26**, 394 (1968).
4. T.A. Karchava and D.G. Sanikidze, *Soviet Phys.—JETP* **31**, 988 (1970).
5. R.P. Wehrum, Thesis, Mainz, 1971 (to be published).
6. D.G. Sanikidze, I.N. Adamenko, and M.I. Kaganov, *Soviet Phys.—JETP* **25**, 383 (1967).
7. H. Wiechert and L. Meinhold-Heerlein, *Phys. Lett.* **29A**, 41 (1969); *J. Low Temp. Phys.* **4**, 273, (1971).
8. K.R. Atkins, *Phys. Rev.* **113**, 962 (1959).
9. K.A. Shapiro and I. Rudnick, *Phys. Rev.* **137**, A1383 (1965).
10. G.L. Pollack and J.R. Pellam, *Phys. Rev.* **137**, A1676 (1965).
11. E. Lifshitz, *J. Phys.* **8**, 110 (1944).
12. H. Wiechert, Thesis, Mainz, 1969 (to be published).
13. H. Wiechert and R. Schmidt, *Phys. Lett.* **40A**, 421 (1972); also this volume.

Developments in the Theory of Third Sound and Fourth Sound

David J. Bergman

Department of Physics and Astronomy, Tel-Aviv University
Tel-Aviv, Israel

and

Nuclear Research Center, Soreq, Israel

Third-Sound in Flat Helium Films

The simplest theory of third sound simply starts out from two-fluid hydro-dynamics in the helium film, ignoring all dissipative processes. It can be summarized by the following equations and Fig. 1:

$$h(x,t)$$

Fig. 1

$$\rho_f \frac{\partial h}{\partial t} = - h\rho_s \frac{\partial v_s}{\partial x}$$

$$\frac{\partial v_s}{\partial t} = - \frac{\partial \mu}{\partial x} = -f \frac{\partial h}{\partial x}$$

where $f \equiv \partial\mu/\partial h$. We can combine these two equations to get

$$\ddot{h} = - h \frac{\rho_s}{\rho_f} \frac{\partial \dot{v}_s}{\partial x} = hf \frac{\rho_s}{\rho_f} \frac{\partial^2 h}{\partial x^2}$$

which gives the third-sound velocity as

$$c_3^2 = hf \frac{\rho_s}{\rho_f}$$

This treatment assumes: (a) no temperature oscillations; (b) no normal fluid motion; (c) no interaction with the surrounding gas. Of these only (b) is justifiable, since the viscous penetration depth for the normal fluid motion, $(\eta_f/\rho_f \omega)^{1/2} = 10^4 - 10^5$ Å, is much greater than the thickness of any film on which experiments are done. Assumption (a) is not obeyed: Accumulation of ρ_s at the peaks inevitably

501

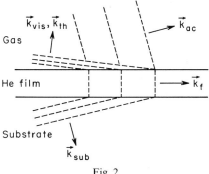

Fig. 2

means lower T. This leads to heat conduction into the surrounding media, thus violating (c). Another cause for violation of (c) is that the changes in h and T both mean that μ oscillates; hence there will also be particles of helium flowing from the film to the gas.

In both the film and the gas there are waves which accompany the one in the film: They travel away from the film and have the same periodicity in the x direction, while in the y direction their amplitude changes exponentially over a certain characteristic distance. The wave in the substrate is a solution of the thermal conduction equation, and the thermal penetration depth is $(\kappa_{sub}/C_{sub}\rho_{sub}\omega)^{1/2} = 0.1$ cm. This is the same depth which characterizes the following property: If the temperature at the surface of a fluid is made to oscillate at the frequency ω, then these oscillations penetrate into the fluid with an amplitude that decays exponentially. The thermal penetration depth is the characteristic distance of this decay.

In the gas (Fig. 2) the situation is more complicated as a result of the fact that the basic equations—the hydrodynamic equations—admit three types of waves in the gas: a thermal wave M_{th} as in the substrate, a viscous wave M_{vis}, and an acoustic wave M_{ac}. In the y direction their amplitudes change exponentially—decreasing for M_{th} and M_{vis} but increasing for M_{ac}.

For low frequencies or thin films $\mathbf{k}_{sub}$, $\mathbf{k}_{th}$, and $\mathbf{k}_{vis}$ are essentially normal to the film and have an imaginary part that is equal to the real part—so that damping is very rapid. The characteristic distances are the appropriate penetration depths:

$$- ik_{th} \equiv q_{th} \cong e^{-i\pi/4}(C_{pg}\rho_g\omega/\kappa_g)^{1/2} \cong (1/0.003) \quad \text{cm}^{-1}$$

$$- ik_{vis} \equiv q_{vis} \cong e^{-i\pi/4}(\rho_g\omega/\eta_g)^{1/2} \cong (1/0.003) \quad \text{cm}^{-1}$$

$$ik_{sub} \equiv q_{sub} \cong e^{-i\pi/4}(C_{sub}\rho_{sub}\omega/\kappa_{sub})^{1/2} \cong (1/0.1) \quad \text{cm}^{-1}$$

Under the same conditions $\mathbf{k}_{ac}$ is essentially parallel to the film and is equal to $\mathbf{k}_f$. Its damping is very small in the direction of propagation, being essentially equal to that of ordinary sound in the gas. But in the y direction it increases exponentially because the attenuation of third sound is stronger than that of sound in the gas: On a wavefront of the acoustic mode points that are further away from the film originated from earlier points on the film where the amplitude was greater. Hence the amplitude must increase in the y direction in the steady state of third-sound propagation.

How do all of these results come about? The third-sound wave in the film excites each of these modes in the surrounding media to a greater or lesser extent depending on the efficiency with which they interact. The interactions are determined by the boundary conditions which ensure that mass, energy, and momentum are all conserved at the interfaces. For example, at the liquid-substrate interface only energy can be exchanged, in the form of heat. Hence we must have

$$- \kappa_{\text{sub}}\, \partial T_{\text{sub}}/\partial y = B_1 (T_{\text{sub}} - T_f)$$

where $1/B_1$ is the Kapitza thermal boundary resistance.

At the gas–liquid interface the boundary conditions are

$$J_M = \rho_g (v_{gy} - \dot{h}) \qquad \text{(mass conservation)}$$

$$J_Q = - \kappa_g\, \partial T_g/\partial y \qquad \text{(heat flux equality)}$$

$$P_g = P_l \qquad \text{(equality of pressures)}$$

$$v_{gx} = 0 \qquad \text{(no tangential slipping of the gas)}$$

The last two are the momentum conservation equations—the gas is assumed not to exchange momentum with the superfluid flow, but to be in equilibrium with the locked normal component. J_M and J_Q are calculated from simple kinetic theory:

$$J_M \left(1 - \frac{\rho_g}{\rho_f}\right) = 2\rho_g \left(\frac{m}{2\pi k T}\right)^{1/2} \left[\mu_f - \mu_g + \left(s_g - \frac{k}{2m}\right)(T_f - T_g)\right]$$

$$J_Q = - \frac{1}{2} \rho_g \left(\frac{kT}{2\pi m}\right)^{1/2} \left[\mu_f - \mu_g + \left(s_g - \frac{9}{2}\frac{k}{m}\right)(T_f - T_g)\right]$$

When the equations are solved[1,2] it is found that for very thin films both J_M and J_Q are very small because both $T_f - T_g$ and $\mu_f - \mu_g$ are very small. For thicker films J_Q is still small but J_M is not. The terms $T_f - T_g$ and $\mu_f - \mu_g$ are no longer small but they must cancel very well in the expression for J_Q: If they did not, then we would have, since $T_f - T_g$ is of the same order as T_g,

$$J_Q \sim \rho_g \left(\frac{kT}{m}\right)^{1/2} \frac{kT}{m} \frac{T_g}{T}$$

On the other hand, since $T_g = T_{g,\text{th}} + T_{g,\text{ac}}$ and since $q_{\text{th}} \gg q_{\text{ac}}$, we have

$$- \kappa_g\, \partial T_g/\partial y \lesssim \kappa_g q_{\text{th}} T_g \cong (\kappa_g \omega C_{pg} \rho_g)^{1/2} T_g$$

The temperature gradient is mostly due to M_{th}, since $q_{\text{th}} \gg q_{\text{ac}}$, and since M_{vis} involves no temperature oscillations at all. Dividing the two results, we find that

$$- \kappa_g \frac{\partial T_g}{\partial y} \bigg/ J_Q \lesssim \left(\frac{\kappa_g \omega}{\rho_g C_{pg} c^2}\right)^{1/2} \ll 1$$

where c is the velocity of sound in the gas. Therefore we must either have $T_f = T_g$ and $\mu_f = \mu_g$, as in thin films, or $T_f - T_g$ cancelled by $\mu_f - \mu_g$, as in thick films.

For thin films we can supplement the original equations as follows: To the mass conservation equation we add a term representing a flow of mass J_M from film to gas,

$$\rho_f \frac{\partial h}{\partial t} + h\rho_s \frac{\partial v_s}{\partial x} + J_M = 0 \tag{1}$$

To the equation for $\dot{v}_s$ we add a temperature term in $\nabla \mu$:

$$\dot{v}_s = -\nabla \mu_f = \bar{S} \frac{\partial T_f}{\partial x} - f \frac{\partial h}{\partial x} \tag{2}$$

To these we add a third equation which accounts for heat flows:

$$h\rho_f C_h \dot{T}_f - T\bar{S}h\rho_s \frac{\partial v_s}{\partial x} = -\kappa_{\text{sub}} \frac{\partial T_{\text{sub}}}{\partial y} + \kappa_g \frac{\partial T_g}{\partial y} - LJ_M \tag{3}$$

The left-hand side is the sum of heat needed to raise the temperature of the film plus the heat needed to keep it from dropping due to the accumulation of superfluid. The right-hand side is the sum of heat flux from substrate to film, plus heat flux from gas to film, plus heat of condensation from gas condensing on the film. In this equation we substitute J_M from Eq. (1), as well as

$$\kappa_g \partial T_g/\partial y \cong -\kappa_g q_{\text{th}} T_g, \qquad T_g \cong T_f$$

The heat conduction equation for the substrate is solved to yield T_{sub} in terms of T_f, leading to

$$-\kappa_{\text{sub}} \partial T_{\text{sub}}/\partial y = -BT_f$$

where

$$1/B \cong (1/B_1) + e^{i\pi/4} (\omega \kappa_{\text{sub}} \rho_{\text{sub}} C_{\text{sub}})^{-1/2}$$

is the total thermal resistance of the interface plus bulk substrate, which is here assumed to be infinite. These substitutions lead to

$$\rho_f \dot{h} - h\rho_f C_h \dot{T}_f - \frac{B + \kappa_g q_{\text{th}}}{L} T_f + h\rho_s \left(1 + \frac{T\bar{S}}{L}\right) \frac{\partial v_s}{\partial x} = 0 \tag{3'}$$

Finally we must use the fact that $\mu_f = \mu_g$, hence $d\mu_f = d\mu_g$, i.e.,

$$-\bar{S}\, dT_f + f\, dh = -s_g\, dT_g + (dP_g/\rho_g) = -s_g\, dT_g$$

The last part of this, where dP_g/ρ_g is ignored, is due to the fact that only M_{th} is appreciably excited and in it the pressure oscillations vanish. Hence we can write

$$f h_1 = -(L/T) T_f \tag{4}$$

Using (4) and (2) to substitute in (3'), we get

$$\ddot{h}\left(1 + \frac{hf\,TC_h}{L^2}\right) + \frac{Tf}{\rho_f L^2}\left[B + e^{-i\pi/4}(\kappa_g \omega \rho_g C_{pg})^{1/2}\right]\dot{h} - \frac{hf\,\rho_s}{\rho_f}\left(1 + \frac{T\bar{S}}{L}\right)^2 \frac{\partial^2 h}{\partial x^2} = 0$$

This leads to the dispersion equation

$$\frac{hf\,\rho_s}{c_3^2 \rho_f}\left(1 + \frac{T\bar{S}}{L}\right)^2 = 1 + \frac{Tf}{L^2 \rho_f}\left[e^{i\pi/4}\left(\frac{\kappa_g \rho_g C_{pg}}{\omega}\right)^{1/2} + \frac{iB}{\omega} + h\rho_f C_h\right]$$

$$\cong 1 + \frac{Tf}{L^2 \rho_f} e^{i\pi/4}\left[\left(\frac{\kappa_g \rho_g C_{pg}}{\omega}\right)^{1/2} + \left(\frac{\kappa_{\text{sub}} \rho_{\text{sub}} C_{\text{sub}}}{\omega}\right)^{1/2}\right]$$

where the last form is a very good approximation for low frequencies and thin films.

Some properties of this third-sound mode are as follows.

(a) $\alpha = \omega \, \mathrm{Im}(1/c_3) \sim \omega^{1/2} h^{-5/2}$. This attenuation coefficient is the result of heat conduction to the surroundings of the helium film.

(b) We have

$$J_M = i\omega\rho_f h_1 - ikh\rho_s v_s = i\omega\rho_f h_1 \left[1 - \frac{hf\,\rho_s}{\rho_f c_3^2} \left(1 + \frac{T\bar{S}}{L} \right) \right]$$

$$= \frac{i\omega\rho_f h_1}{1 + (T\bar{S}/L)} \left\{ \frac{T\bar{S}}{L} - \frac{Tf}{L^2 \rho_f} e^{i\pi/4} \left[\left(\frac{\kappa_g \rho_g C_{pg}}{\omega} \right)^{1/2} + \left(\frac{\kappa_{sub}\rho_{sub}C_{sub}}{\omega} \right)^{1/2} \right] \right\}$$

From this one can see that $\mathrm{Im}\,J_M < \mathrm{Re}\,J_M$ and that $0 < \mathrm{Re}\,J_M$, so that J_M tends to be in phase with h_1, as shown in Fig. 3.

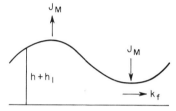

Fig. 3

Since the temperature T_f is out of phase with h_1 [see Eq. (4)], this is perhaps not what one might have expected intuitively. However, it is wrong to try to judge the rate of evaporation by the temperature of the film alone, since (a) it in fact depends on the difference $T_f - T_g$, which is very different both in magnitude as well as phase from T_f, and (b) it also depends on the difference $\mu_f - \mu_g$.

(c) $T_f \cong T_g$, $\mu_f \cong \mu_g$.

In thick films one can ignore the conduction of heat altogether, but not the evaporation. One of the equations is thus $J_Q = 0$ and it leads to a relation between $\mu_f - \mu_g$ and $T_f - T_g$. We can thus write

$$0 = J_Q = -\frac{1}{2}\rho_g \left(\frac{kT}{2\pi m} \right)^{1/2} \left[\frac{L}{T} T_f - \frac{9}{2}\frac{k}{m}(T_f - T_g) + fh_1 \right]$$

$$J_M = \frac{8\rho_g}{T} \left(\frac{kT}{2\pi m} \right)^{1/2} \left(1 - \frac{\rho_g}{\rho_f} \right)^{-1} (T_f - T_g)$$

while the heat flow equation becomes

$$- T\bar{S}h\rho_s(\partial v_s/\partial x) + LJ_M = 0$$

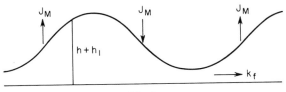

Fig. 4

To this we add mass conservation and the equation for $\dot{v}_s$:

$$\rho_f h + h\rho_s \frac{\partial v_s}{\partial x} + J_M = 0, \qquad \dot{v}_s = -\bar{S}\frac{\partial T_f}{\partial x} + f\frac{\partial h}{\partial x}$$

Solving these equations leads to the following dispersion equation:

$$\frac{hf\,\rho_s}{c_3^2 \rho_f}\left(1 + \frac{T\bar{S}}{L}\right)^2 \cong 1 + \frac{9}{16}\frac{kT}{m}\frac{i\omega\rho_f}{f\rho_g}\left(\frac{2\pi m}{kT}\right)^{1/2}\left(\frac{T\bar{S}}{L}\right)^2\left(1 - \frac{\rho_g}{\rho_f}\right)\left(1 + \frac{T\bar{S}}{L}\right)^{-2}$$

Some properties of this third-sound mode are as follows.

(a) $\alpha = \omega\mathrm{Im}(1/c_3) \sim \omega^2 h^{11/2}$. The attenuation is the result of evaporation and condensation of helium atoms.

(b) $J_M = i\omega\rho_f(T\bar{S}/L)[1 + (T\bar{S}/L)]^{-1}h_1 \sim -\dot{h}$. In this case J_M is in phase with $-\dot{h}$, as shown in Fig. 4. Again, the behavior of J_M is quite different from what one might have expected intuitively.

(c) $T_f \neq T_g$

Third-Sound Resonators

The simple-minded way to look at such a system (Fig. 5) is as though a flat, third-sound film were closed upon itself. In fact, however, the situation is somewhat more complicated. The thickness of the gas is usually of the order of or less than the thermal penetration depth in the gas. The same is true of the substrate. Hence one must take into consideration the interaction between the sound traveling in the two films. This is done by imposing appropriate boundary conditions on the wave in the gas and in the substrate: Instead of boundary conditions at $\pm\infty$, we now take

$$\left.\frac{\partial T_g}{\partial y}\right|_{y=h_g/2} = 0, \qquad J_Q|_{y=-h_{\mathrm{sub}}} = 0$$

and allow wave solutions in the substrate and gas that travel toward as well as away from the film. The physics is the same as in the simple film geometry. The result is

$$\frac{c_{30}^2}{c_3^2} \equiv \frac{hf\,\rho_s}{c_3^2\rho_f}\left(1 + \frac{T\bar{S}}{L}\right)^2 = 1 + \frac{Tf}{L^2\rho_f}\left(\frac{iq_{\mathrm{th}}\kappa_g}{\omega}\tanh\frac{q_{\mathrm{th}}h_g}{2} + \frac{i\tilde{B}}{\omega}\right)$$

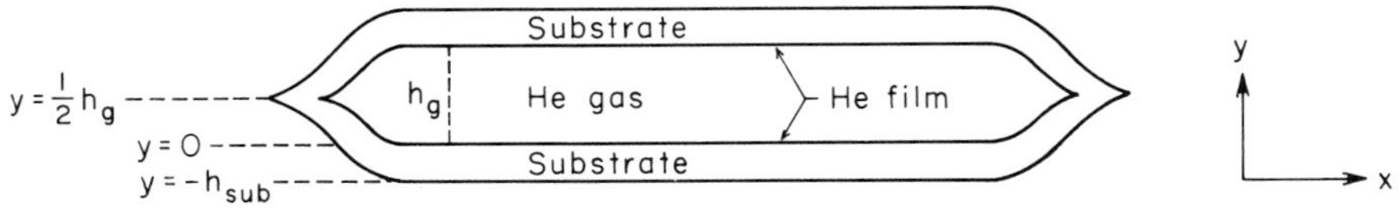

	Substrate	
$y=\frac{1}{2}h_g$ - - - - - -	h_g He gas	He film
$y=0$ - - - - -		
$y=-h_{\mathrm{sub}}$ - - - -	Substrate	

Fig. 5

where

$$\frac{1}{\tilde{B}} = \frac{1}{B_1} + \frac{1}{q_{sub}\kappa_{sub}\tanh q_{sub}h_{sub}} \approx \frac{1}{B_1} + \frac{1}{q_{sub}^2\kappa_{sub}h_{sub}}$$

$$= \frac{q_{sub}^2\kappa_{sub}h_{sub} + B_1}{B_1 q_{sub}^2\kappa_{sub}h_{sub}}$$

because $q_{sub}h_{sub} \ll 1$. Since $q_{sub}^2\kappa_{sub}h_{sub}/B_1$ is usually very small, we can also write

$$\frac{iB}{\omega} = \frac{iB_1 q_{sub}^2\kappa_{sub}h_{sub}}{\omega(B_1 + q_{sub}^2\kappa_{sub}h_{sub})} \approx \frac{iq_{sub}^2\kappa_{sub}h_{sub}}{\omega}\left(1 - \frac{q_{sub}^2\kappa_{sub}h_{sub}}{B_1}\right)$$

Using these results and the fact that $q_{th}h_g \ll 1$, we get

$$\frac{c_{30}^2}{c_3^2} = 1 + \frac{Tf}{L^2\rho_f}\left(\frac{\rho_g C_{pg}h_g}{2} + \frac{i\omega\kappa_g h_g}{2c_3^2} + \frac{i\rho_g^2 C_{pg}^2\omega h_g^3}{24\kappa_g}\right.$$

$$\left. + \rho_{sub}C_{sub}h_{sub} + \frac{i\rho_{sub}^2 C_{sub}^2\omega h_{sub}^2}{B_1} + \frac{i\omega\kappa_{sub}h_{sub}}{c_3^2}\right)$$

$$\approx 1 + i\frac{Tf}{L^2\rho_f}\frac{\rho_{sub}^2 C_{sub}^2\omega h_{sub}^2}{B_1}$$

In the last form we have left the only terms which are important if the gas is made very thin.

Some properties of the third-sound resonator are as follows.

(a) $\alpha = \omega \operatorname{Im}(1/c_3) \sim \omega^2 h^{-5/2}$ and $1/Q = 2\operatorname{Im}(c_{30}/c_3) \sim \omega h^{-4}$. Both of these quantities vary as a rather large power of T:

$$\alpha \sim Q^{-1} \sim TC_{sub}^2/B_1 \sim T^{7-2.4} = T^{4.6}$$

(b) When the gas is thin enough, as is usually the case, the main cause of the attenuation is the Kapitza resistance, through which heat flows into the sink afforded by the heat capacity of the walls.

The last property might make this a useful tool for determining precise values for the Kapitza resistance in thin films of liquid helium.

Fourth Sound

Following our experience with third sound, we may expect that in fourth sound, too, when the channels that hold the helium are sufficiently small so that the normal fluid motion is completely locked out, the only important source of attenuation will be the conduction of heat into the walls of the helium channels.

We look first at a simplified geometry: one thin film of helium bounded by two straight, semiinfinite, parallel walls. The equations of motion are

$$\dot{v}_s = -\nabla\mu = S\nabla T_f - (1/\rho)\nabla P_f$$

$$Th\,\partial(\rho S)/\partial t = -2J_Q = -2BT_f, \qquad \dot{\rho} = -\rho_s\operatorname{div} v_s$$

One needs to use equations of state or thermodynamic relations to express the amplitudes of S and P in terms of the amplitudes of T_f and ρ. The equations then lead to the following dispersion equation for fourth sound[3]:

$$c_4^2 = \frac{\rho_s}{\rho}\left(\frac{\partial P}{\partial \rho}\right)_T + \frac{\rho_s}{\rho}\left(\frac{\partial T}{\partial S}\right)_P S\left[S + 2\rho\left(\frac{\partial S}{\partial \rho}\right)_T\right]\left(1 + \frac{2iB}{\omega\rho hT}\frac{\partial T}{\partial S}\right)^{-1}$$

This looks very similar to the expression originally found by Atkins. However, the derivation and the form of B are different: Whereas Atkins assumed that $1/B$, which is the effective thermal resistance of the wall, is constant, we have calculated it and found it to be a sum of two terms (the same as in the analogous geometry for third sound),

$$1/B = (1/B_1) + e^{i\pi/4}(\omega\kappa_{sub}\rho_{sub}C_{sub})^{1/2}$$

When the wall is infinite it is the second term (due to the bulk of the wall) which is dominant, leading to an attenuation coefficient of the form

$$\alpha = \left(\frac{\rho_s C_s c_4^2}{\kappa_s \omega}\right)^{1/2} \frac{c_2^2}{c_4^4}\frac{\rho_n C_h}{\rho_s C_s}\frac{h\omega^2}{2} \sim \omega^{3/2}h$$

Most fourth-sound experiments are performed on other geometries—where the walls are about as thick as the helium channels and where the geometry is of a more or less random nature; the pores are usually not straight and the waves do not propagate parallel to them. This makes a direct solution of the equations of motion impossible.

The main problem is the attenuation, since the velocity is in practice entirely determined by nondissipative processes to be

$$c_4^2 \cong (\rho_s/\rho) c_1^2$$

For the attenuation we will use a well-known formula which connects the average rate of decline of the available energy in the sound wave to the attenuation coefficient:

$$\alpha = |\bar{\dot{E}}|/2c_4\bar{E} = T|\bar{\dot{S}}|/2c_4\bar{E}$$

The average energy in a fourth-sound wave is almost entirely mechanical energy of the superfluid helium. However, $\dot{E} = T\dot{S}$ is mainly due to thermal conduction processes going on in the helium and through the helium-wall interface. When all of these processes are taken into account one gets a ridiculously low value for α, namely $10^{-9}\,\mathrm{cm}^{-1}$.

Another process in the bulk of the fourth-sound medium which could contribute to attenuation is Rayleigh scattering of the waves by inhomogeneities in the medium. This also is much too weak to explain experimental observations.

We therefore conclude that it must be surface effects in the fourth-sound resonator (e.g., imperfect sealing) which cause the observed attenuation. If that source of attenuation could be eliminated or cut down, we would therefore have a resonator with an extremely narrow resonance.

References

1. D.J. Bergman, *Phys. Rev.* **188**, 370 (1969).
2. D.J. Bergman, *Phys. Rev.* *A***3**, 2058 (1971).
3. D.J. Bergman and Y. Achiam (to be published).
4. K.R. Atkins, *Phys. Rev.* **113**, 962 (1959).
5. L.D. Landau and E.M. Lifshitz, *Fluid Mechanics* (transl. by J.B. Sykes and W.H. Reid), Pergamon Press, New York and Addison-Wesley, Reading Mass. (1966).

Thermal Excitation of Fourth Sound in Liquid Helium II*

H. Wiechert and R. Schmidt

Institut für Physik der Johannes Gutenberg Universität
Mainz, Germany

In narrow channels filled with helium II two wave modes propagate, fourth sound[1-3] and the fifth wave mode.[4-6] According to new results of theoretical studies,[7-8] it has been predicted that it should be possible to excite both wave modes mechanically by vibrating the diaphragm of a condenser microphone as well as thermally by periodically heating the surface of a solid body (e.g., a resistance layer). Shapiro and Rudnick[3] have produced and detected fourth-sound signals mechanically. In the present contribution it will be experimentally verified that it is also possible to excite fourth sound thermally. Since the theory[7,8] gives the result that fourth sound can be generated thermally only if the fifth-wave mode is taken into account, the detection of a thermally excited fourth-sound signal is also a first, indirect proof for the existence of the fifth-wave mode.

The experiments were carried out in an adsorption cryostat.[9] Zeolite was employed as adsorbing substance. The lowest temperature achieved was about $0.75°K$. The temperature of the helium bath above $1.2°K$ was determined from the vapor pressure with the 1958 temperature scale. Below this value the temperature was measured with an Allen-Bradley resistance which was compared with a commercially available calibrated germanium resistance. It was carefully tested that the temperature inside and outside the sound propagation tube was the same during a run within an experimental error of a few millidegrees.

In Fig. 1 a sketch of the sound propagation tube is given. It consists of a thick-walled brass cylinder (3.5 cm long, 1 cm diameter) closed at one end by a chromium resistance layer deposited on a glass plate. This layer serves as a heater to produce fourth sound. The other end is closed by a condenser microphone of the type described by Hofmann *et al.*[10] It is made of a 2.5-μm thick plastic foil metallized with aluminum on one side. The foil is glued to an insulation ring. The counterelectrode is a close-meshed wire gauze (twill) of phosphorbronze ($30~\mu$m wire diameter). The pickup was polarized by a potential of about 250 V. The sound penetrating the membrane was partially absorbed by a porous medium made of Styropor. For the mechanical production of fourth sound the heater was replaced by a condenser microphone of the same type.

Narrow channels were formed by pressing fine-grained powders of different sizes into the cylindrical tube. The space between the particles of the powder yields a complex system of branched capillaries in which the fourth sound propagates.

* Supported by the Deutsche Forschungsgemeinschaft.

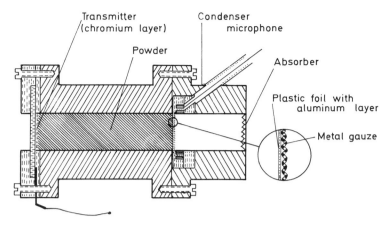

Fig. 1. The sound propagation tube with a chromium resistance layer as transmitter and a condenser microphone as receiver.

Care was taken to get the end areas of the powder flat and aligned with the cylinder ends so that there were no gaps between either the powder and the transmitter or the powder and the receiver.

To produce and detect the fourth-sound waves, two methods were employed: the pulse technique for high frequencies and the standing-wave technique for low frequencies. For the latter method the receiver was replaced by a microphone with a solid counterelectrode.

Figure 2 shows results of our measurements with the pulse method at a frequency of 60 kHz. The powder used to form the channels was Linde Type A aluminum oxide with a grain size of approximately 0.3 μm. The packing pressure of the powder was 15 kg cm^{-2}. The input power to the heater during a pulse was nearly 50 mW cm^{-2}. This was in the upper range of the power inputs which could be used in order to remain in the linear acoustical region. In the figure the relative voltage amplitudes of the thermally and mechanically excited fourth-sound signals are plotted against temperature. The crosses are the data of the thermally generated fourth-sound signals and refer to the left ordinate, whereas the circles are data of mechanically generated fourth sound and refer to the right ordinate. The following features can be seen in the figure.

(1) Fourth sound can be excited thermally as well as mechanically.

(2) The amplitude of the thermally generated sound signal decreases with increasing temperature, vanishes at about 0.95°K, where a phase reversal takes place, rises again at higher temperatures to a maximum near 1.3°K, and decreases to the lambda point.

(3) In contrast to this behavior, the amplitude of the mechanically excited fourth-sound wave decreases monotonically with temperature. It does not vanish at 0.95°K.

For comparison, theoretical curves[7,8] of the ratio of the radiated pressure amplitude $\bar{p}_4'$ to the heat flow density $\bar{q}'$, which passes through the surface of the heater, and the absolute value of the ratio of the radiated pressure amplitude $\bar{p}_4'$ to

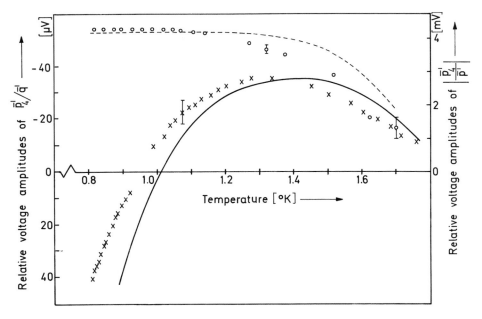

Fig. 2. Temperature dependence of the pressure amplitudes of thermally (×) and mechanically (○) excited fourth sound.

the total pressure amplitude $\bar{p}'$, which is generated by the vibrating membrane of the microphone, are also represented in this figure. The solid line refers to the ratio $\bar{p}'_4/\bar{q}'$, the dashed line to the ratio $|\bar{p}'_4/\bar{p}'|$. The theoretical curves have been normalized to agree with the experimental data at $1.4°$K for the solid line and at $1.1°$K for the dashed line. An essential result of the theory[7,8] is that the quantity $\bar{p}'_4/\bar{q}'$ is determined by the ratio of the coefficient of thermal expansion to the specific entropy α_p/σ.

Therefore the temperature where the thermally generated fourth-sound signal vanishes depends substantially on the values of these quantities. Our calculations are based on experimental data for the coefficient of thermal expansion and of the specific entropy which were measured in the bulk liquid by Harris-Lowe and Smee[11] and Kramers et al.,[12] respectively. Taking into account values of α_p of Kerr and Taylor,[13] the solid line will be shifted nearly 40 m°K toward higher temperatures.

An inspection shows that there is a rough agreement between theory and experiment but that the temperature where $\bar{p}'_4/\bar{q}'$ is zero is about 60 m°K higher for the theoretical curve than for the experimental one. This discrepancy may partly be caused by size effects. It is probable that a purely hydrodynamic theory, on which the theoretical curves are based, may not be sufficient for the description of wave processes in narrow channels at low temperatures, where the mean free path of the elementary excitations may be greater than the channel width. Nevertheless, it is usual to describe such size effects by a hydrodynamic theory in which the thermodynamic quantities are assumed to be functions of the channel width. Rudnick and co-workers (see, e.g., Ref. 14) have, for instance, discovered that the ratio of the superfluid density to the total density ρ_s/ρ in narrow channels is reduced below that of bulk helium. In our case the aspect of size-dependent thermodynamic quantities would lead to the prediction that the temperature where the thermally excited fourth-sound signal vanishes depends on the channel width but is independent of the frequency.

In order to test these predictions, some further measurements have been performed.

1. The influence of the frequency has been investigated. Measurements with the standing-wave technique at a frequency of about 2.8 kHz and with the pulse technique at 30, 60, and 120 kHz have shown, in agreement with theory,[7,8] no noticeable shift of the temperature where $\bar{p}_4'/\bar{q}'$ is zero.

2. In order to study the dependence of this temperature on the thickness of the channels, measurements with powder samples of various grain sizes and under various packing pressures have been carried out. In Fig. 3 the relative voltage amplitudes of thermally produced fourth-sound signals are plotted versus temperature for different powder samples marked by different symbols as indicated in the figure caption. The figure shows that the temperature where the thermally generated fourth-sound signal disappears is decreased by the packing of the sound propagation tube with ever smaller particle sizes under the same pressure. With a given particle size an increase in the packing pressure also results in a lowering of this temperature. Values for this temperature between 0.97° K and $< 0.78^\circ$ K have been obtained. The theoretical curve of $\bar{p}_4'/\bar{q}'$ is also represented in this figure (solid line).

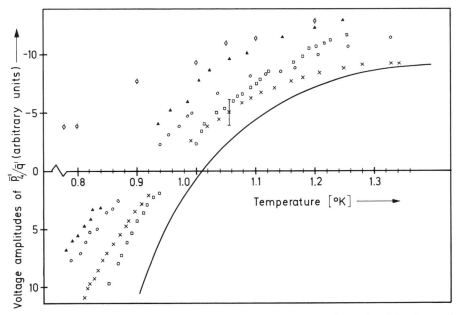

Fig. 3. Temperature dependence of the pressure amplitudes of thermally produced fourth-sound signals for various powder samples. The particle diameters and the packing pressures are: ($\square$) 10,000 Å, 15 kg cm^{-2}; ($\times$) 3000 Å, 15 kg cm^{-2}; ($\bigcirc$) 3000 Å, 190 kg cm^{-2}; ($\blacktriangle$) 500 Å, 15 kg cm^{-2} (all these samples consist of Linde alumina powders); ($\diamondsuit$) 100 Å, 100 kg cm^{-2} (Aerosil, Degussa).

Comparing theory[7,8] and experiment, one can conclude from these measurements that the quantity $u_1^2 \alpha_p / \sigma$ (u_1 is the phase velocity of first sound) is a function of the thickness of the channels. It is possible that this dependence can be interpreted by a modification of the excitation spectrum of superfluid helium near a solid wall.[15] Further measurements have to be performed to investigate this interesting effect.

References

1. J.R. Pellam, *Phys. Rev.* **73**, 608 (1948).
2. K.R. Atkins, *Phys. Rev.* **113**, 962 (1959).
3. K.A. Shapiro and I. Rudnick, *Phys. Rev.* **137**, A1383 (1965).
4. L. Meinhold-Heerlein, *Phys. Lett.* **26A**, 495 (1968); *Z. Physik* **213**, 152 (1968).
5. I.N. Adamenko and M.I. Kaganov, *Soviet Phys.—JETP* **26**, 394 (1968).
6. H. Wiechert and L. Meinhold-Heerlein, *Phys. Lett.* **29A**, 41 (1969); *J. Low Temp. Phys.* **4**, 273 (1971).
7. H. Wiechert, Thesis, Mainz, 1969 (to be published).
8. H. Wiechert and G.U. Schubert, this volume.
9. G. Frommann and K. Keck, *Z. angew. Phys.* **31**, 258 (1971).
10. A. Hofmann, K. Keck, and G.U. Schubert, *Z. Phys.* **231**, 177 (1970).
11. R.F. Harris-Lowe and K.A. Smee, *Phys. Rev.* **A2**, 158 (1970).
12. H.C. Kramers, J.D. Wasscher, and C.J. Gorter, *Physica* **18**, 329 (1952).
13. E.C. Kerr and R.D. Taylor, *Ann. Phys.* **26**, 292 (1964).
14. M. Kriss and I. Rudnick, *J. Low Temp. Phys.* **3**, 339 (1970).
15. C.G. Kuper, *Physica* **24**, 1009 (1956).

Inelastic Scattering from Surface Zero-Sound Modes: A Model Calculation*

Allan Griffin and Eugene Zaremba

Department of Physics, University of Toronto
Toronto, Ontario, Canada

In dealing with density oscillations near a planar boundary, one must make some choice on how one deals with the static part of the problem and the dynamic part of the problem. In the present model calculations we are interested in relatively high-frequency collisionless modes and we argue that one may treat the boundary (whether a wall or a free liquid surface) in terms of a high potential barrier which gives rise to simple specular scattering. For the dynamics we will use the well-known self-consistent field approximation (or RPA) appropriate to a degenerate Fermi system. More precisely, we shall use the semiclassical RPA. Thus we neglect the quantum mechanical interference effects for particles bounding off the surface.[1,2] Formally, our calculations involve extending the usual discussions of surface modes in an electron gas[3] to the case of neutral systems interacting via a short-range Yukawa potential

$$v(\mathbf{R}) = g^2 e^{-\kappa R}/R \tag{1}$$

As a specific system to which our model calculations may be of interest we of course have liquid ^{3}He in mind. In fact, our calculations may be interpreted in terms of Fermi liquid theory for a contact potential and in the strong coupling limit.

We recall that the differential scattering cross section for inelastic scattering of particles from some "blob" is given by (see, e.g., Ref. 4)

$$d^2\sigma/d\Omega\,d\omega \propto v_{\text{ext}}^2(\mathbf{q})\,S(\mathbf{q},\omega) \tag{2}$$

if the Born approximation is valid. Here $v_{\text{ext}}(\mathbf{q})$ is the Fourier transform of the interaction potential between a probe particle and a particle in the system of interest, while the dynamic structure factor of our "blob" is

$$S(\mathbf{q},\omega) = \int d\mathbf{r} \int d\mathbf{r}' \int_{-\infty}^{\infty} dt\,(\exp i\omega t)\,(\exp - i\mathbf{q}\cdot(\mathbf{r}-\mathbf{r}'))\,\langle n(\mathbf{r},t)\,n(\mathbf{r}',0)\rangle \tag{3}$$

Finally, $\hbar\omega$ is the energy transfer and, since we are interested in a system with a single boundary, $\hbar\mathbf{q} \equiv (\hbar\mathbf{q}_{\|}, \hbar q_z)$ is the momentum transfer parallel $(\hbar\mathbf{q}_{\|})$ and normal $(\hbar q_z)$ to the surface. Using the semiclassical RPA and our assumption of specular scattering, we can analytically work out the induced density fluctuation to first order in a weak external field. From this we can obtain the retarded density–density

* Supported by the National Research Council of Canada.

correlation function of our bounded system, which in turn can be related directly to $S(\mathbf{q}, \omega)$. We simply state that for our model problem calculation gives[5]

$$S(\mathbf{q}, \omega) = \frac{2\hbar}{1 - e^{-\beta\hbar\omega}} \left[V \operatorname{Im} \chi_B(\mathbf{q}, \omega) - A\left(\frac{Q}{4\pi g^2}\right) \operatorname{Im}\left(\frac{v^2(q)\,\chi_B^2(\mathbf{q}, \omega)}{D(\mathbf{q}_\parallel, \omega)}\right) \right] \tag{4}$$

Here V is the total volume of the system of interest, A is the area of the bounding surface, $Q^2 \equiv q_\parallel^2 + \kappa^2$, and the bulk density response function is

$$\chi_B(\mathbf{q}, \omega) = \Pi_0(\mathbf{q}, \omega)/[1 - v(\mathbf{q})\Pi_0(\mathbf{q}, \omega)] \tag{5}$$

where

$$v(q) = 4\pi g^2/(q^2 + \kappa^2) \tag{6}$$

and $\Pi_0(\mathbf{q}, \omega)$ is the well-known[4] semiclassical approximation to the Lindhard polarization function for a gas of noninteracting fermions. The denominator of the second (or surface) term in Eq. (4) involves the function

$$D(\mathbf{q}_\parallel, \omega) \equiv 1 + \int_{-\infty}^{\infty} \frac{dq_z}{2\pi} \frac{2Q}{q_z^2 + Q^2} \frac{1}{[1 - v(q)\Pi_0(\mathbf{q}, \omega)]} \tag{7}$$

It can be shown[6] that the dispersion relation for surface zero-sound modes corresponds to the condition $\operatorname{Re} D(\mathbf{q}_\parallel, \omega) = 0$. In particular, if one uses the simple strong coupling approximation $\Pi_0(\mathbf{q}, \omega) \simeq \bar{n}q^2/m\omega^2$, where $\bar{n}$ is the bulk density, then the surface mode is given by

$$\omega_S^2(q_\parallel) = \omega_g^2\, q_\parallel^2/(\kappa^2 + 2q_\parallel^2); \qquad \omega_g^2 \equiv 4\pi g^2 \bar{n}/m \tag{8}$$

which contrasts with the analogous bulk zero-sound mode[4]

$$\omega_B^2(q) = \omega_g^2\, q^2/(\kappa^2 + q^2) \tag{9}$$

The collisionless damping of the surface zero-sound mode in Eq. (8) is very small. In the long-wavelength limit ($q_\parallel \ll \kappa$) we find[5]

$$\frac{\Gamma(q_\parallel)}{\omega_S(q_\parallel)} = \left(\frac{q_\parallel}{\kappa}\right)^2 \frac{q_\parallel}{\sqrt{3}k_{TF}} \ln \frac{\sqrt{3}\kappa}{q_\parallel} \tag{10}$$

where k_{TF} is the Thomas–Fermi wave vector ($k_{TF} \gg \kappa$ in the strong coupling limit).

It is clear from the general form of Eq. (4) that $S(\mathbf{q}, \omega)$ will have two resonances whose positions are given approximately by Eqs. (8) and (9). The total weight of the bulk mode is somewhat reduced as a result of the surface term. As a simple example let us consider a system of electrons moving against a fixed positive background, in which case we must set $\kappa = 0$. In the high-frequency limit, which corresponds to using $\varepsilon_B(\mathbf{q}, \omega) = 1 - (\omega_p^2/\omega^2)$, the expression in Eq. (4) reduces to ($L \equiv V/A$)

$$S(\mathbf{q}, \omega) = \frac{2\hbar}{1 - e^{-\beta\hbar\omega}} V\left(\frac{\pi\bar{n}}{m}\right) \operatorname{sgn}\omega$$

$$\times \left[\left(q_\parallel^2 + q_z^2 - \frac{q_\parallel}{L}\right)\delta(\omega^2 - \omega_p^2) + \frac{q_\parallel}{L}\delta\left(\omega^2 - \frac{\omega_p^2}{2}\right) \right] \tag{11}$$

It can be shown that in this long-wavelength limit the surface and bulk plasmon modes exhaust the f-sum rule.

We have also worked out the generalization of Eq. (4) for a system of degenerate fermions moving in a deep square well potential of width L. One finds that collisionless density fluctuations which arise as a result of the boundaries are now determined by

$$D_{S,A}(\mathbf{q}_{\parallel}, \omega) \equiv 1 + \frac{1}{L} \sum_{q_z}' \frac{2Q}{q_z^2 + Q^2} \frac{1}{1 - v(q) \Pi_0(\mathbf{q}, \omega)} = 0 \qquad (12)$$

For symmetric (antisymmetric) modes the q_z sum is restricted to even (odd) multiples of π/L. A detailed analysis of Eq. (12) is given in Ref. 5. For a given value of $q_{\parallel}$ one has a whole family of mixed mode solutions. These are shifted from the bulk mode frequencies associated with $q_z = n\pi/L$ for $n = 1, 2, 3,\ldots$, i.e., the singularities of the summand in Eq. (12). In addition there is a true symmetric surface mode which is phononlike in the $q_{\parallel} \to 0$ limit. The analogous antisymmetric surface mode transforms into a mixed surface bulk mode for $q_{\parallel} \lesssim (2\kappa/L)^{1/2}$.

References

1. C. Heger and D. Wagner, *Z. Physik* **224**, 449 (1971).
2. D.E. Beck and V. Celli, *Phys. Rev.* **B2**, 2955 (1970).
3. V. Peuckert, *Z. Physik* **241**, 191 (1971).
4. D. Pines and P. Nozières, *The Theory of Quantum Liquids*, Benjamin, New York (1966).
5. A. Griffin and E. Zaremba, *Phys. Rev.* **A8**, 486 (1973).
6. A. Griffin and J. Harris, *Phys. Rev.* **A5**, 2190 (1972).

The Scattering of Low-Energy Helium Atoms at the Surface of Liquid Helium*

J. Eckardt, D.O. Edwards, F.M. Gasparini, and S.Y. Shen

Department of Physics, The Ohio State University
Columbus, Ohio

We have designed an atomic beam experiment to be performed at low temperature in order to study the scattering of helium atoms at the liquid surface. This work has been partly stimulated by the theoretical papers of Anderson[1] and Hyman et al.[2] as well as by the experimental work of Johnston and King[3]. Anderson pointed out that the spectrum of atoms evaporating from the liquid should show a sharp increase in intensity at an energy corresponding to the difference between the atomic binding energy L and the roton minimum Δ. This prediction is made by assuming that inelastic processes at the surface play a minor role in evaporation.

Our experimental cell consists of a metal can about 9 cm in diameter and 17 cm high which is cooled by a dilution refrigerator to $\sim 0.030°K$. The can houses two superconducting motors which can move two graphite film resistors at a radius of 4.1 cm so as to vary their angle relative to the liquid surface. Each resistor can be used interchangeably as an emitter or a receiver. A copper mask screens the receiver from direct radiation from the emitter, except at such positions where the two are visible through a 1.6×1 cm collimating window. A beam of atoms is generated by an electrical pulse of 10–40 μsec duration which evaporates a small fraction of the superfluid film on the emitter. When a resistor is used as the receiver it is part of a positive feedback oscillator circuit. As such, the condition that the loop gain be unity maintains its resistance and hence its temperature constant. When energy is incident on this receiver it causes a transient decrease in the oscillator power, and this is the signal which is recorded.

We have done control experiments in which the cell contained just enough helium to produce a saturated film. Under such condition we have looked at the direct radiation of atoms from the heater to the receiver. A typical signal is shown in Fig. 1, where we have plotted the energy received at our bolometer versus time of arrival. The full curve was calculated by assuming that a Maxwellian distribution of ^{4}He atoms at $T = 0.57°K$ is generated at the emitter and that each atom carries an energy $L + \frac{1}{2}mv^2$ to the receiver. The agreement of the data with this distribution is quite good, except perhaps for some smearing due to the finite response time of our receiver. There is no sign in the data of an enhancement of received power as predicted by Anderson, which, with our receiver–emitter separation, would appear at ~ 1 msec. This result is also in agreement with recently reported measurements by King

* Work supported in part by the National Science Foundation.

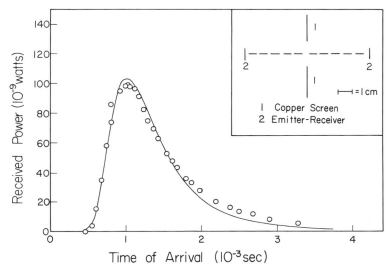

Fig. 1. Power received as a function of transit time for the arrangement shown in the inset. The energy put into the emitter is 0.68 erg. The full curve is a Maxwellian distribution for ^{4}He at 0.57°K.

et al.[4] We have also looked at the angular distribution of atoms emitted from the heater. Figure 2 shows the result of such a measurement. One might have expected that the distribution would obey Lambert's law, in which case the energy received would simply be proportional to the cosine of the angle θ between heater and receiver. On a polar diagram this would appear as a circle. We find instead that more of the energy is being received in the forward direction. On the other hand, with a stationary receiver and a moving emitter we find that the energy received is simply proportional to the cosine of the angle θ. The collimation which we observe is thus definitely connected with the beam generated at the emitter. If one now assumes that the rough heater surface generates an isotropic distribution of excitations in the helium film, and further, if one takes the simplest model for evaporation (i.e., one where inelastic and multiparticle processes are ignored), then one can show that phonons impinging on the liquid vacuum interface will tend to produce a collimated beam of atoms, while rotons will not. This simply follows from conservation of energy and parallel momentum at the liquid surface. This may be the cause of the angular dependence shown in Fig. 2.

Additional control experiments were done to check the effectiveness of the copper mask. On the left-hand side of Fig. 3 we show the result of an experiment where the emitter was fixed at a position $\theta_i = 100°$ while the receiver was scanned from 36° to 102°. The input energy to the heater in this experiment was 0.68 erg. The sharp decrease in total received energy as the receiver goes into the shadow and the extremely small received energy within the shadow region are good indications of the effectiveness of the mask and the absence of serious multiple reflections from within the cell. The theoretical curve shown in this figure was calculated from the experimental geometry by assuming the radiation from the emitter to be Lambertian. This

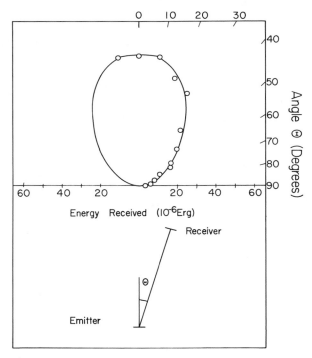

Fig. 2. Total energy received as a function of receiver position. Energy put into the emitter is 0.043 erg and the emitter-to-receiver separation is 3.5 cm.

assumption, although not strictly valid, as we have seen from Fig. 2, is not serious enough to affect the calculation.

To do reflection experiments, the cell is filled with helium until the liquid surface is in the middle of the opening in the copper mask; this coincides with the center of rotation of the resistors. Approximately 13 moles of helium are required to fill the cell. With such a large sample ~ 0.1 ppm of natural ^{3}He impurity is sufficient to cause the ^{4}He to have a substantial fraction of a monolayer of ^{3}He adsorbed on its surface.[5,6] The result of an experiment done under such conditions is shown on the right-hand side of Fig. 3. In this instance the emitter was fixed at an angle of $51°$ and the receiver was scanned from $6°$ through $81°$. The data shows that in the region where the specular part of the reflected beam is expected there is a considerable increase in the amount of energy received. Where the specular component is expected to be zero the signal does not, however, go to zero, but shows a roughly constant value. This is the inelastically scattered part of the incident beam. The results of several experiments at various heater–receiver arrangements show that the inelastic part of the reflected beam is roughly independent of θ_i and θ_r. The theoretical curve for specular reflection involves a calculation which assumes a Lambertian distribution in the atoms emitted from the heater and a $\sin^2 \theta$ dependence for the reflection coefficient on the angle of incidence. This is roughly the angular dependence which we observe in our data. We are not in a position at this point to report on the reflec-

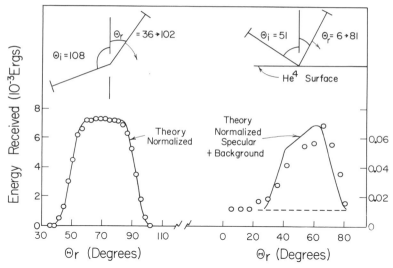

Fig. 3. Left: total energy received as a function of receiver angle for fixed emitter position. Right: the energy received after reflection as a function of receiver angle and for fixed emitter. Energy input is 0.68 and 1.5 ergs, respectively.

tion coefficient as a function of velocity; our measurements do, however, show that it decreases with increasing velocity and that it has a broad minimum in the energy neighborhood of $\Delta - L$. There are no signs, however, of the sharp edge predicted by Anderson.

In addition to the above experiments we have looked at surface excitations which seem to be the principal mode of energy loss from the beam. These experiments have been done with the receiver in the liquid surface and the emitter in the vacuum, in the surface, or below the surface. We have also looked at the reflection of phonons from the liquid–vacuum interface. The results of these experiments, which are being pursued under controlled ^{3}He impurity concentrations, will be reported elsewhere.

References

1. P.W. Anderson, *Phys. Lett.* **29A**, 563 (1968).
2. D.S. Hyman, M.O. Scully, and A. Widom, *Phys. Rev.* **186**, 231 (1969).
3. W.D. Johnston, Jr. and J.G. King, *Phys. Rev. Lett.* **16**, 1191 (1966).
4. J.G. King, J. McWane, and R.F. Tinker, *Bull. Am. Phys. Soc.* **17**, 38 (1972).
5. A.F. Andreev, *Zh. Eksperim. i Teor. Fiz.* **50**, 1415 (1966), [*Soviet Phys.—JETP* **23**, 939 (1966)].
6. H.M. Guo, D.O. Edwards, R.E. Sarwinski, and J.T. Tough, *Phys. Rev. Lett.* **27**, 1259 (1971).

Inelastic Scattering of ^{4}He Atoms by the Free Surface of Liquid ^{4}He

C.G. Kuper

Department of Physics, Technion–Israel Institute of Technology
Haifa, Israel

With the intention of measuring the accommodation coefficient of ^{4}He, i.e., the probability that an atom hitting the surface of the liquid will be captured, Eckardt et al.[1] (EEGS) designed the following experiment (Fig. 1).

A source S generates a beam of ^{4}He atoms at a mean energy $\sim 2°$K. The beam hits the surface of liquid ^{4}He (at temperature $\sim 0.05°$K) at an angle of incidence θ_i. The reflected intensity at angle θ_f is measured by the detector D. A screen B prevents atoms from S reaching D without first being scattered at the liquid surface.

According to Widom et al.[2] and Anderson,[3] the interaction between an atom and the liquid has the character of a tunneling matrix element; if the atom is not captured, the scattering is expected to be elastic. However, the preliminary results of Ref. 1 show nonspecular reflection: The angle of emergence θ_f of the reflected beam is much larger than θ_i and is insensitive to θ_i.

The results of Ref. 1 can be understood in terms of the following hydrodynamic picture. The tangential component of the incident momentum $\mathbf{p}_i$ does not interact with the liquid and is conserved:

$$p_i \sin \theta_i = p_f \sin \theta_f \tag{1}$$

The normal component $p_i \cos \theta_i$ will be changed; indeed, it will change its sign even in specular reflection. The momentum loss

$$P = p_i \cos \theta_i + p_f \cos \theta_f \tag{2}$$

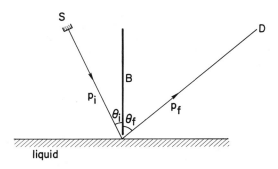

Fig. 1. The experiment of Edwards et al.[1]

must be transmitted to the liquid. It will be assumed that it is entirely carried by the acoustic disturbance which will spread out from the point of impact.

For small amplitudes the total energy carried in both the sound wave and surface wave will be proportional to the square of the impulse P. Hence conservation of energy gives

$$p_i^2/2m - p_f^2/2m = \alpha P^2/2m \tag{3}$$

where α is a dimensionless constant, $\alpha \leq 1$. (In a detailed acoustic model[4] of the collision in which the radius of the radiating "loudspeaker" is taken to be $\sim 10\text{Å}$—the de Broglie wavelength of a ^{4}He atom at energy $2°$K—it can be shown that $\alpha \sim \frac{1}{2}$.) Eliminating P and p_f from Eqs. (1)–(3) gives

$$\tan \theta_f / \tan \theta_i = (1 + \alpha)/(1 - \alpha) \tag{4}$$

which agrees at least qualitatively with the preliminary measurements of Edwards et al.[1]

In the analogous problem of roton scattering at the surface, Eq. (3) is replaced by

$$(p_i - p_0)^2/2m^* - (p_f - p_0)^2/2m^* = \alpha' P^2/2m \tag{5}$$

where p_0 and m^* are Landau's roton parameters, and where α' might naively be expected to be the same as α.

For thermal rotons at temperature $2°$K it is easily shown that

$$|p - p_0|/p_0 \sim 7 \times 10^{-2} \tag{6}$$

whence, from Eq. (5)

$$P/p_0 \lesssim 7 \times 10^{-2}(\alpha' m^*/m)^{-1/2} \tag{7}$$

But from Eqs. (2) and (6)

$$P/p_0 \geq \cos \theta_i \tag{8}$$

The inequalities (7) and (8) are inconsistent (except for very large angles of incidence) unless $\alpha' \ll 1$; i.e., roton scattering must be nearly specular.

References

1. J. Eckardt, D.O. Edwards, F.M. Gasparini, and S.Y. Shen, this volume.
2. A. Widom, *Phys. Lett.* **29A**, 96 (1969); D.S. Hyman, M.O. Scully, and A. Widom, *Phys. Rev.* **186**, 231 (1969).
3. P.W. Anderson, *Phys. Lett.* **29A**, 563 (1969).
4. C.G. Kuper, in: *Collective Phenomena*: (H. Haken and M. Wagner, eds.) Springer Verlag, Berlin (1973), p. 129.

The Scattering of Light by Liquid ^{4}He Close to the λ-Line

W.F. Vinen, C.J. Palin

Department of Physics
University of Birmingham, United Kingdom

and

J.M. Vaughan

Royal Radar Establishment
Malvern, Worcestershire, United Kingdom

Introduction

For several years now hope has existed that the scattering of visible light by liquid ^{4}He would prove a useful tool in the experimental investigation of the λ-transition. In this paper we report some progress that we have made toward the realization of this hope.

Suppose that light of angular frequency ω_0 is incident on a liquid and that we observe the spectrum of light scattered through an angle θ corresponding to a change in optical wave vector of $\mathbf{Q}$. It is well known that in sufficiently simple cases like liquid helium the scattered intensity at frequency $(\omega_o + \Omega)$ is proportional to the dynamic structure factor $S(\mathbf{Q}, \Omega)$ which is obtained by taking the Fourier transform in space and time of the density correlation function $\langle \delta\rho(\mathbf{r}, t)\, \delta\rho(0, 0) \rangle$; $\delta\rho$ is the amount by which the density differs from its average value and $\langle \cdots \rangle$ denotes an equilibrium ensemble average.[1] The light therefore probes the density fluctuations that occur naturally in a liquid in thermal equilibrium, particularly those that occur over a spatial range which is of the order of the wavelength of light. As we shall see, density fluctuations in helium close to the λ-line contain much interesting information. We may note that near the λ-line special interest attaches to the fluctuations in the superfluid order parameter Ψ. At present we cannot probe fluctuations in Ψ directly, but the hope exists that the behavior of these fluctuations is reflected to a significant extent in the directly observable density fluctuations. We remember also[2] that the order parameter fluctuations have associated with them a correlation length ξ_Ψ which diverges as $\xi_0 |\Delta T/T_\lambda|^{-2/3}$ at the λ-point, where $\xi_0 \sim 1$ Å; ξ_Ψ ought therefore to become comparable with the wavelength of light at the quite manageable temperature difference ΔT of about 0.2 m°K, so that light scattering offers us the chance of probing the "critical region" where $Q\xi_\Psi > 1$.

Experimental Method

All the measurements that we have made so far relate to scattering through 90° with light of wavelength 4880 Å derived from an argon-ion laser. The measurements relate therefore to a fixed value of Q, equal to about 1.8×10^5 cm^{-1}. As we shall see, some of the interesting effects are much more pronounced at high pressures than they are near the vapor pressure, so that the scattering cell containing the experimental helium must be designed to withstand pressures corresponding to the upper end of the λ-line. Adequate temperature control, certainly to within 50 $\mu°$K, and correspondingly adequate temperature measurement are also required.

The scattering cell that we have used so far has a stainless steel case and is fitted with silica windows sealed with indium O-rings. The cell contains a closely spaced copper fin system, the fins containing a set of small holes through which the focused laser beam passes. The cell is filled with about 160 cm^3 of clean helium through a fine capillary. Thermal contact between the cell and a main helium bath is provided by a copper link of moderate thermal resistance, and measurements can be made either as the cell temperature drifts very slowly toward a constant bath temperature or with the cell temperature stabilized with a heater. The temperature of the central copper fins is measured with a resistance thermometer with a resolution of about 10 $\mu°$K. The design of the cell is such that in the superfluid phase this temperature should in practice not differ from that of the helium at the point of scattering by more than 10 $\mu°$K, but there is evidence that due to a fault in the cell construction this difference may be as high as 100 $\mu°$K above the λ-point. All temperatures are measured relative to the λ-point, the position of which is obtained from the observed heat capacity of the helium as deduced from the heating and cooling curves of the cell.

We have made two types of optical measurement: the total scattered intensity and the spectrum of the scattered light. In the former case scattered light is focused directly onto the photocathode of a photomultiplier connected to a counter; a "normalized scattered intensity" I is then obtained by dividing the count rate by the output of a photodiode which continuously monitors the intensity of the incident laser beam. In the latter case the scattered light is analyzed with a scanning Fabry–Perot interferometer, which is described in the following contributed paper, and for which the instrumental line broadening is about 6 MHz (fwhh).

Measurements of the Total Scattered Intensity [3]

We have made measurements of the total scattering I as a function of $\Delta T = T - T_\lambda$ along a number of isobars that cross the λ-line, our most detailed measurements having been made at pressures of about 20 bars. Typical results are displayed in Fig. 1.

To understand the significance of these results, we return to the theory of the structure factor $S(Q, \Omega)$. The total scattering I that we measure is clearly proportional to

$$S(Q) = \int_{-\infty}^{\infty} S(Q, \Omega) \, d\Omega$$

We shall assume that all fluctuations with which we are concerned are in the classical limit $\hbar\Omega \ll k_B T$, an assumption that involves a negligible error in the present context. Suppose that the density correlation function has a range ξ_ρ. Then it may be shown that if $Q\xi_\rho \ll 1$, then $S(Q) = \rho_0^2 k_B T K_T$, where K_T is the isothermal compressibility of the fluid.[1] This means that over the small temperature range with which we are concerned $S(Q)$ will be proportional simply to K_T.

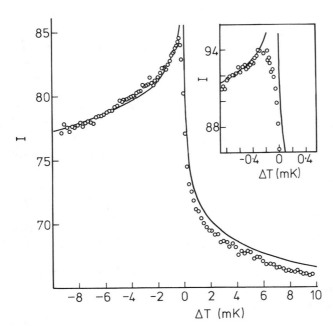

Fig. 1. Normalized scattered intensity I plotted against ΔT at a pressure of 18.80 bars. The solid line is the isothermal compressibility. Inset: detailed behavior very close to the λ-line at a pressure of 19.10 bars. The vertical scales are arbitrary and are not the same in the two plots.

The solid lines in Fig. 1 show the variation of K_T deduced from thermodynamic data[4]; there is a single normalization factor, obtained by fitting the K_T curve to the scattering data at the lowest temperature. We shall ignore the relatively small discrepancy that exists over the whole temperature range above the λ-point. This is probably due to a constant error of about 1% in I (or in K_T) in this temperature range (or to some similar kind of error); it is probably not significant, but it ought to be investigated in future experiments. We see that with this proviso there is generally good agreement between the form of I vs. ΔT and that of K_T vs. ΔT, except in the region very close to the λ-point. In this region K_T appears to diverge logarithmically to an infinite value at T_λ, whereas the scattered intensity has a finite maximum value at a temperature significantly below T_λ. We attribute this difference to a failure of the inequality $Q\xi_\rho \ll 1$. It is tempting to try the assumption that ξ_ρ is the same as ξ_Ψ. If we take $\xi_0 = 1$ Å, say, we find that $Q\xi_\Psi = 1$ in our experiments when $\Delta T = 0.16\,\mathrm{m^\circ K}$. In our experiments departures from proportionality of I to K_T do seem to set in at roughly this value at ΔT, and this fact provides some evidence that our ideas here are correct. As far as the precise form of the I vs. ΔT curve in the critical region is concerned, we have some confidence in our results for $T < T_\lambda$, where the uncertainty in temperature is small, but less confidence for $T > T_\lambda$, where the uncertainty in temperature is larger.

Of course, many more total-scattering experiments of this kind are still required: we wish to improve the precision of the present results; to remove the residual uncertainty in temperature for $T > T_\lambda$; and to establish the Q dependence of the effect. When the Q dependence is known we shall try to use the results to deduce the temperature dependence of ξ_ρ.

It should be emphasized that the anomaly in K_T at the λ-point becomes more marked as the pressure rises, and that it is quite small near the vapor pressure line. We have confirmed that this trend is reflected in the intensity of the scattered light. It is interesting to note that the peak that we have observed in the scattered intensity at the λ-point is, in a sense, analogous to the critical opalescence that is observed at the ordinary critical point of a fluid system.

Measurements on the Spectrum of the Scattered Light in the Hydrodynamic Region

As is clear from the relation between the dynamic structure factor and the density correlation function, the *spectrum* of the light scattered by a simple fluid depends on the way in which a thermally excited density fluctuation evolves in time. For a particular scattering angle the scattered intensity probes those fluctuations with a particular wave vector **Q**. Let us assume that the natural evolution in time of these fluctuations can be described by ordinary local hydrodynamics. We may then distinguish two cases: an ordinary (normal) fluid and a superfluid. In both cases we can, to a good approximation, resolve the density fluctuation into two components: an approximately isentropic pressure wave and an approximately isobaric entropy wave. In both cases the isentropic pressure wave propagates as an ordinary (first) sound wave with speed u_1 and attenuation coefficient α_1. This gives rise to a "Brillouin doublet" in the spectrum with frequency shift $\pm \Omega_1 = \pm 2n\omega_0(u_1/c)\sin(\theta/2)$ and with linewidth $\Delta\Omega_1$ (fwhh) equal to $2u_1\alpha_1$ (n is the refractive index of the liquid). In the case of an *ordinary* liquid the isobaric entropy wave does not propagate but simply decays at a rate depending on the thermal diffusivity $(K/\rho C_p)$ of the medium. This gives rise to an unshifted "Rayleigh line" of width $\Delta\Omega_0 = (K/\rho C_p)Q^2$. With a superfluid, however, the entropy waves do propagate, as second sound with speed u_2 and attenuation α_2, and they therefore lead to a second Brillouin doublet with shift $\pm 2n\omega_o(u_2/c)\sin(\theta/2)$ and linewidth $2u_2\alpha_2$. To a good approximation, the intensity $(2I_1)$ of the first-sound doublet is proportional to the adiabatic compressibility of the liquid, while the total scattered intensity is, as we have seen, proportional to the isothermal compressibility. It follows that if I_0 is the intensity of the Rayleigh component, and $2I_2$ is the intensity of the second-sound Brillouin doublet, then [5-7]

$$I_0/2I_1 = I_2/I_1 = \gamma - 1 \tag{1}$$

where $\gamma = C_p/C_v$.

We emphasize that these predictions hold only as long as the time evolution of the fluctuations can be described by local hydrodynamics, which will be the case only if $1/Q$ is large compared with all relevant correlation lengths.[2] In what follows we shall assume that there is only one such correlation length in the present problem, namely ξ_Ψ. We can say then that our predictions can be valid only in the "hydrodynamic" regions above and below the λ-point, where $Q\xi_\Psi < 1$. We see that within these regions experimental data on light scattering would yield, in principle, the following information: in both phases the speed and attenuation of first sound and the value of γ; in the superfluid phase the speed and attenuation of second sound; and in the normal phase the thermal diffusivity of the medium. All these quantities are of particular interest in the immediate neighborhood of the λ-point.

We now turn to the experimental data. We comment first that experimental

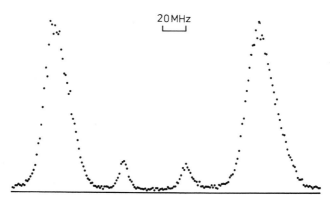

Fig. 2. Typical spectrum obtained with a scanning Fabry-Perot interferometer in the hydrodynamic region below the λ-point; $T = -58.3 \text{ m}^\circ\text{K}$; 20 bars. The smaller peaks are the second-sound doublet and have been shifted by ± 27.4 MHz from the center of the trace. Each larger peak is due to two overlapping first-sound peaks, and these first-sound peaks have been shifted by about 5.5 orders of the spectrum, or by about 1000 MHz.

observation of the first-sound Brillouin doublet in liquid helium near the vapor pressure line was reported several years ago.[8,9] However, at this pressure the value of γ in the superfluid phase is very close to unity, except at temperatures *extremely* close to the λ-point, so that the second-sound doublet is too weak to detect. Well above the λ-point γ rises sharply; our preliminary observation of and measurements on the Rayleigh component have already been reported[10] and were shown to be in agreement with Eq. (1). At high pressures, however, γ is appreciably greater than unity in the superfluid,[7,11] particularly near the λ-point, and, as we shall see, the observation of the second-sound doublet then presents no difficulty.

A typical spectrum obtained in the superfluid hydrodynamic region close to the λ-line at 20 bars is shown in Fig. 2. All our spectra so far relate to this pressure. The two Brillouin doublets are clearly visible. In this paper we shall be concerned only with the second-sound doublet; results on the first-sound doublet will be described in the following contributed paper. Our present spectroscopic measurements have not yet included any systematic work above the λ-point, except for that on first sound.

Values of u_2 and of I_2/I_1 deduced from spectra of this kind for a pressure of 20 bars are shown in Figs. 3 and 4. We see that there is good agreement with the recent data of Greywall and Ahlers[12] on u_2 obtained at audio frequencies, and with values $\gamma - 1$ obtained from thermodynamic data.[4] We emphasize that the frequency of the second sound observed optically is quite high: it ranges from about 30 MHz at $\Delta T = -60 \text{ m}^\circ\text{K}$ to about 4 MHz at $\Delta T = -0.5 \text{ m}^\circ\text{K}$. Our results thus confirm that there is insignificant dispersion in the speed of second sound over the range of frequencies up to these values.*

---

* A similar conclusion was possible in recent experiments on Bragg scattering from injected second-sound waves.[13]

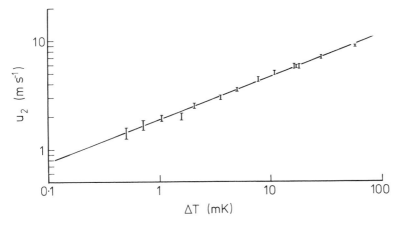

Fig. 3. The speed of second sound u_2 derived from our spectra plotted against ΔT;
20 bars. The solid line is derived from the data of Greywall and Ahlers.

The linewidths that we observe in our second-sound Brillouin doublets are less than the instrumental broadening. The linewidths presumably increase as we approach the λ-point (according to dynamic scaling theory[2,14] and the low-frequency data of Tyson,[15] they ought to increase as $|\Delta T|^{-1/3}$), but even at $\Delta T = -1$ m°K the linewidth is not greater than 3MHz. If we make allowance for the instrumental broadening, our spectra suggest that the linewidth is in fact between 1 and 2 MHz at this temperature, but the relatively large instrumental broadening makes it difficult to be certain. It is interesting that this linewidth is less than it would be at the vapor pressure; Tyson's value[15] for the attenuation of (audio frequency) second sound at the vapor pressure would lead to a linewidth of 7 MHz at $\Delta T = -1$ m°K. This suggests that it would be helpful in this context to extend our measurements to

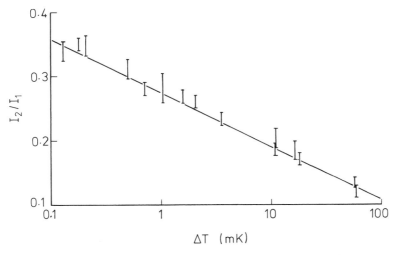

Fig. 4. The intensity ratio I_2/I_1 derived from our spectra plotted against ΔT; 20 bars. The solid line is the value of $\gamma - 1$ derived from thermodynamic data.[4]

lower pressures. A modest improvement in the instrumental broadening and an increase in the scattering angle would also be helpful, and would almost certainly make it possible to make measurements of the second-sound linewidth that are sufficiently precise to be useful.

Measurements on the Spectrum of the Scattered Light in the Critical Region

In the critical region, $Q\xi_\Psi > 1$, the spectrum and intensity of the density fluctuations can no longer be described in the simple manner outlined in Section 4, since the behavior of the liquid is then no longer described by local hydrodynamics. The problem becomes complicated, and we shall not discuss it in any detail, since, as we shall see, the relevant experimental results that we have so far obtained are quite sparse. We must remember, however, that it is in just this region that the thermal fluctuations in the liquid are of great interest. In particular, the fluctuations in the order parameter must become large, so that they begin to play a dominant role in the properties of the system. There is, of course, the problem of the relationship between the fluctuations in the order parameter and the observable fluctuations in density. In the superfluid hydrodynamic region the fluctuations in order parameter and those in entropy are essentially the same, but this is no longer necessarily true in the critical region.[14]

The spectrum that we have observed at $\Delta T = -130\ \mu°K$ is shown in Fig. 5. This temperature is higher than that at which the peak in the total scattering occurs, and we believe therefore that it probably lies in the critical region. We have observed similar spectra at -180 and $-209\ \mu°K$, but have so far made no further study of the critical region.

To attempt a preliminary interpretation of the central peak in the spectrum of Fig. 5, we shall assume, with, e.g., Ferrel et al.,[2] that the dominant contribution has a spectrum of the form

$$I(\Omega) = A\Omega^2/[(\Omega^2 - \Omega_0^2)^2 + \beta\Omega_0^2\Omega^2] \tag{2}$$

In the low-temperature hydrodynamic region $\Omega_0 = Qu_2$ and $\beta = 4\alpha_2^2/Q^2$. In the critical region Ω_0 and β are believed to become independent of ΔT; according to extended dynamic scaling, Ω_0 becomes proportional to $Q^{3/2}$ (with the neglect of logarithmic factors) and β becomes a constant. A reasonable extrapolation into the critical region for our value of Q would lead to $(\Omega_0/2\pi) = 2.4$ MHz. Equation (2) predicts a double peak, but with this value of Ω_0 the two peaks would not be resolved

20MHz

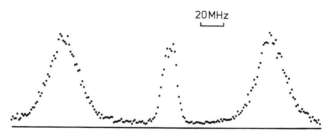

Fig. 5. Spectrum obtained at $\Delta T = -130\ \mu°K$; 20 bars.

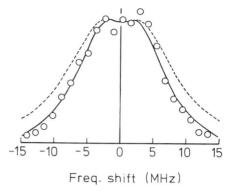

Fig. 6. The spectrum obtained at $\Delta T = -130\mu^\circ K$ and 20 bars compared with two theoretical predictions. The points are derived from the central feature of the spectrum of Fig. 5. The lines have been obtained by combining the function (2) with the Fabry-Perot instrumental function; full line: $\Omega_0/2\pi = 2.4$ MHz $\beta = 1$; broken line: $\Omega_0/2\pi = 2.4$ MHz, $\beta = 4$.

Freq. shift (MHz)

in our experiments. This is illustrated in Fig. 6, where the lines have been obtained by combining Eq. (2) with the known instrumental broadening, for two values of β. We see that in fact there is quite good agreement with experiment if $\beta = 1$, so that the form of Eq. (2) is at least *consistent* with experiment. It is interesting that the agreement is noticeably less good if $\beta = 4$. This value of β corresponds to critical damping of the critical fluctuations, so that we are led to the tentative conclusion that the critical fluctuations are under-critically damped.[2,16]

It should be added that we may expect the intensity ratio I_2/I_1 to become less than $\gamma - 1$ in the critical region; however, our measurements are not yet sufficiently sensitive to reveal this effect.

References

1. H.E. Stanley, *Phase Transitions and Critical Phenomena* (Oxford, 1971), Chapter 13.
2. R.A. Ferrell, N. Menyhàrd, H. Schmidt, F. Schwabl, and P. Szépfalusy, *Ann. Phys.* **47**, 565 (1968).
3. C.J. Palin, W.F. Vinen, and J.M. Vaughan, *J. Phys. C: Sol. State Phys.* **5**, L139 (1972).
4. G. Ahlers, private communication.
5. L.D. Landau and G. Placzek, *Phys. Z. Sowjun.* **5**, 172 (1934).
6. V.L. Ginsburg, *J. Exp. Theor. Phys. (USSR)* **13**, 243 (1943).
7. W.F. Vinen, *J. Phys. C: Sol. St. Phys.* **4**, L287 (1971).
8. R.L. St. Peters, T.J. Greytak, and G.B. Benedek, *Optics Comm.* **1**, 412 (1970).
9. E.R. Pike, J.M. Vaughan, and W.F. Vinen, *J. Phys. C: Sol. St. Phys.* **3**, L37 (1970).
10. E.R. Pike, *J. de Physique* **33**, C1–25 (1972).
11. B.N. Ganguly, *Phys. Rev. Lett.* **26**, 1623 (1971).
12. D.S. Greywall and G. Ahlers, *Phys. Rev. Lett.* **28**, 1251 (1972).
13. S. Cunsolo, G. Grillo, and G. Jacucci, in *Proc. 12th Intern. Conf. Low Temp. Phys., 1970*, Academic Press of Japan, Tokyo (1971), p. 49.
14. B.I. Halperin and P.C. Hohenberg, *Phys. Rev.* **177**, 952 (1969).
15. J.A. Tyson, *Phys. Rev. Lett.* **21**, 1235 (1968).
16. H. Schmidt and P. Szépfalusy, *Phys. Lett.* **32A**, 326 (1970).

Experiments on the Scattering of Light by Liquid Helium

J.M. Vaughan

Royal Radar Establishment
Malvern, Worcestershire, United Kingdom

and

W.F. Vinen and C.J. Palin

Department of Physics
University of Birmingham, United Kingdom

Introduction

The Malvern–Birmingham groups have conducted light scattering experiments on liquid helium in a number of different regimes: Rayleigh and Brillouin scattering from superfluid mixtures of ^{3}He–^{4}He,[1] from normal and superfluid ^{4}He,[2] second-order Raman scattering from ^{4}He above the λ-point,[3] and total intensity measurements close to the λ-line.[4] This paper describes the spectroscopic equipment used in recent studies and the results of attenuation and velocity measurements of first sound close to the λ-line at elevated pressure and a frequency close to 1 GHz. These latter results relate well in their general form with previous measurements at SVP and show a velocity minimum and attenuation maximum several millidegrees below the λ-point.

Instrumentation

The 90° scattering cell that we have used has been described in the preceding paper. The input power of about 70 mW at 4880 Å was derived from a CRL model 52 argon-ion laser. This laser has been operated in a heavy, isolated, external cavity spaced by Invar bars of 2 in. diameter; the cavity is further strengthened by several transverse stiffening rods, and holds an etalon for single mode operation. With this system of passive stabilization the short term ($\sim$ 100 msec) frequency movement, which was previously about 15 MHz, is reduced to less than 1 MHz and no longer dominates the spectral linewidth. The interferometer is a plane Fabry-Perot etalon of spacing over 80 cm giving a free spectral range of 180 MHz. The etalon is scanned repetitively at $\sim$ 6 Hz with piezoelectric elements. Photoelectrons produced at the cathode of the photomultiplier detector (ITT FW130, dark count < 0.5 sec^{-1}) give output pulses that are collected in a multiscaler store; this is swept by a sequence of preset clock pulses. Slow drift in the absolute laser frequency is overcome by starting the multiscaler sweep on the peak of a reference beam which passes through the etalon and is then immediately cut off. After sweeping and recording for approxi-

mately one and a half orders the reference beam is again allowed into the spectrometer and the cycle is repeated. This ensures that each repeated sweep over the
signal adds in the same channels; thus the counts in a particular address of the store
correspond to the intensity of the scattered light at a particular frequency shift.
For the present work the plane etalon offers great flexibility in that the spacing and
free spectral range can be varied to match a particular problem; this, together with
high flare rejection, is to be set against the somewhat greater stability and light
gathering power of the confocal etalon.

A set of spectra taken from the oscilloscope display is shown in Fig. 1. Typical
count rates when scanning have been about 30 sec^{-1}; acceptable recordings are
accumulated in less than 30 min. Spurious signal arising from flare at the incident
laser frequency is found to be less than one in 10^4 of the total spectrum.

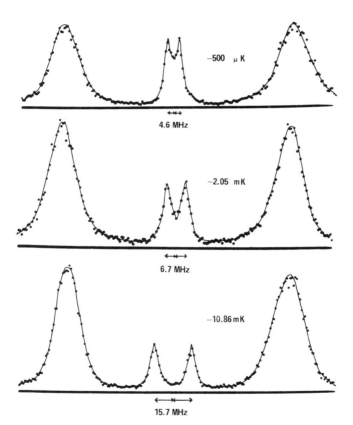

Fig. 1. Spectra obtained with the scanning Fabry-Perot interferometer in the hydrodynamic region at temperatures below the λ-point
and a pressure of 20 bars. Smooth curves have been drawn through the
experimental points. The smaller peaks are the second-sound doublet
shifted from the incident laser frequency at the center. The large peaks
are due to overlapping first-sound scattering which is shifted about
± 5.5 orders (~ 1000 MHz); they are additionally broadened by
k-vector spread.

Measurements of First-Sound Velocity and Attenuation

For these measurements the Brillouin peaks of scattering from first sound were arranged to overlap 5.33 orders of the interferometer and thus were well resolved. A pair of profiles taken at a pressure of 17 bars is shown in Fig. 2. The relative frequency shift within one order is easily measured to better than $\frac{1}{2}\%$. Over five orders the corresponding velocities, from spectra taken at different temperatures, are thus obtained with a relative accuracy of better than one in 10^3, using the usual relationship $\pm\,\Omega_1 = \pm\,2n\omega_0(U_1/c)\sin(\theta/2)$. The error on the absolute value is somewhat larger since knowledge of the absolute velocity is generally limited by the precision of measurement of the scattering angle θ. A set of results is shown in Fig. 3(a).

For the measurements on attenuation the spread in frequency arising from the finite acceptance angle $\Delta\theta$ of the spectrometer (the "k-vector spread") was restricted by measured apertures. In order to fit the experimental spectra, computed folds were made of the known instrumental profile with rectangular functions (corresponding to the k-vector spread) and Lorentzian functions (corresponding to the sound wave attenuation). After small correction for overlap the experimental curves were compared with these profiles and the Lorentzian components extracted. These fits were further estimated against tables of Voigt integrals[5] with which they closely corresponded. Values of the attenuation constant α_1 are shown in Fig. 3(b) derived from the usual expression $\Delta\nu_1 = U_1\alpha_1/\pi$ Hz, where $\Delta\nu_1$ is the Lorentzian width (fwhh) due to attenuation.

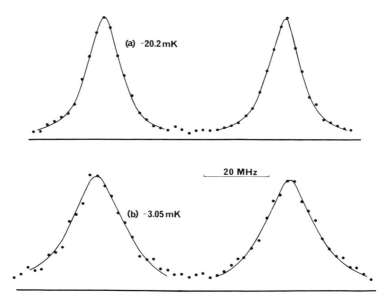

Fig. 2. First-sound peaks at a pressure of 17 bars shifted about $\pm$ 5.33 orders of the interferometer. The instrumental width (fwhh) is about 6 MHz and the k-vector spread is 9 MHz. The Lorentzian components, due to attenuation, derived from these profiles are (a) 5.2 $\pm$ 0.8 MHz and (b) 12.3 $\pm$ 0.8 MHz. The errors are estimated limits.

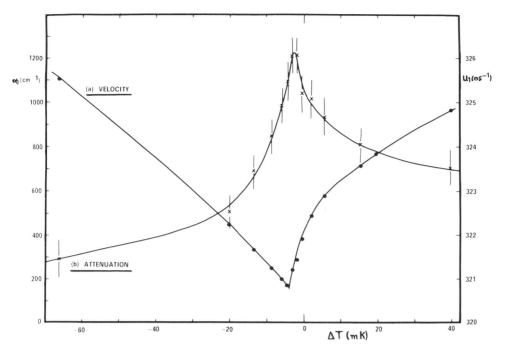

Fig. 3. (a) Velocity and (b) attenuation plotted against ΔT at a constant wave vector of 1.8×10^5 cm^{-1}. Smooth curves have been drawn through the experimental points. The velocity minimum is at $\Delta T = -4.0 \pm 0.5$ m°K and the attenuation maximum appears to be at about -2.5 m°K.

Discussion

In common with previous measurements on first sound we observe, as shown in Fig. 3, a pronounced peak in the attenuation and minimum in the velocity close to the λ-point. The velocity minimum is at $\Delta T = -4.0 \pm 0.5$ m°K, while the attenuation peak appears to be about -2.5 m°K. Additionally, the attenuation is observed to be markedly asymmetric about the λ-point.

An exact comparison between these results and earlier measurements at ultrasonic frequencies is not possible since our observations are made at high pressure. Our values of velocity are close to those from interpolation of the results of Atkins and Stasior[6] using a pulse technique at 12 MHz. At SVP Chase[7] finds a velocity minimum at -0.2 m°K for a frequency of 1 MHz, and Barmatz and Rudnick[8] a minimum at -4 μ°K for 22 kHz. The observed temperature dependence of the attenuation is of the same general form as that obtained by Imai and Rudnick[9] using an acoustical interferometer and by Heinicke et al.[10] using stimulated Brillouin scattering, and also at lower frequencies by Williams and Rudnick.[11] The position of our attenuation minimum is much the same as that suggested by Imai and Rudnick at a very similar frequency of 1 GHz. However, our value of the attenuation peak of ~ 1220 cm^{-1} is smaller by about a factor of two than their measurement. This would suggest an appreciable pressure dependence which requires further investigation.

Acknowledgments

We are indebted to Mr. R. Jones of RRE for his continued assistance with the Fabry-Perot control system and to Dr. E. R. Pike for many discussions.

References

1. C.J. Palin, W.F. Vinen, E.R. Pike, and J.M. Vaughan, *J. Phys. C: Sol. St. Phys.* **4**, L225 (1971).
2. E.R. Pike, J.M. Vaughan, and W.F. Vinen, *J. Phys. C: Sol. St. Phys.* **3**, L37 (1970); E.R. Pike, *J. Physique* **33**, C25 (1972).
3. E.R. Pike and J.M. Vaughan, *J. Phys. C: Sol. St. Phys.* **4**, L362 (1971).
4. C.J. Palin, W.F. Vinen, and J.M. Vaughan, *J. Phys. C: Sol. St. Phys.* **5**, L139 (1972).
5. J. Tudor Davies and J.M. Vaughan, *Astrophys. J.* **137**, 1302 (1963).
6. K.R. Atkins and R.A. Stasior, *Can. J. Phys.* **31**, 1156 (1953).
7. C.E. Chase, *Phys. Fluids* **1**, 193 (1958).
8. M. Barmatz and I. Rudnick, *Phys. Rev.* **170**, 224 (1968).
9. J.S. Imai and I. Rudnick, *Phys. Rev. Lett.* **22**, 694 (1969).
10. W. Heinicke, G. Winterling, and K. Dransfeld, *Phys. Rev. Lett.* **22**, 170 (1969).
11. R.D. Williams and I. Rudnick, *Phys. Rev. Lett.* **25**, 276 (1970).

Brillouin Light Scattering from Superfluid Helium under Pressure*

G. Winterling,† F.S. Holmes, and T.J. Greytak‡

Department of Physics and Center for Material Science and Engineering
Massachusetts Institute of Technology, Cambridge, Massachusetts

We have used Brillouin scattering[1,2] to study both thermally excited first and second sound in pure liquid ^{4}He at a pressure of 25 atm. We have measured the high-frequency velocity and damping of both modes at a wavelength of 4.34×10^{-5} cm and at temperatures between 1.4 and 2.2°K. The ratio of the intensity of the light scattered from second sound to that scattered from first sound has also been determined and is reported here for the temperature region $10^{-5} < T_\lambda - T < 10^{-1}$ °K.

In the past decade laser light scattering has become a powerful tool for probing density fluctuations in dielectric liquids. In normal fluid ^{4}He, as well as in other simple liquids, the spectrum of the scattered light at a fixed scattering angle contains a doublet arising from a propagating sound mode and an unshifted line[3] arising from a nonpropagating heat mode. In superfluid ^{4}He, however, the heat mode propagates as "second sound" and the corresponding spectrum of the scattered light is expected to consist of two doublets.[1]

The second sound, which is mainly an entropy wave, is coupled with the density, and consequently with the light, through the thermal expansion. Under SVP this is very small in He II and, accordingly, the scattering from thermal second sound is very weak; it has not yet been observed. With increasing pressure, however, the thermal expansion coefficient increases considerably along the λ-line. Close to the upper λ-point it becomes so large that the intensity of light scattered from fluctuations of the second sound is expected to become comparable to that scattered from fluctuations of first sound, as pointed out by Ferrell et al.[4] In general, this intensity ratio is given to good approximation by $C_p/C_v - 1$,[4-6] with C_p/C_v the ratio of the principal specific heats.

In our experiment the beam of a single-mode, frequency-stabilized He–Ne laser having a power of 20 mW was passed through a scattering cell filled with liquid ^{4}He under pressure. The spectrum of the laser light scattered at 90° was analyzed with a spherical Fabry-Perot interferometer of a free spectral range of 149 MHz. The total instrumental width, including jitter of the laser frequency, was found to be 3.9 MHz, corresponding to an optical resolving power of 10^8. This high resolving

* Work supported by the Advanced Research Projects Agency under Contract DAHC 15-67-C-0222.
† Supported by a research fellowship of the Deutsche Forschungsgemeinschaft, Bonn, Germany.
‡ Alfred P. Sloan Research Fellow.

power, in combination with a careful design of the scattering cell in order to eliminate any diffuse stray light, enabled us to resolve the second-sound lines even very close to the λ-transition where the frequency shift becomes small. The temperature of the superfluid helium inside the scattering cell was measured with a germanium resistance thermometer within ± 10 μdeg. The lambda temperature was identified by locating the characteristic plateau in the heating curve of the helium and could be determined within ± 15 μdeg. The slope of the λ-curve at 25 atm obtained in this way was found to be in good agreement with the value listed in literature.[7] Although the scattering cell was made out of copper, the temperature determination in the normal fluid phase was found to be much less reliable. Preliminary investigations have shown that the uncertainty of the quoted temperatures $T_\lambda - T$ in He I might be as large as 0.6 mdeg.

Brillouin spectra were taken in the superfluid phase at 25 atm and at temperatures in the range $0.5 > T_\lambda - T \geq 10^{-5}$ °K. Examples of these spectra[8] recorded at $T_\lambda - T = 46.8$ mdeg and $T_\lambda - T = 2.18$ mdeg can be seen in Fig. 2 of Ref. 9. When approaching the λ-transition it was clearly observed that the relative intensity of the second-sound lines grows remarkably and that, on the other hand, the corresponding frequency shift tends to decrease to zero. Due to our narrow instrumental width, we were able to resolve the second-sound lines as a distinct doublet at temperatures as close to T_λ as 0.35 mdeg. It was further observed that the first-sound lines broaden noticeably in the vicinity of the λ-transition, indicating a strong increase in the attenuation of first sound. The peak counting rate in a first-sound line close to T_λ was only 30 counts sec^{-1}. This small number required a very slow pressure scanning of the Fabry-Perot spectrometer. While sweeping through a second-sound doublet the temperature of the scattering cell was stabilized within ± 15 μdeg.

The velocity and attenuation of first and second sound at a wavelength of 4.34×10^{-5} cm were determined from the recorded Brillouin lines by representing the intrinsic spectrum with a pair of Lorentzian lines, convolving them with the instrumental profile and including a k-vector spread due to both the finite collection angle of the spectrometer and the finite laser beam diameter. Figure 1 shows the results on first sound at a frequency of about 800 MHz. The minimum of the first-sound velocity was found to be displaced by 5 ± 1 mdeg below T_λ. Similar, but considerably smaller displacements of the velocity minimum were observed at SVP at much lower frequencies.[10,11] A comparison of our results with low-frequency sound velocities at 25 atm should give valuable information on the dispersion of first sound near T_λ. The observed attenuation of first sound increases strongly in the vicinity of the λ-transition, as shown in Fig. 1. The attenuation maximum was found close to T_λ in the range $0 \lesssim T_\lambda - T \lesssim 2$ mdeg. More experiments are being carried out to establish in more detail the behavior of the attenuation close to T_λ; however, it is already evident—as observed at SVP—that the high-frequency data cannot be explained by the relaxation process which describes the low-frequency results.[11]

Data on the frequency shift and intrinsic linewidth (full width) of second sound for $T_\lambda - T \gtrsim 2$ mdeg are presented in Fig. 2. It can be seen that the measured shifts agree well with the values extrapolated from low-frequency velocity data.[12] Included in Fig. 2 is also the linewidth extrapolated from the low-frequency damping coefficient[13] D using the relation $\Delta v = Dk^2/2\pi$ and taking into account a pressure

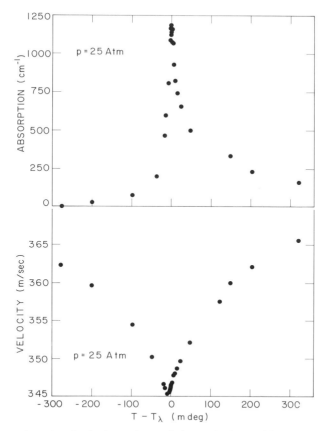

Fig. 1. Amplitude absorption coefficient and velocity of first sound in liquid ^{4}He at a frequency of about 800 MHz, plotted vs. the temperature difference from the λ-transition. The uncertainty of the absorption data is $\pm 50\,\text{cm}^{-1}$; the velocity data are accurate within $\pm 0.1\%$.

dependence of D due to the factor ρ_s/ρ_n. Our data agree well in their magnitude with this hydrodynamic extrapolation. It should be noted that for temperatures close to the transition $T_\lambda - T \lesssim 2$ mdeg, the extrapolated frequency shifts and line-widths become comparable so that the description of the second-sound spectrum as a pair of Lorentzian lines is no longer a good approximation. Therefore it was not appropriate to include the spectra measured closer to T_λ in Fig. 2. We compared the experimental spectrum in this region with the following intrinsic spectrum obtained from superfluid hydrodynamics[4,5]:

$$S(k,\omega) = \frac{1}{\pi} \frac{Dk^2\omega^2}{\left[\omega^2 - (u_2 k)^2\right]^2 + (Dk^2)^2\omega^2}$$

It has been suggested by Ferrell et al.,[4] who considered fluctuations in the phase of the order parameter near T_λ, that the above form for the second-sound spectrum

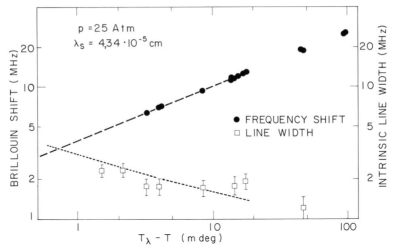

Fig. 2. Frequency shift and linewidth (full width) of second sound at a wavelength of
4.34×10^{-5} cm. The circles and squares represent data obtained at a pressure of
25 atm.; (———) the frequency shift extrapolated from low-frequency velocity data
(see Ref. 12); (- - -) the linewidth expected from a k^2 extrapolation of low-frequency
damping data (Ref. 13).

might be valid even in the critical temperature region; however, the damping coefficient D and the velocity u_2 would then be k dependent. We could not get a satisfactory fit to our experimental traces at any set of u_2 and D for temperatures $T_\lambda -$
$T < 4$ mdeg; in particular, at frequency shift $\omega = 0$ our experimental traces seem
to be consistently higher than those computed with the above expression. We are
now trying other fits, including the possibility of an additional unshifted line[4,5]
which might be present due to fluctuations in the magnitude of the order parameter.

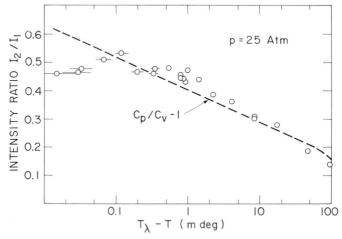

Fig. 3. Intensity ratio of second-sound to first-sound scattering at a pressure
of 25 atm. The dashed line represents data on the ratio of the specific
heats minus one.[14]

Figure 3 shows data on the ratio of the intensity in the second-sound doublet to that in the first-sound doublet. The scatter of the data is typical for our experimental uncertainty of $\pm 7\%$. The dashed line represents data* on $C_p / C_v - 1$, which corresponds to the intensity ratio expected from equilibrium thermodynamics.[4-6] Taking into account the combined experimental uncertainty, the agreement between our data and the expected values can be considered as satisfactory for $T_\lambda - T > 0.1$ mdeg. However, for temperatures $T_\lambda - T < 0.1$ mdeg our data appear to decrease, in contrast to the ratio of the (static) specific heats, which still increases. This discrepancy is connected with entering the critical temperature region. For example, at $T_\lambda - T \simeq 0.1$ mdeg the value for the correlation length ξ obtained in Ref. 11 would result in $k\xi = 1$. A similar deviation from the prediction of equilibrium thermodynamics has been observed in the measurements of the *total* scattered intensity very close to T_λ by Vinen *et al.*[16] This behavior seems to be consistent with an Ornstein–Zernike type of correction which must be applied in the critical temperature region.

References

1. V.L. Ginsburg, *J. Phys.* (*SSSR*) **7**, 305 (1943).
2. R.L.St. Peters, T.J. Greytak, and G.B. Benedek, *Opt. Comm.* **1**, 412 (1970); E.R. Pike, J.M. Vaughan, and W.F. Vinen, *J. Phys. C* **3**, L40 (1970).
3. E.R. Pike, *J. Physique* **33**, C1–25 (1972).
4. R.A. Ferrell, N. Menyhard, H. Schmidt, F. Schwabl, and P. Szepfalusy, *Ann. Phys.* (*N.Y.*) **47**, 565 (1968).
5. W.F. Vinen, in *Physics of Quantum Fluids,* R. Kubo and F. Takano, ed., Syokabo, Tokyo (1971).
6. B.N. Ganguly, *Phys. Rev. Lett.* **26**, 1623 (1971); and *Phys. Lett.* **39A**, 11 (1972).
7. H.A. Kierstead, *Phys. Rev.* **162**, 153 (1967).
8. G. Winterling, F.S. Holmes, and T.J. Greytak (to be published).
9. T.J. Greytak, this volume.
10. C.E. Chase, *Phys. Fluids* **1**, 193 (1958).
11. R.D. Williams and I. Rudnick, *Phys. Rev. Lett.* **25**, 276 (1970).
12. D.S. Greywall and G. Ahlers, *Phys. Rev. Lett.* **28**, 1251 (1972).
13. J.A. Tyson, *Phys. Rev. Lett.* **21**, 1235 (1968).
14. G. Ahlers, private communication.
15. E.R. Grilly, *Phys. Rev.* **149**, 97 (1966).
16. C.J. Palin, W.F. Vinen, and J.M. Vaughan, *J. Phys. C* **5**, L139 (1972).

* Ahlers.[14] The points for $T_\lambda - T > 10^{-2}$°K were computed from the compressibility data of Grilly.[15]

Brillouin Scattering from Superfluid ^{3}He–^{4}He Mixtures*

R. F. Benjamin, D. A. Rockwell, and T. J. Greytak†

Physics Department and Center for Materials Science and Engineering
Massachusetts Institute of Technology, Cambridge, Massachusetts

We are using Brillouin scattering to study thermally excited first sound, second sound, and concentration fluctuations in superfluid ^{3}He–^{4}He mixtures. The spectrum of the scattered light consists of five lines. One pair of lines arises from first sound and is shifted from the incident laser frequency by an amount $v_1 = \pm Ku_1/2\pi$, where u_1 is the velocity of first sound and K is the scattering wave vector. The half-width at half-height of the lines δv_1 is a measure of the amplitude attenuation coefficient α_1 of first sound at that frequency: $\delta v_1 = \alpha_1 u_1/2\pi$. Another pair of lines results from scattering by second sound. The frequency shift of these lines is $v_2 = \pm Ku_2/2\pi$ and their half-width is given by $\delta v_2 = \alpha_2 u_2/2\pi$, where u_2 and α_2 are the velocity and attenuation coefficient of second sound, respectively. The fifth line is due to the presence in the mixture of concentration fluctuations at constant pressure. This line is not shifted in frequency relative to the incident light, indicating that concentration fluctuations do not propagate but merely decay away. The decay time constant is inversely proportional to the width of the unshifted line.

Our ^{3}He–^{4}He mixtures were contained in a copper cell which was cooled by a ^{3}He refrigerator. In order to obtain the resolution required to measure linewidths and to make accurate determinations of velocities, it was necessary to use a single-mode, frequency-stabilized laser. An argon laser was chosen in order to take advantage of the greater scattering and improved phototube response in the blue end of the spectrum. This laser also provided us with several different wavelengths so that we could select values for the scattering wave vector K which would minimize overlap among the five spectral lines. Incident laser power levels ranged from 35 to 65 mW. We took some traces with several different powers at the same temperatures and found no change in the spectra, indicating no "heat flush" problem existed. Light scattered by 90° was focused into a spherical Fabry-Perot interferometer with a 318-MHz free spectral range. Since the shift of first sound is typically 665 MHz, our first-sound lines are shifted by more than two orders of the interferometer. The full width of the instrumental profile was typically 10 MHz and included about equal contributions from the interferometer and the frequency jitter of the laser.

Measurements were made on two different molar concentrations: $x = 0.05$ and 0.135. Temperatures ranged from 0.4 to 1.2°K. Brillouin scattering from ^{3}He–^{4}He

* Work supported by the Advanced Research Projects Agency under Contract No. DAHC 15-67-C-0222.
† Alfred P. Sloan Research Fellow.

mixtures at higher temperatures has been observed previously.[1] Figure 1 shows two typical examples of our spectra. Consider first the trace with $x = 0.05$. At zero frequency shift we see a low peak which contains contributions from both stray light and concentration fluctuations. The pairs of intense peaks are due to first sound. The frequency shift of these lines is measured from an origin two orders away. The apparent width of these lines is that of the instrument broadened by the effect of a finite optical collection angle. The intrinsic linewidth of the first-sound peaks is small compared to the apparent width and has not been measured. Note that the two peaks are of different heights, the intensity of the up-shifted component being proportional to $n(T)$ and that of the down-shifted component being proportional to $n(T) + 1$, where $n(T)$ is the mean occupation number of a phonon state of frequency ν_1. The pairs of lines just rising up over the background correspond to second sound. Their shifts are much smaller than the first-sound shifts, and in this case the intrinsic linewidth is measureable and is on the order of the instrumental width. Comparing the $x = 0.135$ trace with the $x = 0.05$ trace, some important differences appear: The intensities of scattering from concentration fluctuations and second sound have increased and the second-sound linewidth has decreased.

Before considering the detailed experimental results it is helpful to review the theoretical predictions. The appropriate equations of motion for the system are the equations of two-fluid hydrodynamics for ^{3}He–^{4}He mixtures. The general expressions for the quantities we will measure are very complicated. It is found, though, that in the low-concentration limit several simplifications may be made and a very straight-forward physical model may be used to explain the temperature dependence of u_2 and to predict the concentration dependence of $\delta \nu_2$. When the temperature is sufficiently low the normal fluid density arises primarily from the ^{3}He atoms. In this limit one can suppose that the ^{3}He quasiparticles act like a gas of ordinary particles and the effect of the background of ^{4}He atoms is primarily to modify the ^{3}He

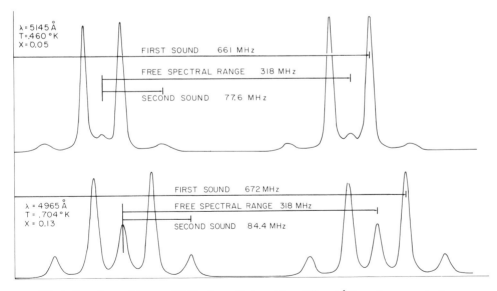

Fig. 1. Experimental traces of Brillouin scattering at two different ^{3}He molar concentrations.

inertial mass. In this model the second-sound velocity takes the form of the ordinary sound velocity in an ideal monatomic gas

$$u_2^2 = \tfrac{5}{3} kT/m_3^*$$
(1)

where m_3^* is the ^{3}He effective mass. Consider this model further: When the density of a classical gas becomes very low the equations of hydrodynamics no longer describe the dynamics of the gas and we must look to a kinetic theory for a correct description. For example, a sound wave begins to lose its collective identity to the random motions of the individual atoms when the mean free path of the atoms becomes comparable to the sound wavelength. This is accompanied by an increase in the sound attenuation as the density of the gas is decreased. This implies in our experiment that the linewidth of second sound should grow as the concentration decreases,

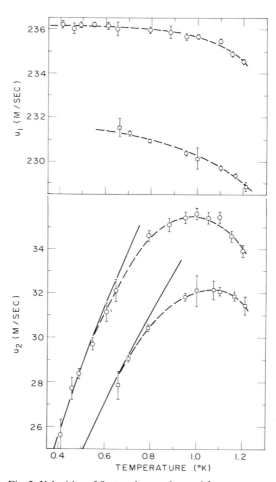

Fig. 2. Velocities of first and second sound for two concentrations. Circles: $x = 0.05$; squares: $x = 0.135$. First-sound velocities are at frequencies ~ 665 MHz; second-sound velocities are at ~ 80 MHz.

as was seen in Fig. 1. In fact, for very low concentrations the central component and the second-sound lines would merge into a single, broad Gaussian line representing the Doppler shift from the Maxwellian distribution of individual quasiparticle velocities.

We would also like to have theoretical values of I_2/I_1 to compare with our experimental measurements, where I_1 and I_2 are the total light intensities scattered by first and second sound, respectively. Again a few physically justified approximations greatly simplify a complicated general expression. If the thermal expansion is neglected and $u_2 \ll u_1$, we have

$$I_2/I_1 \simeq (x^2/16)\,(\rho_s/\rho_n)\,(u_1/u_2)^2 \tag{2}$$

where ρ_s and ρ_n are the superfluid and normal fluid densities, respectively. With the further assumptions that

$$\rho_s \simeq m_4(1-x)\,n \simeq m_4 n \qquad \text{and} \qquad \rho_n \simeq m_3^* x n$$

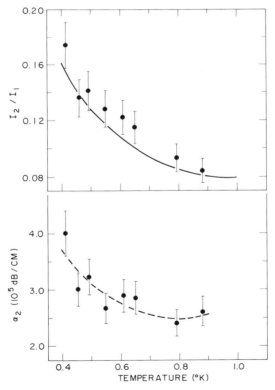

Fig. 3. Temperature dependence of intensity ratios of first- and second-sound lines compared with Eq. (3), and temperature dependence of the second-sound attenuation coefficient. $x = 0.05$.

where n is the total number density and m_4 is the atomic mass of ^{4}He, we find

$$I_2/I_1 = (x/16)(m_4/m_3^*)(u_1/u_2)^2 \qquad (3)$$

This equation agrees with the calculations of Gorkov and Pitaevskii[2] and Vinen[3] but differs slightly from the expression given by Ganguly and Griffin.[4]

In Fig. 2 we have plotted the temperature dependence of the velocities of first and second sound for both concentrations studied. Our results for u_1 and u_2 agree very well with interpolations of previous measurements of first sound in mixtures[5] and in pure ^{4}He,[6] and of second sound in mixtures,[7] all of which were made at lower frequencies. The $T^{1/2}$ dependence of u_2 predicted by Eq. (1) and represented by the solid lines in the figure seems to hold for the lowest temperatures studied. The main contribution to the absolute uncertainty in the velocities arises from small irregularities in pressure-sweeping the interferometer and is generally $< 0.1\%$ for u_1 and $< 1\%$ for u_2. The low intensity of the second-sound peaks also contributes to the error in u_2.

Figure 3 shows the temperature dependence of I_2/I_1 for $x = 0.05$. The solid curve was calculated from Eq. (3) using the temperature-dependent values of the effective mass from Brubaker et al.[8] A better fit to the intensity ratio data would be obtained with a 5% lower value of the effective mass. On the other hand, the low-temperature second-sound velocities for this mixture are best fit with slightly higher values of m_3^*. We are continuing to investigate this point. Also shown in Fig. 3 is the second-sound attenuation coefficient α_2. The dashed curve is not a theoretical fit but merely an aid to the eye.

The results for the two concentrations reported here are very promising. We intend to study several more concentrations, some near the tricritical point, where an exploratory run has shown a greatly enhanced scattering from concentration fluctuations: With $x = 0.57$ and $T = 0.829$ the intensity of the central line is about 35 times greater than that of a single first-sound peak indicating the onset of critical opalescence.

References

1. E.R. Pike, J.M. Vaughan, and W.F. Vinen, *Phys. Lett.* **30A**, 373 (1969); C.J. Palin, W.F. Vinen, E.R. Pike, and J.M. Vaughan, *J. Phys. C* **4**, L225 (1971).
2. L.P. Gorkov and L.P. Pitaevskii, *Soviet Phys.—JETP* **6**, 486 (1958).
3. W.F. Vinen, Physics of Quantum Fluids, in 1970 *Tokyo Summer Lectures in Theoretical and Experimental Physics*, R. Kubo and F. Takano, eds., Syokabo, Tokyo (1970).
4. B.N. Ganguly and A. Griffin, *Can. J. Phys.* **46**, 1895 (1968).
5. T.R. Roberts and S.G. Sydoriak, *Phys. Fluids* **3**, 895 (1960).
6. B.M. Abraham, Y. Eckstein, J.B. Ketterson, M. Kuchnir, and J. Vignos, *Phys. Rev.* **181**, 347 (1969).
7. H.A. Fairbank, *Nuovo Cimento Suppl.* **9**, 325 (1958).
8. N.R. Brubaker, D.O. Edwards, R.E. Sarwinski, P. Seligman, and R.A. Sherlock *Phys. Rev. Lett.* **25**, 715 (1970).

Liquid Structure Factor Measurements on the Quantum Liquids*

Robert B. Hallock

*Department of Physics and Astronomy, University of Massachusetts
Amherst, Massachusetts*

We have completed an analysis of data obtained from recent experiments in which Cu $K\alpha$ (1.54 Å) X rays were scattered from liquid ^{3}He. The liquid structure factor S was determined† as a function of momentum transfer k in two completely independent scattering experiments: (1) $T = 0.36°$K, $0.133 < k < 1.125$ Å^{-1} and (2) $T = 0.41°$K, $0.15 < k < 2.1$ Å^{-1}. As usual (see, e.g., Hallock[2]), the structure factor was determined from the basic expression

$$S = \frac{I\rho_G T_G}{I_G \rho T}\left(\frac{\xi_e + \xi_i}{\sigma_e}\right) - \frac{\sigma_i}{\sigma_e} \tag{1}$$

where I is the observed intensity, ρ is the number density, T is the transmission factor, and σ and ξ are scattering factors for the liquid helium and neon (normalization gas G), respectively.

The existence of these new data allows a detailed comparison with recent liquid structure measurements on ^{4}He.[2,3] In particular, the gentle shoulder or change of slope seen in the structure factor measurements on ^{4}He seems to be evident[4] in the case of ^{3}He as well. This similarity in shape can be examined in more detail by forming the ratio

$$R(k) = S_3(k)/S_4(k) \tag{2}$$

where $S_3(k)$ is the liquid structure factor of ^{3}He, etc. In Fig. 1 we display the results of forming such a ratio. The crosses are the values of R resulting from using explicit values [1,2] of S_3 and S_4. The error bars are due to counting statistics alone. The solid line is a weighted least squares polynomial fit to the values represented by the crosses. The fact that R is reasonably flat demonstrates the basic similarity in shape of the two structure factor functions. These present results support our earlier conclusions that the gentle change in slope of $S(k)$ seen in ^{4}He is not a consequence of the particular quantum statistics involved.

* Supported in part by the National Science Foundation, the Advanced Research Projects Agency through the Center for Materials Research at Stanford University, and the University Computer Center at the University of Massachusetts at Amherst.

† The details of these experiments on ^{3}He appear elsewhere along with tables of values for the structure factor and a comparison with theoretical work.[1]

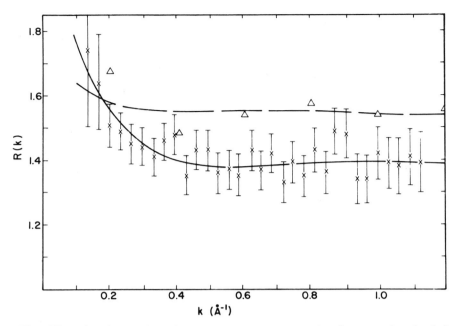

Fig. 1. The ratio $R(k) = S_3(k)/S_4(k)$. The crosses represent the ratio R for our work as detailed in the text. The triangles are computed from the work of Achter and Meyer.[3] The lines are in each case least squares polynomial fits to the computed R values. At small values of k the errors become large due to large window scattering. Also, in the limit $k \rightarrow 0$ the ratio $R \rightarrow \rho_3 T_3 \chi_3/ \rho_4 T_4 \chi_4$, where χ is the isothermal compressibility. This ratio has the value 2.1 for the parameters of our work.[1,2]

If one carries out this same sort of analysis on the data of Achter and Meyer,[3] one finds the values shown in the figure as represented by the triangles. The dashed line results from a least squares fit to their data. As can be seen from the figure, the results of Achter and Meyer also lead to the conclusion that the two structure factors have the same basic shape. It should be emphasized that the two sets of experiments disagree on the value of the shape ratio. The source of this disagreement stems from a relatively uniform stronger disagreement in the ³He results than in the ⁴He results. We do not at present understand the reason for these disagreements.*

Aside from a direct comparison with ⁴He, the detailed form of the ³He structure factor at small momentum transfer is of interest. While there have been several attempts (e.g., Ref. 5) to calculate the liquid structure factor for ³He based on the pair correlation function $g(r)$, these efforts have been unsuccessful at small k due to the extreme difficulty in obtaining $g(r)$ for large r. Arguments based on the Landau theory of oscillations in a Fermi liquid[6] by Widom and Sigel[7] and also a calculation by Tan[8] have aimed more directly at the low-momentum-transfer region of the liquid structure factor. If ³He were an ideal Fermi fluid, its structure factor would be given by (e.g., Ref. 9)

$$S(k) = \tfrac{3}{4}(k/k_F) + \theta(k^2) \tag{3}$$

* A new scattering spectrometer of improved resolution and higher power is presently under construction. When completed perhaps the source of the present discrepancy will become clear.

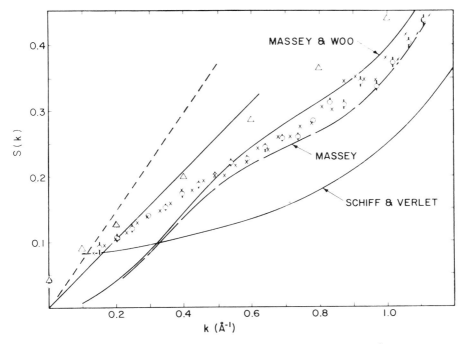

Fig. 2. Plot of $S(k)$ for ^{3}He. The triangles refer to the work of Achter and Meyer[3] while the circles and crosses refer to our work.[1] The straight lines are due to (——) Widom and Sigel[7] and Tan[8] and (- - -) the ideal Fermi theory. The curved lines result from attempts to calculate the entire liquid structure factor of ^{3}He.

Widom and Sigel and also Tan have argued that the structure factor should be given by

$$S(k) \simeq 0.57\,\hbar k/2mc \qquad (4)$$

where m is the mass of a ^{3}He atom and c is the velocity of sound in ^{3}He.

The results of the theoretical work we have mentioned are summarized in Fig. 2 along with both the present results and those obtained earlier by Achter and Meyer. As we have previously indicated,[4] the data necessary to confirm Eq. (4) are not presently available. More experimental work at lower temperatures and perhaps smaller angles is required.

Acknowledgments

The author extends his thanks to the Department of Physics at Stanford University for their kind hospitality during part of this work. He also thanks W. A. Little and L. Goldstein for helpful comments and suggestions and L. Suter for assistance during part of the data collection.

References

1. R.B. Hallock, *J. Low Temp. Phys.* **9**, 687 (1972).
2. R.B. Hallock, *Phys. Rev.* **A5**, 320 (1972).
3. E.K. Achter and L. Meyer, *Phys. Rev.* **188**, 291 (1969).

4. R.B. Hallock, *Phys. Rev. Lett.* **26**, 618 (1971).
5. W.E. Massey and C.-W. Woo, *Phys. Rev.* **164**, 256 (1967); D. Schiff and L. Verlet, *Phys. Rev.* **160**, 208 (1967).
6. D. Pines and P. Nozières, *Theory of Quantum Liquids,* Benjamin, New York (1966).
7. A. Wisdom and J.L. Sigel, *Phys. Rev. Lett.* **24**, 1400 (1970).
8. H.-T. Tan, *Phys. Rev.* *A***4**, 256 (1971).
9. E. Feenburg, *Theory of Quantum Fluids,* Academic Press, New York (1970), p. 14.

The Functional Forms of $S(k)$ and $E(k)$ in He II As Determined by Scattering Experiments*

Robert B. Hallock

*Department of Physics and Astronomy, University of Massachusetts
Amherst, Massachusetts*

The detailed form of the dispersion relation in He II has recently been the subject of intense interest.[1-5] The present discussion† was motivated by the work of Molinari and Regge[2] in which they suggested that the dispersion curve in He II might possess a term quadratic in the momentum. Their work was based on theoretical efforts as well as on computer fits to the data of Cowley and Woods[1] as shown in Fig. 1. In the work we shall describe here we have carried out fits to the data of Cowley and Woods, but in our case we have restricted attention to the small-angle data ($k \leq 1.0 \, \text{Å}^{-1}$).

As a basis for the present comments, let us introduce the following expansions for the dispersion relation $E(k)$ and the liquid structure factor $S(k)$:

$$E(k) = \hbar k c \left[1 + \sum_{j=1}^{p} \alpha_j (\hbar k/mc)^{nj} \right] \qquad (1)$$

$$S(k) = (\hbar k/2mc) \left[1 + \sum_{j=1}^{p} A_j (\hbar k/mc)^{nj} \right] \qquad (2)$$

The present discussion is concerned with whether or not the existing scattering experiments support the conclusions that: (1) the dispersion curve has a small upward concavity[3] and (2) the above expansions contain odd[7] powers of k (i.e., can $n = 1$?). The very recent work of Roach et al.[5] seems to establish beyond doubt that for $n = 1$ the coefficient α_1 in Eq. (1) is either absent or else is very small. If the dispersion curve can in fact be represented by an analytic expression such as Eq. (1), then the results obtained by Roach et al. can be expected to remain valid for larger values of the momentum. To maintain perspective, however, it should be remembered that the measurements of Roach et al. at 90 MHz correspond to phonon energies of $4 \times 10^{-3} \, °\text{K}$ and k values of $2 \times 10^{-4} \, \text{Å}^{-1}$. These are to be compared with the neutron results of interest here, which span energies up to $12°\text{K}$ and momentum transfer values as large as $1 \, \text{Å}^{-1}$ (the smallest being $0.1 \, \text{Å}^{-1}$).

Based on our fits to the neutron data at small momentum transfer ($k \leq 1 \, \text{Å}^{-1}$)

* Supported in part by the National Science Foundation and the University Computation Center of the University of Massachusetts.
† A preliminary discussion of the comments presented here has been given previously.[6] A more complete account of this work is in preparation.

551

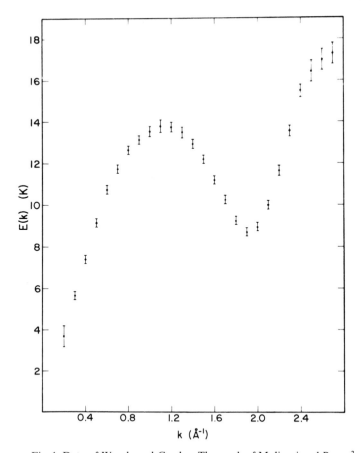

Fig. 1. Data of Woods and Cowley. The work of Molinari and Regge[2]
was confined to the region $k \geq 1.1$ Å^{-1}. We have focused attention on the
region $k \leq 1.0$ Å^{-1}.

we find that: (1) These data are not of sufficient accuracy to determine whether or
not one should take $n = 1$ or $n = 2$ in Eq. (1); (2) if one *demands* an expansion of
the form of Eq. (1) with $n = 2$, then the neutron data favors $\alpha_1 > 0$; and (3) if one
demands an expansion of the form of Eq. (1) with $n = 1$ but with $\alpha_1 = 0$, then $\alpha_2 > 0$
and again the dispersion is positive.

These conclusions can be seen rather clearly by reference to the tables. Table I
displays the relative errors of the various fits for several representative sets of the
data. The table can be most easily understood by an example. Focus attention on the
three-parameter fits for $E(k)$. Take the case of $M = 25$. The entries indicate that
an even-power expansion $(2, 4, 6)$ is preferred, an expansion of mixed powers starting
with $k^2 (2, 3, 4)$ is not as good, and the $(1, 2, 3)$ expansion is the worst of the three.
The entries are the sum square deviations of the data from the given fit relative to
the best fit for the given number of parameters and M value under consideration.
The larger the value of M in the table, the more data points were fit (see Fig. 2) and

Table I. Relative Errors of a Representative Sample of the Closed-Form Fits

M	Two parameters			Three parameters			Four parameters		
	1,2	2,3	2,4	1,2,3	2,3,4	2,4,6	1,2,3,4	2,3,4,5	2,4,6,8
$E(k)$									
10	1	1	1	1	1	1	1	1	1
15	1	1.02	1.04	1.05	1.02	1	1	1	1.01
20	1.01	1	1.03	1	1.01	1.02	1	1	1
25	1.08	1	1.12	1.02	1.01	1	1.02	1.01	1
30	1	1.18	1.40	1.10	1.03	1	1.01	1	1
$S(k)$									
10	1	1.01	1.46	1	1.17	1.35	1	1.06	1.15
15	1	1	1.23	1	1.29	1.54	1	1.12	1.32
20	1.09	1	1.11	1	1.31	1.51	1	1.21	1.49
25	1.34	1	1.01	1	1.41	1.57	1	1.28	1.60
30	1.58	1	1	1	1.26	1.27	1	1.27	1.59

[a] The fits to the elementary excitation spectrum show no definite preference for a given leading term in the fit. The fits to the structure factor data show a slight preference for the quadratic term. The columns headed by parameters indicate the number of coefficients in the fit. M refers to the number of data points used in the fit. For $M = 30$ nearly all of the data $k \leq 1$ Å^{-1} are used. The numbers in the first row refer to the powers of $(\hbar k/mc)$ used in the fit in the column below.

consequently, the larger the values of the momentum transfer which were included. It is our view that the table supports the conclusion that the present neutron data (at small momentum transfer) makes no unambiguous statement about the presence or absence of the quadratic term. Table II, is a highly schematic portrayal of the preferred sign of the term $(\hbar k/mc)^2$ in the various fits. If we focus attention on those fits that are of the form $1.0 + \alpha_2(\hbar k/mc)^2 + \cdots$, the conclusion is that the present neutron data tend to support the results of Maris.

To reach these conclusions, we have carried out a series of weighted least squares polynomial fits to the data of Cowley and Woods.[1] We used their 33 data points as presented in their plot of velocity vs. k^2 (see Fig. 9 of Ref. 1) for the momentum transfer range $k \leq 1.0$ Å^{-1} rather than the nine points in this range as presented in their Table 3. Figure 2 displays these points translated into the more familiar format of the dispersion curve. Our reason for adopting this procedure was that the values presented in Table 3 of Ref. 1 are "best values" and not actual data points.[8] We have also carried out several fits using the "best values" data. In the cases where k_{max} coincided with that used by Molinari and Regge, our fits produced essentially the same coefficients as theirs.

Several comments on the details of the fitting are in order. Our fits were to closed-form polynomial expressions of the sort given by Eq. (1). We have carried out fits with maximum values of P (see Fig. 2) ranging from two to five. For each value of P we have carried out fits over increasing ranges of the data starting with the lowest six ($M = 6$) data points and working up to the full 33 ($M = 33$) points which completely cover the momentum transfer range $k \leq 1$ Å^{-1}. In most cases

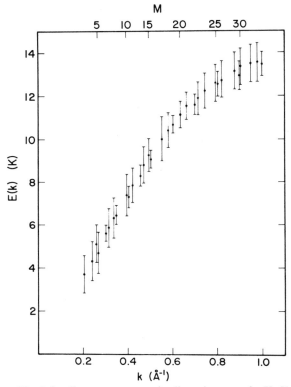

Fig. 2. Small momentum transfer dispersion curve for He II. The data shown here are those which we have used in our various fits. They were obtained from Ref. 1 in the manner described in the text. M refers to the number of points used in a given fit. In all of our fits we have set E equal to zero at $k = 0$.

the resulting coefficients were highly correlated. These coefficients represent a set of numbers too large to be published here.

Complementary to the fitting of the neutron data of Cowley and Woods we have carried out the same sort of fitting on the X-ray $S(k)$ measurements of Hallock.[9]

Table II. Sign of the $(\hbar k/mc)^2$ Term in the Various Fits for $E(k)$

Two parameters	Three parameters	Four parameters
1,2 ($-$)	1,2,3 ($\sim$)	1,2,3,4 ($+$)
2,3 ($+$)	2,3,4 ($+$)	2,3,4,5 ($+$)
2,4 ($+$)	2,4,6 ($+$)	2,4,6,8 ($+$)

[a] This table is highly schematic and represents the sign of the $(\hbar k/mc)^2$ term favored by the fits with M above 25 in the various cases. As can be seen, the neutron data in this range are consistent with positive dispersion.

In the case of the liquid structure factor data we have used from six to 33 points, covering in the same manner as before the momentum transfer span $k \leqq 1.125\,\text{Å}^{-1}$. As can be seen in Table I the results of this effort suggest that a linear term and odd powers for $S(k)$ are slightly preferred.

Finally, we have also carried out a series of small momentum transfer fits to determine the suitability of terms[4] like $k^4 \ln(1/k)$ in the expressions for both $E(k)$ and $S(k)$. We find that for $k \leqq 1\,\text{Å}$ neither the energy data nor the structure factor data are accurate enough to allow a determination as to whether $k^4 \ln(1/k)$ terms are present.

References

1. R.A. Cowley and A.D.B. Woods, *Can. J. Phys.* **49**, 177 (1971).
2. A. Molinari and T. Regge, *Phys. Rev. Lett.* **26**, 1531 (1971).
3. H.J. Maris, *Phys. Rev. Lett.* **28**, 277 (1972).
4. H. Gould and V.K. Wong, *Phys. Rev. Lett.* **27**, 301 (1971).
5. P.R. Roach, B.M. Abraham, J.B. Ketterson, and M. Kuchnir, *Phys. Rev. Lett.* **29**, 32 (1972).
6. R.B. Hallock, *Bull. Am. Phys. Soc.* **17**, 610 (1972).
7. E. Feenberg, *Phys. Rev. Lett.* **26**, 301 (1971).
8. A.D.B. Woods, private communication.
9. R.B. Hallock, *Phys. Rev. A***5**, 320 (1972).

9
^{3}He$-^4$He Mixtures

Effective Viscosity of Liquid Helium Isotope Mixtures

D.S. Betts, D.F. Brewer, and R. Lucking

School of Mathematical and Physical Sciences
University of Sussex, Brighton, United Kingdom

Recently results were reported[1] from this laboratory on the effective viscosity of dilute mixtures ($X_3 < 10\%$) between 20 m°K and 1°K. In the present work we have extended the ranges of concentration up to $X_3 = 100\%$ and of temperature up to 2°K so that all accessible areas of the phase diagram have now been investigated. Only in the region of the tricritical point ($X_3 = 67\%$; $T = 0.87°K$) have serious difficulties been encountered, and further work there will be necessary to improve the technique.

To derive the viscosity, we measured the logarithmic decrement of torsional oscillations of a cylindrical quartz crystal immersed in the liquid. From classical hydrodynamics

$$\eta_e \rho_n = (M/S)^2 \, (f/\pi)(\Delta - \Delta_0)^2 \qquad (1)$$

where η_e is an effective viscosity, ρ_n is the normal density (equal to the total density at points above the λ-line), M, S, and f are respectively the crystal mass, surface area, and resonant frequency $f = 40\,\text{kHz}$, Δ is the logarithmic decrement, and Δ_0 is the value of Δ in vacuum. The fit to a truly exponential decay was found to be good to about 1%. The method involves the propagation of a heavily damped transverse wave by the crystal surface, the penetration depth into the liquid being given by $\delta = (\eta_e/\pi f \rho_n)^{1/2}$. Clearly the penetration depth should comfortably exceed any relevant mean free paths in the liquid and only in the most dilute (i.e., low X_3) mixtures did this margin become close. The crystal and indeed all the viscometric, CMN-thermometric, and refrigerating equipment were the same as those used by Bertinat *et al.*[1]

The series of runs began with pure ^3He ($X_3 = 100\%$) and this was successively diluted by the carefully controlled addition of ^4He and, as the total volume of mixture rose, by the removal of mixture to keep the liquid surface inside the cell. It is believed that even when the lowest concentration was reached, it was known to be $5\% \pm 0.2\%$ and it is reassuring to observe the good agreement between our 5% results and those of Bertinat *et al.*,[1] who began at the opposite end of the phase diagram with $X_3 = 0$. At the lowest temperatures our final results for pure ^3He are adequately described by $\eta T^2 \approx 2.2 \,\mu\text{p deg}^2$, thus showing agreement with Bertinat *et al.*[1] and with Wheatley's result[2] derived from sonic absorption methods. This agreement serves as a double check on our viscometry and thermometry. It is interesting to note here that tiny quantities of ^4He impurity were found to cause a significantly different result at temperatures below the appropriate phase separation temperature. We believe that the reason for this is the formation of a ^4He layer at the crystal surface, further evidence for which is presented below. We found, however, that the effect was entirely removed by overfilling the cell so that the free surface was in the filling capillary. This behavior is consistent with the observations of Laheurte and Keyston[3] and suggests that when the free surface is warmer than the crystal, ^{4}He impurity is "siphoned" away from the crystal by superfluid film flow.

All concentrations other than $X_3 = 100\%$ were done with the free surface inside the cell. When phase separation occurs ^{3}He-rich floats on ^{4}He-rich and for $95\% > X_3 > 20\%$ the interface is in contact with the crystal. For other concentrations the dead spaces above or below the crystal contained the minor phase. Bearing these points in mind and looking at Fig. 1, we see that for X_3 close to 100% the phase separation does not produce a reversion to the 100% readings at the lowest

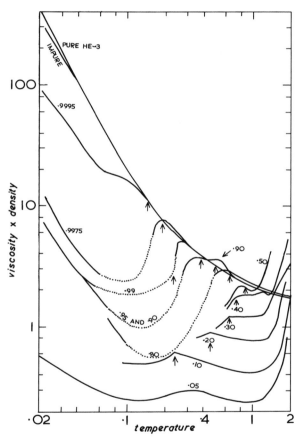

Fig. 1. The smoothed data, plotted as $\eta_e \rho_n$ [according to Eq. (1)] in units of μp g cm^{-3}. Continuous lines indicate regions of low scatter ($\lesssim 1\%$) and the dotted sections indicate regions where day-to-day variations were noticeable (see text). The numbers on the figure are values of X_3 for each curve and the vertical arrows indicate appropriate phase separation temperatures.

temperatures, thus suggesting that the crystal sees not pure ^{3}He at its surface, but rather a ^{4}He-enriched film having a much reduced viscosity. This view is strengthened by the observation that in the low-temperature ($T < 0.07°$K), phase-separated region the curves of $\eta_e \rho_n$ vs. T all coalesced for $X_3 = 99\%, 95\%, 90\%$ (and probably 80%). One can deduce an effective viscosity η_e from this section of the curves if one chooses a suitable value for ρ_n. We find that η_e is remarkably close to the results of Black *et al.*[4] and Kuenhold *et al.*[5] for a saturated solution of ^{3}He in He II if one uses values of $\rho_n(X_3, T)$ appropriate to such a saturated solution. Figure 2 illustrates this point and suggests that the film covering the crystal may be a saturated solution of ^{3}He in He II ($X_3 \sim 6.4\%$) with a thickness of $\gtrsim \delta$, the penetration depth of the excited transverse waves which is of order 10,000 Å at $0.02°$K.

These concentrations ($99\%, 95\%, 90\%, 80\%$) all gave very similar results even though the interface was at very different levels in each case. In fact, even for

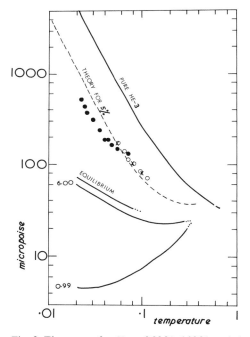

Fig. 2. The curves for $X_3 = 0.99\%, 6.00\%,$ and the equilibrium concentration (6.4% as $T \to 0$) are from Ref. 1, confirmed by the present work. The curve for pure ^{3}He has also been confirmed. The dashed curve is from theory[6] and crosses[5] and open circles[4] indicate experimental points by other workers on the equilibrium concentration. The filled circles are from the present work obtained from mixtures with $X_3 = 90\%, 95\%,$ and 99% below 70 m°K using ρ_n appropriate to the equilibrium concentration ($X_3 = 6.4\%$). It seems possible, therefore, that the crystal sees a layer of equilibrium concentration in these circumstances (see text).

a particular fixed initial concentration the interface level will vary with temperature as determined by the shape of the phase diagram, and this may well account for the regions of hysteresis indicated on the figure by dashed lines. This local hysteresis took the form of day-to-day variations in otherwise consistent results. For concentrations $X_3 \leqq 50\%$ we paid particular attention to the no-phase-separation areas.

References

1. M.P. Bertinat, D.S. Betts, D.F. Brewer, and G.J. Butterworth, *Phys. Rev. Lett.* **28**, 472 (1972).
2. J.C. Wheatley, *Progress in Low Temperature Physics,* C.J. Gorter, ed., North-Holland, Amsterdam (1970), Vol. 6, p. 77.
3. J.P. Laheurte and J.R.C. Keyston, *Cryogenics* **11**, 485 (1971).
4. M.A. Black, H.E. Hall, and K. Thompson, *J. Phys. C.* **4**, 129 (1971).
5. K.A. Kuenhold, D.B. Crum, and R.E. Sarwinski, *Phys. Lett.* **41A**, 13 (1972).
6. G. Baym and W.F. Saam, *Phys. Rev.* **171**, 172 (1968).

The Viscosity of Dilute Solutions of ^{3}He in ^{4}He at Low Temperatures*

K.A. Kuenhold, D.B. Crum, and R.E. Sarwinski

Department of Physics, The Ohio State University
Columbus, Ohio

Measurements of the viscosity of dilute solutions of ^{3}He in ^{4}He made by different groups show little agreement. Viscosities measured by Black *et al.*[1] and later by Fisk and Hall[2] are quite different from those of Webeler and Allen[3] and Bertinat *et al.*[4] Webeler and Allen and Bertinat *et al.* used the torsional crystal method to measure lower viscosities, those of Bertinat *et al.* being an order of magnitude lower at low temperatures. The reason for this discrepancy is not known.

We have measured viscosities of dilute solutions having concentrations X from 0.1% to 7.0% ^{3}He in ^{4}He over the temperature range 0.03–0.9°K. Our measurements were made using the more conventional technique of allowing the liquid to flow between two thermally connected reservoirs through a capillary of known impedance Z (see Fig. 1). The capillary was a 74.6-cm-long tube of 0.028 cm i.d. having a low-temperature impedance of 61.1×10^8 cm^{-3}. The level in one of the reservoirs could be changed by adding or removing ^{4}He liquid on the outside of a bellows located in that reservoir.[5] After a level change we waited a sufficient time for the bellows to come to equilibrium and the ^{4}He to flow to equalize its chemical potential between the reservoirs. We then measured the liquid height h_1 above its final equilibrium value by measuring the capacitance of a vertical parallel plate capacitor as a function of time. The rate of change of h_1 can be described by the equation[6]

$$-\frac{dh_1}{dt} = \frac{h_1 \rho g (1 + ah_1)}{ZA^* \eta (1 + kV_1 V_2/A^* V)(1 + bh_1)} \tag{1}$$

where $A^* = (A_1 A_2)/(A_1 + A_2)$ and $V = V_1 + V_2$, the subscripts 1 and 2 on the areas A and volumes V refer, respectively, to the reservoir containing the capacitor and to the other reservoir. The areas of the reservoirs are constant and the volumes are those at the final equilibrium value. The mass density of the liquid is ρ and g is the gravitational constant. The quantity $k = \rho g/(\partial \pi/\partial X)_T X$, where π is the osmotic pressure.† In Eq. (1) $a = A_1 k(V_1 - V_2)/2A^* V$ and $b = 4a/(1 + kV_1 V_2/VA^*)$. The quantities ah_1 and bh_1 are always small compared to one and take into account to first order the concentration difference between reservoirs due to the flow of ^{4}He to equalize its osmotic pressure. The capacitor was calibrated to measure h_1 and the

* Work supported by National Science Foundation Grant No. GP-13381.
† Values of $(\partial \pi/\partial X)_T$ were provided by Disatnik.[7]

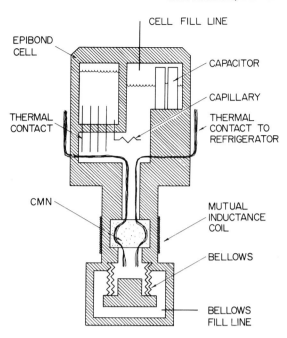

Fig. 1. A schematic drawing of the experimental cell.

data were fit to the integrated form of Eq. (1) to determine the average viscosity η. This equation has the form

$$\left[\frac{1 + a\alpha(C - C_F)}{1 + a\alpha(C_i - C_F)}\right]^\beta \frac{C - C_F}{C_i - C_F} = e^{-B(t - t_i)/\eta} \tag{2}$$

where $\beta = (b - a)/a$ and $\alpha = (C - C_F)/h_1$ are constants for a given X and T, C_i is the capacitance at the starting time t_i, C_F is the final capacitance at $t = \infty$, C is the capacitance at time t, and $B = \rho g/[ZA^*(1 + kV_1V_2/A^*V)]$. For a typical data point the level in one reservoir was lowered 2–3 mm and then the mixture level (capacitance) monitored for about three time constants. The measured time constants ranged from 300 to 6000 sec.

At all concentrations and temperatures the ^{3}He mean free path was at least 1000 times smaller than the capillary diameter, so no slip correction was necessary. We found no corrections necessary at high temperatures due to vapor pressure differences between the reservoirs because of the changing volumes or because of the concentration difference in the vapors. Surface tension effects were accounted for in our method of calibration and all measurements were taken in the region of laminar flow.

Figure 2 shows the viscosity for different values of X as a function of temperature. The viscosity of the 7% solution below 0.1°K is along the phase separation curve and is in excellent agreement with the measurements of Black et al.[1] At 50 m°K, the viscosity measured by Wheatley[8] is 7% lower than an extrapolation of our data

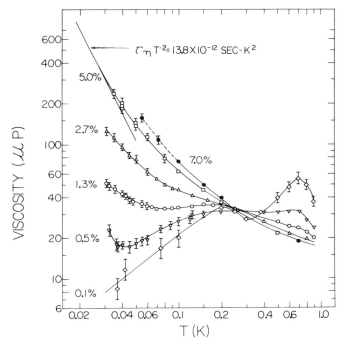

Fig. 2. The viscosity of ^{3}He–^{4}He solutions as a function of temperature for different molar fractions X of ^{3}He in ^{3}He. Below 0.1°K for $X = 7\%$ the broken line indicates the viscosity along the phase separation curve.

along the phase separation curve. An extrapolation of our data at all concentrations to high temperatures gives good agreement with that of Staas *et al.*[6]

At low temperatures the viscosity of the 5% solution is consistent with, but does not define, the T^{-2} behavior expected of a degenerate Fermi gas. One method of extrapolating to this T^{-2} behavior is to extract the relaxation time for viscosity T_η from η and fit it to an equation of the form[9]

$$\tau_\eta T^2 = A[1 + B(T/T_F)^2] \tag{3}$$

where T_F is the Fermi temperature of 0.331°K for a 5% solution. We calculated τ_η from $\tau_\eta = \eta/P(T)$, where $P(T)$ is the pressure of an ideal Fermi gas of the same effective mass and number density as the ^{3}He in solution. We find $A = (13.8 \pm 1) \times 10^{-12}$ sec °K^2 and $B = 9 \pm 1$. This compares with $A = 17.8 \times 10^{-12}$ sec °K^2 predicted by Roach[10] using the original effective interaction potential of Bardeen *et al.*[11] A value for A of 22.5×10^{-12} sec °K^2 was calculated by Baym and Ebner[9] from a theory to fit the ultrasonic sound attenuation data of Abraham *et al.*[12]

For $X \geq 5\%$ and $T > 0.5$°K the viscosity is independent of X, indicating a semiclassical region with little or no phonon contribution to the viscosity. Fitting $T^{1/2}/\eta$, we find

$$T^{1/2}/\eta = (0.29 + 5.8T) \times 10^4 \text{ °K}^{1/2} \text{ P}^{-1} \tag{4}$$

over the temperature range 0.5–0.9°K.

At high temperatures and $X < 5\%$ we find that the viscosity increases as X decreases. When the 5–7% data are subtracted from the data for lower X we find the difference in viscosities proportional to X^{-1} for temperatures above 0.8°K. We find this difference $\eta_{ph} = 3 \times 10^{-3} X^{-1} \mu P$ at about 0.9°K.

In Fig. 3 we compare theory and measurements of viscosities by a number of experimenters with our own work for $X \sim 1.3\%$ and 5%. The theoretical predictions of Baym and Ebner[9] are about 18% higher than our measurements. The high-temperature portions of these curves were calculated using Ebner's[13] expansion to high momentum of an effective interaction potential fit to measurements of the spin diffusion coefficient.[14] This semiclassical calculation predicts viscosities about 18% too high, but this is within the accuracy claimed by the theory.

Our data and that of Fisk and Hall[2] are in excellent agreement at low temperatures but we have no explanation for the disagreement at high temperatures. The data of Bertinat et al.[4] and Webeler and Allen[3] show better agreement with our data and with each other at high X and T. Their anomalously low viscosities at low temperatures cannot be accounted for by slip corrections according to Fisk and Hall, but Betts et al.[15] suggest that perhaps a thick ^{4}He-rich film covering the crystal is involved in understanding the low-temperature data of Bertinat et al.[4] By comparison with our own work, we see that this effect would be small at high X and T. We conclude that at low X and T the torsional crystal method does not measure the hydrodynamic viscosity.

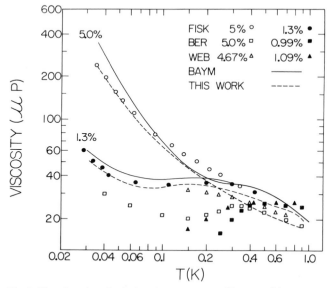

Fig. 3. The viscosity of solutions having $X \sim 1.3\%$ and 5.0% from the experimental work of Fisk and Hall,[2] Webeler and Allen,[3] and Bertinat et al.,[4] and the theoretical work of Baym and Ebner[9] compared to this work.

References

1. M.A. Black, H.E. Hall, and K. Thompson, *J. Phys. C: Sol. St. Phys.* **4**, 129 (1971).
2. D.J. Fisk and H.E. Hall, this volume.
3. R.W.H. Webeler and G. Allen, *Phys. Lett.* **29A**, 93 (1969).
4. M.P. Bertinat, D.S. Betts, D.F. Brewer, and G.J. Butterworth, *Phys. Rev. Lett.* **28**, 472 (1972).
5. H.M. Guo, Ph.D. Thesis, Ohio State Univ. 1971, unpublished.
6. F.A. Staas, K.W. Taconis, and K. Fokkens, *Physics* **26**, 669 (1960).
7. Y. Disatnik, private communication.
8. J.C. Wheatley, R.E. Rapp, and R.T. Johnson, *J. Low Temp. Phys.* **4**, 1 (1971).
9. G. Baym and C. Ebner, *Phys. Rev.* **164**, 235 (1967); G. Baym and W.F. Saam, *Phys. Rev.* **171**, 172 (1968).
10. W.R. Roach, Ph.D. Thesis, Univ. of Illinois, 1966, unpublished.
11. J. Bardeen, G. Baym, and D. Pines, *Phys. Rev.* **156**, 207 (1967).
12. B.M. Abraham, Y.E. Eckstein, J.B. Ketterson, and J.H. Vignos, *Phys. Rev. Lett.* **17**, 1254 (1966).
13. C. Ebner, *Phys. Rev.* **156**, 222 (1967).
14. A.C. Anderson, D.O. Edwards, W.R. Roach, R.E. Sarwinski, and J.C. Wheatley, *Phys. Rev. Lett.* **17**, 367 (1966).
15. D.S. Betts, D.F. Brewer, and R. Lucking, this volume.

The Viscosity of ^{3}He–^{4}He Solutions

D.J. Fisk and H.E. Hall

The Schuster Laboratory
University of Manchester, United Kingdom

Experimental Method

We have measured the viscosity of 1.3%, 3.5%, and 5% solutions of ^{3}He in ^{4}He from 25 m°K to 0.5°K with a vibrating wire viscometer of the type used previously[1] for measurements on pure ^{3}He, operating at a frequency of the order of 200 Hz. The major change to the apparatus for the present measurements was that a pill of powdered CMN immersed in the solution was used as a primary thermometer. This powdered salt was contained in an Epibond 100A appendix to the viscometer cell, connected to it by a tube largely filled with a brush of fine copper wires. Speer resistance thermometers immersed in the liquid were used to monitor the temperatures of the viscometer and salt chamber, and thus to ensure that measurements were made only in thermal equilibrium.

As was done previously,[1] we measured the half-height points of the wire resonance by the "45° null" method. Although in principle both viscosity and normal density can be deduced from the resonance width and frequency shift, we did not attempt this, for two reasons. First, these parameters are rather insensitive to normal density; and second, the vacuum resonance frequency changed slightly each time the cell was filled, so that reliable absolute measurements of frequency shift could not be made. We therefore used normal densities calculated from the experimental inertial masses of Brubaker *et al.*[2] to compute viscosities from the experimental half-height points. The computer program also deduced a "vacuum resonance frequency" from the observed half-height points; the constancy of this frequency during a run provided a check that the observed resonance widths and frequency shifts were consistent, which will prove of importance later.

Results

Our results are compared with the theory of Baym and Saam[3] in Fig. 1; they agree very well for 1.3% and rather less well for 5%. The theoretical curves are for ^{3}He viscosity, since phonon viscosity should not be detected at our frequency; experiment seems to confirm this.

The results of Bertinat *et al.*[4] obtained with a torsional crystal at 40 kHz are an order of magnitude lower. Our first thought was that this discrepancy might be explained by specular reflection of ^{3}He at the solid surface with consequent surface slip. Such an effect might be expected because the pressure dependence of μ_{40} and E_3[5] means that van der Waals forces effectively repel ^{3}He in a solution from a solid

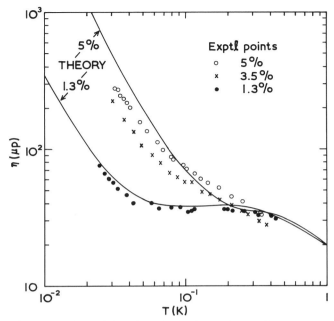

Fig. 1. Experimental viscosity, calculated assuming no slip, compared
with theory.

boundary. We therefore solved the hydrodynamic equations with a slip boundary
condition for the relevant geometries and deduced a slip length and viscosity consistent with our results and those of Bertinat *et al*. This slip hypothesis gave slip
lengths of order 100 μm at the lowest temperatures and raised the viscosities of Fig.
1 by amounts of the order of 50% at the lowest temperatures. A similar procedure
with the results of Webeler and Allen[6] at 11 kHz gave consistent slip lengths. These
large slip lengths imply that, on this interpretation, the reflection of ^{3}He at the solid
boundary is over 99% specular at the lowest temperatures. The magnitude and
temperature dependence of the fraction of diffuse reflection deduced in this way is
in order of magnitude agreement with numerical estimates of the penetrability of
the repulsive van der Waals potential barrier at the wall, referred to above.

    However, fitting our observations to hydrodynamic equations with slip leads
to a temperature-dependent "vacuum resonance frequency," whereas the equations
without slip give a constant vacuum resonance frequency. This is shown for some of
the 5% results in Fig. 2. Systematic trends in period of order 0.3 μsec could arise
from phase-setting errors, but the observed variation exceeds this. We therefore
conclude that hydrodynamic equations without slip give a significantly better fit
to our observations, and the discrepancy between our results and those of Bertinat
et al. cannot be explained by slip of the form we have assumed. If slip is not the
explanation, the correlation between fraction of diffuse reflection and barrier penetrability that we find on the slip hypothesis must be coincidental. It remains possible,
however, that specular reflection of ^{3}He occurs in the experiment of Bertinat *et al*.
but not in ours, for their crystal had an optically polished surface, whereas our wire

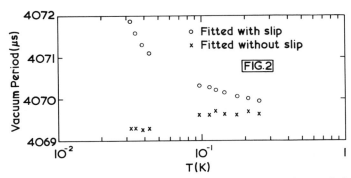

Fig. 2. Temperature dependence of "vacuum resonance frequencies" deduced from the experimental half-height points on various assumptions.

showed slight but visible drawing marks under a microscope. The only alternative to this type of explanation is a dependence of the viscosity on frequency or wave number that is not contained in current theory.

To decide definitively between these possibilities, a fuller investigation is required in which both the damping and inertial components of the viscous drag on a surface would be measured as a function of frequency.

References

1. M.A. Black, H.E. Hall, and K. Thompson *J. Phys. C* **4**, 129 (1971).
2. N.R. Brubaker, D.O. Edwards, R.E. Sarwinski, P. Seligmann, and R.A. Sherlock, *Phys. Rev. Lett.* **25**, 715 (1970).
3. G. Baym and W.F. Saam, *Phys. Rev.* **171**, 172 (1968).
4. M.P. Bertinat, D.S. Betts, D.F. Brewer, and G.J. Butterworth *Phys. Rev. Lett.* **28**, 472 (1972).
5. C. Ebner and D.O. Edwards, *Phys. Report* **2C**, 78 (1971).
6. R.W.H. Webeler and G. Allen, *Phys. Lett.* **29A**, 93 (1969).

Thermodynamic Properties of Liquid ^{3}He–^{4}He Mixtures Near the Tricritical Point Derived from Specific Heat Measurements*

S.T. Islander

Department of Technical Physics, Helsinki University of Technology
Otaniemi, Finland

and

W. Zimmermann, Jr.

School of Physics and Astronomy, University of Minnesota
Minneapolis, Minnesota

In a recent publication we presented the results of some measurements of the specific heat of liquid ^{3}He/^{4}He mixtures at saturated vapor pressure near the interesting junction of the lambda and phase-separation curves which occurs at a temperature $T_0 \cong 0.87°$ K and a ^{3}He mole fraction $x_0 \cong 0.67$.[1] The purpose of the present report is to indicate some of the possibilities for determining a variety of thermodynamic information about the junction region from the specific heat data.†

The junction is a point at which a first-order phase transition goes over into a second-order transition with increasing temperature. Griffiths has argued that when the additional thermodynamic dimension associated with the superfluid order parameter and its conjugate field is included the junction point represents the intersection of three critical lines, the λ-curve and two new ones, and has therefore identified the junction as a tricritical point.[3]

Our thermodynamic calculations for the liquid mixtures are based on the expression

$$dg = -s\,dT + v\,dP + \phi\,dx$$

where g is a molar free energy, s is the molar entropy, T is the temperature, v is the molar volume, P is the pressure, ϕ is the difference $\mu_3 - \mu_4$ between molar chemical potentials, and x is the mole fraction of ^{3}He. Because of the occurrence of phase separation at temperatures below T_0, it turns out to be possible in principle to obtain a complete thermodynamic picture of the junction region at saturated vapor pressure almost entirely from measurements of the molar specific heat $c_{P,x}$ of the liquid in that region by themselves. Such a determination would not be possible for a

---

* We wish to acknowledge the support given this work by the U.S. Atomic Energy Commission.
† For a more complete account of this work see Ref. 2.

liquid which remained everywhere homogeneous. As a result, we have been able to make numerical calculations of the quantities $(\partial s/\partial T)_{P,x} = c_{P,x}/T$, $(\partial\phi/\partial T)_{P,x} = -(\partial s/\partial x)_{P,T}$, s, and $\phi - \phi_0$ in the junction region as functions of T and x at saturated vapor pressure. The only additional data needed were two constants and the dependence of molar volume and saturated vapor pressure upon T and x. The constants, $(d\phi/dT)_{\sigma;0}$ and $x_0(d\mu_3/dT)_{\sigma;0} + (1 - x_0)(d\mu_4/dT)_{\sigma;0}$, were evaluated using low-temperature data discussed in Ref. 1. The path of differentiation indicated by the subscript σ here is the phase-separation curve; the subscript 0 indicates the value at the tricritical point. Because of the smallness of saturated vapor pressure in the junction region, the terms involving the vapor pressure and the molar volume represent corrections to the main terms of the order of only 1%.

It would have been of considerable interest to complete the picture by computing the derivative $(\partial\phi/\partial x)_{P,T}$ throughout the junction region. This derivative was, in fact, already known along the phase separation curve from the work of Ref. 1. However, because of the accumulation of error resulting from the succession of differentiations and integrations of the data involved, it did not seem likely that enough detail could be preserved to make the process worthwhile.

The results of our calculations can be used to discuss a number of features of the critical behavior of the mixtures near the λ-curve and the tricritical point. At the λ-curve our measurements indicated that $c_{P,x}$ is finite and continuous but that it has an infinite slope at values of x not too close to x_0.[1] These findings are in agreement with those of Gasparini and Moldover at lower values of x.[4] The question which immediately arises is whether the complementary specific heat $c_{P,\phi}$ diverges at the λ-curve as does c_P for pure ^{4}He. If so, the behavior of $c_{P,x}$ would be an example of the renormalization of the "ideal" behavior of $c_{P,\phi}$ discussed by Fisher.[5]

We have investigated this question indirectly. At the tricritical point it can be shown that the vanishing of $(\partial\phi/\partial x)_{P,T}$ at that point indicated by the results of our specific heat measurements[1] implies the divergence of $c_{P,\phi}$ there. Along the λ-curve it can be shown that $c_{P,\phi}$ diverges if and only if $(\partial s/\partial T)_{P,x} = c_{P,x}/T$ attains its upper bound of $(ds/dT)_{P,\lambda} + (d\phi/dT)_{P,\lambda}(dx/dT)_{P,\lambda}$. Here the path of differentiation indicated by the subscripts lies in the λ-surface at constant pressure. A comparison between our determinations of $(\partial s/\partial T)_{P,x}$ and its upper bound is shown in Fig. 1. The curves for $(\partial s/\partial T)_{P,x}$ terminate at ± 1 m°K away from the λ-curve. From this plot we see that it is plausible, but by no means established by our data, that $(\partial s/\partial T)_{P,x}$ reaches its upper bound at values of x away from x_0. Hence, although it seems likely, we cannot be sure from our data that $c_{P,\phi}$ diverges all along the λ-curve, as it does at $x = 0$ and $x = x_0$.

Near the tricritical point our results are consistent with the following asymptotic behavior:

$$(\partial s/\partial T)_{P,x;0} - (\partial s/\partial T)_{P,x} \sim (T - T_0)^1 \quad \text{along } x = x_0 \text{ for } T > T_0 \quad (\alpha_u = -1)$$

$$|x - x_0| \sim (T_0 - T)^1 \quad \text{along } \sigma_\pm \text{ for } x \lessgtr x_0 \quad (\beta_\pm = 1)$$

$$(\partial\phi/\partial x)_{P,T} \sim (T_0 - T)^1 \quad \text{along both } \sigma_\pm \text{ for } T < T_0 \quad (\gamma'_\pm = 1)$$

$$(d\phi/dT)_P \sim |T - T_0|^1 \quad \text{along both } \sigma\text{- and } \lambda\text{-curves} \quad (\phi'_c, \phi_c = 2)$$

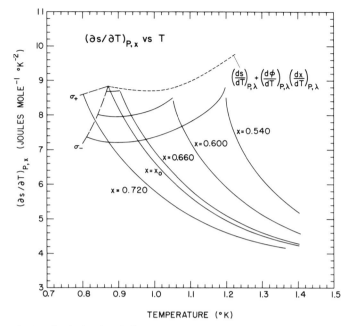

Fig. 1. The derivative $(\partial s/\partial T)_{P,x} = c_{P,x}/T$ at saturated vapor pressure. The continuous curves show $(\partial s/\partial T)_{P,x}$ at five representative values of x, including $x_0 = 0.6735$. The long-dashed curves show $(\partial s/\partial T)_{P,x}$ along the two arms of the phase-separation curve. The short-dashed curve shows $(ds/dT)_{P,\lambda} + (d\phi/dT)_{P,\lambda}(dx/dT)_{P,\lambda}$ along the lambda curve.

$$|\phi - \phi_0| \sim |x - x_0|^2 \qquad \text{along } T = T_0 \text{ for } x \lessgtr x_0 \quad (\delta_\pm = 2)$$

$$(\partial s/\partial T)_{P,\phi} \sim |T - T_0|^{-1/2} \qquad \text{along } \phi = \phi_0 \text{ for } T \lessgtr T_0 \quad (\alpha_t', \alpha_t = \tfrac{1}{2})$$

Here σ_+ and σ_- stand for the ^{3}He-rich and ^{3}He-poor branches of the phase-separation curve, respectively. The exponents in parentheses follow the notation convention recently proposed by Griffiths, with the exception that we introduce two crossover exponents ϕ_c' and ϕ_c in place of his single exponent ϕ to allow for possible differences between the $T < T_0$ and $T > T_0$ cases, respectively.[6] The subscript c is added in order to distinguish the crossover exponents from our variable ϕ.

Of these six relationships the first three come most directly from experimental data. The first of these was determined from data having a resolution of ~ 1 m°K. Although our preferred value for α_u is -1.0, a value as low as -0.8 is possible for a somewhat unlikely alternate interpretation of the data. The second relationship was determined from the phase diagram and the third from the jump in $c_{P,x}$ at the phase-separation curve. Because of the relatively large gaps in x value at which data were taken, the uncertainties in our determinations of β_+, β_-, γ_+', and γ_-' are perhaps several tenths.

The fourth relationship was determined by calculations of $(d\phi/dT)_P$ along the σ- and λ-curves. The crossover exponents ϕ_c' and ϕ_c are known from our work only with considerable uncertainty. The evidence for the behavior described by the fifth

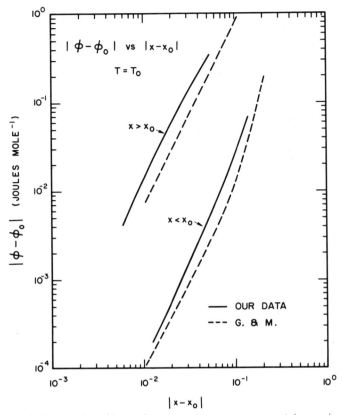

Fig. 2. The quantity $\ln|\phi - \phi_0|$ at saturated vapor pressure vs. $\ln|x - x_0|$ for $T = T_0 = 0.871°K$. The continuous curves show our results, while the dashed curves show the results of Goellner and Meyer.[7]

relationship is shown in Fig. 2, where the continuous curves represent our determinations of $|\phi - \phi_0|$ vs. $|x - x_0|$ along the critical isotherm and the dashed curves represent the determinations of Goellner and Meyer from vapor pressure measurements.[7] The two determinations agree reasonably well with an exponent of 2, but our values of $|\phi - \phi_0|$ are consistently larger in magnitude than theirs by up to a factor of two. Finally, the evidence for the behavior described by the sixth relationship is shown in Fig. 3. Here $(\partial s/\partial T)_{P,\phi}$ has been constructed along the curve of constant ϕ passing through the tricritical point. Note that the effective temperature resolution here is greater by more than an order of magnitude than the actual experimental temperature resolution available for the measurement of $c_{P,x}$. This increase is due to the flatness of the curve $\phi = \phi_0$ near the tricritical point in the T, x plane, particularly for $x < x_0$.

We note that $\alpha'_t, \alpha_t = 1/2$ represents a stronger divergence of $(\partial s/\partial T)_{P,\phi}$ at the tricritical point than the near-logarithmic divergence seen at the other end of the λ-curve. This breakdown of universality along the λ-curve, at least near the tricritical point, appears to be in accord with recent proposals for scaling near the

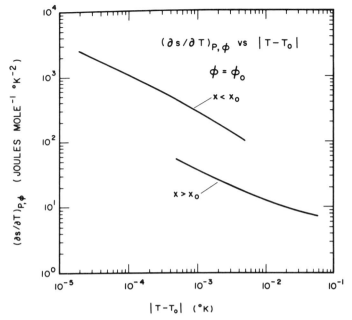

Fig. 3. The quantity $\ln(\partial s/\partial T)_{P,\phi} = \ln(c_{P,\phi}/T)$ at saturated vapor pressure vs. $\ln|T - T_0|$ for $\phi = \phi_0$. The condition $x < x_0$ corresponds to $T < T_0$, while $x > x_0$ corresponds to $T > T_0$.

tricritical point.[6] Furthermore, we note that all of the above exponents are in agreement with Griffiths's scaling relations for the tricritical point. Finally, we note that α_0 satisfies the Fisher renormalization relation with respect to α_t.[5]

References

1. T.A. Alvesalo, P.M. Berglund, S.T. Islander, G.R. Pickett, and W. Zimmermann, Jr., *Phys. Rev.* **A4**, 2354 (1971).
2. S.T. Islander and W. Zimmermann, Jr., *Phys. Rev.* **A7**, 188 (1973).
3. R.B. Griffiths, *Phys. Rev. Lett.* **24**, 715 (1970).
4. F. Gasparini and M.R. Moldover, *Phys. Rev. Lett.* **23**, 749 (1969).
5. M.E. Fisher, *Phys. Rev.* **176**, 257 (1968).
6. R.B. Griffiths, *Phys. Rev.* **B7**, 545 (1973).
7. G. Goellner and H. Meyer, *Phys. Rev. Lett.* **26**, 1534 (1971).

Dielectric Constant and Viscosity of Pressurized ^{3}He–^{4}He Solutions Near the Tricritical Point*

C.M. Lai

Brookhaven National Laboratory, Upton, New York
and
State University of New York, Stony Brook

and

T.A. Kitchens

Brookhaven National Laboratory, Upton, New York

Measurements of the dielectric constant ε and the product of the normal density ρ_n and the viscosity η have been made on ^{3}He–^{4}He solutions at pressures up to the solidus. In this experiment both $\rho_n\eta$, determined from the damping on a vibrating wire, and ε, determined by the frequency shift of a tunnel diode oscillator, can be measured simultaneously in both the ^{3}He-rich and the ^{4}He-rich phases below the separation. Data taken for 63.6–75.1 % ^{3}He for temperatures between 0.45 and 1.2°K are presented. Viscosity and $\rho_n\eta$ results near the tricritical point and the resulting phase diagram for elevated pressures obtained from the measured dielectric constant are given.

As shown by Graf *et al.*,[1] the second-order superfluid transition in ^{3}He–^{4}He liquid solution appears to end in the first-order phase separation. In fact, they found the superfluid transition temperature to be a linearly decreasing function of the ^{3}He concentration and the phase boundaries for both the superfluid ^{4}He-rich and the normal ^{3}He-rich solutions to also be linear. This behavior indicates that these rather different types of ordering are related, indeed strongly related, at the point of intersection. Griffiths[2] has investigated some of the thermodynamic features of this rather unusual situation and has dubbed this point the tricritical point. Here we report an experiment to investigate the region about this point in which a static property, the dielectric constant, and a dynamic property, the effective viscosity, can be simultaneously measured for ^{3}He–^{4}He liquid solutions under pressure. These measurements will be reduced to determine the general features of the phase diagram and to determine the variation of the viscosity as the transition is approached.

The experimental apparatus is capable of simultaneous measurement of both the dielectric constant and the effective viscosity of both the ^{3}He- and the ^{4}He-rich solutions below the phase separation. The technique and the apparatus has been

* Work supported by the U.S. Atomic Energy Commission.

described previously[3] and will be only briefly reviewed here. The sample cell was an integral part of a vapor-pressure-regulated ³He refrigerator. The cell has symmetric upper and lower sections each containing a carbon resistance thermometer, a vibrating bent-wire viscometer equivalent to that used by Black et al.,[4] and a low-volume capacitor in the extreme upper (lower) position. These capacitors, as well as a third capacitor located in the center of the cell, where there was a large excess volume, were in the tank circuits of three independent oscillators driven by tunnel diodes located at 4.2°K. These oscillators were stable to one part in 10^7, which is equivalent to less than 0.05% change in ³He concentration in these liquid solutions. The temperature of the cell was determined by a calibrated germanium thermometer and could be regulated by a controller using one of the carbon resistance thermometers and driving a heater attached to the ³He refrigerator body.

The 80-μm-diameter tungsten wire viscometers, which constituted arms of ac bridges, were in a uniform 6–kOe magnetic field produced with a persistent super-conducting solenoid in the main ⁴He bath of the cryostat. The wires were driven to a maximum amplitude of about 4 μm by current oscillators synchronized with the natural frequency of the wires. The decay or increase in the amplitude of oscillation of each wire is proportional to the error signal of the bridge and, after suitable amplification, was observed on an oscilloscope.

The samples were pressurized by a cryogenic toepler technique; the sample gas was condensed in a container of minimum volume filled with Zeolite cooled to 4°K; the Zeolite container was then warmed to room temperature, thereby pressurizing the sample cell. The ³He concentration of the gas recovered after the completion of the measurements was determined. Corrections due to the small volumes at high temperatures in the sample system were much less than the uncertainties in the isotopic gas analyses, which were typically 1%. Comparison of the measured phase separation temperature with that expected from Graf et al.[1] for saturated vapor pressure (SVP) were also within 1%.

Analysis of Dielectric Constant Data

From an analysis of the dielectric constant measured in each of the three capacitors it is not only possible to determine the temperature of phase separation for the given ³He concentration of the sample but also to determine the complete phase diagram for lower temperatures. We have assumed the molar volume of the solution V_s may be expressed by $V_s = xV_3 + (1 - x)V_4 - x(1 - x)\Delta$,[1] where V_3 and V_4 are the molar volumes of pure ³He and ⁴He respectively, and are known functions of pressure and temperature.[5] The form of the last term is consistent with measurements at SVP by Kerr[6]; from these measurements we know that $\Delta = \Delta(p \sim 0, T)$ for T above 1.2°K. It is positive, yet decreasing with decreasing temperature. At higher pressures Δ can be determined from the data of Watson et al.[7] for $T \rightarrow 0$. Here Δ is found to be small and positive at low pressures, turning slightly negative at high pressures. We have roughly accounted for Δ in the following analysis with resultant phase separation curve shifts of less than 0.5%.

The data were analyzed assuming that the tank circuit was composed of a capacitor C_0 filled with the sample solution and a parallel capacitor C_s, due to stray capacitance in the circuit, which is not filled with sample. The dielectric constant of the solution is related to the number densities N_3 and N_4 of ^{3}He and ^{4}He, respectively, through the Clausius–Mossotti equation

$$(\varepsilon - 1)/(\varepsilon + 2) = (4\pi/3)(\alpha_3 N_3 + \alpha_4 N_4)$$

where α_M is the atomic polarization for the isotope with atomic mass M. Our analysis follows that of previous work[5,7] except that we have accounted for the fact that α_3 and α_4 are not equal and depend on the density as recently shown by Kerr and Sherman.[8] Here we have used $\alpha_3 \,(\text{mole cm}^{-3}) = 0.12341 - (0.0068/V_s)$ and α_4 0.03% higher.

The analysis[7] contains two parameters f_0 and $\gamma = C_0/(C_0 + C_s)$ for each of the oscillator circuits, where f_0 is the frequency when the sample capacitor is void of sample and γ can be determined from the frequency measured above 1.2°K and the known ^{3}He concentration of the sample. The results of such an analysis on a run at SVP is shown in Fig. 1 and the agreement with Graf et al.[1] is excellent. A compilation of results at other pressures agrees with those obtained by Johnson and Graf.[9]

Analysis of Viscosity Data

From an analysis of the decay in the amplitude of vibration of the bentwire viscometers the product of the normal density and the viscosity $\rho_n \eta$ can be obtained near the top and bottom of the cell provided (1) the vibrational amplitude is a small fraction of the wire diameter a, (2) the viscous penetration depth $\lambda \ll a$, and (3) the

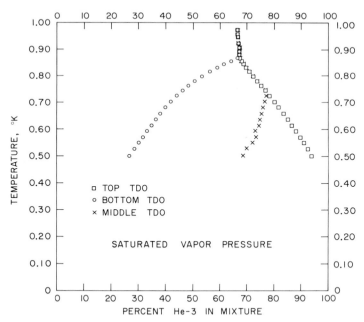

Fig. 1. The phase diagram determined from dielectric constant data from the three tunnel diode oscillators in the experimental cell. The initial solution was 67% ^{3}He in ^{4}He at saturated vapor pressure.

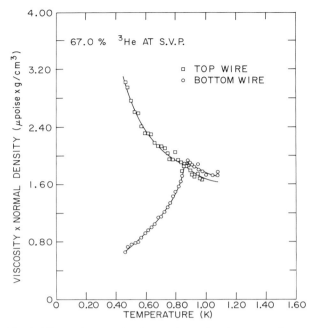

Fig. 2. The product $\rho_n\eta$ for the two bent-wire viscometers for an
initial solution of 67 % ³He at saturated vapor pressure.

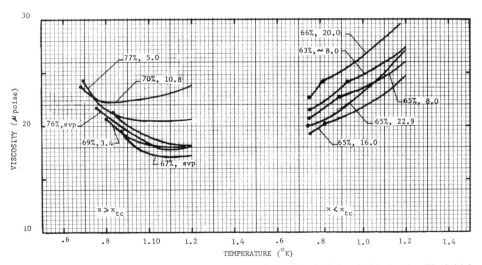

Fig. 3. The qualitative features of the viscosity in the neighborhood of the tricritical point. The initial
concentration of ³He and the pressure in atm label each curve. The terminal points indicate phase
separation and the intermediate point in the $x < x_{tc}$ curves designates the superfluid transition.

nonlinear terms in the fluid velocity are negligible in the Navier–Stokes equation.[10,11] Under these assumptions, well satisfied in this investigation,

$$\rho_n \eta = (1/4\pi f)(\rho_\omega a/\tau)^2$$

within 0.5%, where τ is the decay time constant, ρ_ω is the wire density, and f is the frequency of the oscillation.

In addition to a measure of $\rho_n \eta$ as the phase separation is approached and for both the ^{3}He-rich and the ^{4}He-rich phases below the phase separation, it is possible to determine the superfluid transition $T_\lambda = T_\lambda(P, x)$ and $\rho_n \eta$ near T_λ for an initial ^{3}He concentration x less than the tricritical value x_{tc}. Such data are illustrated in Fig. 2.

Finally, the viscosity in the normal fluid has been calculated from these data near the tricritical point. The data illustrated qualitatively in Fig. 3 are not precise enough to obtain accurately the exponent derived from the scaling hypothesis, but it appears that it is rather more sensitive to concentration than to pressure. However, the curves corresponding to 67% ^{3}He solution at SVP and 65% ^{3}He solution at 22.9 atm terminate very near the tricritical point, yet have quite different slopes.

Acknowledgments

We wish to acknowledge many helpful conversations with M. Vannesste, D. F. Brewer, and P. P. Craig and the technical support rendered by T. Oversluizen.

References

1. E.H. Graf, D.M. Lee, and J.D. Reppy, *Phys. Rev. Lett.* **19**, 417 (1967).
2. R.B. Griffiths, *Phys. Rev. Lett.* **24**, 715 (1970).
3. C.M. Lai, P.P. Craig, T.A. Kitchens, and D.F. Brewer, in *Proc. 12th Intern. Conf. Low Temp. Phys.*, *1970*, Academic Press of Japan, Tokyo (1971), p. 167.
4. M.A. Black, H.E. Hall, and K. Thompson, *J. Phys. C: Sol. St. Phys.* **4**, 129 (1971).
5. C. Boghosian, H. Meyer, and J. Rives, *Phys. Rev.* **146**, 110 (1966); C. Boghosian and H. Meyer (private communication).
6. E.C. Kerr, *Low Temperature Physics and Chemistry*, J.R. Dillinger, ed., Univ. of Wisconsin Press, Madison (1958), p. 158.
7. G.E. Watson, J.D. Reppy, and R.C. Richardson, *Phys. Rev.* **188**, 384 (1969).
8. E.C. Kerr and R.H. Sherman, *J. Low Temp. Phys.* **3**, 451 (1970).
9. R. Johnson and E. Graf, private communication.
10. J.T. Tough, W.D. McCormick, and J.G. Dash, *Rev. Sci. Instr.* **35**, 1345 (1964); *Phys. Rev.* **132**, 2373 (1963).
11. R.G. Hussey, B.J. Good, and J.M. Reynolds, *Phys. Fluids* **10**, 89 (1967); R.G. Hussey and P. Vuja-cic, *Phys. Fluids* **10**, 96 (1967).

Critical Opalescent Light Scattering in ^{3}He–^{4}He Mixtures Near the Tricritical Point*

D. Randolf Watts and Watt W. Webb

School of Applied and Engineering Physics and Laboratory for Atomic and Solid State Physics
Cornell University, Ithaca, New York

Mixtures of the two isotopes ^{3}He–^{4}He display a "tricritical point" at the junction of a "lambda" line of superfluid transitions and the consolute critical point, where a first-order miscibility gap goes continuously into perfect miscibility.[1] The simultaneous occurrence of two transitions in this critical region provides particularly stringent tests of theories of critical points.

We have studied this critical region by carefully measuring the intensity of scattered light.[1] Concentration fluctuations in the mixture cause fluctuations in the dielectric constant in the fluid, which scatter light. Scattering measurements yield detailed data on the "concentration susceptibility" $(\partial x/\partial \Delta)_{TP}$ where x is the ^{3}He mole fraction and $\Delta = \mu_3 - \mu_4$ is the difference in the chemical potentials of the isotopes.

The extinction length for light scattering from the mixture is†

$$h = A(\lambda) \frac{k_{\mathrm{B}}T}{v^2} \left[\frac{1}{v}\left(\frac{\partial v}{\partial x}\right)^2_{TP} \left(\frac{\partial x}{\partial \Delta}\right)_{TP} + \beta_{Tx} \right] \quad \mathrm{cm}^{-1}$$

where

$$A(\lambda) = \left(\frac{2\pi}{\lambda}\right)^4 \frac{(4\pi\alpha N_0)^2}{6\pi}, \qquad \beta_{Tx} = -\frac{1}{v}\left(\frac{\partial v}{\partial p}\right)_{Tx}$$

Here v is the molar volume, αN_0 is the molar polarizability of helium, and λ is the light wavelength. The compressibility β_{Tx} varies only weakly near the tricritical point (x_c, T_c), whereas the susceptibility $(\partial x/\partial \Delta)_{TP}$ diverges and may be fitted to a power law:

$$(\partial x/\partial \Delta)_{TP} = \Gamma |T - T_c|^{-\gamma} \qquad \text{(along critical isochore)}$$

Along the coexistence curve different amplitudes Γ'_+, Γ'_-, and "critical exponents" γ'_+, γ'_- are needed for the normal and superfluid phases, respectively.

The intensities of light of wavelength 632.8 nm scattered at a 90° angle are reported in Fig. 1 for four mixtures, in normal and superfluid phases, above and along

---

* Supported by the National Science Foundation and the Advanced Research Projects Agency through the Materials Science Center at Cornell University.
† This has been checked in a private communication with W. I. Goldburg. See Ref. 2.

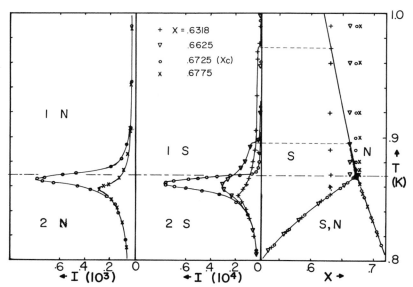

Fig. 1. Scattered intensities as a function of temperature in the single (1) and two-phase (2) regions for normal (N) and superfluid (S) regions. At the right using the *same* temperature scale is the phase diagram near the tricritical point. For mixtures $x_1 = 0.6318$ (+) and $x_2 = 0.6623$ (∇) the scattering is shown for the superfluid phase along the coexistence curve (CC), for $x_3 = 0.6775$ (×) the normal phase along CC, and for $x_4 = 0.6725$ (○) both phases along CC. The single-phase scattering is shown for all phases. Note the different intensity scales for the normal and superfluid phases. Only a small fraction of the data are shown for better clarity.

the coexistence curve and near the λ-line. In order to attain equilibrium conditions in the mixture, temperatures were controlled to better than 1 $\mu°$K stability for hours at a time. The intensity peaks are strikingly asymmetric about T_c, with the opalescence below T_c much greater than above. In the phase-separated region the scattering along the coexistence curve is shown to be always the same, independent of overall composition, as would be expected from the lack of thermodynamic degrees of freedom there. For the critical composition $x_c \approx x_4$ the scattering along the coexistence curve reaches a rounded peak 4 m°K below T_c in the normal phase and 8 m°K below T_c in the superfluid phase. This rounding and displacement from T_c is caused by gravitational effects and by the divergence of the coherence length in the mixture near T_c.* In the single-phase region above the phase separation temperature the scattered intensities decrease smooothly with increasing temperature for mixture x_3 and along the critical isochore (x_4). Mixtures x_1 and x_2 display a sharp decrease in scattered intensity as the temperature increases and crosses the superfluid transition. (A small local maximum in scattering near the lambda temperature of x_1,

---

* When the coherence length ξ becomes so large that $|\mathbf{q}| \xi \sim 1$, where $\mathbf{q}$ is the scattering vector, the scattering becomes concentrated toward small angles. We have studied this effect using two different wavelengths of light and have found the scattering peaks to be shifted differently. This will be discussed in a subsequent paper. See also Ref. 3.

sufficiently far from the tricritical point, has been discovered and investigated, but it is not evident on the scale of Fig. 1.[4]) All this behavior is consistent with thermodynamic calculations of $(\partial x/\partial \Delta)_{TP}$ which may be obtained from recent vapor pressure measurements of Goellner and Meyer[5] (GM) and specific heat measurements of Alvesalo et al.[6] (AZ)

From the scattered intensities $(\partial x/\partial \Delta)_{TP}$ has been calculated and is presented in Fig. 2 for composition $x_4 \approx x_c$. In the single-phase region we find $\gamma = 1.02 \pm 0.06$, and also very good agreement in magnitude with GM. For the normal phase with $T < T_c$ the good agreement with GM and particularly with AZ indicates $\gamma'_+ = 1.12 \pm 0.06$. For the superfluid phase one finds the data to be well fitted by the exponent $\gamma'_- = 1.67 \pm 0.1$; however, the agreement with AZ is good only for $\delta T \gtrsim 20$ m°K, and the disagreement with the GM data is evident. These data contradict present scaling theories of the tricritical point,[7,8] which predict $\gamma = \gamma'_+ = \gamma'_- \sim 1$. Recently Griffiths[9] has suggested the notation employed here, which at least allows for different exponents.

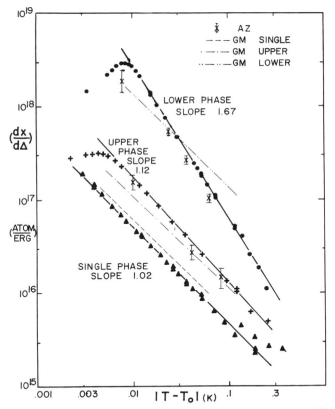

Fig. 2. The power-law divergence of $(\partial x/\partial \Delta)_{TP}$ is demonstrated in this log plot vs. $|T - T_0|$ for both phases along the coexistence curve ($T < T_0 = 0.870$) and for the critical isochore ($x = x_c \sim x_4$, $T > T_0 = 0.8682$). The relative vertical positions of these data are not arbitrary. In addition, values calculated from Refs. 5 (GM) and 6 (AZ) are shown.

The discrepancies reported here are not yet understood; however, a number of possible uncertainties have been eliminated: (1) The system was at equilibrium because relaxed temperature control 3–10 times worse (3–10 $\mu°$K) did not systematically change the scattered intensities. (2) Tests were made for dependence of the scattering upon the laser power and the resulting γ'_{-} is not significantly different. (3) Our experiments have shown that gravitational and coherence length effects for measurements about 2 mm below the interface are important *only* for $T_c - T \lesssim 10$ m°K. One remaining question is whether the system has reached its "asymptotic critical region" for $(\delta T)_{min} \sim 10$ m°K. The following discussion of gravitational effects upon the mixture will help to show the great difficulty in extending measurements much closer to the true tricritical point.

Different exponents γ for different phases would in any case suggest differences in at least some of the other critical exponents $(\delta, \beta, \nu, \eta, \ldots)$ describing thermodynamic behavior near (x_c, T_c).

Next we discuss measurements of concentration gradients in the mixture caused by the dependence of the chemical potentials on gravity. At height z in the liquid $\Delta(z) = \Delta(0) + (m_3 - m_4)gz$, which causes a concentration gradient $\partial x / \partial z = (\partial x / \partial \Delta)_{TP}(m_3 - m_4)g$. When $(\partial x / \partial \Delta)_{TP}$ becomes large near (x_c, T_c) the concentration

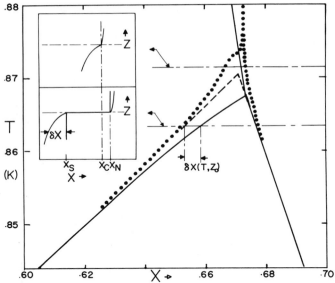

Fig. 3. Concentration changes due to gravity are shown qualitatively in the inset as a function of height z at a temperature in the single-phase and in the two-phase region. The measured change $\delta x(T,z)$ for fixed values of z_0 and 3 mm above and below the interface has been used to correct the phase diagram. The true coexistence curve (——) lies inside the thermodynamic path followed by a point in the fluid a few mm from the interface ($\cdots$). That path never approaches (x_c, T_c), and a straight line extrapolation (- - -) predicts the tricritical point falsely (by several m°K). The normal phase is relatively much less affected because $(\partial x / \partial \Delta)_{TP}$ is much smaller.

changes significantly with height in the fluid. This is indicated qualitatively on the inset of Fig. 3. Berestov *et al.*[10] (BVM) have recently shown that $\delta x \propto z^{1/\delta} \sim \sqrt{z}$ near T_c.

We report here direct measurements of $\delta x(T, z)$ using a modified Jamin interferometer which measures the index of refraction of the mixture as a function of height. These measurements indicate important corrections to the usual phase diagram within 10 m° K of T_c as explained in Fig. 3. The measured correction term $\delta x(T, z)$ indicated is accurate within a factor of two. Rough estimates by BVM predict a value about three times greater. The presently known thermodynamic information on the mixture [in particular, $(\partial x/\partial \Delta)_{TP}$] should suffice for self-consistent numerical calculations to be made of corrections for the gravitational effects on all the thermodynamics near (x_c, T_c). [Hohenberg and Barmatz[11] have recently outlined such a method for a simple fluid which behaves symmetrically about (ρ_c, T_c).]

In summary, the light scattering measurements of $(\partial x/\partial \Delta)_{TP}$, which have been shown to be consistent with much of the data on C_P[6] and P_{sat},[5] have produced highly unexpected critical exponents γ not adequately explained by present theories if these exponents are representative of the asymptotic critical region. Furthermore, *all* experiments that have been done on the ^{3}He–^{4}He mixture near the tricritical point which measure a property in the fluid very close to but not exactly on the interface will be *dominated* by gravitational effects within 5–10 m° K of T_c.

Acknowledgments

We are pleased to acknowledge assistance in the early stages of this project by L. D. Jackel and Dr. W. I. Goldburg.

References

1. D.R. Watts, W.I. Goldburg, L.D. Jackel, and W.W. Webb, *J. Physique* **33**, Cl–155 (1972).
2. L.P. Gor'kov and L.P. Pitaevskii, *Soviet Phys.—JETP* **6**, 486 (1958).
3. M.E. Fisher and R.J. Burford, *Phys. Rev.* **156**, 583 (1967).
4. D.R. Watts (to be published).
5. G. Goellner and H. Meyer, *Phys. Rev. Lett.* **26**, 1534 (1971); *Bull. Am. Phys. Soc.* **17**, 232 (1972); G. Goellner, Ph.D. Thesis, Duke Univ., August 1972, unpublished.
6. T. Alvesalo, P. Berglund, S. Islander, G.R. Pickett, and W. Zimmerman, *Phys. Rev. Lett.* **22**, 1281 (1969).
7. R.B. Griffiths, *Phys. Rev. Lett.* **24**, 715 (1970).
8. E.K. Riedel, *Phys. Rev. Lett.* **28**, 675 (1972).
9. R.B. Griffiths (to be published).
10. A.T. Berestov, A.V. Voronel', and M.Sh. Giterman, *Soviet Phys.—JETP Lett.* **15**, 273 (1972).
11. P.C. Hohenberg and M. Barmatz, *Phys. Rev. A***6**, 289 (1972).

Second-Sound Velocity, Gravitational Effects, Relaxation Times, and Superfluid Density Near the Tricritical Point in ^{3}He–^{4}He Mixtures

Guenter Ahlers and Dennis S. Greywall

Bell Laboratories
Murray Hill, New Jersey

There has been considerable interest recently in the behavior of liquid ^{3}He–^{4}He mixtures near the tricritical point. We wish to report results for the second-sound velocity[1] in this system which yielded information about the superfluid density, the coherence length, the phase diagram, the effect of gravity, and concentration relaxation times.

The measurements were made by determining the frequency of plane-wave resonant modes in a cylindrical cavity with a diameter of 0.6 cm and a height of 0.3 cm. The resonator was located in a well just below a much larger portion of the sample with an additional height of 0.1 cm. This geometry permitted measurements in the superfluid phase along the coexistence curve, as well as along lines of constant concentration between the phase separation temperature T_σ and the superfluid transition temperature T_λ. The results for the velocity are shown as a function of temperature in Fig. 1.

We used 1-μm-Nuclepore membrane filters as the active elements of superleak condenser transducers to generate the second sound. There was a small amount of power dissipation in the sample which depended on the square of the transducer driving voltage. Near the tricritical temperature T_t and the tricritical concentration x_t this was kept as low as 10^{-2} erg sec^{-1} by using only 0.5 V RMS to excite the driven transducer. It was necessary to keep the power dissipation to a minimum in order to avoid excessive temperature and concentration gradients. All data were corrected to zero power, and have a precision of 0.1 %. Near T_t large relaxation times were encountered, and the sample was kept within $\pm 10^{-6}$ °K of a constant temperature for serveral hours before the final velocity measurement was made.

Consistent with the expected thermal diffusivity of the mixtures,[2] the thermal relaxation time τ_T of our system was only about 45 sec and was essentially constant throughout the range of the measurements. Nonetheless, large velocity relaxation times τ were encountered near the tricritical point. We attribute these time constants to concentration relaxation. Along the coexistence curve we found

$$\tau \propto (T_t - T_\sigma)^{-1.0 \pm 0.1}$$

At constant x in the single-phase region, but near T_σ, we also found $\tau > \tau_T$.

Because of the large $(\partial x/\partial \Phi)_{TP}$ near T_t and x_t,[3] one may expect sizable effects of the gravitational inhomogeneity upon the measured properties of the system.

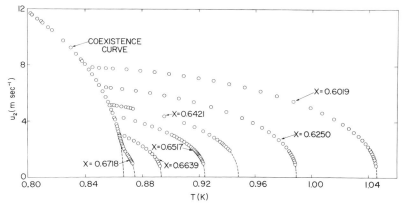

Fig. 1. The second-sound velocity u_2 along the coexistence curve and at six constant concentrations in the vicinity of the tricritical point as a function of the temperature.

For this reason, the height of the sample was kept to a minimum. Nonetheless, we were able to observe explicitly the effect of the gravitational field in the form of "rounding" of the velocity at the junctions of the phase separation curve and the lines of constant concentration. In the absence of a field the velocity, although continuous, would have a discontinuous derivative at these points. The rounding is shown in Fig. 2, which is an expanded plot of u_2 as a function of T. At an average concentration of 0.6718, corresponding to $1 - x_\sigma/x_t \cong 0.004$ and $\varepsilon_t \equiv 1 - T_\sigma/T_t \cong 0.0014$, the gravitational effect is quite pronounced and the measured velocity at T_σ can be estimated to be as much as 4/3 of the gravity free value. At $x = 0.6421$, corresponding to $1 - x_\sigma/x_t \cong 0.04$ and $\varepsilon_t \cong 0.012$, the gravitational rounding, although observable with high-resolution data, is small and affects the velocity by at most 1% or 2%. In analogy to a criterion suggested by Hohenberg and Barmatz[4] for liquid–gas critical points, one can estimate the range of $1 - x_\sigma/x_t$ over which gravitational effects are "important" by equating the gravitational chemical potential difference $(M_4 - M_3) gh$ with $(x_t - x_\sigma)(\partial \Phi/\partial x)_{PT}$ on the phase separation curve. This yields $(1 - x_\sigma/x_t) \cong 0.012h^{1/2} \cong 0.006$ for our sample (h is the height in cm). On the basis of a somewhat different criterion proposed by Berestov et al.,[5]* one obtains the remarkably similar value $1 - x_\sigma/x_t = 0.015h^{1/2}$ for this range. These estimates tend to be in agreement with our experimental result.

From the measurements it was possible to determine T_σ and T_λ with a precision of $2 \times 10^{-4} \,^\circ$K. We believe that this is the case even for T_σ at $x = 0.6718$, where the gravitational effect is strong, although one could argue for a slightly larger uncertainty. The transition temperatures are listed in Table I. For $x \gtrsim 0.64$ all T_σ can be fitted within $10^{-4}\,^\circ$K by a linear function of the concentration. The same is true for T_λ. The intersection of the two lines defines the tricritical temperature and concentration

$$T_t = 0.8671\,^\circ\text{K}, \qquad x_t = 0.6750 \tag{1}$$

* We used $m = M_4 - M_3$.

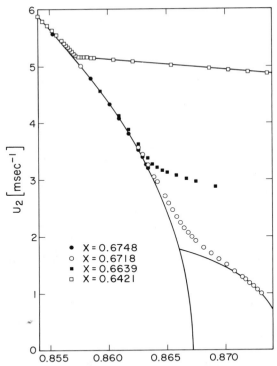

Fig. 2. The second-sound velocity u_2 in the immediate vicinity of the tricritical point as a function of the temperature. Note the "rounding" due to the gravitational field at the junction between the coexistence curve and the line with $x = 0.6718$.

Table I. The Phase Diagram

x	T_σ	T_λ
0.6718	0.8661	0.8749
0.6639	0.8637	0.8936
0.6517	0.8599	0.9245
0.6421	0.8570	0.9481
0.6250	0.8511	0.9895
0.6019	0.8418	1.0460

This phase diagram is in qualitative agreement with previous work,[6,7] but our slope $(1 - x_\sigma/x_t)/\varepsilon_t = 4.19$ of the phase separation curve for $x_\sigma < x_t$ is larger by about a factor of 1.3 than other estimates. This difference is probably attributable to the effect of gravity upon the previous measurements.

We used a relation[1] between u_2 and ρ_s which is given by linear two-fluid hydro-dynamics[8] and combined our velocities with other thermodynamic results[3,7] to obtain the ρ_s/ρ shown in Fig. 3 as a function of ε_t or $\varepsilon_\lambda \equiv 1 - T/T_\lambda$ on logarithmic

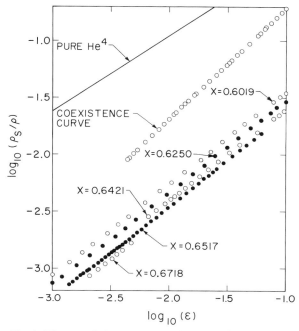

Fig. 3. The superfluid density as a function of ε_t along the coexistence curve and as a function of ε_λ along lines of constant concentration on logarithmic scales.

scales. For comparison, ρ_s/ρ of pure ^{4}He[9] is shown as a solid line. The striking feature of the present results at constant x is that ρ_s/ρ is $1\frac{1}{2}$ orders of magnitude smaller at the same ε_λ than it is in pure ^{4}He. Along the coexistence curve a least-squares fit of the results to a power law yields $\rho_s/\rho = 2.034\varepsilon_t^{1.00}$ with the least-squares-adjusted $T_t = 0.86720°$K. This value of T_t agrees well with Eq. (1), which is based on the phase diagram. At constant concentration ρ_s/ρ does not seem to follow a pure power law; but when singular correction terms to the asymptotic behavior are invoked the data are consistent with $\rho_s/\rho \sim \varepsilon_\lambda^{2/3}$. Higher-order singular contributions are needed also to fit ρ_s in pure ^{4}He under pressure.[9] These results for the mixtures are in agreement with recent theories of tricritical points[10-12] and are discussed in more detail elsewhere.[1]

It is interesting to note that the proportionality between the coherence length ξ and ρ_s^{-1},[13] together with our measurements, implies that the amplitude ξ_0 of ξ becomes extremely large near T_t. For $x = 0.6639$, for instance, one would estimate from the measured ρ_s/ρ that $\xi_0 \cong 10^{-6}$ cm. This yields $\xi \cong 10^{-4}$ cm and equal to the 1-μm hole size of the superleak transducers when $\varepsilon_\lambda \cong 10^{-3}$. Indeed, for smaller ε_λ we have been unable to generate second sound for x near x_t.

References

1. G. Ahlers and D.S. Greywall (to be published).
2. G. Ahlers, *Phys. Rev. Lett.* **24**, 1333 (1970).

3. G. Goellner and H. Meyer, *Phys. Rev. Lett.* **26**, 1534 (1971); and private communication.
4. P.C. Hohenberg and M. Barmatz, *Phys. Rev.* *A***6**, 289 (1972).
5. A.T. Berestov, A.V. Voronel', and M.Sh. Giterman, *Zh. Eksperim. i Teor. Fiz. Pisma* **15**, 273 (1972) [*Soviet Phys.—JETP Lett.* **15**, 190 (1972)].
6. E.H. Graf, D.M. Lee, and J.D. Reppy, *Phys. Rev. Lett.* **19**, 417 (1967).
7. S.T. Islander and W. Zimmermann, Jr., *Phys. Rev.* *A***7**, 188 (1973).
8. I.M. Khalatnikov, *Introduction to the Theory of Superfluidity,* Benjamin, New York (1965).
9. D.S. Greywall and G. Ahlers, *Phys. Rev. Lett.* **28**, 1251 (1972); and to be published.
10. R.B. Griffiths, *Phys. Rev. Lett.* **24**, 715 (1970).
11. E.K. Riedel, *Phys. Rev. Lett.* **28**, 675 (1972).
12. E.K. Riedel and J.F. Wegner, *Phys. Rev. Lett.* **29**, 349 (1972).
13. B.I. Halperin and P.C. Hohenberg, *Phys. Rev.* **177**, 952 (1969).

Excitation Spectrum of ^{3}He–^{4}He Mixture and Its Effect on Raman Scattering

T. Soda

Department of Physics, Tokyo University of Education
Tokyo, Japan

If we add a dilute solution of ^{3}He to liquid ^{4}He and consider the energy spectrum of the mixture, then the original phonon–roton spectrum and the ^{3}He energy spectrum cross each other in the energy spectrum of the ^{3}He–^{4}He mixture if there is no interaction between ^{3}He and ^{4}He. The existence of the interaction between them presents the interesting problem of the two-level crossing. The addition of ^{3}He to liquid ^{4}He gives a shift in the roton-phonon spectrum, but it does not change the second branch of the original ^{4}He excitation spectrum.

Our mixture system consists of bosons and fermions. The total Hamiltonian of the system is given by

$$H = H_B + H_F + H_{FB} \tag{1}$$

where H_B is the Hamiltonian of boson elementary excitations, H_F is the Hamiltonian of fermion elementary excitations, and H_{FB} is the interaction Hamiltonian of boson and fermion density fluctuations. They are given by

$$H_B = \sum_{\lambda,q} \omega_\lambda^q b_\lambda^{q*} b_\lambda^q \tag{2}$$

and

$$H_F = \sum_{\mu,q} \varepsilon_\mu^q a_\mu^{q*} a_\mu^q \tag{3}$$

$$H_{FB} = \sum_q v_q \rho_q^B \rho_{-q}^F \tag{4}$$

Here v_q is the boson–fermion interaction potential and ρ_q^B and ρ_q^F are the density fluctuations of bosons and fermions. We have the following property: $\rho_q = \rho_{-q}^*$. b_λ^{q*} and b_λ^q are the creation and annihilation operators of boson excitations, with λ indicating the different excitations, such as the first-branch phonon–roton excitation and the second-branch excitation. Similarly, a_μ^{q*} and a_μ^q are the operators for the fermion elementary excitations, such as the one particle–hole pair excitation, including a zero-sound excitation, and the two particle–hole pair excitation, etc. We can also express the density fluctuation of Eq. (4) in terms of these operators,

$$\rho_q^B = \sum_\lambda (B_\lambda^{q*} b_\lambda^{q*} + B_\lambda^{-q} b_\lambda^{-q}) \tag{5}$$

$$\rho_q^F = \sum_\mu (A_\mu^{q*} a_\mu^{q*} + A_\mu^{-q} a_\mu^{-q}) \tag{6}$$

Here A and B are the expansion coefficients of the ρ's in terms of these operators.

If we seek the energy eigenvalue E^q for the synthesized system of bosons and fermions by solving a secular equation for the Hamiltonian (1), we obtain the following equation:

$$g(E^{q2}) = 1 - v_q^2 \sum_\lambda \frac{2\omega_\lambda^q |B_\lambda^q|^2}{(E^q)^2 - (\omega_\lambda^q)^2} \sum_\mu \frac{2\varepsilon_\mu^q |A_\mu^q|^2}{(E^q)^2 - (\varepsilon_\mu^q)^2} = 0 \tag{7}$$

From the sum rules[1] we have

$$\sum_n \langle n|\rho_q^* \rho_q|n\rangle \, e^{-\beta E_n} / \sum_n e^{-\beta E_n} = NS_q \tag{8}$$

$$\sum_{n,m} |(\rho_q^*)_{nm}|^2 (E_n - E_0) \, e^{-\beta E_n} / \sum_n e^{-\beta E_n} = Nq^2/2m \tag{9}$$

and

$$\left. \sum_{n,m} \frac{|(\rho_q^*)_{nm}|^2}{E_n - E_m} e^{-\beta E_n} / \sum_n e^{-\beta E_n} \right|_{q \to 0} = \frac{N}{2ms^2} \tag{10}$$

where S_q is the form factor, N is the number of particles, m is mass, s is the isothermal sound velocity, and E_n and E_m are the energies of the nth and mth excited states. The coefficients A and B satisfy the following relations[2]:

$$\sum_\mu |A_\mu^q|^2 \coth(\beta \varepsilon_\mu^q/2) = N^F S_q^F \tag{11}$$

$$\sum_\lambda |B_\lambda^q|^2 \coth(\beta \omega_\lambda^q/2) = N^B S_q^B \tag{12}$$

$$\sum_\mu \varepsilon_\mu^q |A_\mu^q|^2 = N^F q^2/2m_F \tag{13}$$

$$\sum_\lambda \omega_\lambda^q |B_\lambda^q|^2 = N^B q^2/2m_B \tag{14}$$

$$\sum_\mu (|A_\mu^q|^2/\varepsilon_\mu^q)|_{q \to 0} = N^F/2m_F s_F^2 \tag{15}$$

$$\sum_\lambda (|B_\lambda^q|^2/\omega_\lambda^q)|_{q \to 0} = N^B/2m_B s_B^2 \tag{16}$$

where the subscripts F and B stand for fermion and boson, respectively.

The elementary excitations for fermions consist of one particle–hole pair, two particle–hole pairs, etc. However, for the Fermi liquid of high and intermediate density one particle–hole pair contributes predominantly to the density fluctuation, and we assume, therefore, that the index in Eq. (7) mainly consists of one-pair states. Namely, we assume that $\varepsilon_\mu^q = [(\mathbf{p} + \mathbf{q})^2/2m_F] - (q^2/2m_F)$, and $|A_p|^2 = 1$ for $T = 0$ and $|A_p|^2 = f_p$ for finite temperatures, where f_p is the distribution function of a Fermi particle with momentum p.

The second factor of the second term in Eq. (7) takes the following form:

$$N_F I(q, T) = \sum_p \frac{2[(\mathbf{p} \cdot \mathbf{q}/m_F) + (q^2/2m_F)]f_p}{(E^q)^2 - [(\mathbf{p} \cdot \mathbf{q}/m_F) + (q^2/2m_F)]} \tag{17}$$

This amounts to taking a random phase approximation for ^{3}He in the liquid.

For elementary excitations of bosons we consider that they have the two-branch spectrum, namely the first branch of a phonon–roton spectrum given by ω_1^q and the second branch given by ω_2^q. Then the first factor in the second term of Eq. (7) takes the following form,

$$\frac{2\omega_1^q}{(E^q)^2 - (\omega_1^q)^2}|B_1^q|^2 + \frac{2\omega_2^q}{(E^q)^2 - (\omega_2^q)^2}|B_2^q|^2 \tag{18}$$

and we express the quantities B_1^q and B_2^q in terms of experimental values of structure factor, ω_1^q and ω_2^q, from the sum rules (12) and (14). They are given by the following equations[3]:

$$\frac{|B_1^q|^2}{N_B} = \frac{\omega_2^q - \langle H \rangle_q}{\omega_2^q - \omega_1^q}S_q \tag{19}$$

$$\frac{|B_2^q|^2}{N_B} = \frac{\langle H \rangle_q - \omega_1^q}{\omega_2^q - \omega_1^q}S_q \tag{20}$$

Here $\langle H \rangle_q = q^2/2mS_q$ and we choose $\coth(\beta\omega_\lambda^q/2)$ in Eq. (12) as unity because we are mainly concerned with the high-energy portion of the spectrum for low temperatures. We can also choose the magnitude of v_q from the stability condition of the ^{3}He–^{4}He system that the eigenvalue E^q becomes imaginary beyond a certain critical fermion density N_c^F, say 6.4% ^{3}He concentration at absolute zero. This is given by setting E^q equal to zero in Eq. (7) and using the sum rules (15) and (16):

$$v_q^2 = m_B m_F (s_F^c)^2\, s_B^2 / N_c^F N^B \tag{21}$$

where s_F^c is the fermion isothermal sound velocity for the critical density N_c^F. Now Eq. (7) becomes

$$1 - \frac{m_B m_F (s_F^c)^2\, s_B^2}{N^B N_c^F}\sum_p \frac{2[(\mathbf{p}\cdot\mathbf{q}/m_F) + (q^2/2m_F)]f_p}{(E^q)^2 - [(\mathbf{p}\cdot\mathbf{q}/m_F) + (q^2/2m_F)]^2}$$

$$\left\{ \frac{\omega_1^q}{(E^q)^2 - (\omega_1^q)^2}\frac{N^B}{m_B}\frac{\omega_2^q - \langle H \rangle_q}{\omega_2^q - \omega_1^q}\frac{q^2}{\langle H \rangle_q} \right.$$

$$\left. + \frac{\omega_2^q}{(E^q)^2 - (\omega_2^q)^2}\frac{N^B}{m_B}\frac{\langle H \rangle_q - \omega_1^q}{\omega_2^q - \omega_1^q}\frac{q^2}{\langle H \rangle_q} \right\} = 0 \tag{22}$$

We want to calculate the deviation of the first-branch spectrum in a dilute solution of ^{3}He in ^{4}He. We call the deviation Δ_1^q and set

$$E^q = \omega_1^q + \Delta_1^q \tag{23}$$

Then we assume that Δ_1^q is small because the concentration of ^{3}He is small. Then Eq. (22) takes the following form:

$$1 - m_F (s_F^c)^2 (s_B q)^2 (N_c^F)^{-1} \sum_p \frac{2[(\mathbf{p} \cdot \mathbf{q}/m_F) + (q^2/2m_F)] f_p}{(\omega_1^q)^2 - [(\mathbf{p} \cdot \mathbf{q}/m_F) + (q^2/2m_F)]^2}$$

$$\times \left[\frac{1}{2\Delta_1^q} \frac{\omega_2^q - \langle H \rangle_q}{\omega_2^q - \omega_1^q} \frac{1}{\langle H \rangle_q} + \frac{\omega_2^q}{(\omega_1^q)^2 - (\omega_2^q)^2} \frac{\langle H \rangle_q - \omega_1^q}{\omega_2^q - \omega_1^q} \frac{1}{\langle H \rangle_q} \right] = 0 \qquad (24)$$

We first evaluate the function $I(q, T)$ given by

$$N^F I(q, T) = \sum_p \frac{1}{\exp \{ \beta [(p^2/2m_F) - \varepsilon_F] \} + 1}$$

$$\times \frac{2[(\mathbf{p} \cdot \mathbf{q}/m_F) + (q^2/2m_F)]}{(\omega_1^q)^2 - [(\mathbf{p} \cdot \mathbf{q}/m_F) + (q^2/2m_F)]^2} \qquad (25)$$

for temperatures from zero to $kT \lesssim \varepsilon_F = p_F^2/2m$. If we set $\varepsilon_p = p^2/2m_F$ and

$$\frac{dF(\varepsilon_p, q)}{d\varepsilon_p} = 2\pi \frac{m_F^2}{q} \ln \frac{\omega + [(2m_F \varepsilon_p)^{1/2} q/m_F] - (q^2/2m_F)}{\omega - [(2m_F \varepsilon_p)^{1/2} q/m_F] - (q^2/2m_F)}$$

$$\times \frac{\omega - [(2m_F \varepsilon_p)^{1/2} q/m_F] + (q^2/2m_F)}{\omega + [(2m_F \varepsilon_p)^{1/2} q/m_F] + (q^2/2m_F)} \qquad (26)$$

then we can utilize the following Sommerfeld expansion to evaluate Eq. (25):

$$N^F I(q, T) = F(\varepsilon_F, q) + \frac{(\pi T)^2}{6} \frac{d^2 F(\varepsilon_F, q)}{d\varepsilon_p^2} + \frac{7(\pi T)^4}{360} \frac{d^4 F(\varepsilon_F, q)}{d\varepsilon_p^4}$$

$$+ \frac{31(\pi T)^6}{15,120} \frac{d^6 F(\varepsilon_F, q)}{d\varepsilon_p^6} + \cdots \qquad (27)$$

For $kT > \varepsilon_F$ we can approximate $I(q, T)$ in Eq. (25) as follows:

$$I(q, T) \simeq N^{F-1} \sum e^{-p^2/2m_F kT} \left(\frac{1}{\omega - (\mathbf{p} \cdot \mathbf{q}/m_F + q^2/2m_F)} - \frac{1}{\omega + (\mathbf{p} \cdot \mathbf{q}/m_F + q^2/2m_F)} \right) \qquad (28)$$

The shift Δ_1^q is given by

$$\Delta_1^q = \left[\frac{1}{2} m_F (s_F^c)^2 (s_B q)^2 \frac{N_F}{N_c^F} I(q, T) \frac{\omega_2^q - \langle H \rangle_q}{\omega_2^q - \omega_1^q} \frac{1}{\langle H \rangle_q} \right]$$

$$\times \left[1 + m_F (s_F^c)^2 (s_B q)^2 \frac{N^F}{N_c^F} I(q, T) \frac{\langle H \rangle_q - \omega_1^q}{(\omega_2^q - \omega_1^q)} \frac{\omega_2^q}{(\omega_2^q)^2 - (\omega_1^q)^2} \frac{1}{\langle H \rangle_q} \right]^{-1} \qquad (29)$$

We have evaluated the shift due to 6.4% ^{3}He concentration ($\varepsilon_F = 0.8°$K) for temperature $T = 0$, 0.4, and 0.8°K and for a temperature higher than $\varepsilon_F(2°\text{K})$ as examples, assuming that the two-branch spectrum of pure liquid ⁴He at 1.1°K can be used for all the above temperatures and for the liquid of 93.6 % density. The results are shown in Fig. 1.

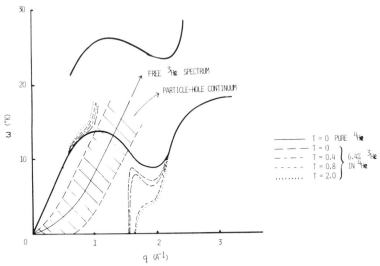

Fig. 1. The roton–phonon spectrum of 6.4% ³He solution in liquid ⁴He in the first approximation to the energy shift for $T = 0$, 0.4, 0.8, and 2.0 K, assuming that the two-branch spectrum of pure liquid ⁴He at 1.1 K can be used for all temperatures and for the liquid of 93.6 % density.

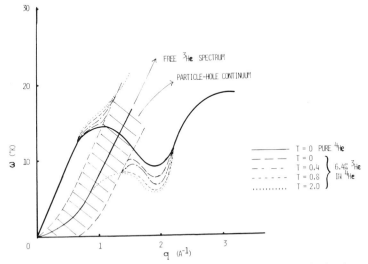

Fig. 2. The roton–phonon spectrum, taking into account both the lowering of the density and a renormalization effect for 6.4% ³He solution in liquid ⁴He.

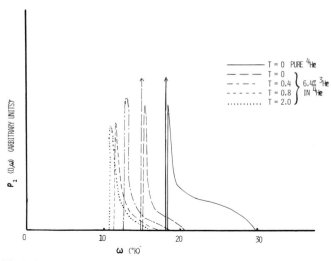

Fig. 3. Schematic picture of the density of two-excitation states of total momentum $K = 0$, $\rho_2(K = 0, \omega)$, for the Raman scattering of a mixture of ^{3}He–^{4}He, taking into account the roton–phonon sepctrum shift and assuming that the light interacts with the bound state of two (synthesized) elementary excitations of mixture if it were formed.

We also take into account the effect of the lowering of the density of liquid ^{4}He. The pure roton energy Δ_0 is increased according to the following empirical formula[4] around the roton minimum:

$$\Delta = \Delta_0[1 - 0.9(\Delta\rho/\rho_0)] \tag{30}$$

Along the phonon part of the spectrum the sound velocity is increased by

$$s_B(\rho) = s_B(\rho_0)[1 - 0.5(\Delta\rho/\rho_0)] \tag{31}$$

where ρ_0 is the density of the pure liquid. We replot the original phonon–roton spectrum by considering Eqs. (30) and (31).

The result of Fig. 1 is only a first approximation for small Δ_1^q. We can go on to the second approximation Δ_2^q by taking the effect of Δ_1^q on the ω_1^q. This is a kind of renormalization procedure. We plot the shift of the phonon–roton spectrum taking into account both effects in Fig. 2. The values for the shift for 6.4% ^{3}He concentration are in good accord with the experimental result of Sobolev and Esel'son,[5] who obtained the averaged value for the shift between T_λ and 0.4°K.

Since the phonon–roton spectrum is separated into two parts having maxima and minima (shown in Figs. 1 and 2), the intensity of Raman scattering is expected to show shifts in location of the peaks (qualitatively shown in Fig. 3), if we assume the light couples with the synthesized density fluctuations of the mixture. But if the light interacts selectively only (in the D state) with the ^{4}He density component of the liquid, the Raman scattering does not show any significant shift in the location of the peak but only shows the increase in the width due to opening of more decay channels, which has been the case so far in the experiment of Surko et al.[6] and Greytak et al.[7]

Acknowledgment

The author acknowledges very stimulating discussions with Professor K. Sawada.

References

1. P. Nozières and D. Pines, *The Theory of Quantum Liquids I,* Benjamin, New York (1966); T. Nagata, T. Soda, and K. Sawada, *Prog. Theor. Phys.* **38**, 1023 (1967).
2. T. Soda, K. Sawada, and T. Nagata, *Prog. Theor. Phys.* **44**, 574 (1970).
3. T. Soda, K. Sawada, and T. Nagata, *Prog. Theor. Phys.* **44**, 860 (1970).
4. J. Wilks, *Liquid and Solid Helium,* Clarendon Press, Oxford (1967), p. 121.
5. V.I. Sobolev and B.N. Esel'son, *Soviet Phys.—JETP* **33**, 132 (1971).
6. C.M. Surko and R.E. Slusher, *Phys. Rev. Lett.* **30**, 1111 (1973).
7. R.L. Woerner, D.A. Rockwell, and T.J. Greytak, *Phys. Rev. Lett.* **30**, 1114 (1973).

The Low-Temperature Specific Heat of the Dilute Solutions of ^{3}He in Superfluid ^{4}He

H. Brucker and Y. Disatnik

Department of Physics and Astronomy, Tel-Aviv University
Tel-Aviv, Israel

The low-temperature specific heat of the dilute superfluid solution of ^{3}He in ^{4}He was measured several years ago by Anderson *et al.*[1] In order to analyze their data, these investigators used the Stoner expression[2] for the specific heat of an ideal Fermi gas. In this work we reanalyze the same data with an improved expression based on the low-temperature equilibrium theory described by us in a recently published study.[3]

The energy density of the dilute superfluid ^{3}He–^{4}He solution is given, within the framework of our equilibrium theory, by the following expression:

$$\varepsilon = \varepsilon_f - \tfrac{5}{2} z_f + \tfrac{3}{2} \mu_f^{\gamma} n_3 \tag{1}$$

The notations used here are the same as those used in Ref. 3. Here ε_f is the energy density of an ideal Fermi gas of number density n_3, temperature T, and fermion mass m^*. The density n_3 is taken to be equal to the number density of the ^{3}He component of the solution, T is the temperature of this system, and m^* is an effective mass given, to first order in the ^{3}He concentration x, by

$$m^* = m_0(1 - 2m_0 \bar{v}_1 n_{40} x + \ldots) \tag{2}$$

where m_0 is the effective mass of a single ^{3}He impurity in a bath of "pure" superfluid ^{4}He system of pressure equal to that of the solution (P); n_{40} is the number density of the same "pure" ^{4}He system, and $\bar{v}_1$ is the "renormalized" expansion coefficient associated with the second-order term in the momentum expansion of the local part of the effective ^{3}He quasiparticle interaction. It should be noted that to second order in the ^{3}He concentration we have

$$n_3 = n_{40} x (1 - \alpha_0 n_{40} x + \cdots) \tag{3}$$

where α_0 is the excess molar volume of a single ^{3}He impurity in the "pure" superfluid system mentioned earlier. Other notations which are used in Eq. (1) are

$$z_f = -\frac{3}{10} \frac{1}{K_B \theta_0} \int d\Omega_p n(p) \left(\frac{p^2}{2m^*} \right)^2 \tag{4}$$

$$\mu_f^{\gamma} = -\frac{5}{6} \frac{\varepsilon_f}{n_3^2 K_B \theta_0} \left(\varepsilon_f - \frac{2}{5} c_f T \right) \tag{5}$$

where K_B is the Boltzmann constant and c_f and $n(p)$ are the fixed-volume specific heat (per unit volume) and the momentum distribution in the ideal Fermi gas defined above. The temperature θ_0 is a characteristic temperature which measures the strength of the p^4 term in the momentum expansion of the ^{3}He quasiparticle energy.[3,4]

The fixed-volume specific heat (per unit volume) c_v is obtained from Eq. (1) in the usual way:

$$c_v = \left(\frac{\partial \varepsilon}{\partial T}\right)_V = c_f - \frac{5}{2}\left(\frac{\partial z_f}{\partial T}\right)_V + \frac{3}{2}n_3\left(\frac{\partial \mu_f^v}{\partial T}\right)_V \tag{6}$$

where V is the volume occupied by the solution. It must be noted that in their work Anderson et al. measured the specific heat at the vapor pressure.[1] Thus it seems useful to get an expression for the fixed-pressure specific heat. In the following we shall denote the fixed-pressure specific heat per unit volume by c_p. In order to obtain an expression for this quantity, we note that

$$c_p - c_v = TK_T\left(\frac{\partial P}{\partial T}\right)_V^2 \cong \left(\frac{2}{3} - w_{40}\right)^2 \frac{c_f T}{n_{40}m_4 S_0^2} c_f \tag{7}$$

where K_T is the isothermal compressibility of the solution, m_4 is the bare ^{4}He mass, and S_0 is the sound velocity of pure superfluid ^{4}He at pressure P and zero temperature. In addition, we have $w_{40} \equiv (n_{40}/m_0)\partial m_0/\partial n_{40}$. The first part of Eq. (7) is a well-known thermodynamic relation.[5] To get the last part of this equation we again apply the equilibrium theory of Ref. 3. A similar result was obtained by Ebner and Edwards.[6] Using the expressions (4)–(7), we can write c_p as follows:

$$c_p = c_{f0}(1 + \delta) \tag{8}$$

$$\delta = \frac{3}{4}\frac{\varepsilon_{f0}}{n_{40}K_B\theta_0 x}\left(-1 + \frac{2}{3}\frac{c_{f0}T}{\varepsilon_{f0}} + \frac{2}{3}\frac{T}{c_{f0}}\frac{\partial c_{f0}}{\partial T} - \frac{25}{6}\frac{\varepsilon_{f0}}{c_{f0}T} - \frac{35}{3}\frac{n_{40}K_B\theta_0 z_f x}{\varepsilon_{f0}c_{f0}T}\right)$$

$$- \left[\alpha_0 + \left(2m_0\bar{v}_1 n_{40} - \frac{2}{3}\alpha_0\right)\frac{T}{c_{f0}}\frac{\partial c_{f0}}{\partial T}\right]x + \left(\frac{2}{3} - w_{40}\right)^2 \frac{c_{f0}T}{m_4 n_{40}S_0^2} + \cdots \tag{9}$$

In deriving the above expressions, we expand the ideal gas specific heat c_f in the small quantities $m^* - m_0$ [Eq. (2)] and $n_3 - n_{40}x$ [Eq. (3)] around its value when these quantities are set equal to zero. In addition, we introduce the notations c_{f0} and ε_{f0}, respectively, for the corresponding zeroth-order values of c_f and ε_f. The quantity δ measures the deviation of c_p from the "Stoner"-like contribution c_{f0}. It can be shown, using the numerical results described below, that δ is roughly of the order of the ^{3}He concentration. Therefore it indeed has to be taken into account promptly when the concentration variation of the low-temperature specific heat is considered. We may add here that the various terms in Eq. (9) for δ come from the following sources: (a) the p^4 term in the expression for the ^{4}He quasiparticle energy used in Ref. 3; (b) the $c_p - c_v$ difference; (c) the effective ^{3}He quasiparticle interaction; and (d) the nonzero excess molar volume of the ^{3}He quasiparticle in the solution. In practice, only some of the terms in Eq. (9) can be accounted for within the framework of the simple Stoner theory. Thus, as we emphasized in Ref. 3, it appears that applications similar to the one made by Anderson et al. of Stoner's expression for the specific heat in studies of the equilibrium properties of the solutions must be regarded with care.

In this work we fitted Eq. (8) to the experimental data of Anderson *et al.* using the "χ-square" test. In general, we obtained a good agreement between theory and experiment. The standard deviation in the fit was about 2%. In the fit we used[7] $n_{40} = 2.18 \times 10^{22}$ cm^{-3}, $S_0 = 238$ msec^{-1}, and $\alpha_0 = 0.284$. In addition, we had[8] $w_{40} = 1.21$. The zero-pressure values of the parameters m_0, θ_0, and $\bar{v}_1$ were determined in the fit:

$$m_0 = 2.38m_3; \qquad \theta_0 = 22°K; \qquad \bar{v}_1 = -3.44°K \qquad (10)$$

where m_3 is the bare ^{3}He mass. To obtain the value of $\bar{v}_1$ in degrees, we multiplied it by the factor $m_4^2 S_0^2 n_{40}/k_B$. It must be noted that c_p depends on θ_0 and $\bar{v}_1$ only through δ. Hence these parameters can affect only a small fraction of the specific heat. Since this fraction is typically smaller than or, at best, of the order of the standard deviation in the fit, we can expect the fitted values of θ_0 and $\bar{v}_1$ [Eq. (10)] to be highly uncertain. The situation is different in the case of the mass m_0. This mass affects c_{f0}, the leading term in expression (8). Hence the error in its fitted value turns out to be only of the order of the standard deviation of the fit.

Using the Stoner expression, Anderson *et al.* have managed to get from their experimental data values for the specific heat effective mass of the solution m^{sh} at two concentrations $x = 1.3\%$ and $x = 5.0\%$. The values they obtain are

$$m^{sh} = 2.38m_3 \quad \text{at} \quad x = 1.3\%; \qquad m^{sh} = 2.46m_3 \quad \text{at} \quad x = 5.0\% \qquad (11)$$

As was mentioned in Ref. 3, m^{sh} is given within the framework of our equilibrium theory by

$$m^{sh} = m_0 \left[1 + \tfrac{3}{5}(T_F/\theta_0) - 2m_0 n_{40}\bar{v}_1 x + \cdots \right] \qquad (12)$$

where T_F is the Fermi temperature of the ^{3}He quasiparticle gas. In order to calculate m^{sh} from this expression for the concentrations mentioned above, we note[1] that:

$$T_F \simeq 0.14°K \quad \text{at} \quad x = 1.3\%; \qquad T_F \simeq 0.33°K \quad \text{at} \quad x = 5.0\% \qquad (13)$$

Hence substituting the values we got for m_0, θ_0, and $\bar{v}_1$ [Eq. (10)] in Eq. (12), we get

$$m^{sh} = 2.40m_3 \quad \text{at} \quad x = 1.3\%; \qquad m^{sh} = 2.45m_3 \quad \text{at} \quad x = 5.0\% \qquad (14)$$

It must be noted that because of the fact that the theory underlying our analysis is not equivalent to the Stoner theory, the above values of m^{sh} differ slightly from those of Anderson *et al.* [Eq. (11)].

To conclude, we should mention that by using the equilibrium theory we applied in this work one can also determine the effective mass m_0 from the velocity of second sound. We have carried out a calculation of this type based on the second-sound data of Brubaker *et al.*,[4] and the results will be described in detail elsewhere.[9] Here we only quote the value we obtained for m_0: $m_0 = 2.26m_3$. The error in this result seems to be less than 1%. Clearly, this value is lower by about 5% than the value obtained for m_0 in the present work ($m_0 = 2.38m_3$). As we mentioned before, the error in the latter is only of the order of 2%. Thus it seems that the difference between the two results is indeed significant and that the second-sound data of Brubaker *et al.* and the specific heat data of Anderson *et al.* actually are inconsistent with one another.

References

1. A.C. Anderson, D.O. Edwards, W.R. Roach, R.E. Sarwinski, and J.C. Wheatley, *Phys. Rev. Lett.* **17**, 367 (1966).
2. E.C. Stoner, *Phil. Mag.* **25**, 899 (1938).
3. Y. Disatnik and H. Brucker, *J. Low Temp. Phys.* **7**, 491 (1972).
4. N.R. Brubaker, D.O. Edwards, R.E. Sarwinski, P. Seligman, and R.E. Sherlock, *J. Low Temp. Phys.* **3**, 619 (1970).
5. L.D. Landau and E.M. Lifshitz, *Statistical Physics,* Pergamon Press, London (1963), p. 51.
6. C. Ebner and D.O. Edwards, *Phys. Reports* **2C**, 77 (1971).
7. G.E. Watson, J.D. Reppy, and R.C. Richardson, *Phys. Rev.* **188**, 384 (1968).
8. D.O. Edwards, E.M. Ifft, and R.E. Sarwinski, *Phys. Rev.* **177**, 380 (1969).
9. H. Brucker and Y. Disatnik (to be published).

Spin Diffusion of Dilute ³He–⁴He Solutions under Pressure*

D. K. Cheng,† P. P. Craig‡

State University of New York, Stony Brook, New York
and
Brookhaven National Laboratory, Upton, New York

and

T. A. Kitchens‡

Brookhaven National Laboratory, Upton, New York

The transport properties of dilute solutions of ³He in ⁴He provide information about the effective interaction between the ³He quasiparticles. Of the transport properties, the spin diffusion coefficient, which can be measured by NMR spin-echo techniques, is most sensitive to these interactions. Here we report such measurements on 4.97% and 1.15% ³He in ⁴He for pressures of 0.04, 9.2, and 20 atm at temperatures between 1.5 and 0.072°K. The effective interaction potentials at the three pressures deduced from these data are presented. From these potentials it appears that the temperature of the expected superfluid transition is highest for the lowest pressure and the smaller concentration studied.

Most of the properties of the dilute mixtures of ³He–⁴He at low temperatures are characterized by the effective interaction among ³He atoms. In order to obtain a fuller knowledge of this interaction, we have made measurements of the spin diffusion coefficient of 4.97% and 1.15% solutions of ³He in ⁴He at 0.04, 9.2, and 20.0 atm between 1.5 and 0.072°K. Spin diffusion is selected because, first, its mean free time relates directly to the interaction; second, because the quantum correction to its classical expression varies as $(T/T_F)^3$ while that for viscosity and thermal conductivity[1] varies as $(T/T_F)^{3/2}$; and third, because there is no direct contribution from the ⁴He background.

Figure 1 is a schematic representation of the low-temperature segment of the cryostat. The refrigeration was provided by a dilution refrigerator which utilizes a capillaries-in-an-annular-space type of countercurrent heat exchanger. The mixing chamber and the sample cell were fabricated from Epibond 100 A. A needle valve was used to seal off the sample mixture at low temperatures and the pressure was generated by compressing a bellows with ⁴He liquid diluted slightly with ³He to

* Work performed under the auspices of the U.S. Atomic Energy Commission.
† Present address: Department of Physics, University of Oregon, Eugene, Oregon.
‡ Present address: National Science Foundation, Washington, D.C.

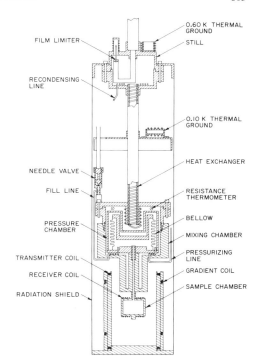

Fig. 1. Low-temperature segment of the cryostat
and sample cell.

reduce its thermal conductivity. The magnetic field was produced by a sixth-order superconducting solenoid. The pulse burst generator which supplied the 90°-180°-180° pulse train at a resonance frequency of 301.1 kHz is described elsewhere.[2] The echo signals were averaged to increase the signal-to-noise ratio with a fast digitizer and a 400-channel analyzer. Temperatures were measured with carbon resistors, which were calibrated against a powdered CMN salt pill both before and after the measurements.

In order to utilize most of the information contained in the echo, the amplitude of the echo was deduced by fitting the entire echo, obtained by averaging several scans at each temperature, to the expected form[3]

$$h_i(t) = h_{0i}\{\sin[A(t - t_0)]\}/A(t - t_0) \tag{1}$$

where h_{0i} is the desired amplitude. The parameter A is determined by the half-height of the sample cell and the total applied magnetic field gradient G, and t_0 is the time at which the maximum amplitude occurs. The diffusion coefficient D was then calculated from the expression[4]

$$D = \frac{12}{\gamma^2 \tau^3 G^2} \ln \frac{h_{02}}{h_{01}} \tag{2}$$

where γ is the gyromagnetic ratio of ³He and τ is the time interval between the two echoes. D was determined from echoes at each temperature with the applied gradient parallel and antiparallel to the uniform magnetic field. Averaging these values reduces the effect of any extraneous gradient field to second order.

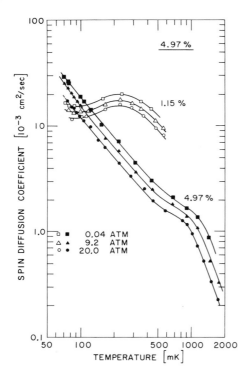

Fig. 2. Experimental results for the spin diffusion coefficient as a function of temperature for 1.15% and 4.97% ^{3}He in ^{4}He.

The results of the measurements are displayed in Fig. 2. The overall uncertainty is about 10–11%, most of which comes from the uncertainty of the baseline and the external gradient value.

In the interpretation of data, we shall rely heavily on the work of Emery, and the equations hereafter can be found in Ref. 1. Starting from a phenomenological Hamiltonian for the dilute mixture, Emery shows that the ^{3}He and ^{4}He subsystems can be essentially decoupled by two canonical transformations. The net result is that the ^{3}He acquire an effective mass from dragging the ^{4}He as they move about and interact with an effective potential resulting from the interaction of the "polarization" clouds around each ^{3}He atom. At low enough temperatures where the excitations of the ^{4}He can be neglected, the effective Hamiltonian of the system becomes

$$H' = \sum_{i=1}^{N_3} \frac{p_i^2}{2m_0^*} + \frac{1}{2} \sum_{i,j}^{N_3} v(r_{ij})$$

where m_0^* is the effective mass of a ^{3}He atom and r_{ij} is the separation of a pair of ^{3}He atoms. Since the ^{3}He quasiparticles form a dilute system, the two-body approximation is used throughout the theory.

The behavior of the spin diffusion as a function of temperature can be divided into three regions. In the semiclassical region ($T \gg T_F$ where T_F is the Fermi temperature) the Boltzmann equation for spin diffusion reduces to the diffusion equation for a binary mixture of gases. The spin diffusion coefficient D is given by

$$\frac{Dx}{T^{3/2}} = \frac{3k_B^{3/2}}{32h^2 N_0 \pi^{1/2}} v(m_0^*)^{1/2}$$

$$\times \left[\int_0^\infty d\gamma (\exp - \gamma^2) \gamma^5 \sum_l (l+1) \sin^2(\delta_{l+1} - \delta_l) \right]^{-1} \tag{3}$$

where x is the concentration, N_0 is Avogadro's number, v is the molar volume, m_0^* is the effective mass of ^{3}He, δ_l is the phase shift for the scattering at the appropriate relative kinetic energy $\gamma^2 k_B T$, and l is the angular momentum.

We note that no expression for the spin diffusion coefficient in the intermediate region ($T \simeq T_F$) has yet been derived. Consequently, no analysis of the present data in that region will be attempted here.

In the quantum mechanical region ($T \ll T_F$) the coefficient can be calculated within the framework of Landau Fermi liquid theory. The resultant expression is

$$D = \tfrac{1}{3}(\chi_0/\chi) v_F^2 \tau \tag{4}$$

where

$$\frac{\chi}{\chi_0} = 1 + \left(\frac{m^*}{h^2 p_F \pi^2} \right) \int_0^{p_F} dp \, p \left[A_E(p, 0) - A_0(p, 0) \right]$$

$$\tau = \frac{\pi^2}{12} \tau_{\vartheta p} + \frac{3}{4}(\tau_A - \tau_{\vartheta p})$$

$$\frac{h}{\tau_A} = \frac{k_B T^2}{T_F} \left(\frac{m^* p_F}{4\pi h^3} \right)^2 \int_0^{2\pi} d\varphi (1 - \cos \varphi) \int_0^{p_F} \frac{dp}{p_F} \frac{(p/p_F)^3}{(1 - p^2/p_F^2)^{1/2}} |A_E + A_0|^2$$

$$\frac{h}{\tau_{\vartheta p}} = \frac{k_B T^2}{2T_F} \left(\frac{m^* p_F}{4\pi h^3} \right)^2 \int_0^{2\pi} d\varphi \int_0^{p_F} \frac{dp}{p_F} \frac{p/p_F}{(1 - p^2/p_F^2)^{1/2}} [3|A_0|^2 + |A_E|^2]$$

and

$$A_E(p, \varphi) = -\frac{4\pi h^2}{m^*} \sum_{(l \text{ even})} (2l + 1) p_l(\cos \varphi) \frac{\tan \delta_l(p)}{p}$$

$A_0(p, \varphi)$ is given by a similar expression with the sum carried over odd values of l. From the above equations it is seen that in the quantum region DT^2 will be temperature independent. This behavior is indeed observed for a 5% solution when $T \lesssim 0.04°$K. For the present data it is necessary to extrapolate into this region. Because of the symmetry between particles and holes, τ is an even function of T. Therefore, accounting for first-order analytic corrections, we have

$$DT^2 = \alpha + \beta T^2$$

This relationship is well obeyed for a mixture with $T/T_F \lesssim 1/3$.[5]

Our analysis proceeds as follows. The 4.97% data are extrapolated from $T \sim 0.1°$K on down to obtain α for the three pressures. Following Emery, we assume the effective potential to be an attractive square well of depth V_0 and range b with a repulsive hard core of radius r_c. In addition, r_c is taken to be 1.8 Å, as determined

Table I. Parameters Deduced from Spin Diffusion Measurements

Pressure, atm	α	β	V_0, K	b, Å	r_c, Å
0.04	115.1 ± 10	6.64 ± 0.7	19.70	0.83	1.8
9.2	97.0 ± 9	5.61 ± 0.6	24.08	0.72	1.8
20.0	67.1 ± 6	5.55 ± 0.5	27.62	0.66	1.8

by Emery for all pressures. Since the core is highly repulsive, it is improbable that the pressure can affect it significantly. Assuming a particular value for b, a curve of DT^2 as a function of V_0 is generated. A range of acceptable values of V_0 can be determined from this curve using the intercept value of DT^2, α. With this value of b and the so determined values of V_0 the semiclassical expression is used to plot a set of curves of $Dx/T^{3/2}$. For the effective mass at different pressures we have used the slope derived from the second-sound experiment,[6] but chose to normalize the zero-pressure value to $2.34m_3$.[7] The molar volume of mixtures is taken from the measurements of Watson et al.[8] The high temperature data are then used to determine the best set of b and V_0. In actual computation, only s- and p-state phase shifts have been included in the degenerate region, since the d-state impact parameter lies outside the range of force. In the semiclassical region the first five phase shifts are used.

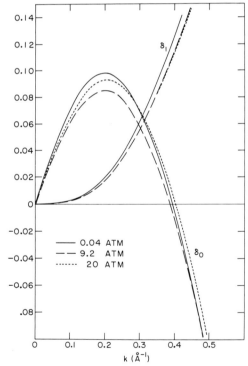

Fig. 3. The s- and p-state phase shifts for the best effective potential determined from spin diffusion data for 0.04, 9.2, and 20 atm.

Also in this region the low concentration ($x < 1\%$) data of Biegelsen *et al.*[9] have proved to be helpful in selecting b and V_0 since they have a smaller quantum correction. The results of the analysis are tabulated in Table I.

As the pressure is increased the effective potential appears to become deeper and to have a shorter range when the core is kept fixed. This trend is qualitatively expected, since an increase in pressure will lead to a decrease of atomic spacing and an accompanying increase in the potential energy. However, from 0.04 to 9.2 atm the change in atomic spacing is only 3%, while the corresponding change is 10% in range and 20% in depth. We believe this is due to the simplicity of the assumed shape of the potential and, as pointed out by Emery, it is more realistic to have a long-range repulsion and an intermediate attractive region. The s- and p-state phase shifts for the three pressures are shown in Fig. 3. They are within about 10% of each other, although DT^2 differs more than 40%. The reason is that DT^2 is proportional to T_F^2/m_0^*, which accounts for most of the pressure dependence.

Since the transition temperature of the ^{3}He subsystem into the superfluid state is the highest when $\delta_l(p_F)$ is at its maximum value, our analysis shows that the more dilute mixture of 1.24% at ^{3}He at vapor pressure has the highest transition temperature of 10^{-8} °K. Under elevated pressure the transition temperature is suppressed.

Acknowledgments

We would like to acknowledge both the encouragement and the assistance given us on this work by Dr. V. J. Emery and Professor J. I. Budnick. The technical assistance given by Mr. T. Oversluizen and also Mr. V. Radeka was essential for the successful completion of this research, and we gratefully thank them for providing it.

References

1. V.J. Emery, *Phys. Rev.* **161**, 194 (1967).
2. V. Radeka, R.L. Chase, M. Petrinovic, and J.A. Glasel, *Rev. Sci. Instr.* **41**, 1766 (1970).
3. H.Y. Carr and E.M. Purcell, *Phys. Rev.* **94**, 630 (1954).
4. H. Hart, Thesis, Univ. of Illinois, 1960.
5. A.C. Anderson, D.O. Edwards, W.R. Roach, R.E. Sarwinski, and J.C. Wheatley, *Phys. Rev. Lett.* **17**, 367 (1966).
6. D.J. Sandiford and H.A. Fairbank, *Phys. Rev.* **162**, 192 (1967); P.V.E. McClintock, K.H. Mueller, R.A. Guyer, and H.A. Fairbank, *Proc. 11th Intern. Conf. Low Temp. Phys., 1968*, St. Andrews Univ. Press, Scotland (1969), Vol. 1, p. 379.
7. W.R. Roach, Thesis, Univ. of Illinois, 1966, unpublished.
8. G.E. Watson, J.D. Reppy, and R.C. Richardson, *Phys. Rev.* **188**, 384 (1969).
9. D.K. Biegelsen and K. Luszczynski, Preprint, Washington Univ., St. Louis, Missouri.

The Spin Diffusion Coefficient of ^{3}He in ^{3}He–^{4}He Solutions

D.C. Chang and H.E. Rorschach

Department of Physics, Rice University
Houston, Texas

Introduction

Nearly all of the previous work on the diffusion of ^{3}He in superfluid ^{3}He–^{4}He solutions has been done on solutions whose molar ^{3}He concentration χ was 5% or less. Husa *et al.*[1] reported measurements for $\chi = 12\%$ for $T < 1.2°$K, and Horvitz and Rorschach[2] reported some indirect estimates for $\chi = 14\%$. This paper reports measurements of the diffusion coefficient by the spin-echo method for solutions with $\chi = 5\%$, 9%, 14%, and 24% for $0.9 < T < 2.5°$K.†

Measurements of the diffusion coefficient by the spin-echo method are a determination of the spin diffusion coefficient D_s. The distinction between spin diffusion and particle diffusion has been carefully elaborated by Fukuda and Kubo.[4] A simplified treatment of the diffusion process based on the Boltzmann equation and the Pomeranchuk model[5] for the solution has been made.[3] In this model the superfluid ^{3}He–^{4}He solution is assumed to consist of a mixture of a gas of ^{3}He atoms (fermions) and a gas of thermal excitations (bosons). D_s depends on the fermion–fermion scattering cross section σ_{FF} as well as on the fermion–boson cross section σ_{FB}. The particle diffusion depends only on σ_{FB}, since the fermion–fermion collisions do not alter the particle current if momentum is conserved. D_s can be expressed in the form[3]

$$\frac{1}{D_s} = \frac{\rho_B}{\rho} \frac{1}{\langle v_{xF}^2 \rangle \tau_{FB}} + \frac{\rho_F}{\rho} \frac{1}{\langle v_{xF}^2 \rangle \tau_{FF}} \tag{1}$$

where ρ_B is the boson mass density; ρ_F is the fermion mass density; $\rho = \rho_F + \rho_B =$ normal fluid density; τ_{FF} and τ_{FB} are the relaxation times for the fermion–fermion and fermion–boson interactions; and $\langle v_{xF}^2 \rangle$ is the average of one component of the velocity squared for the fermion particles.

The relaxation times can be related to the cross sections[6]:

$$\tau_{FF} = \sqrt{2} \frac{m_F}{\rho \sigma_{FF} \langle v_F \rangle} \tag{2a}$$

$$\tau_{FB} = \frac{1}{\rho \sigma_{FB}} \left[\frac{(m_F + m_B)(m_F m_B)^{1/2}}{\langle v_F \rangle \langle v_B \rangle} \right]^{1/2} \tag{2b}$$

* Supported in part by the National Science Foundation and the U.S. Office of Naval Research.
† For a more complete account of this work see Ref. 3.

where m_F is the fermion mass; m_B is the boson mass; $\langle v_F \rangle$ is the average fermion speed; and $\langle v_B \rangle$ is the average boson speed.

In this paper we will show that this description gives a good account of the diffusion process, except for the fermion–roton collisions near the solution lambda point, where the roton density is high and the dilute gas approximation breaks down.

Experimental Methods

The spin diffusion coefficient D_s was measured by the spin-echo method in a spectrometer of conventional design. "Method B" of Carr and Purcell[7] was employed and D_s was determined from the echo decay and its dependence on the magnetic field gradient G. Several measurements were made at each temperature and concentration for both positive and negative values of G. The derived values of D_s for $T \geq 1°$K are estimated to have a maximum error of $\pm 3\%$.

The sample chamber was nylon and the gas handling system was made of stainless steel and glass. The samples were prepared by mixing known volumes of ^{3}He and ^{4}He gas, and the concentration was verified by vapor pressure measurements. The quoted concentrations are accurate to $\pm 1\%$.

The temperature was measured and regulated to within a few millidegrees by use of a precision quartz pressure gauge and servocontrol.

Results and Discussion

The measured values of D_s as a function of χ and T are shown in Fig. 1. The value of D_s for pure ^{3}He as measured by Hart and Wheatley[8] is also shown. The behavior of D_s below T_λ is consistent with the Pomeranchuk model. The spin diffusion coefficient is determined by the ^{3}He–^{4}He and the ^{3}He-thermal excitation scattering.

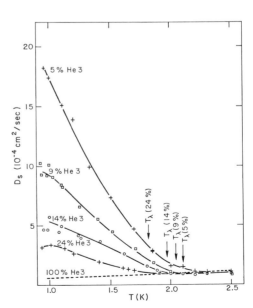

Fig. 1. Spin diffusion coefficient D_s as a function of temperature.

(In the temperature range of these experiments the rotons are dominant.) These two terms are nearly independent of one another, and can be separated by making use of Eq. (1). If σ_{FF} and σ_{FB} are independent of concentration, then Eqs. (2a) and (2b) show that $\rho\tau_{FF}$ and $\rho\tau_{FB}$ are also independent of concentration. It can also be easily shown that it is a good approximation to take $\rho_F \propto \chi$. A plot of $1/D_s$ as a function of χ will thus yield a linear relation, as shown in Fig. 2, from which a value of σ_{FF} may

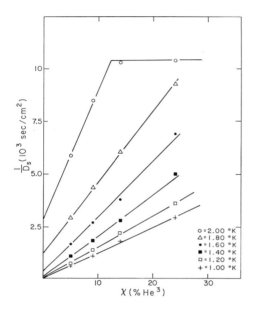

Fig. 2. $1/D_s$ vs. concentration χ.

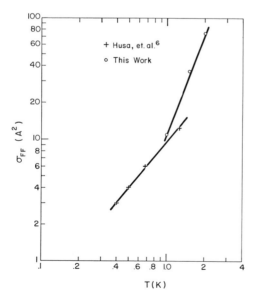

Fig. 3. Fermion–fermion cross section σ_{FF} vs. temperature.

be obtained from the slope by the use of Eqs. (1) and (2a). The values of σ_{FF} obtained in this way are shown as a function of temperature in Fig. 3, along with the values quoted by Husa *et al.*[1] for $T < 1°$K.

A value for σ_{FB} can be obtained from the intercepts of Fig. 2 by use of Eqs. (1) and (2b). Equation (1) may be rewritten as $1/D_s = (1/D_{FB}) + (1/D_{FF})$. The term $1/D_{FF}$ is the contribution to $1/D_s$ due to the fermion–fermion scattering, and it is proportional to T^2 for $T > 1°$K. The term $1/D_{FB}$ is the contribution due to the fermion–boson scattering, and it should be proportional to ρ_B. In the Landau theory $\rho_B \propto e^{-8.6/T}$. The temperature dependence of $1/D_{FB}$ obtained from the intercepts of Fig. 2 is given by $e^{-15.3/T}$. It does not seem reasonable to suppose that σ_{FB} is such a rapidly varying function of temperature, and it appears that the dilute gas transport theory that leads to Eq. (1) does not adequately describe the fermion–roton collisions close to the lambda point where the roton "fluid" becomes very dense.

References

1. D.L. Husa, D.O. Edwards, and J.R. Gaines, in *Proc. 10th Intern. Conf. Low Temp. Phys.*, 1966, VINITI, Moscow, USSR (1967), Vol. 1, p. 345.
2. P. Horvitz and H.E. Rorschach, in *Proc. 9th Intern. Conf. Low Temp. Phys.*, 1964, Plenum Press, New York (1965), Part A, p. 147.
3. D.C. Chang and H.E. Rorschach, *J. Low Temp. Phys.* (to be published).
4. E. Fukuda and R. Kubo, *Prog. Theor. Phys.* **29**, 621 (1963).
5. I. Pomeranchuk, *Soviet Phys.—JETP* **19**, 42 (1949).
6. S. Chapman and T.G. Cowling, *The Mathematical Theory of Non-Uniform Gases*, Cambridge Univ. Press (1970), p. 87 ff.
7. H.Y. Carr and E.M. Purcell, *Phys. Rev.* **94**, 630 (1954).
8. H.R. Hart and J.C. Wheatley, *Phys. Rev. Lett.* **4**, 3 (1960).

Nucleation of Phase Separation in ^{3}He–^{4}He Mixtures*

N. R. Brubaker and M. R. Moldover†

School of Physics and Astronomy, University of Minnesota
Minneapolis, Minnesota

Introduction

We report here preliminary results of experiments designed to study nucleation of phase separation in metastable thermodynamic states of liquid ^{3}He–^{4}He mixtures under a wide range of conditions. In addition to making some interesting qualitative observations we have been able to measure apparent lifetimes of metastable states in six solutions with mole fractions of ^{3}He in the range $0.22 \leq x \leq 0.611$. In a closed container each of these solutions could be supercooled below the temperature at which it would normally phase separate (T_{ps}) until a temperature interval less than $200\mu°$K wide was reached in which the apparent lifetime of the metastable state decreased more than a factor of 25 to less than 10 sec. This temperature interval occurs at much smaller supercoolings than that predicted by a classical model of homogeneous nucleation. Nevertheless, the abruptness of the decrease in lifetimes suggests the onset of some definite, temperature-dependent nucleation process. We speculate that this process is an inhomogeneous one, such as nucleation by negative ions. Our work was stimulated by two earlier experiments[1,2] in which dilute solutions of liquid ^{3}He in ^{4}He were observed to remain in metastable states for times on the order of hours. Both of these earlier experiments were designed to study other phenomena, thus it happened that metastability was observed only below $0.1°$K in solutions only slightly more concentrated than the maximum solubility of ^{3}He in ^{4}He at zero temperature ($x \approx 0.065$).

We will first describe our experimental technique, then some qualitative observations, followed by our lifetime measurements, and finally, some preliminary thoughts on understanding the data.

Experimental Technique

In Fig. 1 we show a schematic diagram and key dimensions of our experimental cell. A guarded capacitor is used to measure the mean number density of helium atoms in a small portion ($\sim 2.5\%$) of the cell volume near the cell's bottom. A germanium thermometer measures the temperature of the cell walls with a resolution of $5 \times 10^{-7}°$K and a servo system is used to program the cell temperature. The noise on the servo system was always smaller than the programmed rate of tempera-

* Work supported in part by National Science Foundation Grant No. GP 14431.
† Present address: Heat Division, National Bureau of Standards, Washington, D.C.

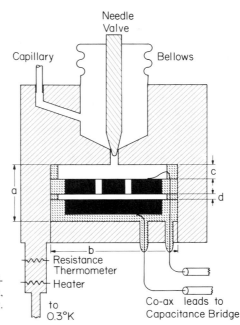

Fig. 1. Schematic diagram of experimental cell. Dimensions are $a = 6.4$ mm, $b = 5.3$ mm, $c = 0.28$ mm, $d = 0.006$ mm. The capacitor plates are 1.0 mm thick. The cell holds 8.0 mm³ of liquid helium.

ture changes. A superfluid-tight valve permits us to fill the cell completely with various mixtures and then to seal it. Thus experiments may be done under conditions of fixed (and measured) total volume and mole fraction. The valve-operating mechanism is detached after use to minimize temperature gradients across the cell. Because of the use of Stycast insulation, the helium in the cell is heated or cooled primarily through the top surface. We estimate the thermal relaxation time for the filled cell to be 2 sec. The mass diffusion relaxation time is more difficult to estimate. Spin diffusion measurements have been made only on very dilute solutions and pure ³He and the diffusion constant varies by orders of magnitude between these two regimes. We suspect one of the main limitations of our apparatus near the tricritical point is the time delay between the occurrence of nucleation and its detection by flow of ³He into or out of the capacitor under diffusive conditions.

For a typical lifetime measurement on the ⁴He-rich side of the tricritical point the cell is filled with a known mixture at a pressure of 200 Torr, at a temperature slightly above the λ temperature of the mixture. Then the valve is closed and the filling capillary is evacuated. The normal phase separation temperature of the mixture in the cell under pressure is measured. To do this, we use the fact that when the cell contains a single phase the density is uniform in the cell (see below for minor exceptions) and the capacitance is essentially independent of temperature. When the cell contains two phases *in equilibrium* the capacitor is filled with the ⁴He-rich lower phase and the capacitance has a temperature dependence reflecting the temperature dependence of the number density of the lower phase. Thus T_{ps} is observed easily with great precision by observing the temperature dependence of the capacitance as the cell is slowly warmed with two phases in it. T_{ps} may be used to determine the mole fraction of the mixture actually in the cell. (We use the saturated vapor pressure phase diagram since the pressure change on closing the valve is known to be small, and the phase diagram is relatively independent of pressure.)

We observe supercooling repeatedly by programming the cell temperature through a cycle consisting of four stages. These stages are: (1) A linear, fast cooldown lasting several hundred seconds from well above T_{ps} (typically tens of m°K) to within several hundred μ°K of the region of interest below T_{ps}. (2) A nearly constant temperature "equilibration interval" lasting several hundred seconds. (3) A linear, slow cooldown through the region of interest at rates from 0.4 to 8.3μ°K sec^{-1}. (4) A warmup to the starting temperature similar to the cooldown but five times faster.

The temperature at which phase separation occurs is detected by monitoring the capacitance. If phase separation is nucleated within the capacitor, the ^{3}He mole fraction in the capacitor will initially increase, leading to a decrease in capacitance. If nucleation occurs outside the capacitor, the capacitance will increase.

Qualitative Observations

Nucleation by Walls. Metastable states were never observed on the ^{3}He-rich side of the tricritical point. This had been expected[3] since it is well known that ^{4}He is preferentially adsorbed on walls and a ^{4}He-rich nucleus can grow continuously out of this adsorbed layer. This is entirely analogous to the nucleation of droplets on the walls of cloud chambers when the fluid in the chamber wets the walls of the chamber. An interesting feature of helium mixtures is that all chemical impurities will be solid at these temperatures (0.5–0.9°K) and therefore should resemble walls in their effects.

Nucleation by Free Surfaces. Metastable states were never observed when the cell was only partly filled, even when the mixture and temperature program were identical to those for which long metastable lifetimes in a filled cell were measured. This is expected[4] since ^{3}He is known to be preferentially adsorbed near the liquid–vapor interface and a ^{3}He-rich nucleus can grow continuously out of the interface.

Nucleation by Heat Flush. A temperature gradient in a cell containing a super-fluid helium mixture will cause ^{3}He to be concentrated at the coldest part of the cell by the so-called "heat flush" effect. We observe concentration gradients in our cell through small capacitance changes even when the cell contains one phase during the fast part of our cooling cycle. We verified that temperature gradients can cause nucleation by first preparing a metastable state in a filled cell at a temperature where the apparent lifetime was quite long. We then induced nucleation by programming the cell temperature to oscillate with a small amplitude and increasing frequency. From this experiment we estimated that the temperature gradients needed to induce nucleation are one to two orders of magnitude larger than those used to measure lifetimes in most cases. Near the tricritical point we are not as confident that temperature gradients are unimportant in lifetime measurements.

We believe we were unable to measure lifetimes longer than about 1000 sec because of heat flush nucleation. Infrequent, large electrical transients are picked up either by the temperature servo system or the cell wiring directly. These transients may produce momentary high-temperature gradients which may cause nucleation.

Lifetime Measurements

Because we have used a technique of repeated, continuous cooling, our data are in the form of a histogram of $N(T)$. Here $N(T)$ is the number of samples prepared in metastable states which have not yet nucleated by the time the samples have been cooled to the temperature T. We fit a polynomial to the histogram and define the "apparent lifetime" τ of a metastable state at the temperature T by the equation

$$\frac{1}{N}\frac{dN}{dT}\dot{T} = -\frac{1}{\tau} \qquad (1)$$

where $\dot{T}$ is the rate of cooling (in $\mu°K\ sec^{-1}$) through the region of interest. To verify that τ is independent of the value of $\dot{T}$ chosen for the experiment without differentiating data, we have chosen to plot the data in a form suggested by integrating Eq. (1), namely

$$\frac{\dot{T}}{\dot{T}_0}\log_{10}\frac{N(T)}{N(T_0)} = \frac{-\log_{10}e}{\dot{T}_0}\int_{T_0}^{T}\frac{dT}{\tau(T)} \qquad vs. \quad T \qquad (2)$$

Fig. 2 shows the data taken at $x = 0.45$ plotted in this fashion with the arbitrary

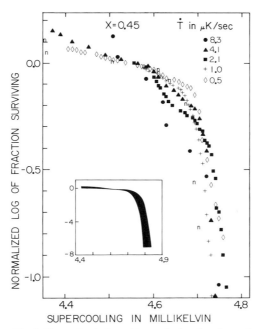

Fig. 2. Normalized logarithm of the fraction of "samples" surviving as a function of supercooling. The normalization is described in the text. Data denoted by n were taken at 4.1 $\mu°K\ sec^{-1}$ in the presence of a neutron source. The shaded area on the inset indicates the temperatures at which all but two of 481 samples phase separated.

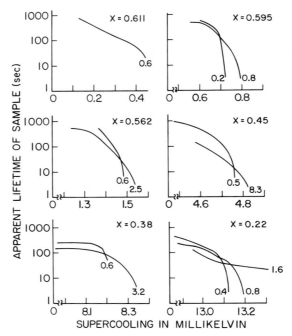

Fig. 3. Apparent sample lifetimes as a function of temperature supercooled below the normal phase separation
temperature. Numbers near curves indicate cooling rate in
$\mu°K\ sec^{-1}$.

normalization choices $\dot{T}_0 = 2.1\ \mu°K\ sec^{-1}$ and $T_0 = 4.55\ \mu°K$ below the normal
phase separation temperature for this mixture. In Fig. 2 data taken at cooling rates
differing by a factor of 16 nearly coincide, thus establishing the independence of
the measured τ from the choice of $\dot{T}$.

Figure 3 shows the apparent lifetimes derived from data for various mixtures.
Each of these mixtures may be supercooled until a narrow temperature interval is
reached in which the apparent lifetime declines abruptly. As one moves away from
the tricritical composition ($x = 0.67$) the amount of supercooling attainable (measured
in units of $(T - T_{tri})/T_{tri}$ increases from 0.018 at $x = 0.61$ to 0.046 at $x = 0.38$ and
then decreases slightly to 0.036 at $x = 0.22$. These values are comparable to but
slightly smaller than those typically observed for pure fluids (e.g., a value of 0.05
is reported for water away from the critical region[5]).

The great reproducibility of the narrow temperature interval in which nucleation
occurs has convinced us we are observing the onset of some definite nucleation process.

Interpretation

We have attempted to compare our data at $x = 0.22$ with the classical theory
of homogeneous nucleation.[6,7] Classical theory predicts nucleation rates J having
a temperature dependence

$$J = \Omega e^{-\Delta F/k_B T} \qquad (3)$$

The prefactor Ω may be crudely estimated for dilute mixtures by ignoring the ^{4}He and considering ^{3}He to be a conventional ideal gas with an increased effective mass. The free energy barrier ΔF may be calculated as a function of T using the measured interfacial tension[4] and published thermodynamic data extrapolated linearly from nearby stable states into the slightly supercooled state. Our estimates suggest we should obtain supercooling an order of magnitude greater than we in fact observed (i.e., $\sim 0.1\,^\circ$K rather than $\sim 0.01\,^\circ$K at $x = 0.22$).

We know that various inhomogeneities are present in our samples, including temperature gradients, pressure waves, walls, vortex lines, positive and negative ions, and perhaps impurities. We speculate that negative ions, which according to the accepted bubble model and experiment[8] are easier to inject into ^{3}He than into ^{4}He, are a likely possibility for reducing the free energy barrier and thus substantially increasing the nucleation rate. A detailed calculation should be made.

We consider our failure to alter the observed lifetime by irradiation of the sample cell with a neutron source to be inconclusive. It is conceivable that we did not change the ambient ion density (resulting from decay of tritium in the sample) in any substantial way with the source available to us. (The source strength was 2000 neutrons sec^{-1} released in 4π ster 7 cm from the cell and the nominal tritium concentration in the ^{3}He component of the gas was 2×10^{-11} mole %.)

Because of the detailed way in which inhomogeneities in liquid helium may be characterized, we consider this system to be a promising one for further experimental and theoretical study of the nucleation process.

References

1. J. Landau, J.T. Tough, N.R. Brubaker, and D.O. Edwards, *Phys. Rev. Lett.* **23**, 283 (1969); J. Landau, The Low Temperature Osmotic Pressure of Degenerate Dilute Liquid ^{3}He–^{4}He Solutions under Pressure, Ph.D. Thesis, Ohio State Univ., June 1969, unpublished.
2. A.E. Watson, J.D. Reppy, and R.C. Richardson, *Phys. Rev.* **189**, 384 (1969).
3. P. Seligmann, D.O. Edwards, R.E. Sarwinski, and J.T. Tough, *Phys. Rev.* **181**, 415 (1969).
4. H.M. Guo, D.O. Edwards, R.E. Sarwinski, and J.T. Tough, *Phys. Rev. Lett.* **27**, 1259 (1971).
5. C.S. Kiang, D. Stauffer, G.H. Walker, O.P. Puri, J.D. Wise, Jr., and E.M. Patterson, *J. Atmos. Sci.* **28**, 1222 (1971).
6. J. Frenkel, *Kinetic Theory of Liquids*, Dover, New York (1955), Chapter VII.
7. A.C. Zettlemoyer (ed.), *Nucleation*, Marcel Dekker, New York (1969).
8. M. Kuchnir, P.R. Roach, and J.B. Ketterson, in *Proc. 12th Intern. Conf. Low Temp. Phys., 1970*, Academic Press of Japan, Tokyo (1971), p. 105.

Renormalization of the ^{4}He λ-Transition in ^{3}He–^{4}He Mixtures*

F.M. Gasparini

Physics Department, The Ohio State University
Columbus, Ohio

and

M.R. Moldover†

Physics Department, University of Minnesota
Minneapolis, Minnesota

The specific heat at constant pressure at the λ-transition of pure ^{4}He is very nearly logarithmically divergent.[1] In the presence of a constant ^{3}He concentration this specific heat is renormalized in such a way that it tends to reach a finite value with an infinite slope.[2,3] The effect of an impurity on the critical point behavior of a pure substance has been discussed by Fisher[4] and Fisher and Scesney.[5] It is pointed out by Fisher that along with the renormalization of C_{px}, the specific heat at constant pressure, and ^{3}He concentration, one expects that $C_{p\phi}$, where $\phi \equiv \mu^3 - \mu^4$, the difference in the molar chemical potential of the two isotopes, will behave as C_p for pure ^{4}He. We have analyzed our data to obtain a more quantitative description for the behavior of C_{px} and $C_{p\phi}$.

Our data covers the concentration range from pure ^{4}He to $x = 0.39$, and extend in temperature to within ~ 5 μ°K of the transition. For all values of x they were taken with a partially filled calorimeter, i.e., along a path of liquid and vapor coexistence. This implies that the pressure, concentration, and amount of liquid sample would change as the temperature was changed. In order to verify that for our experimental conditions the measured heat capacity was C_{px} we made some new measurements at $x = 0.30$ both with a partially filled calorimeter and with a completely filled calorimeter. From these measurements as well as from additional independent estimates we concluded that in the region $t < 10^{-2}$ our data is indeed C_{px} to within the accuracy of our measurements.

The data at all concentrations including pure ^{4}He was least-square-fitted to the function

$$C_{px} = (A/\alpha)(t^{-\alpha} - 1) + B \tag{1}$$

where $t \equiv |(T/T_\lambda) - 1|$. This same function was used for $T > T_\lambda$ as well as for $T < T_\lambda$ with no constraints on the variational parameters. We will denote by primes

* Work supported in part by the National Science Foundation.
† Present address: Heat Division, National Bureau of Standards, Washington, D.C.

the parameters in Eq. (1) for $T < T_\lambda$. T_λ was also allowed to vary, but in practice was well determined. For instance, in the case of pure ⁴He, where the data extend to $t \sim 3 \times 10^{-6}$, T_λ did not vary in the computer fits more than $\sim 3 \times 10^{-7}\,°\text{K}$ if the range of data used in conjunction with Eq. (1) did not give systematic deviations. For all concentrations the data had to be restricted to $t \gtrsim 3.6 \times 10^{-3}$. In order to use data outside this range, additional terms would have been required in the fitted function. The computer fits obtained with this range of data showed a random scatter of $\sim \pm 0.2\%$ in the region where the error was not limited by our experimental temperature resolution of $\sim 0.1\ \mu°\text{K}$.

Figure 1 shows the results for the parameters α and α' as a function of concentration. The data at $x = 0.53$ of Alvesalo *et al.*[6] have also been fitted to Eq. (1) and are shown on Fig. 1 as triangles. The results from these data are in good agreement with the results from our own. For pure ⁴He we note that $\alpha \neq \alpha'$, in violation of scaling laws, and further that $\alpha' < 0$ implies that for $T < T_\lambda$ the heat capacity tends to a finite value. Similar results have also been obtained by Ahlers.[1] With increasing concentration the exponents become more and more negative, retaining, however, the asymmetry observed with pure ⁴He at least up to $x \cong 0.30$. This asymmetry indicates that the branch for $T > T_\lambda$ tends to have a stronger temperature dependence than the branch for $T < T_\lambda$. From renormalization ideas one would expect that the leading temperature-dependent term in the heat capacity would be characterized by an exponent $\alpha(\text{renormalized}) = -\alpha(\text{pure})/[1 - \alpha(\text{pure})]$, where $\alpha(\text{pure})$ is a positive number which gives the divergence in the heat capacity of the pure system. In the

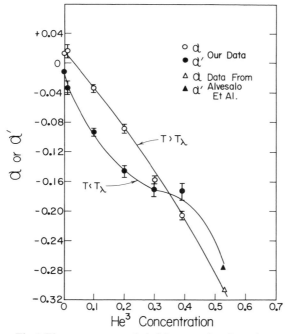

Fig. 1. The exponents α and α' of the power law dependence of C_{px} as a function of ³He concentration.

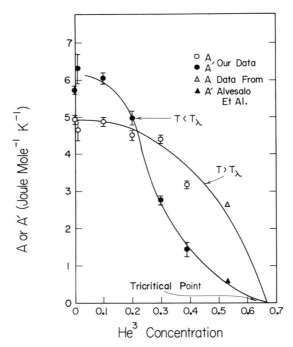

Fig. 2. The parameters A and A' for C_{px} as a function
of ^{3}He concentration.

special case of α(pure) $= 0$ one would expect the leading term in the renormalized
heat capacity to behave as $\sim 1/\log t$. In addition, one expects that the weaker the
divergence, i.e., the smaller α(pure), the greater the importance of higher-order
temperature-dependent terms in the renormalized heat capacity. It seems that it is
precisely because of these higher-order terms that even at a concentration $x = 0.39$,
where one would expect renormalized behavior to be visible, one obtains
α(renormalized) $\cong -0.2$ rather than -0.01, as might be expected in the case when
α(pure) $= +0.01$. In view of this effect of higher-order terms, it seems best to regard
Eq. (1) not as a test for the specific prediction of renormalization but rather as a
useful way of parameterizing the data. In Fig. 2 we show the results for the parameters
A and A'. The tricritical point marked on this figure is the concentration at which
the λ-line meets the two branches of the phase separation curve.[7]* The behavior
of these parameters with concentration strongly indicates that the temperature-
dependent term in the heat capacity vanishes at the tricritical point. The behavior
of the constants B and B' with concentration is also in agreement with this interpretation, since they tend to a value of ~ 8 J mole^{-1} °K^{-1}. This is the measured value
for the specific heat at a concentration very nearly equal to that at the tricritical
point.[3]

In order to investigate the behavior of the conjugate heat capacity $C_{p\phi}$ we

---

* The term "tricritical point" was coined to indicate the unusual nature of the junction of the λ-line
with the phase separation curve.

have used the equation

$$C_{p\phi} = \frac{C_{px}\{T(\partial x/\partial T)_{t,p}\,(\partial\phi/\partial T)_{t,p} - T(\partial s/\partial T)_{t,p}\} + \{T(\partial s/\partial T)_{t,p}\}^2}{T(\partial s/\partial T)_{t,p} + T(\partial x/\partial T)_{t,p}\,(\partial\phi/\partial T)_{t,p} - C_{px}} \tag{2}$$

This is simply an extension to ϕ, x variables of a relation derived by Buckingham and Fairbank[8] between C_p and C_v at the λ-line of pure ⁴He. The partial derivatives of s, ϕ, and x are taken along paths parallel to the λ-line in x, T space. Their temperature dependence for small t is assumed negligible, and their value as computed for $t = 0$ is used in practice. When using Eq. (2) in conjunction with data on C_{px} one obtains $C_{p\phi}(x)$, i.e., $C_{p\phi}$ along paths of constant x. The way Eq. (2) was implemented was to assume the power law behavior of Eq. (1) for $C_{p\phi}$ and then minimize the difference between the right- and left-hand sides of Eq. (2) by varying the parameters α, A, B, α', A', B', and T_λ. Figure 3 shows the results of this analysis for the parameters α and α'. The error bars appearing in this figure represent the variations in these parameters obtained by changing the values of the partial derivatives appearing in Eq. (2) within what was felt was the uncertainty in their estimate. When we compare Fig. 3 with Fig. 1 we note that while the exponents which characterize the temperature dependence of C_{px} become more and more negative, in the case of $C_{p\phi}(x)$ they tend to stay near zero, i.e., $C_{p\phi}(x)$ does seem to retain the behavior of pure ⁴He.

In conclusion, we would say that the effect of ³He on the λ-transition is to

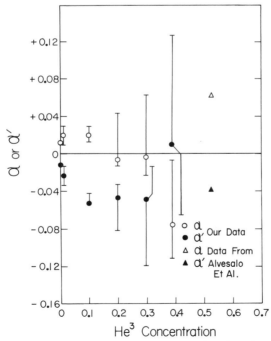

Fig. 3. The exponents α and α' in the power law dependence of $C_{p\phi}(x)$ as a function of ³He concentration.

weaken the temperature dependence and lower the magnitude of the heat capacity until at the end of the λ-line, at the tricritical point, the temperature-dependent term vanishes, leaving only a constant. Regarding renormalization, it would seem that the importance of higher-order terms precludes the extraction of a true renormalized exponent. On the other hand, the conjecture that $C_{p\phi}$ should retain the behavior of the pure system does seem to be true, at least to the extent that $C_{p\phi}(x)$ reflects the behavior of $C_{p\phi}$.

References

1. G. Ahlers, in *Proc. 12th Intern. Conf. Low Temp. Phys., 1970,* Academic Press of Japan, Tokyo (1971), Sect. A, p. 21; G. Ahlers, *Phys. Rev. A***3**, 696 (1971).
2. F.M. Gasparini and M.R. Moldover, *Phys. Rev. Lett.* **23**, 749 (1969); F.M. Gasparini, Ph.D. Thesis, Univ. of Minnesota, 1970.
3. T.A. Alvesalo, P.M. Berglund. S.T. Islander, G.R. Pickett, and W. Zimmermann, Jr., *Phys. Rev. A***4**, 2354 (1971).
4. M.E. Fisher, *Phys. Rev.* **176**, 257 (1968).
5. M.E. Fisher and P.E. Scesney, *Phys. Rev. A***2**, 285 (1970).
6. W. Zimmermann, private communication.
7. R.B. Griffiths, *Phys. Rev. Lett.* **24**, 715 (1970).
8. M.S. Buckingham and W.M. Fairbank, *Progress in Low Temperature Physics,* C.J. Gorter, ed., North-Holland, Amsterdam (1961), Vol. III, p. 83.

The Osmotic Pressure of Very Dilute ^{3}He–^{4}He Mixtures*

J. Landau and R.L. Rosenbaum

Department of Physics, Technion—Israel Institute of Technology
Haifa, Israel

Detailed measurements of the osmotic pressure of dilute ^{3}He–^{4}He mixtures at very low temperatures have been reported recently by Landau et al.[1,2] and Ghozlan et al.[3] which give considerable information on the thermodynamic properties of this Fermi liquid for concentrations greater than 1.5 mole % ^{3}He. The osmotic pressure and other thermodynamic properties of these mixtures have been thoroughly discussed in a review article by Ebner and Edwards.[4] The need for measurement of the osmotic pressure at concentrations substantially less than 1.5% has been discussed in a recent theoretical paper by Disatnik and Brucker.[5] It is in this very low concentration region that the effective interaction theory proposed by Bardeen, Baym, and Pines[6] (BBP) should have its greatest validity. In this paper we present the experimental results from measurements of the osmotic pressure in the range of ^{3}He mole fraction X from 0.08 to 0.6% at the saturated vapor pressure.

The apparatus is shown schematically in Fig. 1. The larger cell, containing the ^{3}He–^{4}He mixture and the powdered cerium magnesium nitrate thermometer, was connected to the pure ^{4}He cell via a porous Vycor glass superleak.[7] The height of the liquid ^{4}He column was measured by a concentric capacitor. The osmotic pressure π which developed across the superleak caused a corresponding difference in liquid height between the two cells and hence a change in the average dielectric constant of the material between the capacitor plates. With a sensitive capacitance bridge we could resolve changes in the liquid height of approximately 1 μm. The 8 cm length of the capacitor fixed the range for measuring π to a maximum of 0.8 Torr. The mixture cell was designed with a cross-sectional area 100 times larger than the annular area between the capacitor plates, thus the change of the liquid level in the capacitor plus a 1% correction for the liquid which rises in the mixture cell equals the osmotic pressure. Both sides of the cell were thermally linked to the mixing chamber of a dilution refrigerator.

The capacitor was calibrated by first making measurements with a mixture $X = 0.04\%$ where π was calculated from the ideal gas law together with the small correction for the Fermi degeneracy.[2] The degeneracy temperature T_F is given by

$$k_B T_F = \frac{h^2}{2m^*}(3\pi^2 n_3)^{2/3} \tag{1}$$

where m^* is the ^{3}He quasiparticle effective mass, determined in the limit of small

* Work supported in part by the Gerald I. Swope Fund and the Batsheva de Rothschild Foundation.

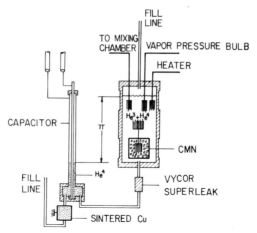

Fig. 1. Schematic drawing of the osmotic pressure apparatus. The CMN thermometer is enclosed in a basket separate from the cell wall to eliminate influence from heat leaks.

X from second-sound measurements[8] to be 2.28 ± 0.04 times the bare ^{3}He mass, and n_3 is the ^{3}He number density. For this calibration concentration T_F is 14 m$^\circ$K, allowing us to calculate π over the temperature range with uncertainty much less than the overall experimental error of 2% of π. One final correction was made to the measured π for the very small but detectable amount of ^{3}He on the capacitor side; this small correction had, in fact, negligible effect on the extrapolation of π to zero temperature which is the principal result to be discussed in this paper.

In this experiment the osmotic pressure at constant X was measured as a function of temperature down to the lowest temperature of 20 m$^\circ$K. The results confirm that the temperature dependence of the osmotic pressure is that of an almost ideal Fermi gas. This is seen in the striking linearity which appears when π is plotted against U_F, the internal energy of an ideal Fermi gas (which is given as a function of T/T_F by Stoner[9]). This agrees with the results of Landau et al.[2] for higher concentrations. Thus we can easily extrapolate π at constant X to the zero-temperature value π_0 with only a small error of extrapolation. The values obtained for π_0 are insensitive to variation of the value of m^*, over the range of experimental uncertainty in m^*, since this just slightly expands or contracts the U_F scale. The results of this extrapolation are shown in Fig. 2.

It is π_0 as a function of X which shows noticeable departure from the behavior of an ideal Fermi gas. This can be seen in Fig. 2, where the results of this experiment are compared to π_{kin}, the predicted osmotic pressure if no effective interaction were present.[2] According to the detailed thermodynamic analysis of Ebner and Edwards,[4]

$$\pi_{kin} = \frac{2}{5} n_3 k_B T_F \left[1 - \frac{10}{7} \chi \left(\frac{P_F}{P_c} \right)^2 \right] \tag{2}$$

where $P_F^2 = 2m^* k_B T_F$ and $P_c/\hbar = 1.5$ Å^{-1}; the term containing χ comes from the p^4 term in the quasiparticle energy spectrum, which Brubaker et al.[8] determined experimentally to be $\chi = 0.14 \pm 0.05$.

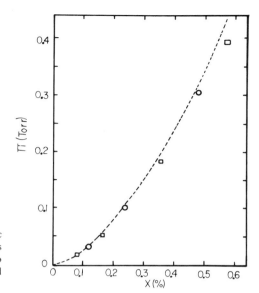

Fig. 2. The zero-temperature limit of the osmotic pressure as a function of concentration. Circles and squares represent values extrapolated from two distinct runs. The dashed curve is the calculated value of π_{kin} from Eq. (2).

It is the difference between these two curves in Fig. 2 which contains the information about the effective interaction $\pi_{\text{int}} = \pi_0 - \pi_{\text{kin}}$. Over the entire range of concentration for the present measurements the experimental π_{int} is smaller than the π_{int} predicted by the BBP theory using the effective interactions $V(k)$ determined by Ebner[11] and Ebner and Edwards [Ref. 4, Eq. (117i)] from transport property measurements.* The disagreement in π_{int} is as large as a factor of 1.5 if one takes a smooth curve through the experimental points. This is to be compared to the higher concentration results of Landau et al.,[1] where the disagreement was 10–20%, in the same direction as the present results.

One can understand the possible sources for this disagreement by examining a plot of $\pi_0 X^{-5/3}$ versus $X^{1/3}$ as is shown in Fig. 3. This graph shows that the present results join smoothly with the results of Landau et al.[1] The solid curve was determined by Disatnik and Brucker[5] by fitting their expression for π_0 to the high-concentration values of the osmotic pressure with the condition that $m^* = 2.28 m_3$. The broken curve represents the results of the dipolar kernel potential which was determined by Ebner and Edwards [Eq. (117ii) of Ref. 4] from the results of transport property measurements. The results of these two theories are almost identical for $X < 2\%$. Both theories include dipolar scattering amplitudes as well as the p^4 term in the quasiparticle energy spectrum. The $X = 0$ intercept is inversely proportional to m^* and the initial slope of the curve (in the limit $X \to 0$) is proportional to V_0, the long-wave length limit of $V(k)$. If one accepts $m^*/m_3 = 2.28 \pm 0.04$, then the experimental values suggest a less negative value of the initial slope (in fact even a positive value seems plausible). This is in contradiction with the Baym expression[12]

$$V(0) = -\alpha^2 m_4 s^2/n_4^0 \tag{3}$$

where m_4 is the mass of a ^{4}He atom and s and n_4^0 are, respectively, the speed of first

* The explicit expression for π_{int} is given by Eq. (30) of Ref. 2.

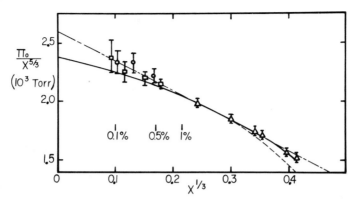

Fig. 3. Comparison between the experimental values of π_0 and the theories of Disatnik and Brucker[5] (solid curve) and Ebner and Edwards[4] (dashed curve). The circles and squares are the results of the present experiment, while the triangles are the results of Landau et al.[1] The straight line intercepts at 2600 Torr corresponding to $m^* = 2.09m_3$.

sound and number density of ^{4}He at $T = 0$. This result supports the conclusion of Eckstein et al.[13] that the term in Eq. (3) is completely cancelled in a more exact calculation of the ^{3}He effective interaction and is replaced by a term of a different form.

Alternatively, one can fit all of the experimental points in Fig. 3 with a straight line whose intercept corresponds to $m^*/m_3 = 2.09$ and whose slope corresponds to a $V(0)$ about double the value given by Eq. (3). This explanation seems unlikely since it requires a value of m^* which disagrees with the results of second-sound and specific heat measurements.

Acknowledgments

We are grateful to Professor D. O. Edwards and to Professor Y. Eckstein for many valuable discussions about the execution of this experiment and useful ideas for the analysis of the data. We thank Dr. M. Black for assistance on the measurements.

References

1. J. Landau, J.T. Tough, N.R. Brubaker, and D.O. Edwards, *Phys. Rev. Lett.* **23**, 283 (1969).
2. J. Landau, J.T. Tough, N.R. Brubaker, and D.O. Edwards, *Phys. Rev.* **A2**, 2472 (1970).
3. A. Ghozlan, P. Piéjus, and E. Varoquaux, *Compt. Rend.* **B269**, 344 (1969).
4. C. Ebner and D.O. Edwards, *Phys. Reports* **2C**, 77 (1971).
5. Y. Disatnik and H. Brucker, *J. Low Temp. Phys.* **7**, 491 (1972).
6. J. Bardeen, G. Baym, and D. Pines, *Phys. Rev.* **156**, 207 (1967).
7. M.F. Wilson, D.O. Edwards, and J.T. Tough, *Rev. Sci. Instr.* **39**, 134 (1968).
8. N.R. Brubaker, D.O. Edwards, R.E. Sarwinski, P. Seligmann, and R.A. Sherlock, *Phys. Rev. Lett.* **25**, 715 (1970).
9. E.C. Stoner, *Phil. Mag.* **25**, 901 (1938).
10. C. Ebner, Ph.D. Thesis, Univ. of Illinois, 1967, unpublished.
11. C. Ebner, *Phys. Rev.* **185**, 392 (1969).   ,
12. G. Baym, *Phys. Rev. Lett.* **17**, 952 (1966).
13. S.G. Eckstein, Y. Eckstein, C.G. Kuper, and A. Ron, *Phys. Rev. Lett.* **25**, 97 (1970).

Pressure Dependence of Superfluid Transition Temperature in ^{3}He–^{4}He Mixtures

T. Satoh and A. Kakizaki

Department of Physics, Faculty of Science
Tohoku University, Sendai, Japan

Introduction

As is well known, the superfluid transition temperature T_λ of pure liquid ^{4}He decreases both with increasing pressure or density and with increasing concentration of ^{3}He impurity.

From the theoretical point of view, many approaches have been made to understand the variation of T_λ, including the Bose gas model,[1-3] the lattice model,[4-6] the phenomenological phonon–roton model,[7] and others.[8,9]

It is curious that at the saturated vapor pressure (SVP) the variation of T_λ with ^{3}He concentration fits rather well to the two-thirds power law of the ideal Bose gas model up to very high concentration of ^{3}He.[1] But this Bose gas model cannot explain the pressure dependence of T_λ.[3] The lattice model qualitatively explains the pressure dependence of T_λ by assuming a change in the number of holes.[4-6]

From the experimental point of view it is desirable to investigate the relation between density and T_λ in mixtures in order to obtain further insight concerning the parameters which determine the superfluid transition.

Although several experiments have been made on the pressure dependence of T_λ in pure ^{4}He[10,11] systematic investigation in ^{4}He–^{3}He mixtures is very rare.[12]

In this paper we report the pressure dependence of T_λ in pure ^{4}He and mixtures of 3.85, 10.0, and 15.0% ^{3}He concentration.

Experiment

The concentration of ^{3}He in mixtures was determined by measuring the pressure and assuming the ideal gas equation of state.

Measurements were made from SVP to ~ 1 atm for each sample. The pressurizing apparatus is very similar to that of Kierstead.[11] The pressure changes were measured by means of a differential helix-quartz Bourdon gauge of Texas Instrument type 3 which was calibrated against a mercury and oil manometer up to 1 atm.

The crystostat used in this work is shown in Fig. 1. This can also be used for specific heat measurements. The λ-point is found by driving a dc heater wound outside of the sample cell and observing the anomaly in the warming rate of the thermometer placed in the sample cell. For pure ^{4}He this anomaly is very clear as

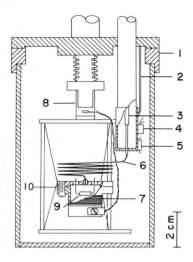

Fig. 1. Schematic drawing of the apparatus: (1) vacuum can; (2) thermal anchor; (3) sample filling copper tubing, 0.7 mm i.d.; (4) manganin heater; (5) thermometer; (6) sample filling copper–nickel tubing, 0.3 mm i.d. and 50 cm length; (7) manganin heater, 1 KΩ; (8) mechanical heat switch; (9) Allen Bradley thermometer, 180 Ω, $\frac{1}{2}$ W; (10) sample cell of 1.5 cm³ volume.

the plateau of the $R(t)$ curve, where R is the resistance of the thermometer and t is time. For mixtures this anomaly is not so clear and only a small change of the slope of the $R(t)$ curve is observed. This thermometer is calibrated before each run of measurements by half filling the sample cell with pure ^{4}He and using the pressure gauge mentioned above. The relative accuracy of our measurements is 10 μdeg in temperature and 0.05 mm Hg in pressure.

The data at SVP were obtained by filling the sample cell to nearly two-thirds of its height, both before and after measuring in the pressurized state.

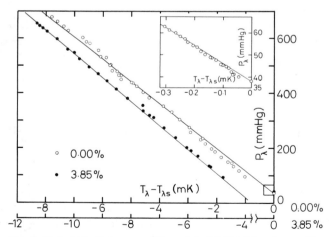

Fig. 2. Plot of $T_\lambda - P_\lambda$ for pure ^{4}He and for 3.85% ^{3}He mixture. The insert is an expansion of the $T_\lambda - P_\lambda$ plot in ^{4}He near saturated vapor pressure.

Results and Discussion

The measured pressure dependence of T_λ in pure ^{4}He and in mixtures is shown in Figs. 2 and 3. Pertinent quantities deduced from these measurements are listed in Table I.

The data on pure ^{4}He give nearly the same value for $(dP/dT)_\lambda$ as that of Kierstead.[11] For mixtures some discrepancies have been noted between the observed T_λ at SVP ($T_{\lambda s}$) and the extraporated $T_{\lambda s}$. This tendency becomes larger for more concentrated ^{3}He mixtures. These may be due to the change in ^{3}He concentration, because the measurements at SVP were made in a partly filled sample cell, as mentioned before.

Although in our method of determining T_λ there is some ambiguity in ^{3}He concentration because of the concentration gradient due to the heat flash effect, the measurements at SVP are reasonable in light of previous results.[13]

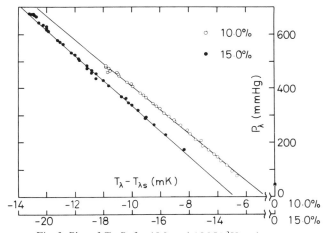

Fig. 3. Plot of T_λ–P_λ for 10.0 and 15.0% ^{3}He mixtures.

Table I. Numerical Values[a] of the Superfluid Transition Temperature at Saturated Vapor Pressure $T_{\lambda s}$ and the Saturated Vapor Pressure at the Superfluid Transition $P_{\lambda s}$

	$T_{\lambda s}$, °K	$P_{\lambda s}$, mm Hg	$(dP/dT)_\lambda$, mm Hg·m°K^{-1}	$T_{\lambda s}^B$, K
Pure ^{4}He	2.173	38.2	−80.8	(2.173)
^{4}He–3.85% ^{3}He	2.122	43.8	−81.4	2.114
^{4}He–10.0% ^{3}He	2.040	44.5	−87.2	2.008
^{4}He–15.0% ^{3}He	1.962	45.0	−94.1	1.894

[a] The listed values of $(dP/dT)_\lambda$ correspond to the slope of the solid lines in Figs. 2 and 3 for each sample. The listed values of $T_{\lambda s}^B$ are the superfluid transition temperatures calculated using the formula $T_{\lambda s}^B = T_{\lambda s}(^4\text{He})(1-x)^{2/3}(1+\alpha x)^{-2/3}$, where $x = N_3/(N_3 + N_4)$ is the mole fraction of ^{3}He and $\alpha = (v_3 - v_4)/v_4$ is assumed to be 0.29.

The tendency of $(dP/dT)_\lambda$ to increase with ^{3}He concentration will be discussed after the completion of the density measurements.

Acknowledgments

The authors would like to thank Prof. T. Ohtsuka for reading the manuscript and for continuous encouragement during the study.

References

1. C.V. Heer and J.G. Daunt, *Phys. Rev.* **81**, 447 (1951).
2. T. Tsuneto, *Sol. St. Phys.* **4**, 652 (1969) (in Japanese).
3. K. Huang, *Studies in Statistical Mechanics,* J. de Boer and G.E. Uhlenbeck, eds., North-Holland, Amsterdam (1964), Vol. 2.
4. T. Matsubara and H. Matsuda, *Prog. Theor. Phys.* **16**, 569 (1956); **17**, 19 (1957).
5. S.G. Brush, *Proc. Roy. Soc.* **A242**, 544 (1957); **A243**, 225 (1958).
6. S. Takagi, *Prog. Theor. Phys.* **47**, 22 (1972).
7. K.R. Atkins, *Liquid Helium,* Cambridge Univ. Press (1959), p. 70.
8. G. Chapline, Jr., *Phys. Rev.* **A3**, 1671 (1971).
9. E. Byckling, *Physica* **38**, 285 (1968).
10. D.L. Elwell and H. Meyer, in *Proc. 10th Intern. Conf. Low Temp. Phys., 1966,* VINITI, Moscow (1967), p. 517.
11. H.A. Kierstead, *Phys. Rev.* **182**, 153 (1967).
12. C. Le Pair, K.W. Taconis, R. de Bruyn Ouboter, P. Das, and E. De Tong, *Physica* **31**, 764 (1965).
13. K.W. Taconis and R. de Bruyn Ouboter, *Progress in Low Temperature Physics,* C.J. Gorter, ed., North-Holland, Amsterdam (1964), Vol. 4.

Thermal Diffusion Factor of ^{3}He–^{4}He Mixtures: A Test of the Helium Interaction Potential

W.L. Taylor

*Monsanto Research Corporation, Mound Laboratory**
Miamisburg, Ohio

The thermal diffusion factor α_T is a particularly useful property for investigating the intermolecular force law because of its enhanced sensitivity to the interaction. Because of this, and because of the lack of experimental data for this property at low temperatures, it was the objective of the present work to measure the composition dependence of α_T at several low temperatures and compare the results to several realistic helium intermolecular potentials. An experimental determination of α_T is usually made by placing a binary mixture of gases (or isotopes) in a temperature gradient and allowing the slight separation to occur, after which the composition is determined in the hot and cold regions. The process is described by the mass flux equation for a binary mixture in a one-dimensional temperature gradient, which must be in the vertical direction in order to eliminate convection. In terms of experimental quantities the thermal diffusion factor is given by

$$\alpha_T = (\ln q)/\ln(T_H/T_C)$$

where q is the separation factor defined by the ratio of composition ratio X_1/X_2 at T_H to X_1/X_2 at T_C. Because the isotopic separation is quite small, a trennschaukel or "swing separator" was used for the present work. The apparatus was constructed so that both the trennschaukel and the bellows pump used to swing the gas could be placed in a cryostat and shielded with liquid nitrogen (see Fig. 1). The trennschaukel consists of seven thin-wall, stainless steel tubes 3.5 cm long and 1.905 cm i.d. with hollow copper caps which were also 3.5 cm long and 1.905 cm i.d. silver-soldered on each end. Each cap contains a heat exchanger consisting of copper wool held in place by copper screen. Successive tubes are connected together top-to-bottom by thin-wall, stainless steel capillaries 0.110 cm i.d. The tubes are soldered to $\frac{1}{4}$-in.-thick circular copper disks to ensure temperature uniformity in the hot and cold regions. The theory of trennschaukel operation has been described in detail by van der Waerden.[1]

The temperature gradient was established by immersing the bottom one-third of the assembly in either liquid neon or liquid helium. A heater was wound around the top block and controlled by a proportional controller. The temperature was sensed by either a calibrated carbon resistor or a platinum resistance thermometer

---

* Mound Laboratory is operated for the U.S. Atomic Energy Commission under Contract No. AT-33-1-GEN-53.

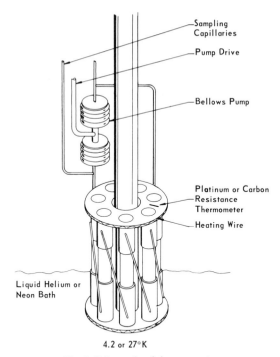

Fig. 1. Schematic of the apparatus.

embedded in the copper disk and constituting one leg of a Wheatstone bridge. The liquid neon bath lasted for the duration of the experiments (~ 10 hr) but the helium was replenished every 2 hr by an "in situ" transfer line.

Duplicate samples were withdrawn from the hot- and cold-end bellows at the end of an experiment after first purging the capillaries leading out of the cryostat by withdrawing a small amount of gas. The separation factor q was obtained by measuring the isotope ratios on a mass spectrometer. A complete description of the analysis procedure is given in Ref. 2.

The composition dependence of α_T has been measured at only one temperature below 200°K and this was by Paul et al.[3] who used a four-tube trennschaukel operating at an average temperature of 160°K. The temperature dependence was measured by Watson et al.[4] down to 136°K using an equimolar mixture in the same device. Van der Valk[5] obtained results down to 12.7°K in a two-bulb apparatus employing a mixture containing 10% ^{3}He in which the cold temperature was varied and the hot bulb was maintained near 300°K. α_T was evaluated by measuring the slope of the logarithmic plot of T_C/T_H vs. the separation factor. The values obtained at the lowest temperatures appeared anomalously high and were never published. More recently Weissman[6] reported α_T values for an equimolar mixture down to 55°K which were found in a manner analogous to that employed by van der Valk.

In order to test the overall performance of the apparatus, the composition dependence was measured at 160°K, the same average temperature used by Paul et al.[3] The reciprocal α_T vs. composition was fitted to a linear function by the method of least squares for both sets of data: for trace ^{3}He, $1/\alpha_T = 14.5$ and 14.8 (present data and Paul et al.,[3] respectively) and similarly for trace ^{4}He, $1/\alpha_T = 14.1$ and 14.0. This agreement to better than a few percent is excellent for thermal diffusion measurements.

The measured composition dependence at the average temperature of $5\frac{1}{2}$ and 30°K is shown in Fig. 2. ΔT values of 3 and 6°K were used, respectively. Duplicate runs were made at four compositions and in three instances reproducibility was excellent. In the fourth instance the values were nearly 20% different. This was traced primarily to the mass spectrometer determination of the isotope ratio in the case where the ratio differs greatly from unity. The end compositions were always more difficult to analyze accurately. The data are well represented by the two linear functions

$$1/\alpha_T = -3.93X(^3\text{He}) + 14.6; \qquad T = 5\tfrac{1}{2}°\text{K}$$

$$1/\alpha_T = -3.58X(^3\text{He}) + 14.4; \qquad T = 30°\text{K}$$

The temperature dependence of an equimolar mixture is shown in Fig. 3. The closed circles are the present data, which were obtained from the linear equations in $1/\alpha_T$. Contrary to the observation of van der Valk, who noted a sharp rise in α_T from 40° down to 14°K, the present results indicate a relatively constant behavior with respect to temperature. Above 40°K the experimental data are essentially in agreement and predict a rather gentle decrease in α_T with increasing temperature.

The experimental data were corrected for trennschaukel operating conditions according to the theory of van der Waerden[1] as described in Ref. 7. Quantum calculations were used in the evaluation of the transport properties required for the

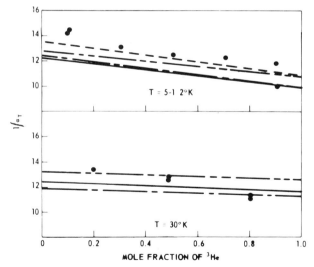

Fig. 2. Composition dependence of the reciprocal thermal diffusion factor; Dots give the present data. The theoretical curves are: (— -) Lennard-Jones (12–6); (——), Beck's; (- - -), Morse V_{DD}; and (— - -) (exp-6).

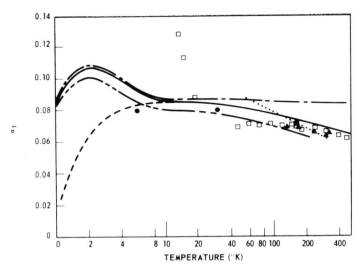

Fig. 3. Temperature dependence of the thermal diffusion factor. The experimental points are: (●) present data; (□) van der Valk; (△) Paul et al.; (▲) Watson et al.; and (···) Weissman. The theoretical curves are the same as in Fig. 2.

correction factors, which were generally less than 6%. The total uncertainty of a given measurement was estimated to be the sum of uncertainties in the temperatures and the sample analysis. The average uncertainty of a single point is estimated to be approximately ±8–10%. For the closed circles in Fig. 3, statistics reduce the uncertainty to approximately ±4–5%.

The theoretical calculations shown in Figs. 2 and 3 are the third Chapman–Enskog approximation to the thermal diffusion factor $[\alpha_T]_3$ calculated quantum mechanically using a procedure described previously.[8,9] The potentials used were the Lennard-Jones (12–6) with de Boer parameters, the modified Buckingham (exp-6) with MR5[10,11] parameters, Beck's, and the Morse V_{DD} potential.

The slope of the theoretical curves in Fig. 2 is very nearly the same for all of the potentials and only slightly less than that of the experimental data. With regard to Fig. 3, the data at $5\frac{1}{2}$°K fit the Morse V_{DD} potential best but at the higher temperatures lie midway between the (exp-6) and the convergent Beck and Morse V_{DD}. The most important conclusion to be drawn is that α_T for ^{3}He–^{4}He mixtures does not rise rapidly with decreasing temperatures as observed by van der Valk but, in both composition and temperature dependences, behaves very nearly as predicted by three of the four intermolecular potentials, the exception being the Lennard-Jones potential. In order to determine whether or not the theoretical maximum in α_T below 4°K is real (Fig. 3), we must await the results of further experiments.

References

1. B.L. Van der Waerden, Z. Naturforsch. **12a**, 583 (1957).
2. W.L. Taylor and S. Weissman, J. Chem. Phys. **55**, 4000 (1971).
3. R. Paul, A.J. Howard, and W.W. Watson, J. Chem. Phys. **43**, 1622 (1965).

4. W.W. Watson, A.J. Howard, N.E. Miller, and R.M. Shiffrin, *Z. Naturforsch.* **18a**, 242 (1963).
5. F. van der Valk, Thesis, Amsterdam Univ. 1963.
6. S. Weissman, *Phys. Fluids* **12**, 2237 (1969).
7. W.L. Taylor, S. Weissman, W.J. Haubach, and P.T. Pickett, *J. Chem. Phys.* **50**, 4886 (1969).
8. J.M. Keller and W.L. Taylor, *J. Chem. Phys.* **51**, 4829 (1969).
9. W.L. Taylor and J.M. Keller, *J. Chem. Phys.* **54**, 647 (1971).
10. J.E. Kilpatrick, W.E. Keller, and E.F. Hammel, *Phys. Rev.* **97**, 9 (1955).
11. E.A. Mason and W.E. Rice, *J. Chem. Phys.* **22**, 187 (1954).

Peculiarities of Charged Particle Motion in ^{3}He–^{4}He Solutions in Strong Electric Fields

B. N. Eselson, Yu. Z. Kovdrya, and O. A. Tolkacheva

Physico-Technical Institute of Low Temperatures
Academy of Sciences of the Ukrainian SSR, Kharkov

The velocity of positive and negative ions in ^{3}He–^{4}He solutions containing 0.22%, 0.75%, and 1.30% ^{3}He has been measured in electric fields up to 1000 V cm^{-1} in the temperature range 0.37–0.67°K by the time-of-flight method. It has been shown that at sufficiently low temperatures where scattering of ions by impurity excitations is predominant the field dependence of the positive-ion drift velocity shifts to the higher-velocity side in strong electric fields.

It has been found that in solutions there are two values of critical field and velocity at which the ion motion is associated with creation of quantized vortex rings. The critical fields in the solutions do not reveal any appreciable temperature dependence and the corresponding critical velocities are somewhat less than in ^{4}He.

The obtained dependence of the vortex ring velocity on electric field shows that the coefficient α_3 which determines the energy losses of the vortex rings in the gas of impurity excitations does not practically depend on temperature. The experimental data qualitatively agree with the calculated values of α_3 obtained, taking into account the attraction and trapping of ^{3}He impurity by the core of vortex line.

First-Sound Absorption and Dissipative Processes in Liquid ^{3}He–^{4}He Solutions and ^{3}He

N. E. Dyumin, B. N. Eselson, and E. Ya. Rudavsky

Physico-Technical Institute of Low Temperatures
Academy of Sciences of the Ukrainian SSR, Kharkov

Absolute measurements of first-sound absorption have been carried out in pure ^{3}He and ^{3}He–^{4}He solutions containing 6.3%, 11.0%, and 19.7% ^{3}He whose frequency was 10 MHz. The absorption has been found to decrease with ^{3}He concentration rise in the solutions below λ-point. Above the λ-point the absorption increases practically according to the linear law. The results obtained are used for calculating the second viscosity coefficient, which sharply decreases with the rise in ^{3}He content; in concentrated solutions it becomes less than the first viscosity coefficient. In this case the main contribution to first-sound absorption is made by dissipative processes associated with impurity particle diffusion and first viscosity. The temperature dependence of sound absorption in ^{3}He has been studied in the range 0.45–2.5°K. The absorption has been shown to be due chiefly to first viscosity and thermal conductivity.

Contents

of Volume 2

2.3 Scattering and Vacancies

3. Helium-3 Nuclear Magnetism

4. Helium Monolayers

5. Molecular Solids

6. Other Topics in Quantum Crystals

Magnetism

7. Plenary Topics

8. Phase Transitions

9. Low-Dimensional Systems

10. Ferromagnetism

14. Small Particles, Heat Capacity, and Paramagnetism

Contents

of Volume 3

Superconductivity

2.3. Dynamics

3. Josephson Effect and Tunneling

3.1. Josephson Effect

3.2. Tunneling

4. Superconductivity Materials

4.1. Elements and Compounds

4.2. Gases, Films, and Granular Materials

5. Phonons

6. Fluctuations

7. Superconductivity Phenomena

Contents

of Volume 4

Electronic Properties

1. Plenary Topics

2. Electron–Electron Interactions

3. Cyclotron Resonance

12. Energy States and Heat

Instrumentation and Measurement

13. Plenary Topics

14. Thermometry

15. Superconducting Instruments

16. Cryogenic Instrumentation and Measurement Techniques

Index to Contributors*

Complete Alphabetical Listing of the Contributors to LT-13, Volumes 1-4

* Volume number appears in parentheses followed by the page number on which the contribution begins.

Clem, J. R. (3) 102 ·
Cochrane, R. W. (2) 427
Cohen, M. L. (3) 619
Cole, M. W. (1) 364
Coleridge, P. T. (4) 9
Coles, B. R. (2) 414, 423
Collan, H. K. (4) 513
Collings, E. W. (3) 403
Collver, M. M. (3) 532
Commins, D. E. (1) 356
Constable, J. H. (2) 223
Conway, M. M. (2) 601, 629
Cook, J. W. Jr. (3) 507
Cooper, A. S. (4) 648
Coqblin, B. (2) 459
Corruccini, L. R. (1) 411
Costa-Ribeiro, P. (2) 520
Cotignola, J. M. (2) 585
Cox, J. E. (3) 480
Crabtree, G. W. (4) 104, 120
Craig, P. P. (1) 602, (4) 385
Creswell, D. J. (1) 163, 195
Crooks, M. J. (2) 95
Crow, J. E. (3) 563, 696
Crum, D. B. (1) 563
Cunsolo, S. (1) 337
Cupp, J. D. (4) 542
Cyrot, M. (2) 551

Daalmans, G. M. (1) 451
Dahm, A. J. (2) 233
Damen, T. C. (4) 310
Danielson, G. C. (3) 408
Danz, R. W. (4) 42
Dash, J. G. (1) 19, 247, (2) 165
Datars, W. R. (4) 335
Date, M. (1) 485
Daunt, J. G. (1) 182
Davis, R. S. (1) 328
Davis, W. T. (3) 507
Day, E. P. (4) 550
Deaton, B. C. (3) 475
Deaver, B. S. Jr. (3) 264
deBoom, C. W. (4) 617
deBruyn Ouboter, R. (3) 772
Decker, W. R. (2) 605
de Graaf, A. M. (3) 238
Deis, D. W. (3) 461
de Klerk, D. (3) 138
de la Cruz, F. (3) 673
Del Castillo, L. (4) 640
DeReggi, A. S. (3) 281
Deroyon, J. P. (1) 96
Deroyon, M. J. (1) 96
Deutscher, G. (3) 573, 603, 692

de Waele, A. Th. A. M. (3) 772
de Wette, F. W. (4) 441
Dietrich, M. (2) 491
Dillon, R. O. (4) 358
DiSalvo, F. J. (3) 417
Disatnik, Y. (1) 598
Doake, C. S. M. (1) 456
Dobbs, E. R. (3) 367, (4) 320
Dolan, G. J. (3) 147
Donnelly, R. J. (1) 57
Douglas, R. J. (4) 335
Douglass, D. H. (1) 414, (3) 468
Dransfeld, K. (2) 53
Dubeck, L. W. (3) 607
Dudko, K. L. (2) 365
Dupas, A. (2) 389
Dupas, C. (2) 355
Dyumin, N. E. (1) 637

Ebisawa, H. (3) 177
Eckardt, J. (1) 518
Eckstein, S. G. (1) 54
Edwards, D. O. (1) 26, 518
Ehrat, R. (3) 134
Ehrenfreund, E. (2) 373, (4) 648
Elgin, R. L. (2) 175
Englehardt, J. J. (4) 445
Eremenko, V. V. (2) 365
Erickson, D. J. (2) 605
Erickson, R. A. (2) 215
Eselson, B. N. (1) 636, 637
Evans, W. A. B. (1) 101

Fairbank, H. A. (2) 85, 90
Fairbank, W. M. (4) 559
Falco, C. M. (4) 563
Falke, H. P. (2) 500, 503
Farrell, D. F. (3) 728
Farrell, H. H. (3) 563
Felsch, W. (3) 543
Fenichel, H. (3) 573
Fert, A. (2) 488
Finnegan, T. F. (3) 268, 272
Finnemore, D. K. (2) 590
Fiory, A. T. (3) 86
Fisk, D. J. (1) 568
Fleury, P. A. (4) 310
Flouquet, J. (2) 448
Flukiger, R. (3) 446
Ford, A. (1) 121
Fournet, G. (3) 232
Fowler, R. D. (3) 377
Fradin, F. Y. (3) 366
Francavilla, T. (3) 750, (4) 287

Franck, J. P. (2) 48
François, M. (1) 96, 319, 324
Fraser, J. C. (1) 253
Frederking, T. H. K. (1) 393
Friedberg, C. B. (4) 177
Friedberg, S. A. (2) 380
Friedlander, D. (1) 54
Friedman, E. J. (3) 238
Frommer, M. H. (3) 306
Frossati, G. (4) 640
Fukuyama H. (2) 537 (4) 50, 54

Gaines, J. R. (2) 223
Gallardo, B. (3) 654
Galleani d'Agliano, E. (2) 459
Gallus, D. E. (3) 182
Gambino, R. J. (3) 387
Gamble, F. R. (3) 438
Gamota, G. (1) 459
Ganguly, B. N. (3) 361
Gardner, W. E. (2) 595
Garito, A. F. (2) 373
Garland, J. C. (4) 399
Gaspari, G. D. (3) 515
Gasparini, F. M. (1) 518, 618
Gauthier, R. (3) 241
Gautier, F. (4) 325
Gavaler, J. R. (3) 558
Geballe, T. H. (3) 411
Genack, A. Z. (3) 69
Gerber, J. A. (4) 622
Gerritsen, A. N. (4) 236
Gershenson, M. (3) 573
Gey, W. (2) 491
Ghozlan, A. (4) 503
Giffard, R. P. (2) 144, (4) 517
Gilabert, A. (3) 312
Gillis, N. S. (2) 227
Ginsberg, D. M. (3) 767
Giorgi, A. L. (2) 605
Girardeau, M. D. (3) 781
Glaberson, W. I. (1) 430
Glasser, M. L. (4) 42
Glover, R. E. III (3) 547, 649
Golding, B. (3) 623
Goldman, A. M. (3) 709
Goldschvartz, J. M. (1) 451
Goldstein, Y. (4) 282
Gomès, A. A. (2) 459, 543
Goodstein, D. L. (1) 243, (2) 175, 180
Gordon, D. E. (3) 475
Gor'kov, L. P. (3) 735
Gorter, C. J. (2) 621
Gossard, A. C. (2) 322, (4) 648

Subject Index

Adsorbed ^{4}He films, 19, 143, 147, 152, 156, 182
   211, 206
Adsorbed ^{3}He and ^{3}He$-^4$He mixture films, 159,
   163, 195, 200

Bose$-$Einstein condensation, 46, 80

Constrained geometries, 167, 172
Correlation length in ^{4}He, 377
Couette flow of superfluid helium, 283
Critical behavior in ^{4}He and ^{3}He$-^4$He mixtures,
   377, 381, 571, 581, 586

Density of ^{4}He, 343
Dissipative heat flux in He II, 324
Dynamic form factor, 61

Electric field effects, 372, 636
Electron bubbles in liquid helium, 469, 482, 485
Elementary excitations, 72, 76
Ellipsometric measurement of film thickness, 185
Equations of state, 87, 381
Evaporation from He II, 364
Excitation spectrum, 67, 493, 591

First sound, 356, 637
Flowing films (Chapter 4)
   dissipation in film flow, 253, 258, 263, 268
   flow rates, 219, 224
   mass transport, 247
   superfluidity, 219, 233
   thermodynamics of superflow, 243
Fourth sound, 324, 497, 501, 510

Heat transfer, 401
Helium atom scattering, 522
He II, general theory, 39, 101
^{3}He, 159, 167, 177, 195, 200, 406, 411, 474
^{3}He$-^4$He mixtures (Chapter 9)
   dissipative processes, 637
   electric field effects, 636
   excitation spectrum, 591
   films, 163, 233
   ^{3}He in He II, 26
   Ising model, 116
   light scattering, 542, 581, 591
   nuclear magnetic resonance, 163
   osmotic pressure, 623

^{3}He$-^4$He mixtures *(cont)*
   pressure dependence of superfluid transition,
     627
   quantum lattice gas model, 112
   rotons, 108, 591
   specific heat, 598
   spin diffusion, 602, 608
   superfluidity, 233
   thermal diffusion, 631
   tricritical point, 571, 581, 586
   viscosity, 559, 563, 568

Ion currents in helium, 426, 456
Ion mobility in helium, 430, 434, 439, 451, 474,
   477, 489
Ising model, 116

Kapitza resistance, 387, 393, 398

Lambda point paradox, 288
Legget-Rice effect in ^{3}He, 411
Light scattering, 9, 377, 524, 532, 537, 542,
   581, 591
Liquid structure factor, 67, 547, 551

Mass transport of ^{4}He films, 247

Neutron scattering, 46, 50, 57, 61, 72
Nuclear magnetic resonance, 163, 167, 195, 200

Orifice flow of helium, 414
Osmotic pressure of ^{3}He$-^4$He mixtures, 623

Persistent currents, 239, 279
Phonons, 50, 54, 84, 591
Pressure effects, 352, 602, 627

Quantum lattice gas model, 87, 112

Rotons, 108, 591

Second sound, 324, 348
Specific heat, 172, 598
Spin diffusion, 177, 602, 608

Date Due

Due	Returned	Due	Returned
DEC 8 1978			
DEC 3 1978			
DEC 16 1978			
NOV 24 87			
APR 23 1981			